Drug Use Trajectories Among Minority Youth

Yonette F. Thomas • LeShawndra N. Price
Editors

Drug Use Trajectories Among Minority Youth

Springer

Editors
Yonette F. Thomas
The New York Academy of Medicine and the American Association of Geographers
Glenn Dale, MD, USA

LeShawndra N. Price
Health Inequities and Global Health Branch
National Heart Lung and Blood Institute, National Institutes of Health
Bethesda, MD, USA

ISBN 978-94-017-7489-5 ISBN 978-94-017-7491-8 (eBook)
DOI 10.1007/978-94-017-7491-8

Library of Congress Control Number: 2016946341

Printed on acid-free paper

This Springer imprint is published by Springer Nature
The registered company is Springer Science+Business Media B.V. Dordrecht

Note: The views expressed are those of the authors and do not necessarily represent the views of the National Institute on Drug Abuse, the National Institute of Mental Health, the National Institutes of Health, the US Census Bureau, the US Department of Commerce, the US Department of Health and Human Services, or the US Government.

Preface

Contrary to much popular opinion, African American and Hispanic/Latino persons in the general population do not consistently show higher rates of substance use disorders than Whites (Table 1). Tobacco use rates are lower for African American and Hispanic/Latino populations than Whites; illicit drug use disorders are higher for African Americans, but alcohol disorders are not higher; and neither illicit drug use disorders nor alcohol disorders are much different for Hispanic/Latino persons compared to Whites in the general population. Yet, a consistent risk factor for substance use and addictive disorders is socioeconomic adversity that is more commonly experienced in these minority populations. Disparate outcomes and pathways reflected in these findings and associated with racial and ethnic status serve as a starting point for this monograph. By tackling many nuances in the social epidemiology of substance use and addictive disorders among persons from diverse racial and ethnic backgrounds, the monograph explores a key question—namely, how to explain the contrasting and similar patterns of substance use and addictive disorders among different racial and ethnic groups.

Research has shown African American and Hispanic/Latino youth to have different pathways or trajectories of drug use compared with White youth, from initiation to development of problem use. Relative to White youth, African American youth are less likely to use alcohol and other drugs as adolescents, begin using in early adulthood, and are more likely to become problem users. Thus, the major risk factor of early adolescent onset of substance use predicting higher rates of addiction does not seem to hold true for African American youth as much as for White youth. As African Americans move into young adulthood and beyond, there appears to be an onset of substance use and addictive disorders more than for White populations. Exploring these trajectories is the goal of this monograph, with some research identifying similar patterns, while others have shown heterogeneity in the developmental course of substance use and substance use disorders across ethnic and racial groups, and this heterogeneity may be of etiological significance.

One of the complications of substance use research is that the behavior involves potential toxicity. That is, research on substance use must take into account the reinforcing/rewarding and brain altering aspects of exposures. This is why early onset is such a potentially important issue. The brain may be particularly vulnerable at different stages of development, and adolescence appears to be such a vulnerable period with heightened risk for establishment of long-term patterns.

The propensity to use drugs is based on both individual predisposition and on environmental influences that interact with one another across human development. Environments that may influence future drug use include the prenatal intrauterine environment, infancy and early childhood environments, and later childhood. Experimentation with psychoactive substances often starts in adolescence, a developmental stage characterized by risk-taking, novelty-seeking, and heightened sensitivity to peer pressure, which might reflect incomplete development of brain regions involved in, for example, executive control, motivation, and decision making. In addition, epidemiological evidence shows that the process of addiction is much more likely to be triggered in an adolescent brain: convergent lines of evidence suggest that exposure to drugs or alcohol during adolescence may result in different neuroadaptations from those that occur during adulthood. For example, recent studies demonstrate that the adolescent period is distinctly sensitive to long-term alteration by chronic exposure to alcohol or nicotine, which may explain the greater vulnerability of young initiates to addiction to alcohol or other drugs. How racial and ethnic status interacts with these fundamental developmental trajectories is a key question.

Addiction Vulnerability

A nuanced way to think about the issue of variations in substance use and addictive disorders in different groups is to conceive these substance use and addictive disorders as explained by a cluster of risk and protective factors. These risk and protective factors include multiple environments from the intraindividual (e.g., genetic predispositions) to family, peer, neighborhood, legal, and cultural domains. Importantly, interactions within and across these domains shape the drug use trajectories and serve as the rich material found within the chapters of this text.

It is estimated that 40–60 % of the vulnerability to addiction is attributable to genetic factors. Animal studies have identified several genes that are involved in drug responses and whose experimental modification markedly affects drug self-administration. In addition, animal studies have also identified candidate genes and genetic loci for alcohol responses, which overlap with genes and loci identified in human studies. Progress in identifying candidate genes for alcoholism and alcohol-related responses continues at a rapid pace. However, identifying the biological function of these new candidate genes has emerged as a major challenge for the next decade. The hope is that a better understanding of the myriad interacting genetic factors and networks that influence addiction risk and trajectory will help increase the efficacy of substance prevention approaches and improve addiction treatment as well.

Table 1 Past 12-month substance use disorders and tobacco use in race and ethnic groups, persons ages 12 and older in 2012

Racial/ethnic group	Illicit drug disorder	Alcohol disorder	Either illicit drugs or alcohol disorder	Tobacco use
Not Hispanic or Latino				
White	2.7 (0.11)	7.0 (0.20)	8.7 (0.22)	34.5 (0.44)
Black or African American	4.1 (0.38)*	6.2 (0.42)	8.9 (0.51)	30.9 (0.97)*
American Indian or Alaska Native	7.6 (1.87)*	17.3 (3.20)*	21.8 (3.39)*	53.8 (3.94)*
Native Hawaiian or other Pacific Islander	2.3 (0.95)	3.7 (0.97)*	5.4 (1.40)*	–
Asian	0.6 (0.17)*	2.9 (0.49)*	3.2 (0.50)*	14.0 (1.32)*
Two or more races	4.2 (0.79)	7.0 (1.25)	10.1 (1.44)	43.7 (3.05)*
Hispanic or Latino	2.8 (0.26)	7.3 (0.42)	8.0 (0.47)	25.0 (0.86)*

Source: Substance Abuse and Mental Health Services Administration, Results from the 2012 National Survey on Drug Use and Health, Detailed Tables. http://www.samhsa.gov/data/NSDUH/2012SummNatFindDetTables/DetTabs/NSDUH- DetTabsTOC2012.htm, Accessed July 5, 2014

*Rate compared to White, $p < 0.05$

One of the best examples of moving from gene identification to biological function is the association between drug metabolizing genes and protection against substance use disorders. These polymorphisms operate by modulating the accumulation of toxic (aversive) metabolites. Therefore, if alcohol or drugs are consumed by individuals who carry variants that convert their substrate at high rates, then the accumulation of toxic metabolites serves as a negative stimulus to prevent further consumption. The ways that these and other genetic factors operate in racially and ethnically diverse populations is a key question.

Environmental factors that have been consistently associated with a propensity to self- administer drugs include low socioeconomic status, poor parental support, within peer group deviancy, and drug availability. Stress might be a common feature in a wide variety of environmental factors that increase the risk for drug abuse and may help explain, for example, why social isolation (which increases anxiety) during a critical period of adolescence increases addiction vulnerability. The ways that racial and ethnic status may alternatively buffer and instigate stress remain an intriguing window into the onset and maintenance of substance use and addictive disorders.

Further, the issue of the overlap of mental illnesses with substance use and substance use outcomes is often under-explored in racial and ethnic subpopulations. Addressing the ways that mental illnesses vary in the ways they shape substance use trajectories in different populations is another challenge for the authors of the chapters in this text.

Interventions

The greater vulnerability of adolescents to experimentation with drugs of abuse and to subsequent addiction underscores why preventing early exposure is such an important strategy to reduce drug addiction. Modifiable risk and protective factors (i.e., targets for prevention interventions) include early aggression, social skills deficits, academic problems, misperceived drug use norms, association with deviant peers, neighborhood drug availability, media glamorization of drugs, and parental monitoring and support. Multiple proven effective interventions target these factors with parent skills training, social skills training, improved self-regulation and impulse control, school tutoring, refusal skills training, and even community policing.

Despite the fact that adolescents are at a stage in their lives when they are more likely to take risks, interventions can decrease the rate of drug use. At present, prevention strategies include not only educational interventions based on comprehensive school-based programs and effective media campaigns and strategies that decrease access to drugs and alcohol but also strategies that provide supportive community activities that engage adolescents in productive and creative ways. However, as we begin to understand the neurobiological correlates that underlie the adverse environmental factors that increase the risks for drug use and for addiction, the hope is to be able to develop interventions to counteract these changes. In addition, as knowledge of how different genes influence a person's vulnerability to taking drugs and to becoming addicted, more targets will be uncovered to tailor interventions for those at higher risk. How each of these interventions and approaches varies according to background factors, including racial and ethnic status, is essential to improving outcomes. In fact, studying variations in response to preventive interventions can provide crucial evidence about the underlying risk trajectories themselves.

Many of the chapters in this book are based on research supported by the National Institute on Drug Abuse (NIDA), and the central themes of the text were built from several years of work following a November 2006 NIDA-sponsored scientific meeting on "Drug Abuse Trajectories Among African Americans." The overall goals of the chapters are to uncover key principles related to the onset and progression of substance use by examining variations across diverse populations. Current and future scientific work depends on such studies helping to explain and elucidate the intersection of drug use exposures and background variation in risk and protective factors.

Deputy Director of the National Institute on Drug Abuse (NIDA), National Institutes of Health (NIH), Bethesda, MD, USA

Wilson M. Compton

Acknowledgments

We thank the many contributors to the drafting of this book. The compilation of this book involved a significant effort by many people—those who are identified as contributors as well as those who are more behind the scenes. Everyone connected with this book is immensely grateful for various contributions while this book was in development. We would also like to acknowledge the substantial contributions by Ms. A. Victoria Lybrand for her editorial and logistical support. Without her dedication, this text could not have been completed. Finally, this book exists because of the talent, experience, and active participation of the contributing authors. We are grateful that they accepted our invitation and perservered through the drafts and editing process. Many, many thanks to each and every one of you.

Glenn Dale, MD, USA — Yonette F. Thomas
Bethesda, MD, USA — LeShawndra N. Price

Contents

Contributors

Fernando H. Andrade was born in Peru where he obtained his bachelor degree in social psychology. He also holds a master of arts in statistics and Ph.D. in education both obtained at the University of Michigan, Ann Arbor. Currently he is a research investigator in the School of Social Work at the University of Michigan. His area of research is focused on understanding the interactive effects of children and adolescents' contexts—such as family and schools—on educational outcomes and substance use.

Margaret Arzon graduated from the University of Miami with a bachelor of arts in 2009 majoring in international relations and Spanish. She went on to complete her master in public health at the University of Miami graduating in 2011 with a focus on global health and nutrition. Her research focuses are on Hispanic adolescent, socio-behavioral intervention programs, HIV and substance use prevention, and global nutrition. Margaret presently aides in clinical trial research for respiratory and monitoring devices.

Jerald G. Bachman is research professor and distinguished research scientist at the University of Michigan's Institute for Social Research and a principal investigator on the Monitoring the Future study since its inception in 1975. In 1965, he initiated the Youth in Transition project and has authored five books and numerous articles based on that research. His scientific publications focus on youth and social issues, and his current research interests include drug use and attitudes about drugs other values, attitudes, and behaviors of youth military plans and experiences and public opinion as related to a number of other social issues.

Lourdes Baezconde-Garbanati is an associate professor of research at the Keck School of Medicine, University of Southern California. She is a member of the directorate of the Institute for Health Promotion and Disease Prevention Research at USC, focusing on community outreach and education, and a member of the USC/Norris Comprehensive Cancer Center. She holds a joint appointment in the Department of Sociology and a courtesy appointment at the Annenberg School of

Communication. Baezconde-Garbanati conducts research, teaches, and provides technical assistance and training on tobacco and cancer control, culturally competent research, gender and ethnic minority health, and international health, emphasizing health disparities and the translation of scientific findings into community initiatives. She is the principal investigator and director of a California Department of Health Services-funded program, the Hispanic/Latino Tobacco Education Partnership, as well as the Southern California Partnership Program Office of the National Cancer Institute.

Cristina B. Bares is an assistant professor at the School of Social Work at Virginia Commonwealth University whose research interests include testing complex etiological models that incorporate emotional health to explain the onset and maintenance of adolescent and early adulthood cigarette use. Using a developmental perspective, she has examined the transactional relationships between psychological, familial, and peer factors influencing internalizing problems of adolescents and young adults. Through both primary data collection studies and analyses of secondary data, she has aimed to understand the multiple influences on smoking initiation and progression to heavier use.

Steven R.H. Beach is a faculty member in the Clinical Psychology program at the University of Georgia and a licensed clinical psychologist with over 25 years of experience investigating the role of family relationships in the development and maintenance of psychopathology, particularly depression and substance use disorders. He currently serves as codirector of the Center for Family Research, an interdisciplinary research center. He is widely recognized as an authority on the link between marital discord and depression and the use of marital interventions in the context of depression. He is also recognized for his research on the role of genetic moderators of intervention response and other contextual variables and the potential role of epigenetic regulation in accounting for contextual and intervention effects.

Gene H. Brody is regents' professor and director of the Center for Family Research at the University of Georgia. Since 1988, he has conducted research on rural African American family life in which more than 3000 families have participated. An author of more than 300 publications, Dr. Brody has explored normative family processes, contextual processes associated with youth adjustment, and factors that protect youth and families from the effects of adversity. He has applied this research to the development and evaluation of preventive interventions tailored to the needs of rural African Americans. Most recently, he has explored gene × environment processes that forecast psychosocial outcomes. Dr. Brody is a fellow of the American Psychological Association and has served on several NIH standing review panels.

Clifford L. Broman is a professor in the Department of Sociology at Michigan State University. His general research area is psychosocial factors in physical and emotional health. Recent research has focused on the role of race/ethnicity in substance use, psychosocial stress, and family formation behaviors and attitudes among

African Americans. Recent publications have appeared in *Drug and Alcohol Dependence*, *Race and Social Problems*, and *Psychological Services.*

C. Hendricks Brown is a professor in the Departments of Psychiatry and Behavioral Sciences and Preventive Medicine in the Northwestern University, Feinberg School of Medicine. He also holds adjunct appointments in the Departments of Biostatistics and Mental Health at the Johns Hopkins Bloomberg School of Public Health as well as the Department of Public Health Sciences at the Miller School of Medicine at the University of Miami. He directs the NIDA-funded Center for Prevention Implementation Methodology (Ce-PIM) for Drug Abuse and Sexual Risk Behavior and an NIMH-funded study to synthesize findings from individual-level data across multiple randomized trials for adolescent depression. Since 1985 he has received NIH funding to direct the Prevention Science and Methodology Group (PSMG), now a national network of over 180 scientists and methodologists who are working on the design of preventive field trials and their analysis and implementation of prevention programs. Recently, his work has focused on the prevention of drug abuse, conduct disorder, depression, and suicide. Brown has been a member of the recent National Academy of Sciences/Institute of Medicine committee on prevention science and serves on numerous federal panels, advisory boards, and editorial boards.

Lindsay Bynum is a currently doctoral student at the University of Illinois at Chicago (UIC). She received her undergraduate degree (B.A.) from Johns Hopkins University in 2009 in psychology and sociology and her master's degree (M.A.) in Community Psychology from UIC in 2013. Her research program investigates how spaces can promote pro-social behavior, with an emphasis on the institutional factors that facilitate or hinder civic engagement. Currently, she is investigating how nonprofit organizations approach volunteer retention and beginning an analysis of service learning environments' effect on nonacademic learning outcomes. Demonstrating her commitment to civic engagement, Lindsay has served as a researcher, teacher, and consultant.

Nicole Cano is the national manager of the National Hispanic Science Network, an international network of over 500 scientists and students dedicated to making significant advances in the field of Hispanic health research. As the national manager, she is responsible for supporting the executive director in the coordination of all national activities and also serves as the program coordinator for the organization's R25 mentoring and training program at Michigan State University. Nicole obtained her MPH in 2011 from the University of Miami Miller School of Medicine and has worked in several prevention research projects targeting parents, newborns, infants, and adolescents and has subsequently served as a coauthor in several publications in these areas. Nicole is currently obtaining an executive MBA in health sector management and policy at the University of Miami to further her career in administration and management.

Alice Cepeda is an assistant professor in the School of Social Work at the University of Southern California. She received her doctoral degree in sociology from the City University of New York, Graduate Center. Her work has focused on the social epidemiology of drug use and the related health risk behaviors that disproportionately affect urban Mexican-origin minority populations including violence, HIV/STI infection risks, and mental health conditions.

Chih-Ping Chou is a professor of Preventive Medicine. His research focuses on the advancement of research methodology and statistical techniques in social and health behavioral research. His research interest falls into three distinct areas: evaluation of prevention intervention of substance use among adolescents, evaluation of substance abuse treatment, and statistical and methodological application and development for prevention research. Dr. Chou is an internationally recognized researcher on structural equation modeling. He has a well-established record on the application and development of statistical models and research methodologies in prevention research and has extensive experience in longitudinal analyses of the effects of health promotion interventions, structural equation modeling, multilevel modeling, growth curve modeling, and psychometrics analysis. Dr. Chou received the Research Scientist Development Award and several research projects from NIH to study advanced statistical methods for prevention research. He has also been serving as the directors of measurement core and statistics core for four NIH-funded transdisciplinary research centers based at USC. Dr. Chou also holds a joint appointment in School of Social Work.

Tammy Chung is an associate professor of psychiatry and epidemiology at the University of Pittsburgh. She received her Ph.D. from Rutgers University in clinical psychology. Her research focuses on assessment, diagnosis, and course of adolescent alcohol and other substance use disorders in community and treatment samples. She is principal investigator of the National Institute on Drug Abuse component of the Pittsburgh Girls Study, as well as a longitudinal study examining changes in treated adolescents' social network characteristics in relation to substance use outcomes. Dr. Chung is a coinvestigator on the National Consortium on Alcohol and Neurodevelopment in Adolescence (Pittsburgh site), a multisite longitudinal neuroimaging study that aims to examine the effects of alcohol on the developing brain.

Wilson M. Compton serves as the deputy director of the National Institute on Drug Abuse (NIDA) of the National Institutes of Health. Dr. Compton's responsibilities include providing scientific leadership in the development, implementation, and management of NIDA's research portfolio and working with the director to support and conduct research to improve the prevention and treatment of drug abuse and addiction. Prior to his current appointment, Dr. Compton served as the director of NIDA's Division of Epidemiology, Services and Prevention Research from 2002 until 2013. Of note, he led the development of a large-scale longitudinal population study to assess the impact of new tobacco regulations in the United States. The project is jointly sponsored by NIDA and the US Food and Drug Administration with yearly data collection from approximately 60,000 study participants, including both surveys and biological assessments of tobacco exposures, risk factors, and health outcomes. Before joining NIDA, Dr. Compton was associate professor of psychiatry and director of the Master in Psychiatric Epidemiology Program at Washington University in Saint Louis as well as medical director of Addiction Services at the Barnes-Jewish Hospital in Saint Louis. Dr. Compton received his undergraduate education from Amherst College. He attended medical school and completed his residency training in psychiatry at Washington University.

Rand D. Conger is a distinguished professor of psychology, human development, and family studies at the University of California, Davis. He received his doctoral degree in sociology from the University of Washington in 1976. He has also been a faculty member at the University of Georgia, the University of Illinois at Urbana/Champaign, and Iowa State University. Dr. Conger's program of research focuses on social, cultural, and individual characteristics that either promote successful development or lead to significant developmental problems. Results of his research with over 500 White families in rural Iowa demonstrated that financial pressures can create severe strains in family relationships which, in turn, increase emotional and behavioral problems for both parents and children. This work also identified individual and family strengths that increase resilience to developmental problems in the face of adversity. His more recent studies suggest that this economic stress process is quite similar for African American and Mexican American families. Other research contributions include the study of intergenerational continuities in family characteristics and varying developmental pathways in the transition from adolescence to adulthood. Dr. Conger has published over 300 books, book chapters, and journal articles, and his research has been supported by a series of federal grants from the National Institutes of Health. In addition, the significance of his scholarly activities has been recognized through several awards from professional organizations.

Michele Cooley-Strickland is a licensed psychologist who joined the UCLA faculty in 2009 and is a research psychologist in the Center for Culture and Health, in the Department of Psychiatry and Biobehavioral Sciences, David Geffen School of Medicine, NPI-Semel Institute. Dr. Cooley-Strickland has also been on faculty in the Department of Mental Health, Bloomberg School of Public Health at the Johns Hopkins University since 1996, and is an adjunct associate professor. She received her bachelor's, M.Ed., and doctorate in clinical child psychology from the University of Virginia. Dr. Cooley-Strickland is a community-based clinical child researcher, preventive interventionist, and teacher. Over the past 20 years, Dr. Cooley-Strickland has served as the principal investigator of multiple grants funded by the National Institute on Drug Abuse (NIDA) and the National Institute of Mental Health (NIMH) designed to study and prevent the emotional, behavioral, and academic outcomes of youth's exposure to community violence. Dr. Cooley has widely disseminated the results of her community violence research projects in presentations, peer-reviewed publications, book chapters, and coauthored clinical treatment intervention manuals. She has been active in the National Institutes of Health (NIH), serving on several advisory committees, multiple standing review committees, and numerous special emphasis panels, as well as serving on many panels and task forces for professional organizations and serving as editorial board members and reviewers of numerous scientific journals.

David Cordova is an assistant professor at the University of Michigan School of Social Work. A substance abuse and mental health services administration minority fellow, David completed his Ph.D. at Michigan State University in the Department of Human Development and Family Studies with a specialization in couple and family therapy. Additionally, he completed a National Institute on Drug Abuse-funded postdoctoral position at the University of Miami Miller School of Medicine in the Department of Public Health Sciences, Center for Family Studies. A recipient of the National Institutes of Health Loan Repayment Program, David's research and clinical and service interests are in adolescent substance abuse prevention and sexual health promotion.

Jorge Delva is professor and associate dean for research at the School of Social Work, University of Michigan. A native of Chile, he conducts research focusing on addressing and reducing health disparities and helping improve the lives of low-income and racial and ethnic minority populations. His research began in Honolulu two decades ago where he worked on SAMHSA-funded projects aimed at improving the health and mental health of Asian and Pacific Islander children and their families. His more recent state and NIH funded projects show his dedication to combating health disparities. His work has served to advance the substance abuse field's understanding of psychosocial-cultural mechanisms associated with substance using behaviors among Hispanic/Latino, African American, and American Indians of lower socioeconomic position in the United States and with disadvantaged populations in Latin America.

Elaine E. Doherty is an associate professor in the Department of Criminology and Criminal Justice at the University of Missouri, St. Louis. Her research focuses on crime and substance use over the life course, especially the impact of adult life events on these behaviors.

J. Mark Eddy is the director of research at Partners for Our Children, a child welfare research, practice, and policy center that is a collaboration between the School of Social Work at the University of Washington, the Washington State Department of Social and Health Services, and private sector funders. He is a prevention scientist and licensed psychologist whose work has focused on the development and testing of multimodal psychosocial interventions for children and families who are living in challenging contexts. He has a particular interest in working with communities to develop culturally appropriate, evidence-informed interventions that are owned by the community and then rigorously tested. Prior to his current position, Dr. Eddy was a senior scientist for 25 years at the Oregon Social Learning Center in

Eugene. He is the principal investigator or coinvestigator on a number of longitudinal randomized controlled trials conducted within social systems of care, including child welfare, schools, juvenile justice, and criminal justice, and focused on key public health outcomes for youth, including aggression, delinquency, academic failure, early sexual behavior, and substance use and abuse. He regularly reviews grant applications for various public and private funders and serves as an associate editor for the flagship journal of the Society for Prevention Research, *Prevention Science*.

Margaret E. Ensminger is a professor in the Department of Health, Behavior, and Society, the Bloomberg School of Public Health, the Johns Hopkins University. Her research has focused on social integration, socioeconomic resources, social context and its influence on individual characteristics, and development over the life course. She has led the Woodlawn Study for several decades.

Estrada received her doctoral degree in counseling psychology in 2012. She is currently a senior research associate in the Department of Public Health Sciences at the University of Miami Miller School of Medicine. She has coauthored over 10 peer-reviewed publications and chapters and is coprincipal investigator of an early career mentored grant which aims to adapt an evidence-based, family-based preventive intervention for Hispanic youth onto the Internet. Her overarching interests involve the role of cultural processes in the prevention of health disparities among the Hispanic population. Specifically, her focus has been on the prevention of substance use and HIV infection among Hispanic adolescents through employment of a family-based intervention, Familias Unidas.

Kate E. Fothergill is an assistant scientist in the Department of Health, Behavior, and Society, the Bloomberg School of Public Health, the Johns Hopkins University. Her areas of research include the development of risk behaviors over the life course, particularly during childhood and adolescence, substance abuse and crime, and individual, social, and neighborhood determinants of behavior and health. She has been part of the Woodlawn Study team for more than a decade.

Deborah Furr-Holden is an epidemiologist with expertise in drug and alcohol dependence epidemiology, psychiatric epidemiology, and prevention science. Her research areas include methodological issues surrounding the design and evaluation of interventions including sampling, program modeling, and innovative statistical and evaluation approaches, measurement of drug and alcohol use disorders, and innovative methods to prevent and reduce health inequalities, with a focus on behavioral health inequalities. Dr. Furr-Holden is also an expert in environmental approaches to violence, alcohol, tobacco, and other drug preventions. She has worked with local and national policy makers to improve data-driven decision making, including "health in all policies." She currently maintains a portfolio of research focused on ameliorating health inequalities in substance abuse and treatment and environmental strategies to prevent and reduce community violence.

Meg Gerrard received her Ph.D. in psychology from the University of Texas. Her major area of research has concerned the prediction and prevention of adolescent health risk behaviors including unprotected sex, substance use, and UV exposure. Much of her research has been devoted to the exploration of social, familial, and neighborhood factors as sources of stress affecting the health and health behavior of African American adolescents and young adults. She is in the Department of Psychology and the Center for Health Intervention and Prevention (CHIP) at the University of Connecticut.

Frederick X. Gibbons received his Ph.D. in social psychology from the University of Texas. He is a health-social psychologist whose research involves applying social psychological theory and principles to the study of health behavior, primarily substance use and risky sexual behavior, especially among African American adolescents and young adults. His research includes laboratory and field methods as well as interventions and preventive interventions. He is currently a professor of psychology and an affiliate of the Center for Health Intervention and Prevention (CHIP) at the University of Connecticut.

Jennifer Goode whose credentials include an M.S.W. from the University of Texas at Arlington and a Ph.D. in sociology from Bowling Green State University in Ohio, conducts research that blends her interest in social structural influences and social psychological processes. Dr. Goode's interest in issues of race is prominent in her research, with particular interest in racial disproportionality and disparity in various social systems and the relative contribution of individual and systemic factors. Most recently, she is coauthor of Racial Differences in Effects of Religiosity and *Mastery on Psychological Distress: Evidence from National Longitudinal Data in Society and Mental Health* (March 2013 3: 40–58). She has published several book chapters on race and macro-/microlevel influences including *Africana Cultures and Policy Studies, Scholarship and the Transformation of Public Policy* (Palgrave Macmillan 2009), and *Teaching City Kids: Understanding & Appreciating Them* (Lang 2006). Dr. Goode has been an invited speaker and panelist on racial issues and has given many presentations on such topics at national and regional conferences.

Kerry M. Green is an associate professor in the Department of Behavioral and Community Health, University of Maryland School of Public Health. Her research focuses on the interrelationship of substance use, mental health, and criminal involvement over the life course. She is particularly interested in childhood and adolescent development and long-term consequences of risk behaviors during these life stages. She began her work with the Woodlawn team over 10 years ago.

Robert S. Griffin is an attorney from Maryland who majored in psychology as an undergraduate at Towson University and graduated magna cum laude in 2003. Prior to earning his law degree, Mr. Griffin served as the program director, data analyst, and field coordinator for several research projects in the Department of Mental Health, Bloomberg School of Public Health, Johns Hopkins University, including the Multiple Opportunities to Reach Excellence (MORE) Project that was funded by the National Institute on Drug Abuse (NIDA). Mr. Griffin has served as the coauthor of multiple peer-reviewed journal articles and book chapters.

Andrew Grogan-Kaylor is associate professor at the School of Social Work, University of Michigan. His substantive interests are in child development, parenting, neighborhoods and social development, and academic achievement. Much of his current research focuses on the effects of schools, neighborhood, and parenting on outcomes for children. A great deal of his current research agenda focuses on corporal punishment, and he has published several pieces using longitudinal data and advanced statistical methods indicating that parental use of corporal punishment is associated with increases in children's antisocial behavior. His research is strongly quantitative and most often makes use of large nationally representative data sets as well as sophisticated longitudinal and multilevel methodologies to examine things like growth and change over time, or community, school, or parent effects on children and families. He has taught classes on statistics and research methodology at the master's level, as well as in the University of Michigan's Joint Doctoral Program in Social Work and Social Science. He has an MSSW in Social Work and PhD in Social Welfare from the University of Wisconsin.

Denise D. Hallfors is a senior research scientist at PIRE and an adjunct professor in Maternal and Child Health at the University of North Carolina, Chapel Hill. Dr. Hallfors holds a masters in psychiatric nursing from Arizona State University and a Ph.D. in social welfare policy from the Heller School at Brandeis University, where she was an NIMH health services research fellow, specializing in the economics of health and mental health policy. Dr. Hallfors specializes in adolescent and young adult health and has led numerous NIH and Robert Wood Johnson Foundation studies to examine mechanisms for preventing risk and promoting health and to test interventions in clinical trials. She has also focused on racial disparities in HIV and other sexually transmitted infections, analyzing data from the National Longitudinal Study of Adolescent Health (Add Health) to better understand underlying paradoxes that might lead to more effective interventions, including structural and contextual interventions. She currently leads an ongoing randomized controlled trial in Zimbabwe to test whether keeping adolescent orphan girls in school can help to prevent HIV infection. She is also coinvestigator and senior advisor on a second similar clinical trial in western Kenya and coinvestigator on a current Add Health study examining the trajectories of sexual behavior from adolescence through young adulthood.

Carolyn Tucker Halpern is professor of Maternal and Child Health (MCH) in the Gillings School of Global Public Health at the University of North Carolina at Chapel Hill (UNC-CH). She is a developmental psychologist, and her research interests center on adolescent and young adult sexuality and reproductive health. She is especially interested in systems approaches to this topic that consider biopsychosocial factors within cultural context. She is the deputy director of the National

Longitudinal Study of Adolescent Health (Add Health) and the principal investigator of an NICHD-funded project that is using Add Health data to examine sexual trajectories from adolescence into adulthood and the implications of those patterns for multiple aspects of young adult well-being. She is a coinvestigator on several NIH-funded projects evaluating interventions to reduce HIV risk in adolescents in sub-Saharan Africa. Halpern directs the doctoral program in the MCH department and teaches courses in adolescent health, sexuality, theory, and grant development. Her most recent publications appear in the *American Journal of Public Health, Journal of Adolescent Health, Journal of Sex Research,* and the *Maternal and Child Health Journal.*

Roderick J. Harrison is a nationally recognized expert on demographic and socioeconomic trends in the United States, and their implications for the nation, states and localities, and private, nonprofit, and public sector actors seeking to understand, reach, employ, or serve increasingly diverse populations. He is a senior research scientist at the Office of the Vice President for Research and Compliance (2009–present) and associate professor in the Department of Sociology and Anthropology (1992–present) at Howard University. Dr. Harrison was also on the faculty of Harvard University (1984–1990). Between 1998 and 2008, he served as director of the Databank, Office of Research at the Joint Center for Political and Economic Studies (1998–2008) in Washington, DC, during which time he received the Roger Herriot Award of the American Statistical Society for Innovations in Federal Statistics. Dr. Harrison served as the special assistant on Special Populations (1997–1998) and as chief of the Racial Statistics Branch, Population Division, of the US Bureau of the Census. His research and publications focus on demographics, employment and earnings, health, education, youth, substance use, housing, and crime. His current research focuses on the development and dissemination of key social indicators for the nation, localities, racial and ethnic groups, and children. Dr. Harrison holds graduate degrees from Princeton University and an A.B. from Harvard University. His research, including landmark studies of trends in residential segregation and in the growing diversity and changing socioeconomic status of racial and ethnic populations, definitive works on measuring race and ethnicity in the federal statistical system, and pioneering models of occupational labor markets, is widely cited.

Abigail A. Haydon is the assistant director of Research at FrameWorks Institute. A public health scholar with training in maternal and child health, developmental science, and demography, she has combined these perspectives in research on adolescent relationships and reproductive health, teen dating violence, and educational and social outcomes among youth with chronic illnesses. Prior to joining FrameWorks, Abigail worked at the *Eunice Kennedy Shriver* National Institute of Child Health and Human Development as an executive branch science fellow sponsored by the American Psychological Association and the AAAS. Her research has appeared in the *Journal of Adolescent Health, Perspectives on Sexual and Reproductive Health*, the *Journal of Youth and Adolescence, Archives of Pediatrics* *and Adolescent Medicine*, and the *American Journal of Public Health*. Abigail received her B.A. from Pomona College and her M.P.H. and Ph.D. from the Gillings School of Global Public Health at the University of North Carolina at Chapel Hill.

Marya Hynes received her master of health sciences from Johns Hopkins University in 1998, with a concentration in the epidemiology of drug abuse and mental disorders. For the past 15 years, she has been working at the Inter-American Drug Abuse Control Commission (CICAD) at the Organization of American States (OAS) where she leads a variety of programs to promote drug research in Latin American countries. Previously, Ms. Hynes was the assistant director of a transitional living program for young adults with mental illness and drug abuse disorders. She is the author of numerous publications on drug use in Latin America and the Caribbean.

Bonita J. Iritani is an associate research scientist at PIRE. She has a master's degree in psychology from Stanford University and a master of social service from Bryn Mawr College. Ms. Iritani has worked as a project coordinator, research associate, and data analyst on a variety of studies primarily pertaining to risk for and prevention of HIV and sexually transmitted infections among adolescents and young adults. The studies have included randomized controlled trials assessing effects of providing school support on sexually transmitted infection, early marriage and school dropout among orphan adolescents in Africa, and research examining data from the National Longitudinal Study of Adolescent Health (Add Health) on sexual behavior, sexually transmitted infections, and intimate partner violence among youth and young adults in the United States.

James S. Jackson is the Daniel Katz distinguished university professor of psychology, professor of Afroamerican and African Studies, and director of the Institute for Social Research, all at the University of Michigan. His research focuses on issues of racial and ethnic influences on life course development, attitude change, reciprocity, social support, and coping and health among blacks in the Diaspora. He is past director of the Center for Afroamerican and African Studies and past national president of the Association of Black Psychologists. He is currently directing the most extensive social and political behavior and mental and physical health surveys on the African American and Black Caribbean populations ever conducted. He is a founding member of the "Aging Society Research Network" of the MacArthur Foundation. He is the president of the Consortium of Social Science Associations and a member of the Institute of Medicine and a fellow of the American Academy of Arts and Sciences.

Marie-Claude Jipguep-Akhtar is an associate professor in the Department of Sociology and Anthropology at Howard University. She currently serves as scholar-in-residence in the Center on Health, Risk and Society at American University. She previously served as social scientist in the Howard University Cancer Center where she was also an assistant professor in the Department of Family and Community Medicine (2004–2006), as well as senior research associate/statistician for Howard University's Research Program in the Epidemiology and Prevention of Drug Abuse (2001–2004). Her research and publications focus on race/ethnicity, "place" disparities to health, and educational and occupational outcomes. Her work emphasizes on identifying possible causal mechanisms that link the social context (social and structural factors) to inequalities (in health and opportunity structure), the life course, by race and gender. Dr. Jipguep-Akhtar holds a Ph.D. in sociology with a specialization in research methodology and social demography.

Lloyd D. Johnston Angus Campbell Collegiate research professor and university distinguished senior research scientist at the University of Michigan's Institute for Social Research and principal investigator of the Monitoring the Future study since its inception in 1975. A social psychologist by training, he has served as advisor to the White House, Congress, and many other national and international bodies and has conducted research on a wide range of issues, including the use of alcohol, tobacco, and various illicit drugs, institutional trust, policy evaluation, the functioning of American high schools, behaviors influencing the spread of HIV, and, most recently, childhood obesity. His research interests also include international comparative studies and the application of survey research to social problems generally. He is the recipient of the 2012 Lifetime Achievement Award from the Community Anti-Drug Coalitions of America (CADCA).

Hee-Soon Juon is a social and behavioral scientist whose research has focused on substance abuse and criminal behavior, minority mental health, and cancer control behaviors among Asian Americans. She has been part of the Woodlawn Study team for more than two decades. She is a professor in the Sidney Kimmel Medical College, Thomas Jefferson University.

Denise Kandel is professor of sociomedical sciences in psychiatry at the College of Physicians and Surgeons at Columbia University and head of the Department of the Epidemiology of Substance Abuse at the New York State Psychiatric Institute. Dr. Kandel received her Ph.D. in sociology from Columbia University. Her major research interests are in the epidemiology, risk factors and consequences of drug use and dependence, comorbidity between substance use and psychiatric disorders, and the intergenerational transmission of deviance. Her current work focuses on the epidemiology, etiology, and consequences of smoking and nicotine dependence in adolescence. Dr. Kandel recently completed a major longitudinal study of the transition from experimental smoking to nicotine dependence in adolescence and early adulthood. Dr. Kandel developed the gateway hypothesis of drug involvement that describes a unidirectional sequence of drug consumption in which the use of tobacco or alcohol precedes the use of marijuana, cocaine, and other illicit drugs. She pioneered in charting the developmental phases of drug use, showed that these stages were present in the United States and abroad, and identified the specific risk and protective factors for adolescent initiation into each of the major stages of drug use.

Charles Kaplan received his Ph.D. from the University of California, Los Angeles, in sociology in 1973. He is currently the associate dean of research and research professor at the School of Social Work, University of Southern California. He is an active member of the NIDA National Hispanic Science Network. His research emphasizes the linking of drug abuse epidemiology with experimental approaches to drug prevention and treatment suitable for emerging special populations from an international comparative perspective. Recently, he is developing exploratory research in emerging patterns of cannabis use and abuse in the changing policy context of legalization and medicalization in the United States and Europe.

Krim K. Lacey earned his doctoral degree in sociology from Wayne State University and completed his postdoctoral training at the Program for Research on Black Americans (PRBA), Institute for Social Research at the University of Michigan. Dr. Lacey currently has research affiliations with Wayne State University, College of Pharmacy and Health Sciences, and University of Michigan's Institute for Social Research. His primary research interest is on intimate partner violence, particularly focusing on minority and immigrant populations. He has also been engaged in research that addresses the influence of social context and cultural factors on the physical and mental well-being of Caribbeans residing in the United States and within Caribbean regions. Dr. Lacey' s research interests further extend to the examination of educational disparities among secondary and postsecondary students.

Rolf Loeber is distinguished university professor of psychiatry and professor of psychology and epidemiology at the University of Pittsburgh, Pittsburgh, Pennsylvania. He is director of the Life History Program and is principal investigator of two longitudinal studies, the Pittsburgh Youth Study, and the Pittsburgh Girls Study. He has published widely in the fields of juvenile antisocial behavior and delinquency, substance use, and mental health problems. He is an elected member of the Koninklijke Academie van Wetenschappen (Royal Academy of Sciences) in the Netherlands and the Royal Irish Academy in Ireland.

Charles R. Martinez is a clinical psychologist and professor and department head in the Department of Educational Methodology, Policy, and Leadership at the University of Oregon, where he also directs the Center for Equity Promotion. He served as the University of Oregon vice president for Institutional Equity and Diversity from 2005 to 2011 and also served as a senior scientist at the nonprofit Oregon Social Learning Center where he founded and directed the center's Latino Research Team since 1999. His substantive interests center on identifying factors that promote healthy adjustment for families and children following stressful life events, (e.g., changes in family structure, socioeconomic status, physical and/or emotional health, acculturation, and immigration status), taking into consideration the cultural contexts in which families operate. Dr. Martinez has led numerous federally and internationally funded research projects designed to examine risk and protective factors involved in linking acculturation to behavioral health outcomes for Latino families and to develop and test culturally specific interventions for Latino families at risk of behavioral health problems. His main areas of published work include substantive and methodological topics related to cultural issues in prevention science, Latino educational and behavioral health disparities, links between acculturation and discrimination on substance use and academic failure prevention for Latino youth, culturally specific parenting interventions, the role of biosocial stress markers on family outcomes, and family process. He is a nationally known consultant on organizational diversity issues, cross-cultural research, and community engagement. He teaches courses in equity, multicultural education, leading for diversity, prevention science, and Latino family health and education.

Niki Matusko works as a research area specialist for the Program for Research on Black Americans at the University of Michigan Institute for Social Research. In this capacity she is responsible for data management and primary analysis of various nationally representative data sets such as the NSAL adolescents, family connections, and the Caribbean cross section. Her educational background and work experience has focused on primary and secondary research areas in both private and public sectors. Ms. Matusko is well versed in advanced statistical procedures such as structural equation modeling, hierarchical linear modeling, time series, and ARIMA forecasting. Additionally, she has extensive experience in compiling and presenting secondary research like market share quantification, in-migration and out-migration patterns, and various other strategic intelligence analyses. Her roles have included client research analyst, health information specialist, and strategic market analyst. Ms. Matusko has a bachelor of science degree in mathematics from the University of Michigan.

Heather H. McClure Jr. is a cultural anthropologist with 15 years of community-based and policy research focused on stress and health, human rights, social networks, and cultural assets with Latino communities in the United States and in Latin America. She is research associate with the Center for Equity Promotion (formerly the Latino Research Team at the Oregon Social Learning Center) and the Anthropology Department at the University of Oregon. Since 2007, she has led the team's work to integrate stress biomarkers and health measures into their research on the effects of acculturation, discrimination, and related stressors for Latino family health. Dr. McClure has led federally funded research projects, including the first household survey of Latino health in Washington, D.C. (2002–2004). Her published work has spanned topics of acculturation, discrimination, stress, and Latino health, methodological issues involved in the integration of biomarkers into Latino health disparities research, childhood behavioral predictors of adult overweight and obesity, and political asylum involving Latin American applicants who are gay, lesbian, transgendered, and/or living with HIV/AIDS. Dr. McClure's courses include social determinants of health, immigration and farm workers in the United States, and intercultural competence (critical race theory grounded in ethnographic field methods).

Ross E. O'Hara is currently a behavioral researcher at Persistence Plus, LLC, an innovative company that engages and motivates students to college completion through a mobile platform that uses transformative behavioral interventions. He earned his Ph.D. in social psychology from Dartmouth College in 2011 and completed a NIAAA-funded postdoctoral fellowship in the Department of Psychiatry at the University of Connecticut Health Center. His research interests focus on how psychological, social, and behavioral variables differentially predict patterns of alcohol and drug use among adolescents of different racial/ethnic backgrounds and identities.

Patrick M. O'Malley Jr. is a research professor at the University of Michigan's Institute for Social Research and coprincipal investigator on the Monitoring the Future study. He received his Ph.D. degree in psychology from the University of Michigan in 1975 and has been associated with the Monitoring the Future project since then. His publications deal with alcohol, tobacco, and illicit drug use and related attitudes and beliefs. His research interests include causes and consequences of drug use, driving under the influence of alcohol or drugs, social epidemiology of drug use, and longitudinal survey data analysis techniques.

Katherine Otte is an adjunct professor of psychology at Howard Community College in Howard County, Maryland. She received her M.S. in psychological and psychiatric anthropology from Brunel University in the United Kingdom in 2011. She has conducted research on the exploitation of the stateless hill tribes living along the border of Thailand and Myanmar as well as the mental health impact of growing up without a nationality in Thailand. Currently, she is assisting with a nationwide study on healthcare in the United States.

Hilda Pantin is professor and executive vice chair of the Department of Public Health Services at the University of Miami Miller School of Medicine. She is also the associate director of the Center for Family Studies. She has over 20 years of experience working in prevention and specializes in developing family-based ecodevelopmental interventions for Hispanic families. Dr. Pantin currently serve as a coinvestigator of a P30 grant and as codirector of the Training Core of that center. Additionally, she has served as an investigator or coinvestigator on nine R01s in the area of prevention science. Dr. Pantin was awarded with the Outstanding Research Publication Award by the American Association for Marriage and Family Therapy in 2002 and received the 2010 Community, Culture and Prevention Science Award from the Society of Prevention Research. She is also the principal investigator and executive director of the National Hispanic Science Network (NHSN), an international network of over 500 scientists and students dedicated to making significant advances in the field of Hispanic health research. As executive director, Dr. Pantin is responsible for the coordination of all national activities including a summer training program at the NIH's Intramural Program for graduate students, a recently funded R25 mentoring and training program, for which she serves as coinvestigator, and a pilot grant funding program for early stage investigators.

Robert A. Philibert is a faculty member at the University of Iowa in the Department of Psychiatry. He is a board-certified psychiatrist with 30 years of experience at the laboratory benchtop. He has a deep abiding interest in the epigenetic processes associated with the development of behavioral illnesses. He is also the founder and chief scientific officer of Behavioral Diagnostics, a recently funded biotechnology company.

Guillermo Prado is the Dean of the Graduate School at the University of Miami. He obtained his doctoral degree in epidemiology and public health in 2005. In addition, he is currently the Leonard M. Miller professor of Public Health Sciences and chief of the Division of Prevention Science and Community Health at the University of Miami Miller School of Medicine. Prado's research focuses on the prevention of substance use and HIV risk behaviors among Hispanic youth. His program of research has been continuously funded by the NIH since the first year of his doctoral program. He is currently the PI or senior mentor of several substance abuse and HIV prevention studies totalling approximately $10 million of NIH and CDC funding. One of these studies is an effectiveness study of Familias Unidas [United Families in English], one of the few preventive interventions found to be efficacious in preventing/reducing substance use and HIV risk behaviors among Hispanic youth. Dr. Prado is also the director of the Behavioral and Social Sciences and Community Outreach Core of the Miami Center for AIDS Research. Additionally, he is the codirector of Training for the Center for Prevention Implementation Methodology for Drug Abuse and Sexual Risk Behavior and director of Training for the South Florida Cancer Health Disparities Center. His research has been recognized by numerous organizations, including the Society for Prevention Research, the Society for Research on Adolescence, and the National Hispanic Science Network on Drug Abuse. He was also selected by the Miami Herald as one of the top 20 business leaders and innovators in South Florida under the age of 40 for his research and community work in the areas of HIV and substance use among Hispanic families.

LeShawndra N. Price is Chief of the Health Inequities and Global Health Branch within the Center for Translation Research and Implementation Science (CTRIS) at the National Heart, Lung, and Blood Institute (NHLBI), National Institutes of Health (NIH). In this position, she supports and fosters a program of innovative research pertaining to health inequities, domestically and globally, including identifying gaps and needs, as well as research opportunities to address them. She also leads the strategic development of "T4" translation research and implementation science in the global arena.

Immediately prior to joining NHLBI, Dr. Price served as Deputy Director for Research and Chief of the Research Scientist Development program in the Office for Research on Disparities and Global Mental Health at the National Institute of Mental Health (NIMH). She also served as Project Scientist for a multi-million dollar initiative involving eight distinct clinical trials designed to improve treatments and expand access to mental health care in more than 20 low- and middle income countries, including Brazil, Peru, Nigeria, South Africa, Ghana, Ethiopia, India, and

Pakistan. She also served as one of the NIH's coordinators of the Mental Health Across the Lifespan Initiative, a collaboration between the NIH and Delta Sigma Theta Sorority, Incorporated to raise awareness about mental health conditions affecting women and their families, including bullying, postpartum depression, and successful aging later in life.

Dr. Price brings a wealth of experience to her work at the NHLBI, as she also has held positions at the National Institute on Drug Abuse, the Carolina Power and Light Energy & Environmental Center, Burroughs-Wellcome (now Glaxo Smith Kline), Research and Evaluation Associates, the University of North Carolina at Chapel Hill, Duke University, George Washington University, and North Carolina Central University. Her leadership, service, and advocacy has been recognized through awards from the NIH Office of the Director, NIMH Director, NIDA Director, American Psychological Association, and the Family Research Consortium.

Dr. Price's academic background includes a B.A. in Psychology from Wake Forest University and an M.A. and Ph.D. in Developmental Psychology from the University of North Carolina at Chapel Hill.

Tanya J. Quille is a licensed psychologist currently in private practice in the Baltimore and Washington metropolitan areas. She was previously an assistant scientist at the Johns Hopkins Bloomberg School of Public Health and research assistant professor at the University of Miami Medical School in the Department of Psychiatry. Dr. Quille received her Ph.D. from the University of Miami. From 1999 to 2003, Dr. Quille was the clinical director of the Center for Treatment Research on Adolescent Substance Abuse. Dr. Quille was coinvestigator, project director, and model developer for multiple research projects investigating substance abuse treatment for adolescents and adults based on family therapy interventions. Dr. Quille was also coinvestigator for projects investigating engagement and retention to treatment for HIV-positive women and children and a study focusing on obesity in African American women. Dr. Quille has also headed numerous dissemination projects incorporating evidence-based treatments into state and local treatment services.

Anamara Ritt-Olson is a research associate at the Keck School of Medicine, University of Southern California. Her research interests currently focus on the impact that family has on the well-being of adolescents and on ways to harness that power to engage youth in optimal development. Her interests also include curriculum development, research methods, and evaluation of programs designed to improve adolescent health.

Judy A. Robertson was a senior research associate in the Department of Health, Behavior and Society, the Bloomberg School of Public Health, the Johns Hopkins University. She was a data analyst and data manager for the Woodlawn data set from 2005 until she retired in 2012. She participated in studies of life course factors relating to alcohol and drug use disorders, physical and mental health problems, criminal behavior, and issues associated with aging.

Richard W. Robins is a professor of psychology at the University of California, Davis. He received his doctoral degree in psychology from the University of California, Berkeley, in 1996. Dr. Robins' program of research focuses on adolescent personality development, risk and resilience processes in Mexican-origin families, and the regulation and expression of self-conscious emotions. He has published over 140 journal articles, book chapters, and books, including the *Handbook of Personality: Theory and Research* and the *Handbook of Research Methods in Personality Psychology*, and his research has been supported by a series of federal grants from the National Institutes of Health. He is the recipient of the *American Psychological Association's* "Distinguished Scientific Award for Early Career Contribution" and the *Society for Personality and Social Psychology's* "Theoretical Innovation Prize" and "Diener Award for Outstanding Mid-Career Contributions." He served as associate editor of the *Journal of Personality and Social Psychology* and the UC Davis director of a training grant in "Affective Science" funded by the National Institute of Mental Health and is currently the codirector of the California Families Project, a longitudinal study of Mexican-origin families funded by the National Institute on Drug Abuse.

Jarron M. Saint Onge is assistant professor in a joint position between the Department of Sociology at the University of Kansas and the Department of Health Policy and Management at the University of Kansas Medical Center. He received his Ph.D. in sociology from the University of Colorado at Boulder. His research interests utilize a social demographic perspective to explore the interaction of health behaviors, social and physical context, and biology on health disparities for marginalized populations

Ninive Sanchez is a doctoral student in the Joint PhD Program in Social Work and Social Science at the University of Michigan. She earned a master of social work from the University of Michigan's School of Social Work and a bachelor of arts in psychology with a minor in public health from the University of Southern California. Her research interests include health and mental health across the life course, particularly among low-income and ethnic minority populations. Ms. Sanchez has also taught classes in advanced research methods in developmental psychology.

John E. Schulenberg is research professor at the University of Michigan's Institute for Social Research and Center for Human Growth and Development, professor of developmental psychology in the Department of Psychology, associate director of the Survey Research Center, and president-elect of the Society for Research on Adolescence. He received his Ph.D. in human development and family studies from Penn State University in 1987. He is coprincipal investigator on the Monitoring the Future study and has been associated with this study since 1991. He has published widely on several topics concerning adolescent development and the transition to young adulthood. His recent research focuses on the etiology and epidemiology of substance use and psychopathology, on the link between developmental transitions and health and well-being, and on the conceptualization and analysis of developmental change.

Hairong Song is an assistant professor in the quantitative program at the University of Oklahoma. She received her doctoral degree in quantitative psychology from the University of California, Davis, in 2009. Dr. Song's research focuses on developing and applying statistical methods for longitudinal data analysis from both Frequentist and Bayesian perspectives. Her research has published in a few top-tier methodological journals. Dr. Song's substantive research concerns various aspects of social-emotional development in adolescence. Her current work involves evaluating effectiveness of the Parent-child Interaction Training program on families with maltreated children as well as long-term changes of perceived discrimination and relational aggression on Mexican-origin adolescents.

Daniel W. Soto is the project manager for three NIH-funded longitudinal research studies that focus on Latino adolescent drug use, social networks, and acculturation. He recently earned a master's degree in public health (MPH) from the Keck School of Medicine, University of Southern California. He has worked with Project Towards No Drug Abuse and Project EX in implementing tobacco, drug, and alcohol use prevention and cessation curricula in Southern California high schools.

Michelle L. Stock earned her Ph.D. in social psychology from Iowa State University. She is currently an associate professor in the Department of Psychology at the George Washington University. Her research focuses on applying social-psychological theories to the study of risky health cognitions and behaviors, including substance use, sexual behavior, and UV exposure. Her experimental and survey research currently focuses on three main areas: (1) the relation between racial discrimination and risky health cognitions and behaviors in minority adolescents, as well as risk and protective factors that may help reduce this relation (2) the relation among risk behavior, social comparison, and perceptions of risk and (3) applying social psychological theory and the prototype-willingness model to health interventions.

Gary D. Stockdale is a senior statistician for the Psychology Department of the University of California, Davis. He received is doctoral degree in quantitative psychology from the University of California, Davis, in 2007 and his master's degree in clinical psychology from Ball State University in 2000. Presently Dr. Stockdale is the data manager and data analyst for the California Families Project, a longitudinal study of Mexican American families who reside in northern California. His present program of research involves precursors to and protectant factors of substance abuse and addiction of children, now teenagers, in the aforementioned study of Mexican American families. Prior to his present academic pursuits, Dr. Stockdale worked as a research biologist for a multinational chemical company for over 30 years where he was named in three US patents for unique experimental and statistical methodology. Dr. Stockdale has published in areas of human development, quantitative methodology, and agricultural experimentation.

Lingqi Tang is a research statistician who has been at the Center for Health Services and Society, University of California Los, since 1998 and has worked on major center projects. She received her Ph.D. in statistics and applied probability from the University of Alberta in 1992. Her interests include study design, finite population sampling, and analysis of incomplete data. She is highly skilled at advanced programming and has extensive knowledge and experience with implementing advanced statistical methods, including multiple imputation methods and hierarchical modeling. She has executed the development of numerous macros to facilitate analyses of Center data sets. She coleads the Biostatistics Design, Analysis, and Innovations Unit for the NIMH Partnered Research Center for Quality Care and Community Partners in Care.

Yonette F. Thomas is science advisor for urban health to the New York Academy of Medicine, a member of the International Society for Urban Health board, and a senior research advisor to the Association of American Geographers. She is founding Board Member and Vice President of the Interdisciplinary Association for Population Health Science (IAPHS). Dr. Thomas is a faculty affiliate of the Maryland Population Research Center at the University of Maryland College Park, and a voluntary associate professor in the Department of Public Health Sciences at the University of Miami Miller School of Medicine where she teaches social epidemiology. She was formerly the associate vice president for research compliance at Howard University. Previously she served as the chief of the Epidemiology Research Branch and program director for the social epidemiology program at the National Institute on Drug Abuse, National Institutes of Health. She has held faculty appointments in the Department of Psychiatry and Behavioral Sciences and in the School of Pharmacy at Howard University. She is a member of the Consortium of Social Science Associations Advisory Committee and the Steering Committee of the National Hispanic Science Network. She served on the National Academy of Sciences Committee on Revisions to the Common Rule for the Protection of Human Subjects in Research in the Behavioral and Social Sciences. Her primary research and publications have focused on the social epidemiology of drug abuse and HIV/AIDS and the link with geography, including edited volumes: Geography and Drug Addiction, Crime, HIV, and Health: intersections of Criminal Justice and Public Health Concerns. She has a Ph.D. in medical sociology and demography, with post graduate training in epidemiology.

Jennifer B. Unger is a professor of preventive medicine at the University of Southern California Keck School of Medicine. Her research focuses on risk and protective factors for substance use and other health-related behaviors among adolescents and young adults across cultural contexts. She is the PI or co-PI of several NIH-funded studies, including a longitudinal study of acculturation patterns and substance use among Hispanic adolescents and young adults in Los Angeles and a two-city study of family acculturation patterns and risk behaviors among recent-immigrant Hispanic adolescents in Miami and Los Angeles. She also studies the effectiveness of entertainment-education approaches to health education, including the development and evaluation of a fotonovela about secondhand and thirdhand smoke among multiunit housing residents and a telenovela about kidney transplantation. She directs the Ph.D. program in Preventive Medicine/Health Behavior Research at USC and serves on an NIH study section and the editorial boards of several journals.

Avelardo Valdez is a professor at the University of Southern California, School of Social Work. He obtained his Ph.D. in sociology at the University of California, Los Angeles. A primary focus of his research is on the relationship between substance abuse and violence and health issues among high-risk groups including youth and adult gang members. He is a recipient of federal grants from the National Institute on Drug Abuse (NIDA), Substance Abuse and Mental Health Service Administration (SAMHSA), the Centers for Disease Control, and National Center for Injury Prevention and Control (CDC). His most recent grant is a NIDA longitudinal study of Mexican American male gang members in San Antonio, Texas.

John M. Wallace Jr. is a professor and holder of the Philip Hallen Chair in Community Health and Social Justice at the University of Pittsburgh's School of Social Work. Dr. Wallace is a principal investigator on the National Institute on Minority Health and Health Disparities-funded Healthy Living, Healthy Learning, Healthy Lives Asthma study and a coinvestigator on the Monitoring the Future study. His current research examines ethnic disparities in adolescent substance use, asthma correlates, consequences and comorbidities among poor African American children, and comprehensive community revitalization initiatives to address food access and insecurity in underserved urban neighborhoods. His work has appeared in numerous professional journals, books, and monographs in social work, sociology, and public health.

Helene Raskin White is a distinguished professor with a joint appointment in the Center of Alcohol Studies and Sociology Department at Rutgers University. She received her Ph.D. in sociology from Rutgers University in 1976. Dr. White's research focuses on the development, causes, consequences, and comorbidity of substance use and other problem behaviors, including crime, violence, mental illnesses, and unsafe sexual behavior. She also develops, implements, and evaluates brief substance use interventions for college students. Dr. White is currently coprincipal investigator on a study funded by the National Institute on Alcohol Abuse and Alcoholism to conduct integrated data analysis of individual-level data from 24 college alcohol intervention studies and coprincipal investigator on a study funded by the National Institute on Drug Abuse to examine racial differences in predictors and consequences of marijuana use from childhood into adulthood. She is also deputy director and research director of the Center for Behavioral Health Services and Criminal Justice Research funded by the National Institute on Mental Health.

Keith F. Widaman is a distinguished professor of psychology at the University of California, Davis. He received his doctoral degree in psychology from the Ohio

State University in 1982. He has also been a faculty member at the University of California, Riverside. Dr. Widaman has pursued research in a number of substantive domains, including cognitive processes underlying mental abilities, the growth and development of mental abilities from childhood through early adulthood, structure and development of adaptive behaviors in persons with intellectual disability, and more recently the personal, family, and cultural influences that promote resilience during adolescent development. Dr. Widaman has also published methodological work on exploratory and confirmatory factor analysis, multitrait-multimethod models, and factorial invariance across groups and across time. Dr. Widaman has published over 200 books, book chapters, and journal articles, and his research has been supported by a series of federal grants from the National Institutes of Health. He is a past president of the Society of Multivariate Experimental Psychology (SMEP), won the Cattell Award from SMEP for early career contributions to multivariate psychology, received the inaugural Tanaka Award from SMEP for best article published in the society's journal *Multivariate Behavioral Research*, has served on the editorial boards of ten professional journals, and is currently the editor of *Multivariate Behavioral Research*.

Chapter 1
Introduction

Yonette F. Thomas and LeShawndra N. Price

Epidemiological data indicates that ethnic minority children have lower rates of substance use in comparison to non-minority children (Wallace and Muroff 2002; Wallace and Bachman 1991; Wallace et al. 2002, 2003, 2009; Galea et al. 2004; Delva et al. 2005; Substance Abuse and Mental Health Services Administration 2009; Belgrave et al. 2009). Research has shown African American and Hispanic youth have different pathways or trajectories of drug use compared with white youth, from initiation to development of problem use. Relative to white youth, Research has consistently found that relative to white youth, African-American youth are less likely to use alcohol and other drugs as adolescents, begin using in early adulthood, and are more likely to become problem users. In fact, as African Americans move into young adulthood and beyond, drug use appears to increase until it is higher than that found for any other ethnic group (Biafora and Zimmerman 1998; Horton 2007; Watt and Rogers 2007; Watt 2008). Research also indicates that Hispanic youth have the highest rates.

This phenomenon begs several questions: Why are African-American youth more likely than their white counterparts to abstain from drug use during adolescence? However, when they begin to use, why are they more likely to become problem users? What is the trajectory of drug use among African-American youth compared with white youth? Which environmental, cultural, and biological factors are actors across the trajectory? Are there gender differences along the trajectory?

The idea behind the NIDA meeting and this resulting volume was triggered by the work of JM Wallace (Wallace 1998; Wallace et al. 1999), in which he called for

Y.F. Thomas (✉)
The New York Academy of Medicine and the American Association of Geographers, Glenn Dale, MD, USA
e-mail: ythomas@yfthomasdr.com

L.N. Price
Health Inequities and Global Health Branch, National Heart Lung and Blood Institute, National Institutes of Health, Bethesda, MD, USA
e-mail: lprice@mail.nih.gov

Y.F. Thomas, L.N. Price (eds.), *Drug Use Trajectories Among Minority Youth*,
DOI 10.1007/978-94-017-7491-8_1

more emphasis on the social epidemiology of drug use among African American youth in adolescence and into adulthood. Specifically, what social environmental and other causative influences were motivating the race differences seen in problem drug use among minority youth compared to their white counterparts. Wallace's original focus was on African-American youth but in the minds of the authors, this warranted a broader discussion about drug use among minority youth in general and African American and Hispanic youth in particular. This curiosity and interest lead to the National Institute on Drug Abuse (NIDA)-sponsored meeting entitled: "Drug Use Trajectories Among Minority Youth." This meeting brought together experts in the field, many whose research were currently supported by NIDA. The purpose of the workshop was to bring together investigators from different disciplines and provide an opportunity to highlight and answer the questions in paragraph two above and engage participants in a focused discussion on how to critically generate prevention interventions that are likely to best serve these populations.

Investigators were asked to address the following questions:

- Do minority, African American and Hispanic youth, begin using drugs later than their white counterparts but are more likely to become problem users?
- What do we know about how trajectories of drug use differ by race/ethnicity?
- What does the current literature say about differences in drug use trajectories?
- What factors are known to influence these trajectories and crossover effects?
- What additional research questions are needed and what research needs to be conducted to help us better understand variations across the groups?
- What prevention interventions might best address these differences?
- What should the next generation of research in this area look like?
- How can the National Institute on Drug Abuse facilitate research to address these questions?
- What are the key disciplines and most appropriate study designs that are likely to yield the best answers?

The original meeting was held in 2006 and subsequent workshops and research panels between 2006 and 2012 were convened at American Psychological Association (APA) and American Sociological Association (ASA) annual meetings to engage a broader group of researchers around these questions.

Previous research has shown that later age onset of drug use leads to a reduced chance of extensive and persistent drug use leads to a reduced change of extensive and persistent drug use into adulthood, yet African Americans who start using drugs later than whites are more likely to continue using into adulthood. Our hope is to find the set of factors that create this "crossover phenomenon" and, in turn, bring light the need for longitudinal studies that follow individuals from initiation of use onward. The chapters in this volume represent the works of the leading researchers in the field who have addressed various aspects of the phenomenon. Following are the abstracts of each chapter.

Our introductory chapters begin with an overall framework within which to consider drug use trajectories, their risk, protective factors, and consequences among African American and Hispanic youth as compared to their white counterparts. This overview is largely based on the presentation given by **Denise Kandel** at the original meeting. In her presentation, Dr. Kandel provided an overall framework within which to consider the drug use trajectories, their risk and protective factors, and their consequences among African Americans in comparison to other racial-ethnic groups, specifically whites. Etiological factors included socio-demographic, biological, drug use history, individual psychosocial characteristics, family environment, role models for drug use, and social environment to determine the amount of risk and protective factors in each individual with the hope of determining which factors predict different trajectories among African Americans and whites.

Kandel, Schaffran, and Thomas discuss alternate conceptualizations of drug use trajectories and outline a strategy for identifying etiological factors and consequences of different trajectories. They investigate the Crossover Hypothesis for developmental trajectories of drug use among African Americans compared with Whites. They implement the third conceptualization of trajectories discussed in the Overview and focus on age-specific developmental drug use patterns. They examine current drug use of selected legal and illegal drugs, especially nicotine.

The age Crossover Hypothesis first suggested by Geronimus et al. (1993) proposes that there is a reversal in the prevalence of drug use among Whites and African Americans at a certain point in the life cycle. In adolescence, the prevalence of lifetime and current use is higher among Whites than African Americans. At some point in adulthood the pattern is reversed; the prevalence becomes higher among African Americans than Whites. With rare exceptions, most of what we know about life cycle differences in patterns of drug use are based on separate adolescent and adult samples.

Subsequent chapters discuss the various risk and protective factors that serve as actors in the drug use patterns exhibited by minority youth and their white counterparts.

Ensminger et al., explore the evolution of patterns of substance use associated with the interaction between age and race/ethnicity known as the *crossover effect*. The crossover effect suggests that although African Americans have the same or less substance use as adolescents compared to White populations, as they age into adulthood, their drug use becomes more problematic than that of White populations. Many studies indicate that African Americans making the transition into adulthood are less likely to be employed, are less likely to marry, and are more likely to be raising children without the support of a second adult. Studies also indicate that being employed and being married are deterrents to drug use and drug problems. One reason for the crossover effect may be the higher rates of unemployment and the lower rates of marriage among African Americans.

They examine the life course trajectories of drug use for a cohort of African Americans (N = 1053) followed from first grade (1966–1967) through mid adulthood (2002–2003). They also compare the social roles of those who do not use drugs in adulthood to those with drug use and/or drug use problems (as indicated by the CIDI drug abuse/dependence indicators). The sample comprises essentially all first grade children and their families living in Woodlawn, a disadvantaged Chicago community, in 1966 and who were invited to participate in the Woodlawn project with four periods of data collection from first grade to age 42.

They hypothesize that higher rates of unemployment and the lower rates of marriage among African Americans may be important contributing factors to the crossover effect. Specifically, they examine whether those who are employed and those who are married are more likely to desist from drug use and less likely to have started drug use in adulthood than those who are unemployed and those who are not married.

While they do not have as explicit a hypothesis about childrearing, evidence suggests that the impact of childrearing on parents' drug use may differ for mothers compared to fathers. Mothers in the Woodlawn population are much more likely to have children in their households than are fathers. They propose that fathers who do have children in their households will be less likely than other men in the study to use drugs or have drug problems. Further, they expect fewer differences in substance use between the women with and without children in the household.

To examine these differences they compare trajectories of marijuana, and cocaine use. They are interested in those who are late starters (after the age of 25) and those who start drug use early and do not desist drug use in early or mid adulthood compared to those who are never users, those who are adolescent only users, and those who quit drug use by age 25. They also compare rates of drug use with those of similar age from national populations. As adolescents, the drug use of the Woodlawn population compared with the Monitoring the Future (MTF) sample assessed at the same time, show no dramatic differences. In fact, the prevalence of drug use in the Woodlawn population is closer to that of White adolescents from MTF than to the African Americans in that survey.

In addition, they compare rates of drug use in the past year at age 32 and at age 42 for the Woodlawn cohort. At age 32, they find lower rates of alcohol use and higher rates of marijuana, cocaine, and heroin use among the Woodlawn cohort compared to those reported in the National Comorbidity Survey (NCS) and the National Household Survey on Drug Abuse (NHSDA), both assessed during the same years as the Woodlawn population. At age 42, they find lower rates of past year alcohol use reported by those in the Woodlawn cohort and the African Americans from the National Epidemiological Survey on Alcohol and Related Conditions (NESARC 2001–2002) and Blacks in the National Survey on Drug Use (NSDUH 2003) compared to Whites of similar ages from the NESARC and NUSDUH. The rates of past year marijuana are higher for those from Woodlawn and the NSDUH Blacks and Whites than for those Whites and Blacks assessed in the NESARC. The rates of past year cocaine and heroin are higher among the Woodlawn cohort and the NSDUH Blacks and lower among the NESARC Whites and Blacks and the NSDUH Whites.

They discuss the implications of these findings for the existence of a crossover effect; whether lower participation in employment and marriage may contribute to the more problematic drug use that African Americans have as adults. Importantly, they describe the strengths and weaknesses of the Woodlawn study for examining these issues.

Jorge Delva, et al. present an agenda for research on substance use among Latino youth.

Hispanic youth and adults have been largely overrepresented among individuals who use substances in the United States (U.S.). This over-representation is particularly severe for youth. Likewise, in Latin America, substance use has been, and continues to be, a serious social and public health problem. The extent of this problem means that millions of individuals and families, and thousands of communities, in North, Central, and South America are dealing with the devastating consequences of substance use and abuse, consequences that are compounded by the socioeconomic problems many Hispanics face, problems that when interacting with emotional and psychological problems, caused by the consumption of substances, further exacerbate substance use and abuse. In this chapter Delva et al. focus on both Hispanic populations in the U.S. and on people from Latin America because of the interconnectedness that exists between these populations.

The extent of the substance abuse problem among Hispanics in the U.S., and increasingly among Latin American populations, is well documented by studies relying on national household and school-based surveys. The science of survey research is well developed and there are a number of studies in the U.S. and Latin American countries describing the epidemiology of the problem. These studies show considerable variations in substance use and abuse patterns between Hispanic groups in the U.S. and between populations across Latin America. More research is needed to understand these differing patterns. However, very few longitudinal studies of substance use have taken place in the U.S. that focus on Hispanics or that include sufficiently large samples. Similarly, with only a few exceptions, in Latin America longitudinal studies are essentially non-existent. The lack of longitudinal studies with Hispanics in the U.S. and in Latin America results in a critical gap in our understanding of the etiology and developmental pathways or trajectories of substance use in these populations. This dearth of research limits the ability to inform policies and interventions that could prevent onset and reduce the deleterious effects of substance use among these populations.

Broman investigates discrimination as possible influence on drug use trajectories among minority youth.

While research is clear that the trajectory of drug use differs across race and ethnicity, there are many discussions about why this is the case. One understudied phenomenon that may bear on this is racial and ethnic discrimination. Broman argues that due to the cumulative impact of social disorganization and stress exposure, derogated racial and ethnic minorities in the US are more likely to turn to drugs over the life course. The chapter uses a macro-level perspective on factors in drug use. This model incorporates systemic and individual-level factors in drug use. Because perceptions of injustice are systematic and widespread in the US, and discrimination is rampant, this is likely a factor in patterns of drug use over the life course among racial minorities. Broman argues that this factor must be considered more strongly in future research on substance abuse over the life course.

Wallace et al., discusses race/ethnicity, religiosity and differences and similarities in African American adolescents substance use behaviors.

Despite the conclusion that religion relates inversely to adolescent substance use, a number of important questions remain to be answered. One of the most important questions concerns the extent to which this broad conclusion generalizes to the millions of non-white young people who comprise America's increasingly racially and ethnically diverse population. Given that the vast majority of the extant research is based upon relatively small, often non-representative, racially homogenous (i.e., white) samples, the answer to this question remains largely unknown. Even past studies that have attempted to address this question are limited in that they have only compared black and white youth (Wallace et al. 2003) or in one instance, black, white and Hispanic youth (Wallace et al. 2007a). According to another recent review of the literature on the relationship between religiosity and adolescent health behaviors like substance use, "no study investigated differences in R/S [religion/spirituality] between Asian Americans or Native Americans and other racial groups" (Rew and Wong 2006). Similarly, a review of studies on religion and health concludes that, "…research has neglected specific subpopulations, such as Hispanics, Asian Americans, Native Americans and groups of low socioeconomic status" (Williams and Sternthal 2007).

In an effort to begin to address this gap in the literature, Wallace uses large, nationally representative samples of white, black, Mexican American, Puerto Rican, Other Latin American, Asian American and Native American young people to examine empirically the relationship between religiosity and adolescent alcohol, cigarette and marijuana use. This chapter replicates and extends our earlier analyses (i.e., Wallace et al. 2007b) that examined the relationship between religiosity and substance use among white, black and Hispanic youth. Consistent with the earlier study, two questions motivate the present investigation: First, "how religious are American youth?" and second, "to what extent does religiosity 'protect' black, Hispanic (including Mexican American, Puerto Rican, Other Latin American), Asian American and Native American young people from drug use, as past research suggests that it protects white youth?"

White, Loeber, and Chung examine racial differences in substance use with longitudinal data.

Despite the fact that several studies have addressed racial differences in trajectories of substance use, there still exist several gaps in understanding trajectories of substance use among African-Americans. In this chapter, the authors focus on three major gaps. The first is the lack of attention to heterogeneity, which includes: (1) differences across drugs, (2) differences across contexts, such as where people live and when they were born, and (3) intraindividual differences, including gender differences and socioeconomic status differences. The second gap has been the failure to adequately explain *why* there are differences in trajectories of substance use for African-Americans compared to Whites. The third gap is the lack of research examining how racial differences in how substance use trajectories relate to other behaviors. They use data from two prospective longitudinal studies, the Pittsburgh Youth Study and the Pittsburgh Girls Study, to address these issues. First, they examine trajectories of different types of substance use among White and African-American young men from early adolescence into early adulthood. They then examine different types of heterogeneity that may affect racial differences including cohort differences, gender differences, and college attendance differences. Next they summarize data from two previous studies. The first examined racial differences in developmental stages of substance use and movement from initiation to regular use. The second attempted to explain late-onset smoking among African-Americans.

Finally, they examine racial differences in the association between substance use and violence. They end the chapter by discussing the implications of the results for developing prevention programs.

Unger et al., describe predictors of growth trajectories of substance use among Hispanic adolescents.

Previous studies have documented cultural risk and protective factors for substance use among Hispanic adolescents, but few studies have investigated these associations longitudinally. This study examined the associations between cultural variables (acculturation, perceived discrimination, acculturative stress, cultural values) and trajectories of substance use from 9th to 11th grade among 1668 Hispanic adolescents in Southern California, using individual growth curve analyses. From 9th to 11th grade, use of cigarettes, alcohol, and marijuana increased significantly. Significant risk factors for substance use in 9th grade included perceived discrimination, fatalism, and low levels of respeto. Significant risk factors for growth in substance use between 9th and 11th grade included male gender, English language usage, low levels of Hispanic acculturation, respeto, fatalism, and low levels of friends' and parents' use. Results indicate that the effects of perceived discrimination, friends' use, and parents' use occur prior to 9th grade, whereas other variables are associated with escalation of substance use during high school. Substance use prevention interventions are needed throughout early and mid-adolescence. Prevention curricula may be more effective if they address the specific risk and protective factors that are most relevant at each stage of development.

Conger et al. addresses developmental trajectories of substance use risk and resilience for youth of Mexican origin.

In this chapter, Conger et al., evaluate increases in substance use and risk for substance use for a cohort of over 300 children of Mexican origin assessed in the fifth and seventh grades. This period represents the developmental transition from late childhood to early adolescence, a time of increasing mental health, behavioral and substance use problems. The children and their families are participants in the California Families Project (CFP), an ongoing study of substance use and its antecedents, correlates and consequences. The CFP is especially concerned with developmental pathways leading to the initiation and escalation of alcohol, tobacco, and other drug (ATOD) use from late childhood through adolescence. A developmental approach is generally considered the best strategy for informing the creation of effective intervention programs that can prevent the early initiation of ATOD use, when it is most likely to have severe long-term consequences. The present analyses examine hypotheses generated from the family stress model, which proposes that acculturation processes, minority experiences, economic hardship, and neighborhood and school risks will affect parent-parent, parent-child, and sibling and peer relationships. These relationships, in turn, are expected to influence the initiation and escalation of ATOD use and related adjustment problems. The family stress model also proposes that a set of cultural (e.g., respect, familism), personal (e.g., ethnic identity, self-control), and social (e.g., effective family problem solving) resources will promote resilience to the hypothesized risk factors. Initial findings reported here during the early phase of this developmental transition from childhood to adolescence are consistent with specific predictions from the model.

Gibbons et al. examine intergroup differences and role of protective factors.

Gibbons et al. review research that examines psychosocial factors that influence trajectories of drug use among African Americans. They focus on perceived racial discrimination (PRD) as a predictor of change in drug use from adolescence through early adulthood. Previous studies, most of them cross-sectional, have linked PRD with drug use and other types of substance use in Black adolescents. Research discussed in this chapter includes studies with longitudinal designs that allow for assessment of the prospective relation between PRD and use, as well as the relations between changes in each during this critical developmental period. These longitudinal designs also allow for assessment of factors that mediate the PRD → use relation, such as type of affect (e.g., anger vs. depression), as well as factors that moderate the relation, such as level of integration of the neighborhood in which the adolescent lives. The studies include field/survey designs or experimental/lab designs, and, in some cases, both types of designs. We will conclude with some discussion of the implications of this research for interventions, as well as an outline for future research directions, including an examination of the role of genetics and, more specifically, Gene x Environment (G x E) interactions in explaining the effects of PRD on substance use.

Cooley-Strickland et al. discuss the influence of neighborhood context on exposure to and substance use among urban African American youth.

Given the comparatively late onset of alcohol, tobacco, and other drug (ATOD) use among African American youth as compared with Whites, there are limited data on risk and protective factors associated with African American children's ATOD use initiation. Prior research has established that individual-level characteristics are insufficient to account for ATOD use among youth; community-level contextual effects are critical to consider in understanding both group and individual-level behavior (Wilcox et al. 2003). In comparison to family and peer contexts, neighborhood factors have been understudied in relation to youth's substance use (Lambert et al. 2004). In the current chapter, Cooley Strickland first reviews the existing relevant literature that addresses these issues. Next, she uses data from a prospective longitudinal study, the Multiple Opportunities to Reach Excellence (MORE) Project, to investigate the risk and protective trajectories associated with alcohol, tobacco, and other substance exposure – and use – among urban primarily African American school children who have been exposed to varying levels (low, moderate, high) of neighborhood violence. Data are from over 400 youth (ages 7–13), their parents and teachers in a longitudinal community-epidemiological sample. She ends the chapter by discussing the implications of this research for prevention and intervention programs for urban minority youth.

Jipguep, et al. investigate academic performance and graduation differentials as actors in trajectories of substance use.

State of the art research (Mosher and Akins 2007; Reardon and Buka 2002) note that African American youth are less likely than white youth to use substances until about late adolescence and young adulthood, when they become more likely than whites to use substances. They consider whether lower academic performance diminishes the academic success and graduation prospects of some African American and whether this increases risks for greater use of substances. They examine this question using the National Educational Longitudinal Study of 1988 (NELS 88), which first interviewed a nationally representative sample of eighth-graders in the spring of 1988, and then re-interviewed them in 1990, 1992, 1994, and 2000. They specifically test whether low academic performance in 8th and 10th grades (scores on reading, social studies, mathematics and science achievement tests) increases the onset and use of substances (cigarettes, alcohol, and drugs) in the 8th, 10th, and 12th grades, and whether these in turn are associated with poor academic performance and higher likelihood of not completing high school. The analysis will compare African American, Hispanic, and White students to determine whether differences in academic performance might contribute to changes and crossovers in their relative use of substances in later adolescence and young adulthood. Also, they test for the effects that higher poverty urban and rural neighborhoods and schools might have on the relationships between academic performance and later substance use.

Structural and social aspects of neighborhoods, including neighborhood poverty, crime, and social disorganization, and adolescents' subjective experience of their neighborhood environments (e.g., perceived neighborhood disorder) have been linked with adolescent substance use outcomes. However, existing literature using aggregate data (e.g., Census-level data) and adolescent perceptions does not provide information about particular locations that might be associated with increased risk for substance use. In addition, analysis of aggregate data may increase problems related to the ecological fallacy, the assumption of individuals in a particular area, such as overlooking heterogeneity within areas. To address these concerns, the authors used spatial analysis to examine geographic clustering of adolescent substance use behaviors, and the degree to which individual, family, and peer factors accounted for the geographic clustering. Participants were a community sample of 585 African American youth originally assessed in first grade and followed at regular intervals during adolescence. Youth reported about their use of tobacco, alcohol, and marijuana, including frequency of use, in each of grades 6–12. Information about individual, family, and peer risk factors for substance use were obtained via adolescent self-report and from teacher and parent report. Youth place of residence (recorded as x, y coordinates) was recorded each year, and information about residence was linked to the individual-level data for the spatial analyses. Spatio-temporal clustering was used to determine whether adolescent use of tobacco, alcohol, or marijuana clustered in geographic space, and whether the degree of geographic clustering changed over time.

Valdez, et al., discuss the influence of informal social control processes in drug use trajectories.

In recent years, existing life course research has found that weak ties to parents, school and conventional peers increase the probability of the initiation of drug use trajectories and delinquent behavior. The emphasis on informal social control is especially salient among subpopulations in disadvantaged minority communities, given the weakness of formal social control mechanisms in these neighborhoods. This chapter is based on data from a NIDA funded study of 150 Mexican American male gang members between the ages of 16–20 years old. The analysis focuses on the extent adolescent informal social control processes mediate the relationship between gang-type and adolescent delinquent behavior in this population. Multivariate analyses are conducted using the following measures for each of the constructs: (1) family and school informal social control and (2) street vs. drug gang type dichotomy and (3) drug use, general delinquency index and violence delinquency index. Findings indicate that such items as parental supervision is inversely related to the frequency of adolescent cocaine use. The chapter concludes that in adolescence social control processes will mediate the effects of specific gang membership and delinquent behavior. Discussed are implications for future research being conducted in explaining the desistance or continuation of antisocial behavior across the life course.

Martinez, et al. focus on immigrant status among Latino youth.

Latinos are the largest, most rapidly growing racial/ethnic subgroup in the U.S. While the U.S. population grew 7 % from 2000 to 2007, the Latino population grew 29 % (Pew Hispanic Center 2009). This demographic trend is particularly striking in 22 states that are sites of emerging immigrant communities (e.g., Oregon, North Carolina, Georgia, Iowa). Such rapid growth, when combined with public health and social service systems that are unprepared to address the needs of a culturally pluralistic population, can contribute to a lack of effective substance use prevention and intervention programs that target Latino adolescents in general and Latino youth in immigrant families in particular. Research to date shows that prevalence rates for Latino youth substance use are not typically higher than that for the general population, though Latino adolescents may initiate use earlier, activating an earlier trajectory toward more severe outcomes, including school dropout and incarceration (Hawkins et al. 1992; Martinez et al. 2006; Wallace et al. 1995). Little is known, however, about the etiology of Latino youth substance use in emerging immigrant communities, and whether there exist unique risk and protective factors from those for Latino youth in states that are established sites of immigrant settlement (e.g. California, Texas, Florida, New York). This chapter investigates the state of knowledge regarding the etiology of Latino adolescent substance use – with particular attention given to the influences of acculturation, stress and parenting – in states with rapidly changing demographics, and considers the implications of this research for effective interventions.

Lacey et al. describe the influence of race and ethnicity on mental and substance use disorders specifically in the case of Caribbean Blacks.

The scientific literature has long held that differences in socioeconomic status, poor living arrangements, and even genetic factors might account for race and ethnic inequalities in health. Prior studies of substance use disorders have generally grouped respondents of African American and Caribbean black ethnicities into one global category of "Black" or African American. This aggregation may obscure important differences in substance use patterns and mental health outcomes. This chapter is based upon national household probability samples of non-institutionalized

African Americans and (Caribbean) blacks from Caribbean countries living in the U.S., conducted between February 2001 and June 2003, with a slightly modified version of the World Mental Health version of the World Health Organization's Composite International Diagnostic Interview. A total of 3570 African Americans and 1621 Caribbean blacks, aged 18 and over were interviewed in the United States and an additional 2000 adults in Guyana and 1800 in Jamaica. In the United States, overall differences in prevalence of substance disorders between the two ethnic groups were not significant. Within both populations, prevalence was higher for men, for those with lower education and incomes, and for the U.S. born. The prevalence rates of substance disorders, however, among African Americans exceeded that of Caribbean blacks among females, those aged 45–59, and the divorced, while prevalence of substance abuse among college-educated Caribbean blacks exceeded that for African Americans. African Americans in major metropolitan areas had higher prevalence rates, and those in the South had lower ones, compared to those living in other areas. Overall, first generation Caribbean blacks are significantly less likely, but second generation are more likely, than African Americans to meet criteria for lifetime alcohol abuse and overall substance disorders. Lacey and Jackson believe that failure to distinguish between individuals of African American and Caribbean black ethnicities and immigration and ancestry status mask important differences in substance use patterns among these black populations. In this chapter they examine in- and out-of country Caribbean blacks at the population and individual levels and how processes of immigrant protection may operate in lowering rates of substance use disorders in second-generation Caribbean blacks, but have less positive influences on third and subsequent generations.

Prado, et al., discuss preventive interventions among Hispanic youth and implications for future research.

Drug abuse preventive interventions for Hispanic youth require an understanding of both drug use initiation patterns and trajectories. Culturally relevant drug abuse preventive interventions for Hispanic adolescents should address those culturally specific ecological determinants associated with drug initiation and escalation. In recognition of the systemic nature of the relationship among the many ecological processes that are likely to precede, be concurrent with, or follow drug use, many developmental theoretical approaches undergirding evidence-based drug abuse prevention interventions target a complex set of interacting antecedents. Understanding the mechanisms by which these drug abuse preventive interventions have their effect and the varying impact of these interventions on different Hispanic subgroups is required to maximize the efficacy/effectiveness of preventive interventions for Hispanic youth. In this chapter the authors provide recommendations for advancing the drug abuse prevention field among this population.

Hallfors, et al. describe patterns of risk behavior change from adolescence to emerging adulthood as implications for disparities in HIV/STDs among minority youth and the link to drug use.

In previous research, the authors documented large racial disparities in both self-reported and test-identified STD/HIV prevalence that could not be explained by contemporaneous patterns of risk behavior (Halpern et al. 2004, 2007). The present study expands these earlier investigations by applying a longitudinal person-centered analysis of sex and drug use behavior patterns to see whether the implications of behavioral risk taking for racial disparities in STI/HIV among young adults vary according to the developmental period in which they begin, and/or trajectory over time. They found that similar proportions (24–25 %) of White and African American men were in the *stable low* risk category. More African American men, however, were in the *decreasing* risk category (29 % versus 9 %) and more White men were in the *increasing* and *stable high* risk categories (40 % versus 26 %, and 27 % versus 20 % respectively). Despite these patterns, African American men were much more likely than White men to have a positive STD/HIV test in young adulthood in all trajectories.

On the other hand, young adult African American women were more likely to be consistently low risk in their behavior compared to White women (41 % versus 31 % in the *stable low* category). Another 34 % of African American women engaged in sexual risk behavior in adolescence but in lower risk behavior by young adulthood; in comparison, only 14 % of White women were in this *decreasing* category. Only 25 % of African American women were in the *increasing* or *stable high* groups, compared to 54 % of White women. Despite a higher prevalence of consistently low risk behavior or desistance from earlier higher risk patterns, the prevalence of STIs among African American women was much higher than any other sex/race group.

These findings confirm conclusions that the current prevention paradigm, to reduce personal risk behavior and increase personal protective behavior, does not seem adequate for racial STD/HIV disparities. Structural and contextual factors that may provide better clues are discussed.

Brody, et al. describe a research program that began in 2005 that was designed to refine etiologic models of drug use and sexual risk behavior, as well as the prevention programs the models inform. The authors were motivated to start this research because inclusive reviews of programs designed to prevent drug use/abuse and sexual risk behavior reveal mixed results. Many prevention programs do not attain their goals, and others are effective for some subgroups but not others (1–4), suggesting that the causes of these risk behaviors are not yet well enough understood for prevention efforts to achieve large and reliable effects. This suggested a need for new approaches to etiologic models of drug use/abuse and sexual risk behavior, particularly greater articulation of the ways in which interactions among genetic, psychosocial, and developmental processes can inform them. Concurrent advances in both knowledge and technology related to basic genetics and epigenetic processes have created unprecedented opportunities for conceptual integration. The authors responded to this challenge and this opportunity by initiating a research program that uses findings from gene-environment interplay, developmental, and epigenetic research conducted with rural African Americans and participants in the Iowa Adoption Studies (IAS) to refine etiologic models in ways that increase their predictive utility.

References

Belgrave, F. Z., Clark, T., & Nasim, A. (2009). Handbook of African American psychology. In *Drug use among African American youth* (pp. 469–481). New York: Plenum Press.

Biafora, F., & Zimmerman, R. S. (1998). Developmental patterns of African American adolescent drug use. In *Drug use and ethnicity in early adolescence* (pp. 149–175). New York: Plenum Press.

Center, P. H. (2009). *Between two worlds: How young Latinos come of age in America*. Washington, DC: The Pew Hispanic Center.

Delva, J., Wallace, J. M., Jr., O'Malley, P. M., Bachman, J. G., Johnston, L. D., & Schulenberg, J. E. (2005). The epidemiology of alcohol, marijuana, and cocaine use among Mexican American, Puerto Rican, Cuban American, and other Latin American eighth-grade students in the United States: 1991–2002. *American Journal of Public Health, 95*, 696–702.

Galea, S., Nandi, A., & Vlahov, D. (2004). The social epidemiology of substance use. *Epidemiologic Reviews, 26*, 36–52.

Geronimus, A. T., Neidert, L. J., & Bound, J. (1993). Age patterns of smoking in US black and white women of childbearing age. *American Journal of Public Health, 83*(9), 1258–1264.

Halpern, C. T., Hallfors, D., Bauer, D. J., Iritani, B., Waller, M. W., & Cho, H. (2004). Implications of racial and gender differences in patterns of adolescent risk behavior for HIV and other sexually transmitted diseases. *Perspectives on Sexual and Reproductive Health, 36*(6), 239–247.

Halpern, C. T., Kaestle, C. E., & Hallfors, D. D. (2007). Perceived physical maturity, age of romantic partner, and adolescent risk behavior. *Prevention Science, 8*(1), 1–10.

Hawkins, J. D., Catalano, R. F., & Miller, J. Y. (1992). Risk and protective factors for alcohol and other drug problems in adolescence and early adulthood: Implications for substance abuse prevention. *Psychological Bulletin, 112*(1), 64.

Horton, E. G. (2007). Racial differences in the effects of age of onset on alcohol consumption and development of alcohol-related problems among males from mid-adolescence to young adulthood. *Journal of Ethnicity in Substance Abuse, 6*(1), 1–13.

Lambert, S. F., Brown, T. L., Phillips, C. M., & Ialongo, N. S. (2004). The relationship between perceptions of neighborhood characteristics and substance use among urban African American adolescents. *American Journal of Community Psychology, 34*(3-4), 205–218.

Martinez, C. R. (2006). Effects of differential family acculturation on Latino adolescent substance use. *Family Relations, 55*(3), 306–317.

Mosher, C., & Akins, S. (2007). *Drugs and drug policy: The control of consciousness alteration*. Thousand Oaks: Sage Publications.

Reardon, S. F., & Buka, S. L. (2002). Differences in onset and persistence of substance abuse and dependence among whites, blacks, and Hispanics. *Public Health Reports, 117*(Suppl 1), S51.

Rew, L., & Wong, Y. J. (2006). A systematic review of associations among religiosity/spirituality and adolescent health attitudes and behaviors. *Journal of Adolescent Health, 38*(4), 433–442.

Substance Abuse and Mental Health Services Administration. (2003). *Results from the 2002-2001 National Survey on Drug Use and Health: National Findings*. Rockville: Office of Applied Studies, NSDUH.

Substance Abuse and Mental Health Services Administration. (2007). *Results from the 2006 National Survey on Drug Use and Health: National Findings* (Office of Applied Studies, NSDUH Series H-32, DHHS Publication No. SMA 07–4293). Rockville: Office of Applied Studies NSDUH.

Substance Abuse and Mental Health Services Administration. (2009) *Results from the 2008 National Survey on Drug Use and Health: National findings* (Office of Applied Studies, NSDUH Series H-36, HHS Publication No. SMA 09–4434) Rockville: Office of Applied Studies, NSDUH.

Wallace, J. M., Jr. (1998). Explaining race differences in adolescent and young adult drug use: The role of racialized social systems. *Drugs & Society, 14*(1–2), 21–36.

Wallace, J. M., Jr., & Bachman, J. G. (1991). Explaining racial/ethnic differences in adolescent drug use: The impact of background and lifestyle. *Social Problems, 38*, 333–357.
Wallace, J. M., Jr., & Muroff, J. R. (2002). Preventing substance abuse among African American children and youth: Race differences in risk factor exposure and vulnerability. *The Journal of Primary Prevention, 22*, 235–261.
Wallace, J. M., Jr., Forman, T. A., Guthrie, B. J., Bachman, J. G., O'Malley, P. M., & Johnston, L. D. (1999). The epidemiology of alcohol, tobacco and other drug use among black youth. *Journal of Studies on Alcohol, 60*(6), 800–809.
Wallace, J. M., Jr., Bachman, J. G., O'Malley, P. M., Johnston, L. D., Schulenberg, J. E., & Cooper, S. M. (2002). Tobacco, alcohol, and illicit drug use: Racial and ethnic differences among U.S. high school seniors, 1976–2000. *Public Health Reports, 117*(Suppl. 1), S67–S75.
Wallace, J. M., Jr., Bachman, J. G., O'Malley, P. M., Schulenberg, J. E., Cooper, S. M., & Johnston, L. D. (2003). Gender and ethnic differences in smoking, drinking and illicit drug use among American 8th, 10th and 12th grade students, 1976–2000. *Addiction, 98*, 225–234.
Wallace, J. M., Yamaguchi, R., Bachman, J. G., O'Malley, P. M., Schulenberg, J. E., & Johnston, L. D. (2007a). Religiosity and adolescent substance use: The role of individual and contextual influences. *Social Problems, 54*(2), 308–327.
Wallace, J. M., Delva, J., O'Malley, P. M., Bachman, J. G., Schulenberg, J. E., Johnston, L. D., & Stewart, C. (2007b). Race/ethnicity, religiosity and adolescent alcohol, cigarette and marijuana use. *Social Work in Public Health, 23*(2-3), 193–213.
Wallace, J. M., Jr., Vaughn, M. G., Bachman, J. G., O'Malley, P. M., Johnston, L. D., & Schulenberg, J. E. (2009). Race/ethnicity, socioeconomic factors, and smoking among early adolescent girls in the United States. *Drug and Alcohol Dependence, 104*(Suppl. 1), S42–S49.
Watt, T. T. (2008). The race/ethnic age crossover effect in drug use and heavy drinking. *Journal of Ethnicity in Substance Abuse, 7*(1), 93–114.
Watt, T. T., & Rogers, J. M. (2007). Factors contributing to differences in substance use among Black and White adolescents. *Youth & Society, 39*, 54–74.
Wilcox, H. C., Neidehiser, J., & Reiss, D. (2003). Genetic and environmental influences on adolescent cannabis use: The role of family relationships. *Behavior Genetics, 33*(6), 723–723.
Williams, D. R., & Sternthal, M. J. (2007). Spirituality, religion and health: Evidence and research directions. *Medical Journal of Australia, 186*(10), S47.

Chapter 2
Overview and Perspectives

Denise Kandel and Yonette F. Thomas

What Does One Mean by Developmental Trajectories?

There are at least four conceptualizations of behavioral trajectories of drug use. Three of the approaches focus on *development within single drug classes.*

In one approach, the focus is on development in terms of *stage or extensiveness of use*. In this descriptive approach, stage of use subsumes onset, current use, perhaps daily use, and quitting. Extensiveness of use subsumes experimental use, number of times used, and dependence. These definitions tend to be categorical. Cross-sectional, as well as longitudinal data, are used to define an individual's stage of drug use.

In a second approach, the focus is on development framed within a *dynamic approach* to identifying developmental patterns of drug use following initiation. There is increasing recognition of the heterogeneity in the developmental course of substance use over time with variation in latency, level of escalation, chronicity, and remission (Jackson et al. 2005, p. 612). Statistical methods developed over the last 10 years make it possible to identify heterogeneous developmental behavior patterns over time. These methods, e.g., growth curve, latent class, growth mixture, hierarchical models (Duncan and Duncan 2004; Muthén 2004; Muthén and Schedden 1999; Nagin 1999; Raudenbush and Bryk 2002; Singer and Willet 2003),

D. Kandel
Department of Psychiatry, College of Physicians and Surgeons and Mailman School of Public Health, Columbia University, New York, NY, USA

New York State Psychiatric Institute, Columbia University, New York, NY, USA
e-mail: dbk2@cumc.columbia.edu

Y.F. Thomas (✉)
The New York Academy of Medicine and the American Association of Geographers, Glenn Dale, MD, USA
e-mail: ythomas@yfthomasdr.com

Y.F. Thomas, L.N. Price (eds.), *Drug Use Trajectories Among Minority Youth*,
DOI 10.1007/978-94-017-7491-8_2

characterize interindividual change and interindividual differences in intraindividual change, providing a dynamic identification of behavioral trajectories. This approach requires longitudinal data and multiple measurement points, although modeling can be based on age at which different behaviors have occurred. The distinguishing feature of these models is the identification of heterogeneous developmental patterns that characterize individuals rather than variables. The measures may be the same as those used in the first approach. Over the last 10 years, there has been an explosion of such studies with respect to smoking behavior, with the outcome of interest being frequency/quantity of cigarettes smoked over the prior 30 days (Abroms et al. 2005; Audrain-McGovern et al. 2004; Brook et al. 2006a, b; Chassin et al. 2000; Colder et al. 2001; Karp et al. 2005; Orlando et al. 2004; Stanton et al. 2004; Vitaro et al. 2004; White et al. 2000, 2002, 2004). Studies of other substances are fewer. There are studies of alcohol (Muthén and Muthén 2000; Oesterle et al. 2004; Scheier et al. 2000; Toumbourou et al. 2003), marijuana (Brook et al. 2000; Brown et al. 2004; Ellickson et al. 2004; Schulenberg et al. 2005; Windle and Wiesner 2004), as well as two or more substances in parallel (Flory et al. 2004; Jackson et al. 2000; Patton et al. 2007; Walden et al. 2007). Most studies report three or four trajectory classes. The smoking classes typically include a non-user class; an early onset/rapid escalator class; and varying intermediate patterns, such as late onset stable users or quitters. Trajectories that have been described for marijuana include early high users; stable light users; steady increasers; and occasional light users (Ellickson et al. 2004).

The third approach on development within a simple drug class simply *charts rates of use* across specific ages over the life span (Geronimus et al. 1993).

In a fourth approach, the focus is on *development across drug classes*. The interest is in characterizing sequential stages of progression across multiple drug classes. This is the approach underlying the Gateway Hypothesis (Kandel 2002) in which the use of legal drugs, i.e., alcohol and/or cigarettes, has been shown to precede the use of marijuana, which in turn precedes the use of cocaine and other illicit drugs.

Most of the chapters in this volume conceptualize trajectories as per the second and, especially, the third approach.

Etiological Factors-Risk and Protective Factors

A wide range of potential risk and protective factors are relevant to understanding patterns and trajectories of drug use: sociodemographic, biological, drug use history, psychosocial characteristics, family environment across the life cycle (e.g., parent-child interactions, spouse/partner interactions), role models for drug use, and the broader social environment ranging from proximal contexts, such as school or neighborhood, to more distant contexts, such as social policies (See Table 2.1). No single study can encompass all these factors.

As stressed by Wallace and Muroff (2002), in considering the importance of specific risk and protective factors across racial/ethnic groups, one needs to take

Table 2.1 Classes of risk and protective factors

Sociodemographic
Age
Gender
Education
Income
Biological
Genetic
Drug metabolism
Initial sensitivity to the substance used
Sensation seeking
Prenatal exposure
Drug use history
Age of onset
Individual psychosocial characteristics
Personality
Problem behaviors
Psychiatric problems
Academic orientation/performance
Attitudes
Religiosity
Family environment across the life cycle
(a) Childhood/Adolescence
Parent-child interactions
Closeness
Discipline
Family conflict
Parental psychopathology
Parental attitudes and norms
(b) Adulthood
Spouse/partner interactions
Role models for drug use
Parents
Siblings
Peers
Spouse/partner
Broader social environment
School
Neighborhood
Work
Drug availability
Media and advertising
Norms
Social policies

into account exposure, i.e., the levels of risk or protective factors in specific racial/ethnic groups, and vulnerability, i.e., the effect of a particular factor on drug behavior in each group. The effects may be the same or different across different groups. We would emphasize in addition the importance of considering the individual's stage in the life cycle. Effects of the same variables may differ in adolescence, early adulthood, or middle adulthood. Certain variables are only relevant to certain stages in the life cycle, such as puberty in adolescence or marital status in adulthood. Thus, any explanatory scheme must address the specific drug class; the stage/trajectory of use; and the individual's stage in the life cycle. The ultimate goal is to fill out the cells in the table below on the basis of results from multiple studies.

Data Requirements and Availability

Ideally, one needs longitudinal studies that follow individuals of different race/ethnicity from initiation of use, predominantly in adolescence, through the periods of major use and desistance, at least to the late forties. Such studies do not exist, although *Monitoring the Future* is the closest approximation (Merline et al. 2004). Nor do we have studies on more limited age spans based on different birth cohorts drawn from the same population so as to approximate a cohort-sequential design.

The majority of existing studies focus on two phases of the life cycle: adolescence or early adulthood. Furthermore, most of the studies that have compared different stages/trajectories of drug use among ethnic groups have examined only cigarette use. Most studies have examined a single ethnic group or an undifferentiated multiethnic sample.

Acknowledgement We wish to acknowledge the contributions of Christine Schaffran, Pamela Griesler, and Mei-Chen Hu. Work on this manuscript was supported by Grant DA09110 and Research Scientist Award K05-DA0081 from the National Institute on Drug Abuse to D. Kandel.

References

Abroms, L., Simons-Morton, B., Haynie, D. L., & Chen, R. (2005). Psychosocial predictors of smoking trajectories during middle and high school. *Addiction, 100*, 852–861.

Audrain-McGovern, J., Rodriguez, D., Tercyak, K. P., Cuevas, J., Rodgers, K., & Patterson, F. (2004). Identifying and characterizing adolescent smoking trajectories. *Cancer Epidemiology, Biomarkers and Prevention, 13*, 2023–2034.

Brook, J. S., Whiteman, M., Finch, S. J., Morojele, N. K., & Cohen, P. (2000). Individual latent growth curves in the development of marijuana use from childhood to young adulthood. *Journal of Behavioral Medicine, 23*, 451–464.

Brook, J. S., Ning, Y., & Brook, D. W. (2006a). Personality risk actors associated with trajectories of tobacco use. *American Journal on Addictions, 15*, 426–433.

Brook, J. S., Pahl, K., & Ning, Y. (2006b). Peer and parental influences on longitudinal trajectories of smoking among African Americans and Puerto Ricans. *Nicotine & Tobacco Research, 5*, 639–651.

Brown, T. L., Flory, K., Lynam, D. R., Leukefeld, C., & Clayton, R. R. (2004). Comparing the developmental trajectories of marijuana use of African American and Caucasian adolescents: Patterns, antecedents, and consequences. *Experimental and Clinical Psychopharmacology, 12*, 47–56.

Chassin, L. A., Presson, C. C., Rose, J. S., & Sherman, S. J. (2000). The natural history of cigarette smoking from adolescence to adulthood in a midwestern community sample: Multiple trajectories and their psychosocial correlates. *Health Psychology, 19*, 223–231.

Colder, C. R., Mehta, P., Balanda, K., Campbell, R. T., Mayhew, K. P., Stanton, W. R., Pentz, M. A., & Flay, B. R. (2001). Identifying trajectories of adolescent smoking: An application of latent growth mixture modeling. *Health Psychology, 20*, 127–135.

Duncan, T. E., & Duncan, S. C. (2004). Special series: Advances in latent variable analysis: Applications to clinical research. An introduction to latent growth curve modeling. *Behaviour Therapy, 35*(2), 333–363.

Ellickson, P. L., Martino, S. C., & Collins, R. L. (2004). Marijuana use from adolescence to young adulthood: Multiple developmental trajectories and their associated outcomes. *Health Psychology, 23*, 299–307.

Flory, K., Lynam, D., Milich, R., Leukefeld, C., & Clayton, R. (2004). Early adolescent through early adulthood alcohol and marijuana use trajectories: Early predictors, young adult outcomes, and predictive utility. *Development and Psychopathology, 16*, 193–213.

Geronimus, A. T., Neidert, L. J., & Bound, J. (1993). Age patterns of smoking in US black and white women of childbearing age. *American Journal of Public Health, 83*, 1258–1264.

Jackson, K. M., Sher, K. J., & Wood, P. K. (2000). Trajectories of concurrent substance use disorders: A developmental, typological approach to comorbidity. *Alcoholism: Clinical and Experimental Research, 24*, 902–913.

Jackson, K. M., Sher, K. J., & Schulenberg, J. E. (2005). Cojoint developmental trajectories of young adult alcohol and tobacco use. *Journal of Abnormal Psychology, 114*(4), 612–626.

Kandel, D. B. (2002). *Stages and pathways of drug involvement*. New York: Cambridge University Press.

Karp, I., O'Loughlin, J., Paradis, G., Hanley, J., & DiFranza, J. (2005). Smoking trajectories of adolescent novice smokers in a longitudinal study of tobacco use. *Annals of Epidemiology, 15*, 445–452.

Merline, A. C., O'Malley, P. M., Schulenberg, J. E., Bachman, J. G., & Johnston, L. D. (2004). Substance use among adults 35 years of age: Prevalence, adulthood predictors, and impact of adolescent substance use. *American Journal of Public Health, 94*, 96–102.

Muthén, B. O. (2004). Latent variable analysis: Growth mixture modeling and related techniques for longitudinal data. In D. Kaplan (Ed.), *Handbook of quantitative methodology for the social sciences* (pp. 345–368). Newbury Park: Sage Publications.

Muthén, B. O., & Muthén, L. K. (2000). The development of heavy drinking and alcohol-related problems from ages 18 to 37 in a U.S. national sample. *Journal of Studies on Alcohol, 61*, 290–300.

Muthén, B., & Shedden, K. (1999). Finite mixture modeling with mixture outcomes using the EM algorithm. *Biometrics, 55*, 463–469.

Nagin, D. S. (1999). Analyzing developmental trajectories: A semiparametric group based approach. *Psychological Methods, 4*, 139–157.

Oesterle, S., Hill, K. G., Hawkins, J. D., Guo, J., Catalano, R. F., & Abbott, R. D. (2004). Adolescent heavy episodic drinking trajectories and health in young adulthood. *Journal of Studies on Alcohol, 65*, 204–212.

Orlando, M., Tucker, J. S., Ellickson, P. L., & Klein, D. J. (2004). Developmental trajectories of cigarette smoking and their correlates from early adolescence to young adulthood. *Journal of Consulting and Clinical Psychology, 72*, 400–410.

Patton, G. C., Coffey, C., Lynskey, M. T., Reid, S., Hemphill, S., Carlin, J. B., & Hall, W. (2007). Trajectories of adolescent alcohol and cannabis use into young adulthood. *Addiction, 102*, 607–615.

Raudenbush, S. W., & Byrk, A. S. (2002). *Hierarchial linear models: Applications and data analysis methods* (2nd ed.). Thousand Oaks: Sage.

Scheier, L. M., Botvin, G. J., Griffin, K. W., & Diaz, T. (2000). Dynamic growth models of self-esteem and adolescent alcohol use. *Journal of Early Adolescent, 20*, 178–209.

Schulenberg, J. E., Merline, A. C., Johnston, L. D., O'Malley, P. M., Bachman, J. G., & Laetz, V. B. (2005). Trajectories of marijuana use during the transition to adulthood: The big picture based on national panel data. *Journal of Drug Issues, 35*, 255–280.

Singer, J. D., & Willet, J. B. (2003). *Applied longitunal data analysis: Modeling change and event occurence*. New York: Oxford University Press.

Stanton, W. R., Flay, B. R., Colder, C. R., & Mehta, P. (2004). Identifying and predicting adolescent smokers' developmental trajectories. *Nicotine & Tobacco Research, 6*, 843–852.

Toumbourou, J. W., Williams, I. R., Snow, P. C., & White, V. M. (2003). Adolescent alcohol-use trajectories in the transition from high school. *Drug Alcohol Review, 22*, 111–116.

Vitaro, F., Wanner, B., Brendgen, M., Gosselin, C., & Gendreau, P. L. (2004). Differential contribution of parents and friends to smoking trajectories during adolescence. *Addictive Behaviors, 29*, 831–835.

Walden, B., Iacono, W. G., & McGue, M. (2007). Trajectories of change in adolescent substance use and symptomatology: Impact of paternal and maternal substance use disorders. *Psychology of Addictive Behaviors, 21*, 35–43.

Wallace, J. M., & Muroff, J. R. (2002). Preventing substance abuse among African American children and youth: Race differences in risk factor exposure and vulnerability. *Journal of Primary Prevention, 22*, 235–261.

White, H. R., Johnson, V., & Buyske, S. (2000). Parental modeling and parenting behavior effects on offspring alcohol and cigarette use. A growth curve analysis. *Journal of Substance Abuse, 12*, 287–310.

White, H. R., Pandina, R. J., & Chen, P.-H. (2002). Developmental trajectories of cigarette use from early adolescence into young adulthood. *Drug and Alcohol Dependence, 65*, 167–178.

White, H. R., Nagin, D., Replogle, E., & Stouthamer-Loeber, M. (2004). Racial differences in trajectories of cigarette use. *Drug and Alcohol Dependence, 76*, 219–227.

Windle, M., & Wiesner, M. (2004). Trajectories of marijuana use from adolescence to young adulthood: Predictors and outcomes. *Development of Psychopathology, 16*, 1007–1027.

Chapter 3
Developmental Trajectories of Drug Use Among Whites and African-Americans: Evidence for the Crossover Hypothesis

Denise Kandel, Christine Schaffran, and Yonette F. Thomas

The age Crossover Hypothesis, first suggested by Geronimus et al. (1993) in a study of smoking by pregnant women, describes the reversal in the prevalence of drug use among whites and African Americans over the life cycle. In adolescence, the prevalence of lifetime and current use is higher among whites than African Americans. At some point in adulthood, the pattern is reversed: prevalence becomes higher among African Americans than whites. This pattern has been repeatedly observed (Biafora and Zimmerman 1998; Ellickson et al. 2003; Feigelman and Lee 1995; Flint et al. 1998; Kandel 1991, 1995; King et al. 2004; Pampel 2008). However, with rare exceptions, most of what is known about racial/ethnic differences in life cycle differences in patterns of drug use is based on separate adolescent and adult samples.

A sociological perspective rooted in socialization theory that emphasizes role incompatibility, role selection and socialization on patterns of drug use over time (Kandel and Yamaguchi 1987; Yamaguchi and Kandel 1985a, b) may help account in part for the crossover phenomenon. An individual's progression through the life course

---

Reprinted in part from Drug and Alcohol Dependence vol 118, issues 2–3, Denise Kandel, Christine Schaffran, Mei-Chen Hu, Yonette Thomas: Age-related differences in cigarette smoking among whites and African-Americans: Evidence for the crossover hypothesis, pages 280–287, 2011 with permission from Elsevier.

D. Kandel (✉)
Department of Psychiatry, College of Physicians and Surgeons and Mailman School of Public Health, Columbia University, New York, NY, USA

New York State Psychiatric Institute, Columbia University, New York, NY, USA
e-mail: dbk2@cumc.columbia.edu

C. Schaffran
New York State Psychiatric Institute, New York, NY, USA

Y.F. Thomas
The New York Academy of Medicine and the American Association of Geographers, Glenn Dale, MD, USA

Y.F. Thomas, L.N. Price (eds.), *Drug Use Trajectories Among Minority Youth*,
DOI 10.1007/978-94-017-7491-8_3

is marked by successive participation in different social roles. Participation in the traditional roles of adolescence and adulthood tends to be associated with traditional values, attitudes and behavior in various realms including political ideology, religious participation and the use of drugs. Being a student is an important role in adolescence; being a worker, a spouse, and a parent are important roles in adulthood. Cross-sectional and longitudinal data suggest that conventional social roles and drug use are incompatible. In cross-sectional samples, the use of drugs is consistently negatively related to being married, being a parent and working but positively related to being absent from school, being unemployed, divorced or living with a partner (Bachman, O'Malley, and Johnston 1984; Clayton and Voss 1977; Kandel 1984; Kandel et al. 1986; Miller and Cisin 1980). Education is strongly negatively related to tobacco use (Eikemo et al. 2007; House 2002; Robert and House 2000; Schaap et al. 2008; Schnittker and McLeod 2005; U.S. Department of Health and Human Services 2007).

Prior longitudinal analyses focused mainly on marijuana use and documented incompatibility between marijuana use and conventional family and work roles. Incompatibility that was resolved either by role selection or role socialization. In the first case, individuals selected roles that were compatible with their drug use; in the second case, they modified their drug behavior to be compatible with a particular role. Thus, current marijuana use predicted postponement in marriage; continued marijuana use during marriage was associated with increased separation or divorce and increased job instability and unemployment (Kandel and Yamaguchi 1987; Yamaguchi and Kandel 1985b). By contrast, marriage increased the rate of stopping marijuana use, and divorce increased the rate of initiating or resuming marijuana use (Yamaguchi and Kandel 1987). Because African Americans have lower education attainment and are less likely than whites to be married and to be employed full time, these disparities could account for the fact that rates of selected drugs do not decline after the mid-twenties as strongly for African Americans as for whites. In an analysis of black-white differences in aging out of substance use and abuse, implemented in the National Comorbidity Study, Yuan (2009) found that work, economic conditions and family roles explained part of the cross-over between whites and African Americans, especially as regards alcohol use and abuse. In that study, the unique effect of education was not investigated separately from the effect of other covariates.

In this chapter, we investigate the age Crossover Hypothesis at the national level in the United States. In the absence of longitudinal data, we examine age-specific rates of the drug use in the population. In order to confirm the hypothesis, we describe patterns of age-specific rates of current use of selected legal and illegal drugs, especially cigarettes, from age 12 onward among whites and African Americans. We then attempt to identify the specific ages at which crossover appears for cigarettes, when rates are equal in the two groups, and the factors that may account for the differential age related patterns that are observed. Successful identification of relevant factors will shift upward the age at which crossover is estimated to occur.

Methods

The Data

We examined the Crossover Hypothesis for five drug classes: cigarettes, alcohol, marijuana, cocaine and non-prescribed psychoactive drugs using data are from the National Survey on Drug Use and Health (NSDUH). The NSDUH, a successor since 2002 to the National Household Survey on Drug Abuse (NHSDA), consists of a series of annual cross-sectional national surveys of the US population 12 and over and provides data for individuals at different stages of the life cycle. The target civilian non-institutionalized population represents over 98 % of the total population, including persons living in non-institutionalized group quarters, such as homeless shelters, rooming houses and college dormitories. Individuals on active military duty, in jail or drug treatment programs, and the homeless not in shelters are excluded. The samples are very large, making it possible to compare racial/ethnic groups, even though, as we will see, there are still too few cases for analyzing cocaine users among African-American adolescents. The same measures have been used for adolescents and adults. The whole range of licit and illicit drugs has been ascertained.

Most of our analyses are based on data from the NSDUH 2006 (Substance Abuse and Mental Health Services Administration, 2007). Youths (12–25 years old) were over-sampled. Respondents were administered computer assisted structured personal household interviews and asked about the use of tobacco and 11 other drug classes. The completion rate was 67.2 %. Weights take into account the stratified multistage cluster sampling design and correct for over-sampling and non-response rates so that the resulting weighted sample is representative of the U.S. population.

Statistical Analysis

To explain the crossover phenomenon, we applied Ross and Bird (1994)'s statistical approach, which they developed to account for age related differences in perceived quality of health between men and women and to estimate the crossover point at which perceived health is equal in the two genders. This method was also implemented by Yuan (2009). The method makes it possible to specify the age at which crossover actually occurs in the particular population under study and to estimate the crossover age when potential explanatory factors are taken into account. By including each covariate in a sequential order in the model, this strategy identifies the relative importance of covariates and specifies the ages at which the crossover in the event of interest, in this case current smoking, would occur if the two groups were equal on the selected variables. At the crossover age, the rates of current smoking would be the same in both racial/ethnic groups. Adjusting for relevant covariates moves the crossover point to older ages, if not eliminate it.

A series of hierarchical logistic regressions were estimated to predict current smoking among lifetime smokers. In a first step, to establish the occurrence of a crossover and to estimate the specific age at which it appeared, the predictors included only the variables of age, age squared (to reflect declines in smoking after a certain age) and an interaction term between age and ethnicity to test systematically the hypothesis that the age related decline in smoking differed between African Americans. In successive steps, individual factors that could affect the crossover age were added. These factors indexed age of onset into smoking and the interaction between age of onset and race/ethnicity, gender, socioeconomic resources (education) and social role participation (work and marital status). Control for these factors made it possible to estimate the crossover point assuming that the two racial/ethnic groups were equal on the variables of interest. The impact of specific factors is reflected in a higher predicted age at which crossover would occur. Effective controls would reduce or eliminate the crossover.

Results

Racial/Ethnic Specific Patterns of Drug Use

For each drug, the lifetime rates of use in the total population are consistently higher for whites than African-Americans. In 2006, among individuals aged 12 and over, the proportions having ever smoked cigarettes are 72.0 % for whites and 54.7 % for African-Americans. The respective proportions for alcohol are 87 % and 74.6 %; for marijuana 43.6 % and 37.5 %; for cocaine 16.3 % and 9.1 %; and for non-medical use of prescription-type (Rx) psychotherapeutic drugs 22.6 % and 12.9 %. All the differences are highly statistically significant at p<001.

We examined patterns of use for each of the five drug classes in four age groups ranging from adolescence to age 49, ages 12–17, 18–25, 26–34, and 35–49, the periods of highest drug use. As illustrated in Fig. 3.1, the same differences between African-Americans and whites occur for lifetime use of each drug class among each of the four age groups as in each age undifferentiated ethnic group. At each age the prevalence of use are consistently higher among whites than African Americans. At no time do the rates among African Americans exceed those among whites.

The age-related patterns in the two racial/ethnic groups, however, are quite different for selected drugs when the drug behavior under consideration shifts from lifetime to current use among those who have ever used each drug class. These conditional rates of current use, defined as any use in the last 30 days, index persistence of use. Crossover appears but only for two substances: cigarettes, and to a lesser extent for marijuana (Fig. 3.2), and does so at a later age for cigarettes (between ages 26–34) than for marijuana (between ages 18–25). Prior to these ages, the prevalences are higher for whites than African Americans; as of these ages, the prevalences are higher for African Americans than whites. The curves converge at

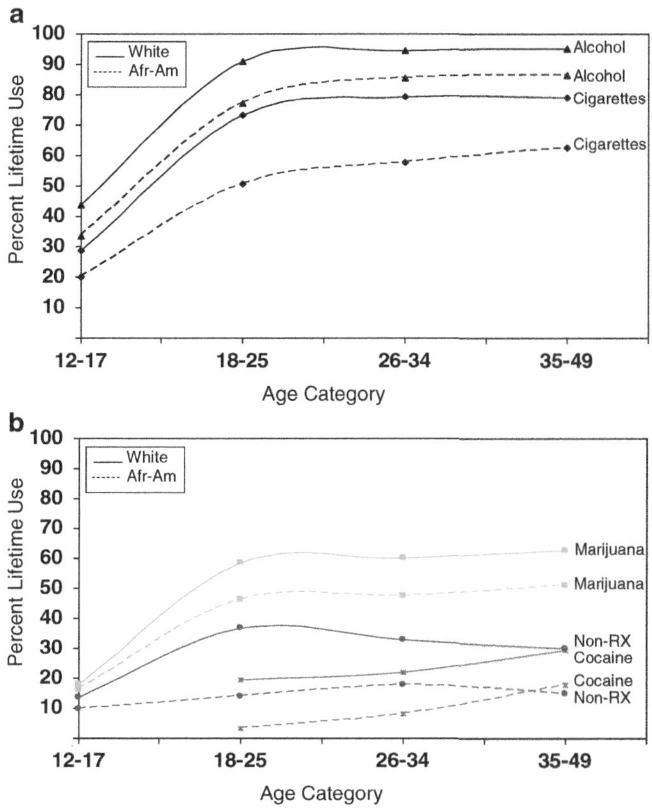

Fig. 3.1 Lifetime Prevalence of use of alcohol, cigarettes and three illicit drugs by age among Whites and African Americans (NSDUH 2006). (**a**) Lifetime prevalence of use of alcohol and cigarettes, (**b**) Lifetime prevalence of use of three Illicit drug classes

ages 35–49 for non-medical use of RX medicines. There does not appear to be any crossover for alcohol nor cocaine as of age 18 (Fig. 3.2). Even in such a large sample as the NSDUH, there are too few cases of current cocaine use (N = 10) among African Americans adolescents aged 12–17 to include them in the analysis.

How Real a Phenomenon Is Crossover in Smoking among Whites and African Americans?

These age comparisons may not reflect true maturational changes, since cross-sectional age differences confound developmental changes with cohort differences and historical factors. The optimal documentation would be based on longitudinal panels spanning adolescence to late adulthood.

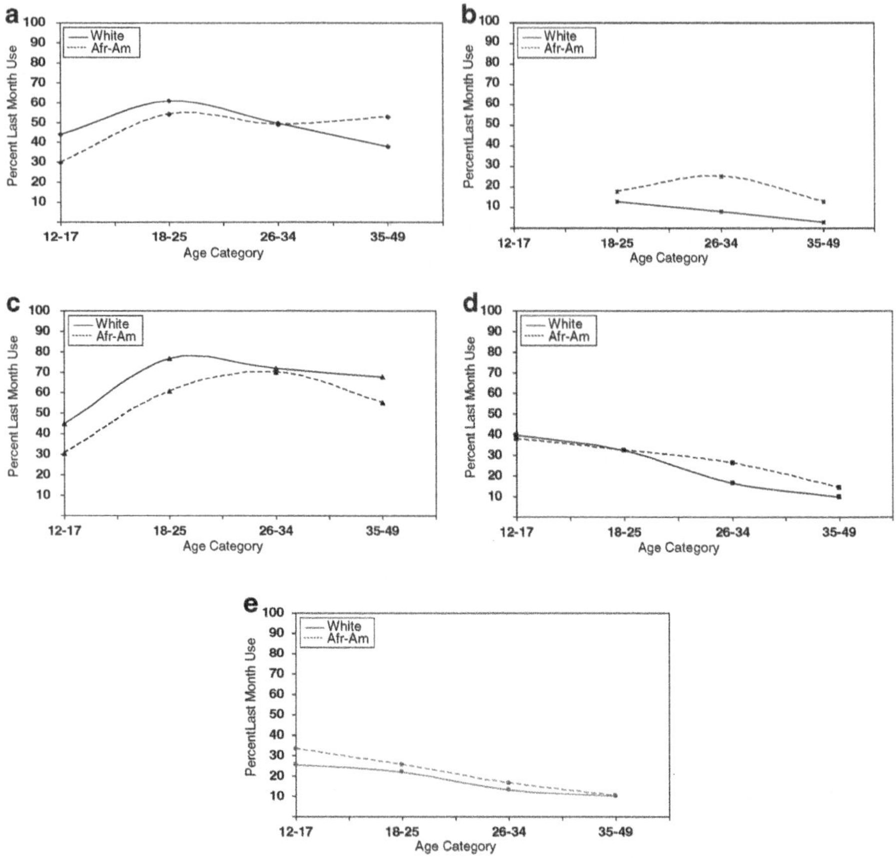

Fig. 3.2 Current use of five drug classes by age among lifetime users of each drug class among Whites and African Americans (NSDUH 2006). (**a**) Current cigarette use among lifetime users. (**b**) Current alcohol use among lifetime users. (**c**) Current marijuana use among lifetime users. (**d**) Current cocaine use among lifetime users. (**e**) Current non-medical use of RX psychotherapeutics use among lifetime users

We approximated such a longitudinal panel by analyzing data from multiple waves of the NHSDA/NSDUH, where the behavior of the same national birth cohorts could be followed over time since each annual sample is a random subsample of the same birth cohorts. We focused on cigarette use, which exhibits a strong crossover effect. We calculated the rate of current use among lifetime smokers at four different ages (12–17, 18–25, 26–34, 35+) for the aggregated 1975–1980 birth cohorts across 15 surveys from 1992 to 2006. These six birth cohorts correspond approximately to the range of ages included in the grouped age categories presented in the published reports and available in the public use data for ages above 20. We used age-specific published prevalences of use for the years 1992–1998 and determined the rates of use from the publicly released data for the years 1999–2006 (See

Table 3.1 Last month cigarettes use among lifetime smokers by age in 2006 and in six birth cohorts among whites and African Americans in NHSDA/NSDUH 1990–2006

Birth		Age 12-17		Age 18-25		Age 26-34		Age 35+	
Cohorts	Survey Year	White	Afr-Amer	White	Afr-Amer	White	Afr-Amer	White	Afr-Amer
75-80	1992	30.5	14.7	48.0	38.8	44.7	51.3	31.3	44.2
	1993	29.1	18.9	44.9	35.0	39.6	45.7	29.0	39.3
	1994	52.4	45.8	51.2	49.3	42.0	47.8	33.2	44.6
	1995	53.1	47.8	52.3	47.1	46.2	50.6	33.3	43.6
	1996	51.7	46.7	57.5	51.9	47.4	50.9	33.1	47.9
	1997	51.8	52.8	61.5	55.3	46.2	45.1	35.2	46.8
	1998	51.6	52.5	62.6	56.7	40.9	49.6	31.1	45.0
	1999	42.3	30.4	60.2	47.4	44.9	41.0	30.8	38.8
	2000	42.0	25.7	59.3	50.9	43.2	42.7	30.5	42.3
	2001	41.7	26.7	59.7	47.3	43.8	45.2	30.0	42.8
	2002	43.5	24.2	61.0	48.9	45.1	60.5	29.6	43.3
	2003	44.1	30.1	60.2	51.4	47.7	45.3	28.7	43.7
	2004	45.1	27.1	61.0	53.5	46.9	48.8	29.6	37.6
	2005	44.0	32.1	61.1	53.4	49.2	43.6	29.5	43.0
	2006	44.1	29.9	60.9	54.2	49.5	48.8	28.9	42.5

1975-1980 birth cohorts

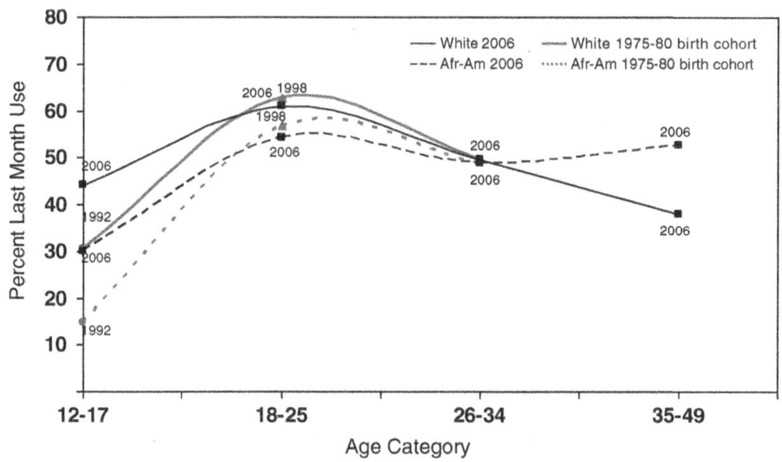

Fig. 3.3 Current cigarette use by age among White and African American lifetime smokers in 2006 and in the 1975–1980 birth cohorts in 1992 and 1998 (NHSDA/NSDUH 1992, 1998, 2006)

annual reports published by NHSDA/ NSDUH). The rates of current smoking among lifetime smokers in the six aggregated birth cohorts at the four different ages are presented in Table 3.1.

The crossover phenomenon in persistence of smoking appears not only in the age-specific cross-sectional data for the year 2006 but for the 1975–1980 birth cohorts when followed at different ages from 1992 to 2006. Crossover in conditional rates of current use appears at ages 26–34, as displayed in Fig. 3.3.

Table 3.2 Mean ages of onset into cigarettes, alcohol and illicit drugs among whites and African-Americans 18–49 years old (NSDUH 2006)

	Cigarettes	Alcohol	Marijuana	Cocaine	Non-RX
Whites					
Mean age (yrs)	15.0***	15.8***	16.3***	19.4***	19.1*
s.d.	3.4	3.1	3.1	3.8	5.4
N's	(14,973)	(18,286)	(11,736)	(4285)	(6558)
African-Americans					
Mean age (yrs)	15.9	17.1	16.7	21.3	19.7
s.d.	3.7	3.5	3.3	4.8	6.1
N's	(2227)	(3242)	(1903)	(287)	(587)

*Differences between whites and African-Americans significant at p<05; ***p<001

Paradox

The crossover pattern in conditional rates of current smoking observed among white and African-American adults is paradoxical. Indeed, the ages of onset into various drug classes are consistently higher for African Americans than whites (See Table 3.2). Among those aged 18–49 who have gone through the major period of risk for initiating the use of different drugs, African Americans initiate the use of cigarettes about 11 months later than whites, alcohol about 15 months later, marijuana about 5 months later, cocaine about 2 years later, and non-medical use of Rx drugs about 7 months later. Later onset into drugs is thought to be protective. Later onset into drugs is related to lower risk of persistent and extensive use, lower risk of becoming dependent on the drug as well as lower risk of using other substances (Giovino 1999; Kandel 2003; King and Chassin 2007; Substance Abuse and Mental Health Services Administration, 2005; Yamaguchi and Kandel 1984). Yet, African Americans, who initiate the use of cigarettes later than whites, are more likely than whites to continue to smoke in adulthood.

The paradox is partially resolved when the association between age of onset into cigarette use and persistence of smoking, i.e., current smoking among lifetime smokers, is examined separately for whites and African Americans (Fig. 3.4). Among white adults, persistence of smoking **decreases** with increasing age of onset. By contrast, among African American adults, persistence of smoking **increases** with increasing age of onset. As of onset age 14 and higher, the rates of current smoking among lifetime smokers are consistently higher among African Americans than whites. Age of onset into cigarettes has opposite effects on persistence of use among African American and white cigarette smokers.

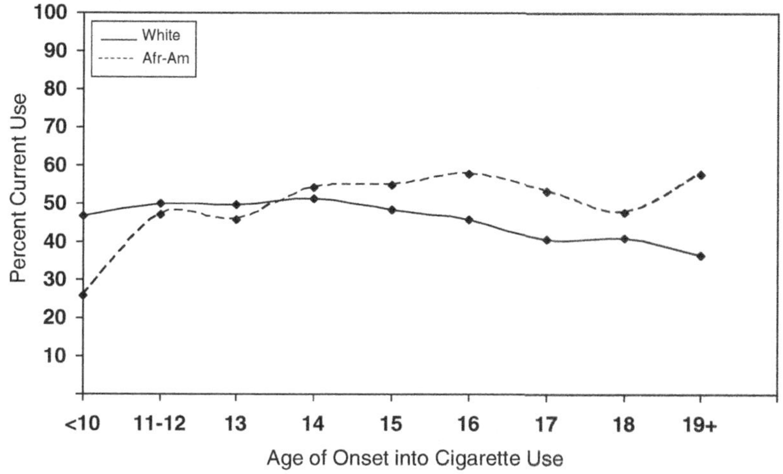

Fig. 3.4 Current cigarette use by age of onset into cigarette use among White and African American lifetime cigarette users 18–49 years old (NSDUH 2006)

Age of Onset and Persistence of Marijuana Use

The inverse pattern between age of onset and persistence of use among whites and African Americans is much weaker for marijuana than for cigarettes and is not statistically significant.

Explaining the Crossover Phenomenon

The crossover curves for smoking illustrate very strikingly that while African American youths start with an advantage, this advantage fails to persist later in adulthood. The factors that come into play for whites do not appear to do so for African-Americans. One factor, as discussed above, is the reverse impact of age of onset on persistence of smoking in the two groups. The opposite relationship between later age of onset and persistence of smoking among whites and African Americans may be one factor that explains the crossover phenomenon observed with respect to smoking.

In addition, differences in social role participation between whites and African-Americans may also contribute to the age related differential patterns of use observed in the two groups. The social role participation of whites and African Americans in the NSDUH sample reflect traditional racial/ethnic patterns (Table 3.3). Twice as many whites as African-Americans were college graduates and were married. The proportion working was higher among whites than African Americans.

Table 3.3 Education, work, and marital status among whites and African-Americans (NSDUH 2006, ages 12–49)

	Whites (%)	African Americans (%)
Education		
Less than high school	8.6	13.0*
High school graduate	26.1	31.2
Some college	24.2	23.8
College graduate	25.9	13.1
Age 12–17	15.2	18.9
Working last week		
Working, has job	72.3	64.2**
Keeping house full-time	5.5	3.7
In school/training	5.7	8.0
Disabled/retried/no job	9.1	15.2
Age<=14	7.4	8.9
Marital status		
Married	45.3	24.2**
Div/sep/wid	10.3	11.8
Never married	37.1	55.2
Age<=14	7.4	8.9
Total N	(30,862)	(6564)

*Square test significant at *p<05; **p<01

The combined impact of the inverse relationship between age of onset and persistence of smoking and the disparity in social roles between whites and African Americans could explain the crossover phenomenon. To test this hypothesis, hierarchical logistic regressions were estimated, as described in the methods section. Results from the hierarchical models are displayed in Table 3.4. The first regression (Model 1) displays the effects of race/ethnicity, age, age squared, and the interaction of race/ethnicity and age squared on current smoking. Model 1 indicates that African Americans are less likely to be current smokers than whites, as reflected in the negative coefficient for race/ethnicity. But this advantages decreases with age, as reflected in the significant positive coefficient for age, the negative effect of age squared, (indicating that smoking initially increases with age but as of a certain age starts to decrease), and the positive race/ethnicity interaction with age squared. At younger ages, African Americans are less likely to smoke than whites but after a certain age they are more likely to do so. The age at which the rates of smoking for whites and African Americans are the same is 30.4 years.

Model 2 adds gender to the model. Gender has a negative impact on current smoking but does not affect the race/ethnic differential.

Model 3 adds age of cigarette onset and the interaction with race/ethnicity. Age of cigarette onset is negatively related to persistence of smoking while the interaction is positive, indicating that as age of smoking onset increases African Americans are more likely to smoke than whites. Controlling for age of onset into smoking and

Table 3.4 Logistic regressions of current cigarette moking (last month) among white and African-American lifetime cigarette smokers aged 12–49 years on race, age, and their interaction (Model 1), controlling for gender (Model 2), age of cigarette onset and interaction with race/ethnicity (Model 3), education (Model 4), work status (Model 5) and marital status (Model 6). (NSDUH 2006, N=20,881)

		Model 1		Model 2		Model 3		Model 4		Model 5		Model 6	
	Univariate	Multivariate	SE	Multivariate	SE	Multivariate	SE	Multivariate	SE	Multivariate	SE	Multivariate	SE
Race/ethnicity (=black)	0.179**	–0.653***	0.118	–0.655***	0.1186	–1.978***	0.2686	–2.018***	0.2865	–2.088***	0.290	–2.104***	0.288
Age	–0.026***	0.100***	0.017	0.100***	0.0167	0.106***	0.0173	–0.002	0.030	0.008	0.031	0.084*	0.0337
Age ^{2}	0.000***	–0.002***	0.000	–0.002***	0.000	–0.002***	0.000	–0.001	0.000	–0.008	0.000	–0.002***	0.001
Race/ethnicity * Age ^{2}		0.001***	0.000	0.001***	0.000	0.001***	0.000	0.001***	0.000	0.001***	0.000	0.001***	0.000
Gender (=female)	–0.088*			–0.089*	0.037	–0.084*	0.0366	–0.030	0.040	–0.003	0.043	–0.001	0.0451
Age of cig onset	–0.015**					–0.030***	0.0072	–0.008	0.007	–0.008	0.007	–0.010	0.0068
Race/ethnicity *Age of cig onset						0.090***	0.0168	0.072***	0.0174	0.072***	0.017	0.073***	0.0173
Education (vs less than high school)													
High school grad	–0.708***							–0.674***	0.091	–0.634***	0.091	–0.634***	0.0891
Some college	–1.203***							–1.209***	0.074	–1.157***	0.076	–1.169***	0.080
College grad	–2.091***							–2.025***	0.099	–1.960***	0.099	1.928***	0.0988
Age 12–17	–1.314***							–1.874***	0.119	–1.661***	0.120	1.415***	0.1242
Work status (vs working/has job)													
Keeping house full time	–0.112									–0.308***	0.088	–0.108	0.0867
In school/training	0.091									–0.082	0.080	–0.159*	0.0812
No job/retired/ disabled	0.797***									0.549***	0.066	0.410***	0.0713

(continued)

Table 3.4 (continued)

		Model 1		Model 2		Model 3		Model 4		Model 5		Model 6	
	Univariate	Multivariate	SE	Multivariate	SE	Multivariate	SE	Multivariate	SE	Multivariate	SE	Multivariate	SE
Age ≤ 14	–0.702***									–0.655***	0.125	–0.033	0.137
Marital status (vs married)[a]													
Divorced/ separated/ widowed	1.118***											1.000***	0.0779
Never married	0.909***											0.692***	0.0651
Age ≤ 14	–0.270**												
Estimated Crossover Age (years)		30.36		30.36		32.12		37.19		37.84		41.74	
χ^2 (df)		470.08(4)		478.68(5)		559.40(7)		2180.89(11)		2343.63(15)		2850.62(17)	

[a]category for age ≤14 omitted in multivariate logistic regressions

*p<05; **p<01; ***p<001

the interaction with race/ethnicity increases the estimated age of onset by about 2 years to age 32.12.

Model 4 adds education to the model. Education is significantly negatively related to current smoking. Compared with not having graduated from high school, each increasing educational level, from high school graduation, to some college, and to college graduation, almost doubles the reduction in levels of current smoking. The estimated crossover age, assuming that the educational levels of African Americans would match those of whites, would be 37.19 years. Adjusting for education (after adjusting for age of smoking onset) raises the estimated crossover age by 5 years.

Model 5 adds the work status variables to the model. Compared with those in the labor force, those who are keeping house full time are less likely to be currently smoking while those with no job or who are retired or disabled are more likely to be smoking currently. Adding the work status variables to the model affects the estimated crossover age relative little, by about 7 months.

Model 6 adds the marital status variables to the model. Compared with those who are married, those who are separated or divorced or those who never married are more likely to be current smokers. Additional control for marital status has a relatively large impact. The expected crossover age increases by 3.9 years to age 41.74.

With each additional control, the negative effect of race/ethnicity becomes larger. Overall, controlling for age of smoking onset, the interaction between age of smoking onset and race/ethnicity, education, and social role participation in work and marriage increases the crossover age by 11.4 years.

Conclusion

The crossover phenomenon in patterns of drug use in which African Americans have lower rates of drug use than whites in adolescence and early adulthood but higher rates in adulthood is observed selectively in the United States population. It is not observed for lifetime use of legal and illegal drugs but it is observed for persistence of use, i.e., current use among those who have ever used the drug, and mainly for cigarettes. The age at which crossover occurs for smoking cigarettes is 30.4 years. Prior to that age, whites have higher rates of persistent smoking than African Americans; after that age, the pattern is reversed.

Several factors account for this reversal. The impact of these factors on the crossover age was assessed by applying a statistical technique (Ross and Bird 1994) that estimates crossover ages predicated on standardizing white and African American smokers on explanatory variables of interest. Three classes of factors are important. One factor is age of onset into smoking. Unexpectedly, the relationship between persistence of smoking (i.e., current smoking among lifetime smokers) and age of smoking onset is opposite for white and African American smokers. For whites, the earlier the age of smoking initiation, the higher the rates of persistence; for African Americans, however, the earlier the age of smoking initiation, the lower the rate of persistence. Among African Americans aged 18–49, smoking initiation takes place about 11 months later on average than among whites. Were the ages of initiation the same among both groups, the crossover age would increase by about 2 years.

The second important factor, and perhaps the most important, is education. African Americans have lower education than whites. Equal educational achievement among whites and African Americans would delay the crossover age by 5 years, after controlling also for age of smoking onset. The third important factor is marital status. African Americans have much lower rates of marriage than whites. Controlling for age of onset and educational level, similar rates of marriage among the two racial/ethnic groups would delay the crossover age further by 3.5 years. Yuan (2009), who investigated potential crossover in last year use of alcohol, illicit drugs and prescription drugs for non medical purposes observed crossover only for alcohol use. Cigarette smoking was not investigated.

The effect of education was statistically significant but its impact on crossover age was not estimated separately from that of other factors. Similar to this study, however, there were no significant effects of work on the age-by-race crossover while family related variables, such as being married and having good relationship with children, increased crossover for alcohol use by 5 years. The observed association between participation in social roles, such as schooling and marriage and individual behaviors, such as smoking, reflect the dynamic interactions of two processes: role selection, i.e., the influence of preexisting behaviors on the timing and sequencing of social role participation, and role socialization, i.e., the subsequent influence of social role participation on individual behaviors. A limitation of the present analyses is their reliance on cross-sectional data which do not permit these two processes to be disentangled.

Aspects of smoking history, education and social role participation together would postpone crossover in current smoking by more than 11 years. Since control for these factors does not eliminate the observed crossover, there obviously are other relevant factors that we have not taken into account. These factors remain to be identified and may include broader contextual community and societal factors than the individual factors that we examined. The role of age of onset is particularly puzzling. To the best of our knowledge, this is the first report in the literature to highlight the potentially opposite impact of age of onset on the natural history of smoking among whites and African Americans. Why age of onset should have different consequences for different groups needs to be investigated further.

However, the present results provide some insights into the factors that contribute to the observed crossover in rates of current smoking among whites and African Americans. They highlight the one factor that needs to be the major focus of intervention- namely schooling and education. Prevention and intervention efforts targeted at retaining youths in school and encouraging the pursuit of educational goals, while not explicitly targeted to reducing the use of drugs, would in fact do so and have many beneficial effects for the individual and society besides the effect on reducing persistent smoking. Not only would these individuals have better economic prospects and contribute to the social good, but they would also be more likely to stop smoking with the additional health benefits that this would entail.

Acknowledgment A revision of a presentation at the conference on "Drug Use trajectories among African Americans" November 13, 2006. National Institute on Drug Abuse, Bethesda, MD in part of this chapter were published as; "Age-related differences in cigarette smoking among whites and African-Americans: Evidence for the crossover hypothesis." (2011) *Drug & Alcohol Dependence*, 118; 280–287.

Supported by Research Scientist Award K-5 DA00081 from the National Institute on Drug Abuse.

References

Bachman, J. G., O'Malley, P. M., & Johnston, L. D. (1984). Drug use among young adults: The impacts of role status and social environment. *Journal of Personality and Social Psychology, 47*, 629–645.

Biafora, F., & Zimmerman, R. S. (1998). Developmental patterns of African American adolescent drug use. In W. A. Vega (Ed.), *Drug use and ethnicity in early adolescence* (pp. 149–176). New York: Plenum Press.

Clayton, R. R., & Voss, H. L. (1977). Shacking up: Cohabitation in the 1970s. *Journal of Marriage and the Family, 39*, 273–283.

Eikemo, T. A., Huisman, M., Bambra, C., & Kunst, A. E. (2007). Health inequalities according to educational level in different welfare regimes: A comparison of 23 European countries. *Sociology of Health and Illness, 30*, 565–582.

Ellickson, P. L., Perlman, M., & Klein, D. J. (2003). Explaining racial/ethnic differences in smoking during the transition to adulthood. *Addictive Behaviors, 28*, 915–931.

Feigelman, W., & Lee, L. (1995). Probing the paradoxical pattern of cigarette smoking among African-Americans: low teenage consumption and high adult use. *Journal of Drug Education, 25*, 307–320.

Flint, A. J., Yamada, E. G., & Novotny, T. E. (1998). Black-white differences in cigarette smoking uptake: progression from adolescent experimentation to regular use. *Preventive Medicine, 27*, 358–364.

Geronimus, A. T., Neidert, L. J., & Bound, J. (1993). Age patterns of smoking in US black and white women of childbearing age. *American Journal of Public Health, 83*, 1258–1264.

Giovino, G. (1999). Epidemiology of tobacco use among US adolescents. *Nicotine & Tobacco Research, 1*, S31–S40.

House, J. S. (2002). Understanding social factors and inequalities in health: 20th century progress and 21st century prospects. *Journal of Health and Social Behavior, 43*, 125–142.

Kandel, D. B. (1991). The social demography of drug use. *Milbank Quarterly, 69*, 365–414.

Kandel, D. B. (1995). Ethnic differences in drug use: patterns and paradoxes. In G. J. Botwin, S. Schinke, & M. A. Orlandi (Eds.), *Drug abuse prevention with mutiethnic youth* (pp. 81–104). Thousand Oaks: Sage.

Kandel, D. B. (1984). Marijuana users in young adulthood. *Archives of General Psychiatry, 41*, 200–209.

Kandel, D. B. (2003). The natural history of smoking and nicotine dependence. In *Proceeding of symposium on addiction: Impact on Canada*. Ottawa/Ontario, Royal Society of Canada.

Kandel, D. B., & Yamaguchi, K. (1987). Job mobility and drug use: An event history analysis. *American Journal of Sociology, 92*, 836–878.

Kandel, D. B., Davies, M., Karus, D., & Yamaguchi, K. (1986). The consequences in young adulthood of adolescent drug involvement: An overview. *Archives of General Psychiatry, 43*, 746–754.

King, K. M., Polednak, A., Bendel, R. B., Vilsaint, M. C., & Nahata, S. B. (2004). Disparities in smoking cessation between African-Americans and whites 1990-2000. *American Journal of Public Health, 94*, 1965–1971.

King, M. K., & Chassin, L. (2007). A prospective study of the effects of age of initiation of alcohol and drug dependence. *Journal of Studies on Alcohol and Drugs, 68*, 256–265.

Miller, J. D., & Cisin, I. H. (1980). *Highlights from the national survey on drug abuse: 1979. ADL-801032*. Rockville: National Institute on Drug Abuse.

Pampel, F. C. (2008). Racial convergence in cigarette use from adolescence to the mid-thirties. *Journal of Health & Social Behavior, 49*, 484–498.

Robert, S. A., & House, J. S. (2000). Socioeconomic inequalities in health: An enduring sociological problem. In A. Bird, P. Conrad, & A. Fremont (Eds.), *Handbook of medicinal sociology*. London: Sage.

Ross, C. E., & Bird, C. E. (1994). Sex stratification and health lifestyle: Consequences for men's and women's perceived health. *Journal of Health and Social Behavior, 35*, 161–178.

Schaap, M. M., Van Agt, H. M. E., & Kunst, A. E. (2008). Identification of socioeconomic groups at increased risk for smoking in European countries: Looking beyond educational level. *Nicotine & Tobacco Research, 10*, 359–369.

Schnittker, J., & McLeod, J. D. (2005). The social psychology of health disparities. *Annual Review of Sociology, 31*, 75–103.

Substance Abuse and Mental Health Services Administration. (2005). *Results from 2004 National Survey on Drug Use and Health: National Findings* (Office of Applied Studies, NSDUH Series H-28, DHHS Publication No. SMA 05-4062). Rockville: Office of Applied Studies NSDUH.

Substance Abuse and Mental Health Services Administration. (2007). *Results from the 2006 National Survey on Drug Use and Health: National Findings* (Office of Applied Studies, NSDUH Series H-32, DHHS Publication No. SMA 07–4293). Rockville: Office of Applied Studies NSDUH.

U.S. Department of Health and Human Services. (2007). *Preventing smoking and exposure to secondhand smoke before, during, and after pregnancy*. Atlanta: U.S. Department of Health and Human Services/Centers for Disease Prevention and Health Promotion.

Yamaguchi, K., & Kandel, D. B. (1984). Patterns of drug use from adolescence to young adulthood: III predictors of progression. *American Journal of Public Health, 74*, 673–688.

Yamaguchi, K., & Kandel, D. B. (1985a). Dynamic relationships between premarital cohabitation and illicit drug use: A life event history analysis of role selection and role socialization. *American Sociological Review, 50*, 530–546.

Yamaguchi, K., & Kandel, D. B. (1985b). On the resolution of role incompatibility: Life event history analysis of family roles and marijuana use. *American Sociological Review, 90*, 1284–1325.

Yamaguchi, K., & Kandel, D. B. (1987). Drug use and other determinants of premarital pregnancy and its outcome: A dynamic analysis of competing life events. *Journal of Marriage and the Family, 49*, 257–270.

Yuan, A. S. V. (2009). *Black-white differences in aging out of substances use and abuse*. Blacksburg: Virginia Polytechnic Institute and State University (unpublished).

Chapter 4
The Crossover Effect: The Influence of Social Roles in an African American Cohort

Margaret E. Ensminger, Kate E. Fothergill, Elaine E. Doherty, Kerry M. Green, Judy A. Robertson, and Hee-Soon Juon

Despite some stereotypes that may exist, substance use among African American adolescents is generally comparable to or somewhat less than use among White populations. However, evidence of a pattern shift emerges as they age into adulthood, when African Americans, compared to Whites, are more likely to initiate drug use and develop problem use, and are less likely to terminate their drug use (French et al. 2002). Graphic representation of this pattern of use for Whites would show a curve that peaks in early adulthood, and then steadily declines in midlife. For African Americans, the drug use curve begins its upward climb after adolescence but, unlike use data for Whites, continues its upward trajectory to an older age before beginning to decline. These age/ethnicity curves intersect as drug use for African Americans continues to rise, while that for Whites tapers off, illustrating the *crossover effect*.

In this chapter, we evaluate the crossover effect in a cohort of African Americans studied from ages 6 to 42 from the Woodlawn Study. We begin with a comparison of the Woodlawn cohort's drug use trends with that of national samples. Further, we evaluate whether social role performance can explain the patterns of increase or decrease in their drug use over the life course. We hypothesize that one reason for

M.E. Ensminger (✉) • K.E. Fothergill • J.A. Robertson
Department of Health, Behavior and Society, The Bloomberg School of Public Health, The Johns Hopkins University, Baltimore, MD, USA
e-mail: mensming@jhsph.edu

E.E. Doherty
Department of Criminology and Criminal Justice, University of Missouri – St. Louis, St. Louis, MO, USA

K.M. Green
Department of Behavioral and Community Health, School of Public Health, University of Maryland, College Park, MD, USA

H.-S. Juon
Sidney Kimmel Medical College, Thomas Jefferson University, Philadelphia, PA, USA

Y.F. Thomas, L.N. Price (eds.), *Drug Use Trajectories Among Minority Youth*,
DOI 10.1007/978-94-017-7491-8_4

higher drug use among African American adults compared to Whites relates to their lower likelihood of participating in the expected central social roles of mid life, especially, marriage and employment.

The Crossover Effect

National data show lower rates of marijuana, alcohol, and cocaine use among African American adolescents than among Whites. The Monitoring the Future study shows that compared to White teens, African American teens have lower rates of lifetime, annual, 30-day, and daily prevalence rates for nearly all licit and illicit drugs (Johnston et al. 2009). However, by early adulthood, there are higher rates of illegal drug use among African Americans compared to Whites (Kandel et al. 1997; SAMSHA 1998, 2001). For example, according to National Household Survey of Drug Abuse (NHSDA) estimates of past-year use, among those age 12–25, Whites had higher rates of marijuana and cocaine use than African Americans, but among those older than 25, African Americans had higher rates of cocaine and marijuana use than Whites (Ma and Shive 2000).

The crossover effect can also be seen in the higher rates of drug abuse and dependence among African Americans. For example, the NHSDA surveys show that although past-year dependence rates among white substance users began to decline at age 26, they increased among African American users at this age (Kandel et al. 1997). In looking specifically at cocaine use and dependence reported in the NHSDA, Chen and Kandel (2002) found that even though African Americans and Whites had similar rates of lifetime and past-year cocaine use, African American users were twice as likely as Whites or Hispanics to report past-year dependence. More specifically, 25.5 % of African American cocaine users reported past-year cocaine dependence compared to 11.4 % of Whites and 16.7 % of Hispanics. This race difference held for both males and females after controlling for socioeconomic status (Chen and Kandel 2002).

Studying the differential patterns of African American drug use is particularly important given that the adverse consequences of drug and alcohol use are most common among those who continue use in adulthood (Brunswick et al. 1991; Newcomb and Bentler 1988). The consequences of substance abuse and dependence are numerous and diverse, including health, cognitive, and emotional problems at the individual level as well as economic and criminal problems at the societal level. African Americans suffer higher levels of drug-related morbidity and mortality. For example, they are about three times more likely than are Whites to be in treatment for drug related problems (NIDA 1986). According to the Centers for Disease Control data, deaths among African Americans for drug or alcohol related causes are at least twice as high as among Whites for all adult age categories (CDC 1997). The excess rates of disease, disability and death have been linked to the high involvement of African Americans in illegal drug use and trafficking (Murray and Lopez 1996; NIDA 1998, 2003). African Americans have disproportionately higher rates of drug-related homicides (BJS 1996) and imprisonment for drug-related offenses (NIDA 2003). Thus, a better understanding of the mechanisms driving the crossover effect could inform interventions and policies as to how to alleviate some of these negative consequences.

The Importance of Social Roles

One potential reason for the crossover effect may be the higher rates of unemployment and the lower rates of marriage among African American adults compared to White adults. We know from many studies that African American adults are less likely to be employed, are less likely to marry, and are more likely to be raising children without the support of a second adult. Recent research has shown an association between employment, divorce, and other social roles with adult onset of drug use (Agrawal and Lynskey 2009; Green et al. 2010).

Social bonds and social roles are important over the life course. The Life Course Social Field Theory (Cicchetti and Schneider-Rosen 1984; Kellam et al. 1975; Kellam and Ensminger 1980; Kellam and Rebok 1992) suggests the importance of key social roles across the life course and the continuing impact of adaptation to these social roles for later success. In adolescence, key social bonds are with family, school, and peers. Failure to bond with family, peers, or school can lead to deviant behavior, which can affect subsequent chances for success in adult social roles. The transition from adolescence to adulthood is one of great change as the roles of spouse, employee, and parent become key, and partially depend on the role performance in the earlier stages of the life course.

Maladaptation in any of these adult social roles could lead to further difficulties in other social contexts and increased deviant behaviors. In mid adulthood, continuing maladaptation or nonparticipation in these central roles may escalate the isolation and disconnection from mainstream society with clear implications for deviant behavior, including health risk behaviors. This is similar to a Social Control Theory perspective (Hirschi 1969; Sampson and Laub 1990) that suggests that a lack of social roles leads to poor attachment to larger society and its norms. Those without strong social ties are less likely to abide by prescribed norms and more likely to engage in deviant behavior (Ensminger et al. 1983, 1984; Hawkins et al. 1992; Mirkowsky and Ross 1986; Seeman 1959; Sieber 1974). In contrast, those with strong bonds are constrained by norms and social obligations from engaging in deviant activities. For example, Umberson (1992) has argued that spouses regulate the behavior of their partners both directly and indirectly. Therefore, those in the role of spouse are more likely to internalize healthy norms that allow for adequate functioning in this role.

Social Roles and Drug Use

The research on drug use over the life course supports the tenets of life course social field theory. Studies have found that failure to succeed in key social roles at each stage of life is related to substance use and disorders. For example, studies of adolescents have found that problems with school, family, and friends are related to an increased risk of drug use initiation (Chilcoat and Breslau 1996; Hawkins et al. 1992; Ensminger 1990). One study, for instance, found that a failure to engage in the social role of school (i.e., truancy) has been associated with an onset of drug use (Henry and Huizinga 2007). Also, educational attainment has had a consistent inverse relationship with drug use and drug use problems (Agrawal et al. 2005; Brunswick and Titus 1998; Crum and Anthony 2000; Fothergill and Ensminger 2006; Lewis et al. 1985; Mensch and Kandel 1988; Warner et al. 1995).

Research with adults has found that being an employee, spouse, and parent is associated with less substance use in mid adulthood, declines in substance use, and less likelihood of initiation after young adulthood (Bachman et al. 1997; Green et al 2010; Horwitz and White 1991; Kandel and Yamaguchi 1987; Merline et al. 2004; Newcomb and Bentler 1988; Yamaguchi and Kandel 1985). Similarly, Sampson and Laub (1990) found that poor marital attachment and job instability increased the risk for adult onset of excessive drinking.

In addition to the line of research and theory that suggests that social roles facilitate desistance from drug use and that the lack of social role participation can lead to drug use and escalation, a complementary line of work show that lack of social role participation can result from drug use (Green and Ensminger 2006; Newcomb and Bentler 1988). Thus, it is important to recognize the reciprocal and complex association between drug use and social roles.

The Current Study

Although much has been learned about the association between social roles and substance use, few studies have examined patterns of drug use over time and how these patterns vary according to social roles relevant to each life stage and how social role participation may inform the crossover phenomena. Using a community population of African Americans, we explore patterns of crack/cocaine use from adolescence to mid adulthood. We hypothesize that in this population we may find higher rates of later onset, greater persistent use through mid adulthood, and later desistance than typically found in White populations. We expect that patterns of cocaine use over time to be associated with social role success in adolescence, early adulthood, and mid adulthood. For example, we expect failure in the adolescent social roles of school and family to be associated with drug use at that age. We also hypothesize that those who are employed and those who are married are more likely to desist from drug use and less likely to have started drug use in adulthood than those who are unemployed and those who are not married. While we do not have as explicit a hypothesis about childrearing, we propose that parents who do have children in their households will be less likely than parents without children in the house to use drugs or have drug problems.

The Woodlawn Project

In this section, we first provide a detailed description of the Woodlawn study design and population. We then compare the drug use of the Woodlawn population with data on drug use from national surveys conducted at the same time as data were collected from the Woodlawn population. These comparisons will provide some perspective on how Woodlawn drug use compared to the nation's at difference stages of the life course. We also examine the participation in social roles from the national surveys by Whites and African Americans, again focusing on the age categories that correspond to those of Woodlawn, and how these social roles relate to trajectories of cocaine use.

Study Design and Population

The Woodlawn study is an epidemiological, prospective study focusing on a cohort of African American children and their families. Most first grade children (N = 1242) in 1966–1967 in Woodlawn, a neighborhood community on the South Side of Chicago, were included in the study (13 families did not participate). Woodlawn is one of 76 community areas in Chicago. In the 1960s when the study began, it was overcrowded and one of the five poorest areas in Chicago, with high rates of unemployment, poverty, and welfare participation: 23 % of Woodlawn families were receiving aid in 1969 compared to 7 % for the city of Chicago as a whole (Council for Community Services 1975). Despite the overall rates of poverty in the community, there was variation in the community, with some blocks having high rates of employment, home ownership, and high levels of education.

The children were assessed in first grade in the nine public and three parochial schools in the community. Teachers and mothers (or mother surrogates) reported on the children's social adaptational status, their mental health, and the family and classroom contexts. The children and their mothers were later assessed when the children were adolescents. Mothers (or mother surrogates) were again interviewed in 1975 about themselves, their families, and the study child. With the mothers' permission, the teenagers were assessed using questionnaires presented on slides and audio tape to control for reading differences (Petersen and Kellam 1977). The adolescents were asked about their psychological well-being, their alcohol and drug use, their delinquency, their family and peer relationships, and their participation in school, church, and other activities. In 1992–1994, the adult children were traced and re-interviewed at ages 32–34. They were assessed on psychiatric symptoms and alcohol and drug abuse/dependence using the Michigan version of the Composite International Diagnostic Interview (CIDI) (Kessler et al. 1994). We also asked them about their living arrangements, family relationships, education and employment histories, health, social support, participation in church and other associations, criminal activities, and their economic situation. The mothers were reinterviewed in 1997–1998. In 2002–2003, we located and reinterviewed the cohort at ages 42–43, using an assessment very similar to the one in 1992–1993. We also have collected school records, death records, and criminal justice records.

Several significant historical and political events occurred as the cohort aged. Most of the cohort was born in 1960 at the end of the migration of African Americans from the rural south to northern cities such as Chicago. About half of the mothers had been born in the South (mostly Mississippi) and then moved to Chicago. The Civil Rights movement was at its height during the 1960s—concerns about voting and school integration were high on the political agenda. Martin Luther King's assassination occurred in 1968 when the children were in 3rd or 4th grade. During the time of the cohort's childhood and adolescence, Woodlawn had very high rates of delinquency and crime (Council for Community Services 1975). The street gang the African Americanstone Rangers, based in Woodlawn, came to the attention of the local and national press during the late 1960s and the 1970s.

The Blackstone Rangers evolved into the El Rukins which continued as a youth gang and matured into an adult gang. Chicago elected its first African American mayor Harold Washington in 1983—cohort members would have been about 23 years old, and this election may have been the first election in which many of them voted. Crack cocaine became widely available during the 1980s. HIV became the leading cause of death for African Americans during the 1990s. Welfare reform was passed in 1996 when the cohort would have been about 36 years old. Barack Obama, an African American Chicago resident from the Southside of Chicago, was elected President of the United States in 2008 when the cohort would have been 48 years old. These events cannot be studied because they do not vary for the cohort; they do provide the context in which cohort members have matured to adulthood.

Attrition

Like all prospective studies, the Woodlawn Study has experienced attrition. About 85 % of the study population was assessed in adulthood at one or both time-points. Mortality by the time of the mid adult assessment was 86 (7.0 %). Because most of the outcomes are derived from the adult data, we tested for attrition biases by comparing those who had at least one adult interview (N = 1054) with those who did not (N = 188). We found few differences on the relevant variables to this study. Those interviewed in adulthood were also more likely to have graduated from high school and less likely to be in poverty in first grade or adolescence. Those who had died by the midlife assessment were more likely to be male, to have dropped out of high school, and to have smoked heavily during adolescence. We also compared those who were missing in adolescence with those who had an adolescent interview. Missingness during adolescence was not related to any of the variables relevant to this study. Those missing in adolescence were more likely to be missing an adult interview (Fothergill and Ensminger 2006).

Trends in Woodlawn Drug Use and Social Roles: Background

Our previous analyses have shown that while drug use declined in general in the decade of the 20s for the Woodlawn cohort as would be expected, there were cohort members who did not desist in their drug use, who even initiated drug use, and who had later adult-onset drug use disorders (Doherty et al. 2008; Green et al. 2010). We have also found that unemployment and overall social integration in early adulthood related to later onset of cocaine/heroin in mid adulthood and unemployment and being unmarried predicted later adult-onset drug use disorders (Green et al. 2010). We did not find any relationship between the number of children in the household in early adulthood and later incidence of substance use or substance use disorders. However, in a person-oriented analysis, we found that those men who were married and had more children living in their households (12.5 % of the males) had the lowest percentage of heavy marijuana use, heavy alcohol use, and incarceration in young adulthood, indicating the strong association between the combined social roles of marriage and parenting and substance use and deviant behavior (Ensminger and Juon 1998).

We also found consequences of adolescent marijuana use for later social roles (Green and Ensminger 2006). For example, heavy adolescent marijuana use (defined as 20 or more times by age 16) led to unemployment, lower participation in marriage, and an increased likelihood of having children outside of marriage for both men and women in young adulthood, 15 years later (Green and Ensminger 2006). We also found (Stuart and Green 2008), that effects of heavy adolescent marijuana use on social role participation continued to persist in mid-life, particularly for women. Heavy adolescent female marijuana users were significantly more likely to report having an unemployment bout between ages 32 and 42 than those who used marijuana less frequently or not at all during adolescence. Thus, in understanding the long-term lack of participation in adult social roles, it is important to consider early risk behaviors.

Comparison of Woodlawn and National Drug Use

Data from national surveys corresponding to the Woodlawn assessments (i.e., adolescence (1975), early adulthood (age 32–33 in 1992), and mid adulthood (age 42 in 2002) provide evidence of the crossover effect for this cohort. We first compare the data for the African American and White respondents of each national survey to see if the data support a crossover effects, and then we compare the Woodlawn data to these national data sets to provide some perspective on the level of drug use of Woodlawn population compared to data from their age level peers collected at about the same time.

As shown in Table 4.1, in 1978, the Monitoring the Future (MTF) survey shows lower rates of all substances among African American adolescences compared to White adolescents (Johnston et al. 2009). In 1992, the National Household Survey of Drug Abuse (NHSDA) (U.S. DHHS 1992) shows that African Americans aged 30–34 had higher rates of cocaine and heroin use than Whites of the same age group, and African Americans had slightly higher marijuana use. Cigarette and alcohol use was lower among African Americans. In 2003, the National Survey of Drug Use in Households (NSDUH) shows that African Americans aged 35–49 continued to have higher rates of cocaine and heroin use than Whites (U.S. DHHS 2004).

With respect to the Woodlawn cohort, for the adolescent assessment, we make comparisons with the Monitoring the Future Study (MTF). In 1975–1976, the year of the Woodlawn adolescent assessment, the MTF was assessing high school seniors. The Woodlawn cohort was in the tenth grade at that time. We used the 1978 MTF assessment of seniors as that is the year that the Woodlawn study participants would be seniors in high school if they were on track and not held back in school. For the young adult and midlife assessments we compare the Woodlawn data with the NHSDA from 1992, which was later renamed the National Study of Drug Use in Households (NSDUH) and NSDUH 2003.

As shown in Table 4.1, the Woodlawn adolescent reports of drug use in the last 2 months were similar to the White adolescents surveyed in MTF (the MTF report was for past month, the Woodlawn report was past 2 months). The reports of use of alcohol and marijuana were very similar to the reports of the Whites in the MTF, and both were considerably higher than the African American reports in that same MTF assessment. Cocaine use was slightly lower (6.4 %) among the Woodlawn adolescents than the MTF White adolescents (8.3 %) with African American MTF adolescents showing the lowest use (4.6 %). According to these 1978 data, then, Woodlawn adolescents did not have lower use of the commonly used drugs than the national sample of MTF white adolescents while both had higher use than the MTF African American adolescents. Given that only seniors in the MTF were surveyed during this year, some of the differences may reflect sample selection in terms of who remained in school until their senior year. Woodlawn adolescents were not assessed in a school setting as we surveyed all those in Chicago we could locate whether they were in school or not.

At the time of young adulthood, evidence emerges supporting the crossover effect in the Woodlawn population. Use of marijuana, crack/cocaine, and heroin was higher among the Woodlawn population than among the either the Whites or the African Americans in NHSDA, and the problem use was considerably higher among the Woodlawn group. Use of alcohol, however, was lower in the Woodlawn population (53 %) compared to both the Whites (80 %) and African Americans (71 %)

Table 4.1 Prevalence of substance use in adolescence and in young- and mid-adulthood in the Woodlawn Study and in National samples by race

	Adolescence			Young adulthood						Mid-adulthood					
		MTF N=17,800				NHSDA						NSDUH			
	Woodlawn N=705	African American	White	Woodlawn N=916		African American N=2,715,640		White N=18,294,013		Woodlawn N=815		African American N=7,574,181		White N=44,838,833	
				Past year		Past year		Past year		Past year		Past year		Past year	
Substance	Past 2 months	Past 30 days or past year	Past 30 days or past year	Use	Problem use	Use	Problem use	Use	Problem use	Use	Problem use	Use	Problem use	Use	Problem use
Cigarettes	77.1	32.7[a]	37.6[a]	49.6	–	36.6	1.3	37.6	4.3	42.6		41.6	–	42.6	–
Alcohol	75.6	48.7[a]	75.0[a]	53.1	11.0	70.8	3.9	80.4	10.3	42.9	2.7	60.6	7.1	78.8	3.8
Marijuana	53.2	39.6[b]	50.1[b]	19.9	4.6	13.8	0.4	13.2	1.5	11.4	2.7[c]	9.8	0.9	10.9	0.5
Cocaine/crack	6.4	4.6[b]	8.3[b]	9.8	5.0	6.2	1.8	4.1	0.7	4.6		5.3	2.2	2.3	0.2
Heroin	1.6	0.6[b]	0.8[b]	1.4	0.7	0.4	–	0.1	–	1.6		0.5	0.3	0.1	0

Note: In adolescence, Woodlawn African American males and females were ages 15–16 (1975–1976). Weighted data from Monitoring the Future are for those in grade 12 in 1978, the expected 12th grade-year for Woodlawn cohort members. Adult Woodlawn data are for African American non-incarcerated males and females ages 32–33 in young adulthood (1992–1993), and 42–43 in mid-adulthood (2002–2003). NHSDA is the 1992 National Household Survey of Drug Abuse, ages 30–34, weighted data. NSDUH is the 2003 National Survey on Drug Use and Health, ages 35–49, weighted data. Problem use is abuse or dependence except NHSDA, in which case problems use is defined as endorsing one of eleven problems attributed to using a specific substance

[a]Past 30 days

[b]Past year

[c]Woodlawn mid-life interview problem use was not drug-specific

in the NHSDA survey. Nonetheless, the problem use of alcohol in the Woodlawn group (11.0 %) was similar to the problem use in the White population (10 %), and both were higher than the problem use among national African Americans (4 %).

By the time of mid adulthood (ages 42–43), the Woodlawn population had lower rates of alcohol use and lower rates of problem alcohol use than the other two populations. The rates of marijuana use were similar for the three populations. Woodlawn and the NSDUH African Americans had similar rates of crack/cocaine use and were about twice as high as the White rates. Woodlawn rates of heroin (1.6 %) were higher than either the African Americans (0.5 %) or the Whites (0.1 %). The problem use seemed to be similar for the Woodlawn population (2.7) and the NSDUH African Americans (2.2–3.4 %) and lower for the NSDUH Whites (0.5–0.7). (Since the problem rates for Woodlawn are based on drugs, in general, the range for the NSDUH groups includes a range that, on one hand, assumes all those with drug abuse or dependence for one drug also have it for the other drugs and, on the other hand, that the rates are all independent). There are many problems with the above comparisons: the measures are not exactly the same, the interviewers were different, and the reasons for the studies were different. Hence, our primary conclusion is that the rates seem to support the crossover idea that African Americans have higher rates of marijuana, cocaine, and heroin use in the adult years than the Whites. There is also evidence to support that they have more problem use, especially by the mid adulthood period.

Comparisons of Social Roles between African Americans and Whites

We examine the participation in social roles from the national surveys by White and African Americans, again focusing on the age categories that correspond to those of Woodlawn assessments. We are hypothesizing that one reason that the Woodlawn population (or other African American populations) have more drug use and/or drug problems as adults is because their participation in employment and marriage (and perhaps, childrearing) is less than the participation of White Americans.

Comparing African American and White respondents in the 1992–1994 NHSDA survey and the 2002 NSDUH survey, we see lower participation in social roles among African American adults. As shown in Table 4.2, among NHSDA respondents aged 30–34 and among NSDUH respondents aged 35–49, African Americans are less likely to have completed high school, to be employed, and to be married. At age 30–34, African Americans are more likely than Whites to have children in their households, but this reverses at ages 35–49.

We compare these national surveys with the Woodlawn data to see how the Woodlawn population compares in social role participation. As Table 4.2 shows, the populations do not differ dramatically in the high school completion rates, although Whites at the mid adulthood assessment have slightly higher rates than do the

Table 4.2 Percent of population involved in social roles at young- and mid-adulthood in the Woodlawn study and in National Samples by Race

	Young adulthood			Mid-adulthood		
		NHSDA			NSDUH	
Social roles	Woodlawn N=916	African American N=2,715,640	White N=18,294,013	Woodlawn N=815	African American N=7,574,181	White N=44,838,833
High school completion	80.4	81.9	83.8	81.8	82.7	90.9
Employed	65.1	68.9	77.9	74.8	78.8	85.0
Married	28.3	49.9	71.0	35.3	44.8	69.6
Children in household	57.1	77.1	68.0	56.9	51.4	59.7

Note: Woodlawn data are for African American non-incarcerated males and females ages 32–33 in young adulthood (1992–1993) and 42–43 in mid-adulthood (2002–2003). NHSDA is the 1992 National Household Survey of Drug Abuse, ages 30–34, weighted data. NSDUH is the 2003 National Survey on Drug Use and Health, ages 35–49, weighted data others. In terms of employment, those in the Woodlawn sample are less likely to be employed at both times compared to the Whites, but the Woodlawn rates are comparable to the rates for African Americans in the national surveys. Rates of marriage differ rather substantially, with Woodlawn study participants having much lower rates of current marriage compared to both the Whites and the African Americans in the national surveys. In early adulthood, the Woodlawn participants were less likely to have children in the household than either the African American or White NHSDA respondents; in mid adulthood, the Woodlawn cohort was less likely than the NHSDA Whites (but more likely than the African Americans) to have children in the household. These data are all consistent with the observation that African Americans participate less in the primary social roles of adulthood.

Measures

We focus on cocaine use for several reasons. First, this cohort entered their 20s at the time of increased use of cocaine in the United States; therefore, cocaine might reasonably be the drug that would best demonstrate the crossover effect. In an earlier review of ethnic differences in drug use, Kandel (1995) found that past year cocaine use among those younger than 26 was higher among Whites than African Americans, but for those older than 26, the rates were higher among African Americans. This study was in the early 1990s, closely corresponding with the Woodlawn young adult assessment. Furthermore, cocaine use is an indicator of serious substance use given its illegal nature and association with dependence and negative health effects. Finally, the prevalence of cocaine use in the Woodlawn Study is sufficient to demonstrate a reliable crossover effect and distinguish patterns of use by social roles.

Cocaine Use Trajectories

The cocaine use information is drawn from the 1053 people who were interviewed at age 32 and/or age 42. Each person was asked whether he or she had used cocaine and if so, the age of first and last use. Among the Woodlawn cohort, cocaine was first used, on average, in the mid-twenties (age 23.6) and last used, on average, at age 31.1. Although about 70 % of the cocaine users terminated their drug use by age 42, 30 % of cocaine users were still using into their 40s. This information for each Woodlawn case was then annualized such that a person is coded as a cocaine user in each year between the age of first use and the age of last use. This coding procedure assumes that a person uses in every year between the ages of first and last use, which may overestimate the continuity of use. However, based on theories of continuity (Farrington 1991; Gottfredson and Hirschi 1990; Huesmann et al. 1984; Jessor et al. 1991; Nagin and Paternoster 2000) and empirical evidence supporting these theories (e.g., Brook et al. 1996; Caspi et al. 1989; Fothergill et al. 2009; Hamil-Luker et al. 2004), the assumption is not an unreasonable one and allows us to establish general trajectories of cocaine use over the life course. Moreover, we assess the validity of this assumption in the results section.

Social Roles

The social roles data are drawn from the adolescent, young adulthood, and mid-adult interviews to assess the engagement in several age-appropriate social roles. Peers, family, and school are salient social institutions in the life of an adolescent. The Woodlawn data does not allow an investigation into deviant peer bonds, which may be the most important aspect of peer bonding; thus, we focus here on family and school. To assess participation in the social roles of family and school, we use self-reports from mother and/or the child (either prospectively or retrospectively) of running away from home more than once (13 %, n = 1118), adolescent self-reports of skipping school three or more times in the past 3 years (27 %, n = 702), and retrospective adult self-reports of dropping out of high school (28 %, n = 1052). In young and mid-adulthood, marriage, employment, and parenthood become primary social roles. At age 32 (n = 916), 28 % are currently married, 57 % have a child living in the household, and 65 % are employed. By age 42 (n = 819), 35 % are currently married, 57 % have a child living in the household, and 75 % are employed. For the young and mid-adulthood social role, we also use a sum of the number of social roles to capture the overall level of social integration in conventional society. In young adulthood, 18 % participated in all three social roles, 25 % in two, 41 % in one, and 16 % in no social roles. In mid-adulthood, 22 % were participating in all three roles, 32 % in two roles, 34 % in one role, and 12 % participating in no age-appropriate social roles.

Analysis

Cocaine Use Trajectories Using a semiparametric mixed logit model, cocaine trajectories from ages 7 to 42 are estimated to identify the long-term patterns of cocaine use for males and females separately. Each individual has their own trajectory of cocaine use, and this model assumes that groups of individuals displaying similar cocaine use trajectories can be identified. Thus, this method allows a disaggregation of the trajectories of cocaine use over time by assigning each individual to the cocaine trajectory group to which he or she is most likely to belong based on the posterior probability of membership. More technically speaking, each developmental trajectory assumes a cubic relationship that links age and cocaine use. Briefly, using SAS Proc Traj (Jones et al. 2001), an incrementally larger number of groups are estimated and the optimal model is assessed using the Bayesian Information

Criterion (BIC) along with other model diagnostics such as population estimates for each group, average posterior probabilities of assignment, and odds of correct classification (see Nagin 2005). We used each of these diagnostic tools as well as the BIC when deciding on the optimal trajectory model. The final result is a number of different groups comprised of individuals who demonstrate similar patterns of cocaine use.

Cocaine Trajectories and Social Roles A series of bivariate chi-square and ANOVA tests are conducted to determine the relationship between participation in adolescent, young adulthood, and mid-life social roles and cocaine trajectory group, for males and females separately.

Results

Cocaine Use Trajectories

Woodlawn Males' Trajectories

Five groups best describe the cocaine patterns of the Woodlawn men (n=503), depicted in Fig. 4.1. Although the BIC statistic continued to increase past five groups, the five group model was selected on the basis of parsimony and the model diagnostics (e.g., the average posterior probabilities were high (.97 to .99), and

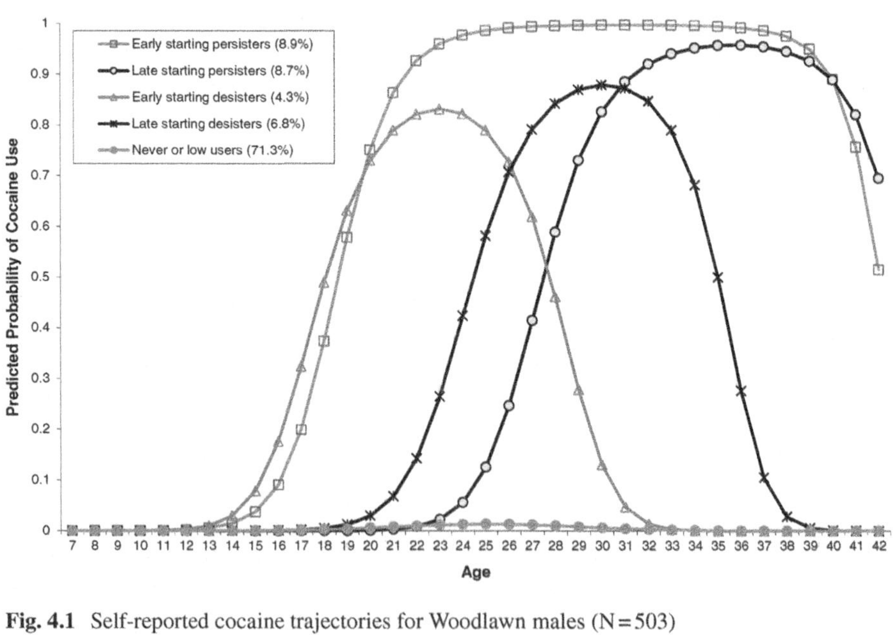

Fig. 4.1 Self-reported cocaine trajectories for Woodlawn males (N=503)

evaluation of the odds of correct classification, which were well above the recommended number five) (see Nagin 2005 for more details on these diagnostics).

The majority of the men (71 %) either never used cocaine or was likely to be experimental users having a very low likelihood of use with a short "career." Two groups of men began their cocaine use in adolescence, yet one group desists in their use beginning in their mid-twenties ("early-starting desisters," 4.3 %) while the other continues through mid-adulthood ("early-starting persisters," 8.9 %). Similarly, while the final two groups begin their use later in adulthood, one desists beginning in their early 30s ("late-starting desisters," 6.8 %) while the other continues into their 40s ("late-starting persisters," 8.7 %).

The trajectories concur with the findings on mean age of onset and the longer duration of use with the finding of the two late-starting groups, that is the majority of those in a cocaine use trajectory was classified into a late-starting trajectory. In addition, those who are in the early-starting persister group are the most likely to have drug problems in both young adulthood (61 %) and mid adulthood (54 %), with those in the never/low using group the least likely to have drug problems in young or mid-adulthood (8 % and 1 %, respectively).[1] This finding may also inform the crossover pattern of cocaine use with a substantial proportion of users beginning their use in adulthood and another proportion of problem users continuing into adulthood.

Woodlawn Females' Trajectories

For the females, both the BIC statistic and the additional model diagnostics indicate that the five-group model is the optimal model. The five groups identified by the trajectory analysis show similar patterns of cocaine use for the Woodlawn women (n=549), with some differences in magnitude and timing (see Fig. 4.2). Again, the vast majority is a never or low-level user (81 %) with two early using groups and two late using groups. All of these groups are smaller in size than the male groups, but the trends are similar. For instance, 8.7 % of the women are early starters, yet 5.2 % persist in their use ("early-starting persisters") while a smaller proportion (3.5 %) desist beginning in their mid-twenties ("early-starting desisters"). An equal proportion of the late starters persist as desist with 5.1 % being classified as "late-starting persisters" and 5.4 % as "late-starting desisters."

[1] This finding also provides evidence that the assumption of cocaine use in each year between the first and last age of use is valid. Although it is possible that a person does not use cocaine in each year, the general distinction between persistent and serious users versus non-users as well as the timing of first and last use seems valid and is consistent with theories of continuity, which have been supported by numerous empirical studies (eg, Hamil-Luker et al., 2004; Caspi and Moffitt 1993).

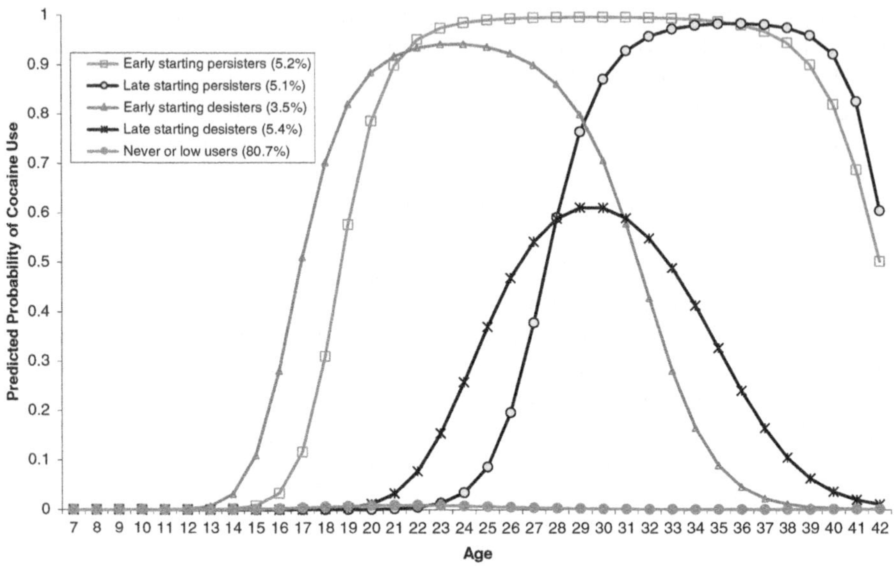

Fig. 4.2 Self-reported cocaine trajectories for Woodlawn females (N = 549)

Cocaine Trajectories and Social Roles

Association with Social Roles for Woodlawn Men

As Table 4.3 shows, many of the social roles are significantly related to cocaine trajectory membership for the Woodlawn men. In adolescence, those males who were currently using cocaine (the early-starting desister and early-starting persister groups) were the most likely to have been runaways (30.0 % and 31.0 %, respectively) and to have dropped out of high school (35.0 % and 38.6 %, respectively). Similarly, the men who were currently using cocaine in young adulthood (i.e., the early-starting persister group and the two late-starting groups) were the least likely to be married or employed. For instance, only 11.6 % of the men in the late-starting persister group were married, followed by the late-starting desister (18.8 %) and early-starting persister group (19.5 %). This is considerably lower than the never using group with 37.3 % married. Similarly, only 34.9 % of the late-starting persister group was employed, followed by the early-starting persisters (51.2 %) and the late desisters (59.4 %), compared with 69.9 % of the never users. The lack of engagement in social roles is also reflected in the fact that the late-starting persisters engaged in significantly fewer roles in young adulthood on average (0.67) than their never user and early desister counterparts who engaged in 1.45 social roles, on average.

By age 42, marriage is the only social role that significantly distinguishes the trajectory groups for the men. Following a similar pattern to that seen at young adulthood, 18.4 % of the late-starting persisters were married compared with 41.9 %

Table 4.3 Bivariate Relationship between Social Roles and Cocaine Trajectories for Woodlawn Males and Females

Woodlawn males						
Social role	Cocaine trajectory group					Significance
	Never (N=360)	Early-starting desisters (N=20)	Early-starting persisters (N=45)	Late-starting desisters (N=34)	Late-starting persisters (N=44)	p-value
Adolescence						
Skip school (N=305)	29.4 %	46.2 %	37.5 %	29.0 %	33.3 %	<0.667
Runaway (N=484)	10.7 %	30.0 %	31.0 %	15.2 %	15.9 %	<0.002
HS dropout (N=475)	22.4 %	35.0 %	38.6 %	25.0 %	38.6 %	<0.034
Young adulthood						
Married (N=455)	37.3 %	35.0 %	19.5 %	18.8 %	11.6 %	<0.001
Employed (N=455)	69.9 %	70.0 %	51.2 %	59.4 %	34.9 %	<0.001
Child in household (N=455)	37.9 %	40.0 %	34.1 %	18.8 %	20.9 %	<0.063
Number of roles (N=455)	1.45	1.45	1.05	0.97	0.67	<0.001
Mid-adulthood						
Married (N=372)	41.9 %	58.3 %	22.9 %	33.3 %	18.4 %	<0.008
Employed[a] (N=371)	74.9 %	100.0 %	80.0 %	74.1 %	60.5 %	<0.071
Child in household (N=372)	41.2 %	41.7 %	25.7 %	33.3 %	23.7 %	<0.139
Number of roles (N=371)	1.58	2.00	1.29	1.41	1.03	<0.008
Woodlawn females						
Social role	Cocaine trajectory group					Significance
	Never (N=443)	Early-starting desisters (N=18)	Early-starting persisters (N=30)	Late-starting desisters (N=30)	Late-starting persisters (N=28)	p-value
Adolescence						
Skip school[a] (N=329)	21.1 %	30.8 %	46.7 %	31.3 %	25.0 %	<0.180
Runaway[a] (N=525)	9.8 %	38.9 %	34.5 %	10.0 %	17.9 %	<0.001
HS dropout (N=514)	17.1 %	27.8 %	66.7 %	21.4 %	32.1 %	<0.001
Young adulthood						

(continued)

Table 4.3 (continued)

Married[a] (N=496)	28.2 %	16.7 %	6.9 %	14.3 %	3.7 %	<0.003
Employed (N=496)	68.3 %	66.7 %	24.1 %	32.1 %	22.2 %	<0.001
Child in household[a] (N=496)	74.6 %	77.8 %	72.4 %	64.3 %	81.5 %	<0.664
Number of roles (N=496)	1.71	1.61	1.03	1.11	1.07	<0.001
Mid-adulthood						
Married[a] (N=458)	35.3 %	33.3 %	8.0 %	19.0 %	13.0 %	<0.007
Employed[a] (N=459)	75.5 %	73.3 %	44.0 %	76.2 %	56.5 %	<0.005
Child in household[a] (N=459)	71.2 %	80.0 %	68.0 %	57.1 %	73.9 %	<0.602
Number of roles (N=458)	1.82	1.87	1.20	1.52	1.43	<0.001

Note: All analyses are conducted using chi-square tests with 4 degrees of freedom except for the young adult and mid-adult number of social roles which is conducted using ANOVA
[a]One or more cells for this analysis has less than the recommended five cases, therefore the results should be interpreted with caution

of the never users and 58.3 % of the early-starting desisters. By mid-adulthood, the late-starting persisters remained the least socially integrated as indicated by the finding that, on average, this group of men participated in one social role compared with the never using group and early-starting desister groups who participated in 1.58 and 2.00 social roles on average, respectively.

Association with Social Roles for Woodlawn Women Table 4.3 also shows that there is a significant relationship between involvement in social roles and cocaine use among the women as well. The early-starting persister women are the least likely to engage in age-appropriate social roles throughout the life course. These women are most likely to drop out of high school in adolescence (66.7 %) and least likely to be married at any time in adulthood (6.9 % in young adulthood and 8.0 % in mid-life) and have some of the lowest rates of employment (24 % in young adulthood and 44 % in mid-life). In contrast, the early-starting desisters, while likely to have been runaways in adolescence, were more likely to graduate from high school (27.8 % dropout) and find success in adulthood, with 33.3 % married and 73.3 % employed by age 42.

During adolescence, the late-starting persister group fared better than their early-starting persisting counterparts on all three school variables, yet they did not continue to engage in social roles in young and mid adulthood, with only 3.7 % and 13.0 % married, respectively and 22.2 % and 56.5 % employed, respectively, in young and mid-adulthood. Overall, the early-starting persisters and the two late-starting groups actively used cocaine in adulthood and engaged in considerably fewer social roles in both young adulthood and mid-life than the never users or early-starting desisters.

Conclusion

In this chapter we focused on the patterns of drug use of a community cohort of African Americans followed prospectively into midlife and compared their substance use and abuse to national data. We explored in-depth whether their trajectories of cocaine use were consistent with the crossover effect and whether their social roles participation might offer some explanation of crossover. A notable finding of our trajectory analysis was the identification of two groups who began cocaine use later and two groups who desisted from cocaine use late in life compared to what is typically seen among White populations.

By comparing rates of drug use among the Woodlawn cohort and examining rates of African Americans and Whites of similar age in national surveys administered at times corresponding to the three Woodlawn assessments, we demonstrated some evidence of cross-over. Although we did not find that drug use among Woodlawn adolescents was lower than Whites in the MTF national survey, we did show that national African American adolescents had lower rates of use of all substances than White adolescents in 1978. We hypothesized that the somewhat higher rates among the Woodlawn cohort compared to national African Americans (but comparable to national Whites) was due in part to different assessment timing, measures, and techniques. It also could be related to the Woodlawn cohort living in a relatively disadvantaged, urban community, although one would speculate that African Americans in general at this time would be more likely to live in disadvantaged urban communities than Whites.

We did show that rates of marijuana, cocaine, and heroin use were higher among the Woodlawn cohort than Whites in the national surveys in early adulthood and continued into mid adulthood. Within the national surveys, the crossover also became prominent in young adulthood (ages 30–34) for the "harder" drugs of cocaine and heroin. By midlife, rates of abuse of all substance for national African Americans surpass those of Whites. This suggests that the crossover is evident for both use and abuse though the timing of the crossover may vary by whether we examine substance use or the problems associated with use. The cross-over phenomenon also seems to vary by substance, with alcohol being an exception to the crossover effect. Overall, these observations and comparisons from the MTF, NHSDA, NSDUH, and Woodlawn are generally consistent with the premises of a crossover effect: lower use of drugs for African Americans during adolescence; increasing use (relative to Whites) over adulthood; and higher reported abuse in adulthood. However, if we consider alcohol, it is not entirely consistent—the crossover effect only seems to apply to abuse since Whites report higher rates of alcohol use.

We next investigated the rates of social role involvement by race in national surveys and compared this to our Woodlawn data. Essentially, our premise is that a reason for the crossover effect is the lower participation of African Americans in the major social roles of adulthood—namely, employment, marriage, and parenthood. We explored whether according to national survey data, African Americans of the same age cohort as the Woodlawn population had lower employment rates and lower marriage rates than Whites. They did for both males and females; rates of marriage were dramatically lower for African Americans compared to Whites. Woodlawn men and women had even lower participation in these roles than national African Americans, which may be due to a variety of reasons, including limited opportunities in their disadvantaged neighborhoods. While we speculated that presence of children in the households of African Americans and White Americans might differ, we found no consistent differences, with relatively consistent rates around 50 % for all three groups by midlife.

We then presented within the Woodlawn data patterns of cocaine use over the life course and whether these patterns were related to social role involvement in adolescence, young adulthood, and mid adulthood. We found five trajectories of use for both the males and the females. For males and females, the largest groups consisted of never users. About equal numbers of men started using cocaine in adolescence (13 %) and after adolescence (15 %). The patterns then were further defined by whether the users continued in their use or stopped. About nine percent of the males started cocaine use early and then continued well into adulthood. An additional nine percent started using later, but persisted in their use into adulthood. Based on these trajectories, about 18 % of the cohort were still using in their 30s and 40s, more so than is typically reported in other populations (e.g., Chen and Kandel 2002; Johnston et al. 2009). The patterns were similar for the females, at lower levels of use.

These findings somewhat diverge from the theories of stability and change in deviant behavior over the life course (Moffitt 1993; Robins 1978; White et al. 1990) that focus on adolescent initiation without much attention to those who initiate later. The conventional wisdom has been that late initiators are rare. The trajectory groups of users do show an adolescent experimentation group (early desisters) and an adolescent persister group (early persisters), but they also indicate that initiation into cocaine use can come later in the life course (the late persisters and desisters). This later initiation may be much more prevalent in African Americans than among Whites. This, in fact, may be the most important aspect of the crossover effect. By identifying two late starting groups among a community population of African Americans, we are providing a clear picture of late onset, a major tenet of crossover.

When we examine these groups with the social roles data, we clearly see the close relation between the pattern of use and the social roles at the different stages of life. For example, those who are early users of cocaine are more likely to have been runaways and high school dropouts in adolescence. All those who were in a cocaine-using trajectory were more likely to be a school dropout than those who had never used cocaine. It may be that in this cohort, lacking a high school credential may be too big a hurdle to overcome in terms of drug use and in terms of later social role performance.

The association between adult social roles and the cocaine patterns of use is also very substantial. Those who started early but desisted by adulthood (early desisters) are similar in the adult roles of employment and marriage to those who were never users. This suggests that the early-starting desisters may be more representative of adolescent experimenters who are engaged in a relatively healthy phase of experimentation that subsides naturally with the transition to adult social roles. Marriage and employment may have steered them away from their cocaine use, that is, those who were married and/or working were much more likely to discontinue their cocaine use than those who were not involved in these social roles. They were among those who looked "worst" with regard to social role participation during adolescence.

The late starters did not have difficulties with social roles in adolescence—fewer ran away from home and fewer were high school dropouts than the early starters. Did something happen in young adulthood? The two late-starting groups had low rates of employment compared to the nevers and the early desisters, suggesting that employment may have had an important influence on their cocaine initiation. Perhaps the early-starting desisters were able to stop because they achieved employment, but the late starters failed to gain employment and thus turned to drug use to cope. Similarly, comparing the late-starting desisters and persisters, we see that the desisters are more likely to get married in early adulthood (a possible turning point) and more likely to be employed in early adulthood (although still at relatively low rates) and mid adulthood (highest rate of all groups). Again, we cannot know the direction of effects, but it is consistent with the possibility that the adoption of these adult social roles is a powerful turning point.

What does the Woodlawn data add to the nationally representative data that has been so important in identifying the crossover effect? At the initiation of the study, Woodlawn was a community with high rates of disadvantage, crime, and presumably drug use. It is important to understand how the drug data from this community compares with drug data from nationally representative populations and informs cross-over. At all three data collection points the Woodlawn data diverged from the nationally collected data. During adolescence the drug use was much higher in Woodlawn than among African Americans in MTF. During the two adult data collection points, the Woodlawn data and the African American survey data were more similar, but in general, the Woodlawn population showed higher rates of drug use and less participation in social roles. Whether these differences mirror differences with other urban community populations, as we expect, is important to explore.

In general, there has not been much interest or speculation in the literature about whether or why prevalence rates from community populations do or do not differ from those reported in national populations. Do these differences reflect mainly the "real" differences in use that exist? Could they indicate that nationally representative samples that are based on households do not obtain data from those not well connected to households, and is this an important shortcoming? In the United States, this problem is further compounded by the large percentage of people that comprise the prison population. While the Woodlawn Study conducts interviews in prisons and jails, we excluded these individuals from these analyses to be better able to equate populations. However, it is well known that drug use rates are very high among those involved in the criminal justice system, and our past work has shown this to be the case among Woodlawn cohort members (Ensminger et al. 1997). Therefore, when national studies omit prisoners, they are underestimating national prevalence of substance use and abuse.

The nationally representative population has been considered the "gold standard" in research, and populations that are more specialized or local have needed to explain why their findings may differ. Given that interventions and policies are often designed for more local applications in the United States, data from community populations may be appropriate for the assessment of needs and policy formation. In addition to the overall picture that the national data provides, the variation that is shown in community population data reminds us that locality is still very important in terms of drug use and social role participation. This deserves more research attention, so that we can better evaluate the information from both kinds of studies and understand why the crossover effect is happening and how we can intervene.

The results from our analyses with the Woodlawn data and observations of the nationally available data suggest that the social role hypothesis may partially explain the crossover effect. Many questions remain. Is there a crossover effect with ethnic groups other than African Americans that are disadvantaged with regard to social role performance in adulthood? While the results are suggestive that early social roles may influence later social role adoption and later drug use, we have not examined this in detail.

The implications of our study are that drug use is related to one's social role participation and performance. Those policies that influence employment and marriage opportunities may also influence the trends in drug use if this association is causal. Many have commented on the lower rates of marriage within the African American community, often attributed at least in part to the high incarceration rates, the low employment rates and the decreased life expectancy of African American males (Wilson 1987). Many have also observed that African Americans have been the last hired and the first fired in economic recessions. Presumably, the low rates of employment would influence the drug trends. While we expect deviant behavior to decrease as youth age into adulthood, this may be less likely in populations with low rates of employment and marriage.

References

Agrawal, A., & Lynskey, M. T. (2009). Correlates of later-onset cannabis use in the National Epidemiological Survey on Alcohol and Related Conditions (NESARC). *Drug and Alcohol Dependence, 105*, 71–75.

Agrawal, A., Gardner, C. O., Prescott, C. A., & Kendler, K. S. (2005). The differential impact of risk factors on illicit drug involvement in females. *Social Psychiatry and Psychiatric Epidemiology, 40*, 454–466.

Bachman, J. G., Wadsworth, K. N., O'Malley, P. M., Johnston, L. D., & Schulenberg, J. E. (1997). *Smoking, drinking, and drug use in young adulthood: The impacts of new freedoms and new responsibilities*. Mahwah: L. Erlbaum Associates.

Brook, J. S., Whiteman, M., Finch, S. J., & Cohen, P. (1996). Young adult drug use and delinquency: Childhood antecedents and adolescent mediators. *Journal of the American Academy of Child and Adolescent Psychiatry, 35*, 1584–1592.

Brunswick, A. F., & Titus, S. (1998). Heroin patterns and trajectories in an African American cohort. In L. Harrington & J. Inciardi (Eds.), *Heroin in the age of crack-cocaine* (pp. 77–108). Newbury Park: Sage Publications.

Brunswick, A. F., Lewis, C., & Messeri, P. (1991). A life span perspective on drug use and affective distress in an urban African American population. *Journal of Community Psychology, 19*, 123–135.

Bureau of Justice Statistics. (1996). *Violence-related injuries treated in hospital emergency departments: A special report from the Bureau of Justice Statistics*. NCJ-156921. Washington, DC: U.S. Government Printing Office.

Caspi, A., & Moffitt, T. E. (1993). The continuity of maladaptive behavior: From description to understanding in the study of antisocial behavior. In D. Cicchetti & D. Cohen (Eds.), *Manual of developmental psychopathology*. New York: Wiley.

Caspi, A., Bem, D., & Elder, G. (1989). Continuities and consequences of interactional styles across the life course. *Journal of Personality, 57*, 375–406.

Centers for Disease Control. (1997). *Death count; 15 to 54 years; all races; both genders: 1994; by race-age; the United States*. Centers for Disease Control Webpage [On-line]. Available: http://wonder.cdc.gov/WONDER/anon/ANON913/MORT.00.ex

Chen, K., & Kandel, D. (2002). Relationship between extent of cocaine use and dependence among adolescents and adults in the United States. *Drug and Alcohol Dependence, 68*, 65–85.

Chilcoat, H. D., & Breslau, N. (1996). Alcohol disorders in young adulthood: Effects of transitions into adult roles. *Journal of Health and Social Behavior, 37*(Dec), 339–349.

Cicchetti, D., & Schneider-Rosen, K. (1984). Toward a transactional model of childhood depression. In D. Cicchetti & K. Schneider-Rosen (Eds.), *Childhood depression: A developmental perspective* (pp. 5–28). San Francisco: Jossey-Bass.

Council for Community Services in Metropolitan Chicago. (1975). *Community analysis project*. Report No. 1: Chicago problem analysis. Chicago: Council for Community Services in Metropolitan Chicago.

Crum, R. M., & Anthony, J. C. (2000). Educational level and risk for alcohol abuse and dependence: Differences by race-ethnicity. *Ethnicity & Disease, 10*(1), 39–52.

Doherty, E. E., Green, K. M., Reisinger, H. S., & Ensminger, M. E. (2008). Long-term patterns of drug use among an urban African-American cohort: The role of gender and family. *Journal of Urban Health, 85*, 250–267.

Ensminger, M. E. (1990). Sexual activity and problem behaviors among African American, urban adolescents. *Child Development, 61*, 2032–2046.

Ensminger, M. E., & Juon, H. S. (1998). Transition to adulthood among high-risk youth. In R. Jessor (Ed.), *New perspectives on adolescent risk behavior* (pp. 365–391). New York: Cambridge University Press.

Ensminger, M. E., Kellam, S. G., & Rubin, B. R. (1983). School and family origins of delinquency: Comparisons by sex. In K. T. Van Dusen & S. A. Mednick (Eds.), *Prospective studies of crime and delinquency* (pp. 73–97). Boston: Kluwer-Nijhoff Publishing Co.

Ensminger, M. E., Brown, C. H., & Kellam, S. G. (1984). Social control as an explanation of sex differences in substance use among adolescents. NIDA research monograph 49. In L. J. Harris (Ed.), *Problems of drug dependence, 1983. Proceedings of the 45th annual scientific meeting*, The Committee on Problems of Drug Dependence. Rockville, MD: NIDA.

Ensminger, M. E., Anthony, J. C., & McCord, J. (1997). The inner city and drug use: Initial findings from an epidemiological study. *Drug and Alcohol Dependence, 48*, 175–184.

Farrington, D. P. (1991). Antisocial personality from childhood to adulthood. *The Psychologist, 4*, 389–394.

Fothergill, K. E., & Ensminger, M. E. (2006). Childhood and adolescent antecedents of drug and alcohol problems: A longitudinal study. *Drug and Alcohol Dependence, 82*, 61–76.

Fothergill, K. E., Ensminger, M. E., Green, K. M., Robertson, J. A., & Juon, H. S. (2009). Pathways to adult marijuana and cocaine use: A prospective study of African Americans from age 6 to 42. *Journal of Health and Social Behavior, 50*, 65–81.

French, K., Finkbiner, R., & Duhamel, L. (2002). *Patterns of substance use among minority youth and adults in the United States: An overview and synthesis of national survey findings.* Fairfax, VA.:U. S. Department of Health and Human Services, Substance Abuse and Mental Health Services Administration, Center for Substance Abuse Treatment/Caliber Associates/National Evaluation Data Services.

Gottfredson, M. R., & Hirschi, T. (1990). *A general theory of crime*. Stanford: Stanford University Press.

Green, K. M., & Ensminger, M. E. (2006). Adult social behavioral effects of heavy adolescent marijuana use among African Americans. *Developmental Psychology, 42*(6), 1168–1178.

Green, K. M., Doherty, E. E., Reisinger, H. S., Chilcoat, H. D., & Ensminger, M. E. (2010). Social integration in young adulthood and the subsequent onset of substance use and disorders among a community population of urban African Americans. *Addiction, 105*(3), 484–493.

Hamil-Luker, J., Land, K. C., & Blau, J. (2004). Diverse trajectories of cocaine use through early adulthood among rebellious and socially conforming youth. *Social Science Research, 33*, 300–321.

Hawkins, J. D., Catalano, R. F., & Miller, J. Y. (1992). Risk and protective factors for alcohol and other drug problems in adolescence and early adulthood: Implications for substance abuse prevention. *Psychological Bulletin, 112*, 64–105.

Henry, K., & Huizinga, D. (2007). Truancy's effect on the onset of drug use among urban adolescents placed at risk. *Journal of Adolescent Health, 40*, 358.e9–358.e17.

Hirschi, T. (1969). *Causes of delinquency*. Berkeley: University of California.

Horwitz, A. V., & White, H. R. (1991). Marital status, depression, and alcohol problems among young adults. *Journal of Health and Social Behavior, 32*, 221–237.

Huesmann, L. R., Eron, L. D., Lefkowitz, M. M., & Walder, L. O. (1984). Stability of aggression over time and generations. *Developmental Psychology, 20*, 1120–1134.

Jessor, R., Donovan, J. E., & Costa, F. M. (1991). *Beyond adolescence: Problem behavior and young adult development*. New York: Cambridge University Press.

Johnston, L. D., O'Malley, P. M., Bachman, J. G., & Schulenberg, J. E. (2009). *Monitoring the future national survey results on drug use, 1975–2008. Volume I: Secondary school students* (NIH publication no. 09-7402). Bethesda: National Institute on Drug Abuse.

Jones, B. L., Nagin, D. S., & Roeder, K. (2001). A SAS procedure based on mixture models for estimating developmental trajectories. *Sociological Methods & Research, 29*, 374–393.

Kandel, D. (1995). Ethnic differences in drug use: Patterns and paradoxes. In G. J. Botvin, S. Schinke, & M. A. Orlandi (Eds.), *Drug abuse prevention with multiethnic youth* (pp. 81–104). Thousand Oaks: SAGE Publications, Inc.

Kandel, D. B., & Yamaguchi, K. (1987). Job mobility and drug use: An event history analysis. *American Journal of Sociology, 92*, 836–878.

Kandel, D., Chen, K., Warner, L. A., Kessler, R. C., & Grant, B. (1997). Prevalence and demographic correlates of symptoms of last year dependence on alcohol, nicotine, marijuana and cocaine in the U. S. population. *Drug and Alcohol Dependence, 44*, 11–29.

Kellam, S. G., & Ensminger, M. E. (1980). Theory and method in child psychiatric epidemiology. In F. J. Earls (Ed.), *Studies of children* (pp. 145–180). New York: Prodist.

Kellam, S. G., & Rebok, G. W. (1992). Building developmental and etiological theory through epidemiological based preventive intervention trials. In J. McCord & R. E. Tremblay (Eds.), *Preventing antisocial behavior: Interventions from birth through adolescence* (pp. 162–195). New York: Guildford Press.

Kellam, S. G., Branch, J. D., Agrawal, K. C., & Ensminger, M. E. (1975). *Mental health and going to school: The Woodlawn program of assessment, early intervention and evaluation*. Chicago: The University of Chicago Press.

Kessler, R. C., McGonagle, K. A., Zhao, S., Nelson, C. B., Hughes, M., Eshleman, S., Wittchen, H. U., & Kendler, K. S. (1994). Lifetime and 12-month prevalence of DSM-III-R psychiatric disorders in the United States. *Archives of General Psychiatry, 51*(Jan), 8–19.

Lewis, C. E., Robins, L., & Rice, J. (1985). Associations of alcoholism with antisocial personality in urban men. *Journal of Nervous and Mental Disease, 173*, 166–174.

Ma, G., & Shive, S. (2000). A comparative analysis of perceived risks and substance abuse among ethnic groups. *Addictive Behaviors, 25*, 361–371.

Mensch, B. S., & Kandel, D. B. (1988). Dropping out of high school and drug involvement. *Sociology of Education, 61*, 95–113.

Merline, A. C., O'Malley, P. M., Schulenberg, J. E., Bachman, J. G., & Johnston, L. D. (2004). Substance use among adults 35 years of age: Prevalence, adulthood predictors, and impact of adolescent substance use. *American Journal of Public Health, 94*, 96–102.

Mirkowsky, J., & Ross, C. E. (1986). *Social causes of psychological distress*. New York: Aldine de Gruyter.

Moffitt, T. E. (1993). Adolescence-limited and life-course-persistent antisocial behavior: A developmental taxonomy. *Psychological Review, 100*, 674–701.

Murray, J. L., & Lopez, A. D. (1996). *Global burden of disease and injury series*. Geneva: World Health Organization.

Nagin, D. S. (2005). *Group-based modeling of development over the life course*. Cambridge: Harvard University Press.

Nagin, D., & Paternoster, R. (2000). Population heterogeneity and state dependence: State of the evidence and direction for future research. *Journal of Quantitative Criminology, 16*, 117–144.

National Institute on Drug Abuse. (1986). *Demographic characteristics and patterns of drug abuse treatment programs in selected states: Annual data 1983*. Rockville: NIDA, Division of Epidemiology and Statistical Analysis.

National Institute on Drug Abuse. (1998). *Drug use among racial/ethnic minorities* (DHHS Pub. no. NIH 98-3888). Rockville: National Institute on Drug Abuse.

National Institute on Drug Abuse. (2003). *Drug use among racial/ethnic minorities* (DHHS Pub. no. NIH 03-3888). Rockville: National Institute on Drug Abuse.

Newcomb, M. D., & Bentler, P. M. (1988). *Consequences of adolescent drug use: Impact on the lives of young adults*. Newbury Park: Sage.

Petersen, A. C., & Kellam, S. G. (1977). Measurement of psychological well-being of adolescents: The psychometric properties and assessment procedures of the How I Feel. *Journal of Youth and Adolescence, 6*, 229–247.

Robins, L. N. (1978). Sturdy childhood predictors of adult antisocial behavior: Replications from longitudinal studies. *Psychological Medicine, 8*, 611–622.

Sampson, R. J., & Laub, J. (1990). Crime and deviance over the lifecourse: The salience of adult social bonds. *American Sociological Review, 55*, 609–627.

Seeman, T. E. (1959). Social ties and health: The benefits of social integration. *Annals of Epidemiology, 6*, 442–451.

Sieber, S. D. (1974). Toward a theory of role accumulation. *American Sociological Review, 39*, 567–578.

Stuart, E. A., & Green, K. M. (2008). Using full matching to estimate causal effects in nonexperimental studies: Examining the relationship between adolescent marijuana use and adult outcomes. *Developmental Psychology, 44*(2), 395–406.

U.S. Department of Health and Human Services, Substance Abuse and Mental Health Administration. (1998). *Prevalence of substance use among racial/ethnic subgroups in the United States, 1991–1993*. SAMHSA Fact Sheet: National Clearinghouse for Alcohol and Drug Information. Rockville, MD.

U.S. Department of Health and Human Services, Substance Abuse and Mental Health Services Administration, Office of Applied Studies. (1992). *National household survey on drug abuse, 1992 [Computer file]. ICPSR06887-v2*. Ann Arbor: Inter-university Consortium for Political and Social Research [distributor], 2008-08-18. doi:10.3886/ICPSR06887

U.S. Department of Health and Human Services, Substance Abuse and Mental Health Services Administration, Office of Applied Studies. (2001). National *household survey on drug abuse, 2001*. http://www.oas.samhsa.gov/NHSDA/2k1NHSDA/vol2/appendixh_1.htm#tableh.2. Accessed 1 Dec 2009.

U.S. Department of Health and Human Services, Substance Abuse and Mental Health Services Administration, Office of Applied Studies. (2004). *National survey on drug use and health, 2003 [Computer file]*. ICPSR version. Research Triangle Park, N.C.: Research Triangle Institute [producer].

Umberson, D. (1992). Gender, marital status and the social control of health behavior. *Social Science & Medicine, 34*(8), 907–917.

Warner, L., Kessler, R., Hughes, M., Anthony, J., & Nelson, C. (1995). Prevalence and correlates of drug use and dependence in the United States, results from the National Comorbidity Survey. *Archives of General Psychiatry, 52*, 219–229.

White, J. L., Moffitt, T. L., Earls, F., Robins, L., & Silva, P. A. (1990). How early can we tell? Predictors of childhood conduct disorder and adolescent delinquency. *Criminology, 28*, 507–528.

Wilson, W. J. (1987). *The truly disadvantaged: The inner city, the underclass, and public policy*. Chicago: University of Chicago Press.

Yamaguchi, K., & Kandel, D. B. (1985). Dynamic relationship between premarital cohabitation and illicit drug use: An event history analysis of role selection and role socialization. *American Sociological Review, 50*, 530–554.

Chapter 5
An Agenda for Longitudinal Research on Substance Use and Abuse with Hispanics in the U.S. And with Latin American Populations

Jorge Delva, Andrew Grogan-Kaylor, Fernando H. Andrade, Marya Hynes, Ninive Sanchez, and Cristina B. Bares

Introduction

Estimates from the 2010 United States (U.S.) Census indicate that there are nearly 50.5 million individuals of Hispanic background, constituting approximately 16 % of the total U.S. population, not including individuals from Puerto Rico (Humes et al. 2011). In fact, the only country in Latin America to have more Spanish-speaking origin populations than the U.S. is Mexico. (Note: Brazil also has a larger population but they are not considered Hispanics due to their Portuguese ancestry and language). Not only have Hispanics become the largest racial or ethnic minority group but also "More than half of the growth in the total population of the United States between 2000 and 2010 was due to the increase in the Hispanic population" (Humes et al. 2011, p 3).

As a result of the fast growing Hispanic population and the overrepresentation of substance use, including the social, political, and economic risk factors they experience, there is an urgent need for more research to be conducted to better understand the etiology of substance use across developmental stages and different population subgroups. Some of these subgroups may be women versus men from the various Hispanic populations, individuals of diverse sexual identities, pre-adolescents versus adolescents, immigrants versus U.S. born, those at risk of HIV/AIDS, and

J. Delva (✉) • A. Grogan-Kaylor • F.H. Andrade • C.B. Bares
School of Social Work, University of Michigan,
1080 S. University, Ann Arbor, MI 48109, USA
e-mail: Jdelva@umich.edu

M. Hynes
Inter-American Drug Abuse Control Commission, Organization of American States,
Washington, DC, USA

N. Sanchez
School of Social Work, University of Missouri, 712 Clark Hall, Columbia, MO 65211, USA

Y.F. Thomas, L.N. Price (eds.), *Drug Use Trajectories Among Minority Youth*,
DOI 10.1007/978-94-017-7491-8_5

school dropouts, among others. Also, there is a need to design and test interventions aimed at preventing any use, preventing individuals' transition into more deleterious use or into using new substances, and rehabilitation services. In this chapter we focus on the first goal, to better understand the etiology and trajectories or developmental pathways of substance use among Hispanics through a detailed discussion of longitudinal studies. Our discussion of longitudinal research focuses mainly on conceptual issues and advantages surrounding longitudinal research but also contains discussion of the statistical methodologies used in this type of work.

Our main goal in writing this chapter is to encourage researchers and their agency and community partners, community stakeholders, families, and policy makers to better understand the type of knowledge that can be gained from longitudinal studies. In turn, with this additional understanding of longitudinal research we hope to build a convincing argument to encourage more longitudinal research with Hispanic youth and their families to develop a more comprehensive body of knowledge about substance use among these populations, knowledge that, we argue, can better inform the field of prevention and treatment.

The chapter is organized into five sections. First, we provide an overview of the epidemiology (distribution) of substance use among Hispanic youth. Second, we describe selected demographic trends that have an impact on Hispanic families and their children. Third, we present a summary of longitudinal studies already completed or underway that were conducted with Hispanic youth or that included sufficient samples to describe findings about these populations. Fourth, we describe and discuss longitudinal research designs that can help shed light into important research areas that could benefit from prospective investigation such as the ways by which acculturation, enculturation, identity development, religiosity, among many others, explain onset and trajectories of substance use. In this section we also discuss statistical issues that need to be considered when conducting this type of research. Fifth, we conclude the chapter with a summary and suggestions for future research.

Before proceeding, we note that in this chapter we use the term Hispanic to refer to the multiple Spanish-speaking origin populations from Latin America that reside in the United States. In other work we use the term Latino, actually a preferred choice among the authors. However, for this chapter we elected to use the term Hispanic to be consistent with the term used by the Federal Government.

Epidemiology of Substance Use among Hispanic Youth in the United States and Among Youth in Latin America

The prevalence of substance use is disproportionately high among Hispanic youth (Delva et al. 2005; Johnston et al. 2009). As shown in Table 5.1, in 2008, a larger percent of Hispanic 8th graders had consumed (in the past year) most substances studied with some exceptions (i.e., amphetamines, OxyContin), a pattern that has

Table 5.1 Drug consumption by Hispanic, White, and African American 8th graders in the U.S.: Results of the 2008 Monitoring the Future Study[a]

Substance	Hispanics	Whites	African Americans
Cigarettes			
% smoked in past 30 days	7.0	7.2	4.6
% smoked daily in past 30 days	2.5	3.3	1.9
Alcohol			
% been drunk in past 30 days	6.7	5.5	3.8
% drank 5 or more drinks in a sitting in past 2 weeks	12.3	7.8	5.7
Marihuana			
% used in past year	13.2	9.6	10.6
Inhalants			
% used in past year	10.2	8.5	5.6
Cocaine			
% used in past year	3.5	1.6	0.8
Crack			
% used in past year	2.5	1.0	0.5
Tranquilizers			
% used in past year	2.9	2.6	0.9
Amphetamines			
% used in past year	4.0	4.9	1.8
Hallucinogens			
% used in past year	2.5	2.1	0.8
Heroin			
% used in past year	1.3	0.8	0.4
Ecstasy (MDMA)			
% used in past year	2.1	1.6	0.7
OxyContin			
% used in past year	1.5	2.3	1.4
Vicodin			
% used in past year	2.1	3.2	1.6
Any Illicit Drug			
% used in past year	16.1	13.0	12.3

[a]Data are for from the unpublished report of the 2008 Monitoring the Future study (Johnston et al. 2009), a national representative study of school-attending youth in the U.S
Report can be found at: http://monitoringthefuture.org/pubs/monographs/vol1_2008.pdf

held quite steady for nearly two decades (Johnston et al. 2009). Findings are similar for lifetime and current (past 30 days) use.

Interestingly, by 12th grade, the percentage of Hispanic youth who use drugs is lower than that of Whites (Johnston et al. 2009) but this is thought to occur because of the disproportionately high dropout rates Hispanics experience (Delva et al. 2005). However, data from the National Household Survey on Health and Drug Use do not point to higher rates of substance use among adult Hispanics without a high school education when compared to non-Hispanic adults without such education (Substance Abuse and Mental Health Data Archive, www.icpsr.umich.edu/icpsrweb/SAMHDA/). These findings suggest that despite the disproportionately higher school dropout rates with its associated social and economic consequences there may be a number of important factors that either mediate or moderate the link between dropping out from school and substance use among Hispanics. Given the heterogeneity that exists among Hispanic populations based on country of origin, region of residence, generational status, acculturation, enculturation, religiosity, and socioeconomic status, among others, significantly more research is needed to understand substance use patterns and etiological factors according to these heterogeneous factors.

Key Demographic Trends Among Hispanics in the U.S.

Quite clearly, the high prevalence of substance use among Hispanic youth should be of concern not only because of direct deleterious effects but also because Hispanics experience a host of socio-economic conditions that can only serve to exacerbate the problems associated with drug involvement. For example, relative to the general U.S. population, Hispanic youth have substantially higher school dropout rates, a high proportion of families living in poverty, and a high proportion of births among young women (15–19 year olds) (Pew Hispanic Center 2009).

The Hispanic population is characterized by a very young population whereby they account for approximately 22 % of children younger than 18-years-old and have a median age of 27.7 years compared to 36.8 for the general population (U.S. Census 2009). These trends indicate that a very large percent of Hispanics are in the greatest risk period of using and abusing substances given that drug use tends to manifest most strongly among younger individuals.

Further contributing to the risk of using and abusing substances are socioeconomic trends faced by Hispanics. For example, the median net worth of Hispanic households in 2002 was only 9 % that of non-Hispanic Whites ($7932 vs. $88,651) (Kochar 2004, 2005). Furthermore, unemployment among Hispanics rose to 6.5 % in the first quarter of 2008, a substantially higher rate than the 4.7 % observed for non-Hispanics (Kochar 2008). The continued growth of the Hispanic population is likely to be maintained by its considerably young age structure and continued immigration from Latin America. Unfortunately, the economic gains Hispanics had experienced in the past two decades have been overturned due to the current economic downturn. Furthermore, and perhaps because of their recent gains and increasing visibility, it seems that Hispanics have been experiencing an increase in targets of hate crimes (Federal Bureau of Investigation. November 2009). The stressors and frustrations individuals and families experience as a result of these social and economic hardships can lead to increased substance use as a coping mechanism. This should be of concern given there is increasing evidence linking experiences of discrimination with increased substance use (Gee et al. 2007; Delva et al. 2009).

Longitudinal Studies of Substance Use with Hispanic Youth

Considerable research has been undertaken to identify pathways associated with substance use initiation and patterns of use and to understand progression to abuse and dependence. However, the knowledge base remains deficient among racial and ethnic minorities and Hispanics in particular. There is an important need to conduct longitudinal studies with Hispanic youth to identify and better understand pathways for substance use in these populations. The lack of studies that can prospectively shed light into the mechanisms that increase the risk of substance use among Hispanic youth is problematic because of the overrepresentation of Hispanic youth presently using drugs and the disproportionate overall burden of disease in this population. As Vega and Gill (1998) stated, *"The past reliance on White non-Hispanics as a source of reference against which other groups were compared, along with the lack of cross-ethnic comparative studies, highlights our need for additional research designed specifically to extend our understanding of the role that risk factors play in substance use over the course of the maturational processes among ethnic groups"* (p. 74). More than a decade later, and despite Hispanics becoming the largest minority population in the U.S., research on Hispanics, particularly longitudinal research, has not grown proportionately.

For the purpose of this study we conducted an extensive search for peer-reviewed publications of substance use among Hispanics that indicated a longitudinal design had been utilized during the 1996–2009 period. We used the following databases to conduct our literature review: Chicano database, ERIC, PsycINFO, ProQuest, Sociological Abstracts, Social Sciences Abstracts, ProQuest Research Library, Social Sciences Citation Index and Social Work Abstracts. The key words used in the search were "longitudinal", "Hispanic", "latino", "latina", "alcohol", "drinking", "illicit drug", "cigarette", "smoke", "marijuana", "substance use". The last update of the search was September 20, 2009.

We conducted the initial search based on the combination of the terms "longitudinal" with any of the substance use keys words ("alcohol", "drinking"... see above). This initial search yielded 1117 studies. Based on these 1117 studies, we refined the search filtering the records using the words "Hispanic" or "Latina" or "Latino" and "longitudinal". This step filtered most of the 1117 articles resulting in 40 studies. Subsequently, we filtered these 40 studies based on the study's sample size (over 400) and that some random selection procedure was utilized in selecting participants to allow for stronger generalizability (.e.g., random selection of schools or random selection of students from schools). We did not consider studies with a small sample of Hispanic participants (less than 400) and studies that did not describe the sampling procedures. This search resulted in 17 published papers based on longitudinal studies involving Hispanic population in the U.S. The search resulted in several important studies that serve as an illustration of longitudinal studies conducted in the past decade with Hispanics and point to further areas of research (see Table 5.2).

Table 5.2 Longitudinal studies of substance use with Hispanic samples of at least 400 participants

Paper code	year of publication	Topic	Study used in paper	Population of longitudinal study	Sample used in paper			Waves used in paper	Analyses	Obs
					total	Hispanic	Representative			
1	1996	Alcohol, cigarette and illicit drugs by ethnicity	Longitudinal study of substance use among adolescents residing in south Florida	Hispanic, African American and Non Hispanic White Adolescents (12–14 years old)	5370	3403	Dade county Florida	Fall 1990 Fall 1992 Spring 1993	ANOVA ANOVA repeated measures	No multilevel No covariates
2	2004	Alcohol, acculturation and intimate partner violence	General Population Household Survey Longitudinal Data Collection	Couples 18+years old	387 couples	387 couples	48 cont states	1995 and 2000	3 way cross tabulation and Chi square taking into account sample design	No multilevel No covariates No trajectories
3	2006	Onset of alcohol drinking	National Longitudinal Survey of Youth NLSY	Youth 14–22 years old	8165	474	Nation wide	1982, 1983, 1989, and 1994[a]	Multinomial logistic regression	No multilevel No trajectories
4	2008	Binge drinking and acculturation	National Longitudinal Study of Adolescent Health (Add Health)	Adolescents (7th -12th grade in 1994)	2035	2035	Nation wide Hispanic	First and second	logistic and multiple regression, weighted	No multilevel No trajectories

5	2004	Drinking, alcohol problems and intimate partner violence	General Population Household Survey Longitudinal Data Collection	Couples 18 +	793 couples	387 couples	48 cont states	1995 and 2000	Pathway analyses	No multilevel No trajectories
6	2005	Family protection and perception of alcohol use	National Cross-site Evaluation of High Risk Youth Program	Youth 9–17 years old	2742	2742	Hispanic females over represented Hispanic males under represented[b]	Program entry, program exit, 6 months after exit and 18 months after exit	HLM	No covariates No trajectories
7	1996	substance use patterns among Hispanic adolescents	Longitudinal study of substance use among adolescents residing in south Florida	Hispanic, African American and Non Hispanic White Adolescents (12–14 years old)	893	400	4 middle schools	Fall 1990 Fall 1992 Spring 1993	Ttest, ANOVA, Repeated measure analyses of variance (ANOVAs), one way ANOVA	No multilevel No covariates
8	2008	Adolescent alcohol use behavior based on segmented-assimilation-theory	National Longitudinal Study of Adolescent Health (Add Health)	Adolescents (7th -12th grade in 1994)	1504	1504	Mexican, Puerto Rican of Cuban Ancestry[c]	First and second	Logistic Analyses	No multilevel No trajectories

(continued)

Table 5.2 (continued)

Paper code	year of publication	Topic	Study used in paper	Population of longitudinal study	Sample used in paper			Waves used in paper	Analyses	Obs
					total	Hispanic	Representative			
9	2005	Evidence -based evaluation of prevention program: "keepin' it REAL"	Keepin' it REAL	35 public middle schools in Phoenix metropolitan area	3402	3402	Mexican American, Mexican, and Chicano	Pre and post intervention	Multiple imputation NORM, Multilevel regression	Problems linking pre test and post test students. No trajectories
10	2009	The effect of conflicting cultural differences between adolescents and their parents (Acculturation discrepancy theory) on substance use	Reteniendo y Entendiendo Diversidad para la Salud (RED)	Hispanic/Latino adolescents in South California. 7 schools in Los Angeles area	1683	1683	seven schools from Los Angeles area	9th and 10th grade	Multilevel logistic regression	No covariates No trajectories
11	2007	Patterns of progression in substance use	Service Use, Need and Outcome Study (SUNO). National Longitudinal Study of Youth (NLSY)	Hispanic youth 13–17	663 Puerto Rico; 1445 USA	663 Puerto Rico; 1445 USA	Puerto Rico and USA	NLSY wave 2 and 3 and first wave from SUNO	Latent transition analysis (LTA)	No multilevel No covariates No trajectories

12	1997	Prevalence, incidence and stability of drinking problems among adolescents	The National Alcohol Survey	Adults 18 +	1616	446	48 cont states	1984, 1992	Cross tab and chi square	No multilevel No covariates No trajectories
13	2007	Smoking cessation and first marriage	National Longitudinal Survey of Youth NLSY	14–22 in 1979	4050		Nation wide	1992, 1994, 1998	EHA, discrete time hazard model	
14	2009	School context effect on alcohol misuse	National Longitudinal Study of Adolescent Health (Add Health)	Adolescents (7th -12th grade in 1994)	10,574 students and 128 schools	1247	Nation wide	time 1 and time 2	Multilevel multinomial logistic models HLM	No trajectories
15	2008	Parental warmth and control on alcohol use	National Longitudinal Study of Adolescent Health (Add Health)	Adolescents (7th -12th grade in 1994)	1887	1887	Nation wide Hispanic	time 1 and time 2	SEM	No trajectories No multilevel
16	2009	Impact of adolescent drug use on socioeconomic outcomes in young adulthood	National Longitudinal Study of Adolescent Health (Add Health)	Adolescents (7th -12th grade in 1994)/young adults 18–29	w1:6504, w2: 4881	w1:780, w2: 586	Nation wide	wave 1, wave 3	Regression OLS and logistic regression	No multilevel No trajectories

(continued)

Table 5.2 (continued)

Paper code	year of publication	Topic	Study used in paper	Population of longitudinal study	Sample used in paper			Waves used in paper	Analyses	Obs
					total	Hispanic	Representative			
17	2009	effect of nativity on substance use mediated by ecological processes	National Longitudinal Study of Adolescent Health (Add Health)	Adolescents (7th -12th grade in 1994)	742	742	Nation wide Hispanic	wave 1, wave 2	CFA, SEM and Post hoc mediation analyses	No multilevel No trajectories

[a]This longitudinal study measured yearly from 1979 to 1994 and every 2 years after 1994
[b]Hispanic females are over represented because 5 out of 7 programs were target only to females
[c]Authors selected only participants who had Mexican, Puerto Rican or Cuban ancestry

As mentioned earlier, the articles found were published between 1996 and 2009, all in peer-reviewed journals. We did not restrict the search to high impact journals. The sample sizes for the Hispanic individuals for these publication rage from 446 to 3403 participants. Most of the studies target adolescents living in the U.S. Only two studies sampled adult couples (18 years and older) and one study sampled adults (at least 18 years old). Nine out of the 17 studies were representative of adolescents living in the US. Another study was representative of adolescents of Mexican and Puerto Rican or Cuban ancestry living in the US. One study was representative only of Hispanic females who participated in the National Cross-site Evaluation of High Risk Youth Program.

There were four studies with non-national representative samples. From these four studies one was conducted in Los Angeles, another in the metropolitan area of Phoenix and two of them from the same study in South Florida. From the two studies in south Florida, one was representative of Dade County and the other has no representation other than the four schools from where the participants were selected. The study in Phoenix is representative of Mexican American, Mexican and Chicano adolescents and the study in the Los Angeles area appeared to be representative of the seven schools from where the students were sampled. The remaining three studies were representative of adults living in the 48 contiguous states.

We classified the longitudinal studies used in the 17 publications into four categories: National longitudinal studies, household surveys, small regional or small longitudinal studies and programs. There were three national longitudinal studies; these three studies are: the "National Longitudinal Study of Adolescent Health (Add Health)", the "National Longitudinal Survey of Youth NLSY", and the "National Alcohol Survey". There were two household surveys: the "General Population Household Survey Longitudinal Data Collection" and "Service Use, Need and Outcome Study (SUNO)" in Puerto Rico. There were two regional longitudinal studies: "Longitudinal study of substance use among adolescents residing in south Florida" and "Reteniendo y Entendiendo Diversidad para la Salud (RED)" in Los Angeles. There were two programs: the "National Cross-site Evaluation of High Risk Youth Program" and "Keepin' it REAL".

In addition to the above studies, we are aware that there are other longitudinal studies with substantial number of Hispanics such as the Project on Human Development in Chicago Neighborhoods and the National Educational Longitudinal Study (NELS) that potentially could be used to produce knowledge about these populations using longitudinal data. Unfortunately, we did not find any peer-reviewed publications about substance use among Hispanic populations using these two studies. Furthermore, we also contacted several research teams in the country that we were aware of that had conducted longitudinal studies to inquire about any findings they may have had concerning Hispanics. We learned that most of these teams either did not have sufficient numbers of Hispanics to identify as separate categories in their analyses or had not entertained the idea of focusing on Hispanics. Also, some teams did not respond to our requests for more information and therefore are not able to ascertain the extent to which Hispanics were included in these other studies.

Going back to the studies we selected for this paper, we find that most of the studies (12/17) used in their analyses only two waves of measurement. Three of the 17 had three waves, and the remaining two studies used four waves of measurement. From the five studies with more than three waves of measurement only two of them analyzed trajectories of substance use. The main analytic approach ranged from simple Chi-square with two- or three-way cross-tabulation, analysis of variance and t-tests, to more sophisticated techniques such as longitudinal hierarchical linear modeling analysis, structural equation modeling and latent transition analysis. Four of these publications used multilevel techniques to account for the nested structure of the data. Some of the 17 publications did not consider necessary to adjust their analysis by covariates or confounders.

Almost all the publications are related to alcohol use, one is only about smoking, and some of the publications used several outcomes under the term substance use. The type of drug consumption ranged from unspecified use to binge drinking and abuse of substances.

There were several topics related to the consumption of substances. One paper focused on the onset of alcohol use while another paper concentrated on smoking cessation. Five papers examined the association between substance use and acculturation and/or immigration status. Two papers explored the effect of family factors (family protection and parental warmth/parental control). Two other papers related the use of alcohol to intimate partner violence. Three papers explored patterns or prevalence of substance use. One paper was specifically about substance use and ethnicity and the last two topics were about the effect of substance use on socioeconomic outcomes and the last topic was an evidence-based evaluation of a prevention program.

All the papers show significant results between their covariates and outcomes (substance use). Our review of these studies suggests that some of these results could have been improved on in one of two ways. One of these is theoretical, including the use of covariates as a way to control for possible confounding factors and the second is methodological, taking into account, when possible, the sampling design and the nested structure of the data. Some of the papers used strong theoretical backgrounds and generated models to be tested while other papers did not show much of a theoretical framework to address their research questions. Missing data was a common problem in all of the papers. Interestingly, most of the papers presented a small attrition analysis concluding that missing was at random and therefore not a big concern. One paper used multiple imputation to deal with the missing data problem.

Due to time constraints, we acknowledge that this search is not the most comprehensive that could have been undertaken as a more comprehensive review is beyond the scope of this chapter. We recognize that we could have acquired a few more studies if we had relaxed some of our inclusion criteria. For example, we could have included studies conducted prior to 1996 and longitudinal studies with smaller sample sizes and that relied on convenience samples. It is also possible that we did not identify studies that included Hispanics but failed to highlight findings relevant to these populations or studies with key words that did not include any one of the substances/drugs used in our search criteria. Despite these limitations, we are not convinced that the additional number of published studies would be much larger and that the findings from less rigorous studies would considerably add more significant information to this review.

As illustrated by the present review, more longitudinal studies are needed to generate additional knowledge about substance use among the diverse Hispanic populations. It is unfortunate that despite the availability of numerous longitudinal data sets in the U.S., a substantial body of knowledge about substance use among Hispanics is clearly lacking. The general problem of not including sufficiently large samples of Hispanics in many studies and the underutilization of these secondary datasets severely limits the knowledge that could be obtained about substance use among racial and ethnic minorities in particular. As we discuss next, conducting longitudinal studies can serve to fill the knowledge vacuum that exists about substance use among Hispanics.

Longitudinal Statistical Methods

When compared with cross-sectional data, longitudinal data offer a number of advantages. Some of these advantages are readily apparent while other advantages of longitudinal data are only revealed after some reflection. In this section of the chapter, we focus our discussion on the specific advantages that are present when one has collected data with *repeated measures* of specific constructs of interest. To better illustrate the use of longitudinal methods to answer questions about growth and change of certain behaviors over time, we focus on the need to prospectively examine the potential link between dropping out from school and substance use among Hispanics. As we describe in the introduction section, discussed elsewhere as well (Gil and Vega 2010), the higher lifetime prevalence rates of substance use among Hispanics in lower grades (8th & 10th) but not in higher grades (12th) have been attributed to higher school dropout rates among Hispanics. Longitudinal methods offer advantages in helping us understand if, how, for whom, and under what circumstances, dropping out of school may be associated with substance use.

Repeated Measures and Growth Trajectories

Repeated measures data are a specific kind of longitudinal data in which, at a minimum, data on a dependent variable of interest are collected at least at two different time points. As an example, one might imagine a study in which the intent was to examine the predictors of cigarette use. For this to be a repeated measures study, data would have to be collected on the dependent variable of cigarette use at multiple points in time.

Within the repeated measures literature, the simple plot of a dependent variable against some measure of time is termed a trajectory or growth trajectory (Raudenbush

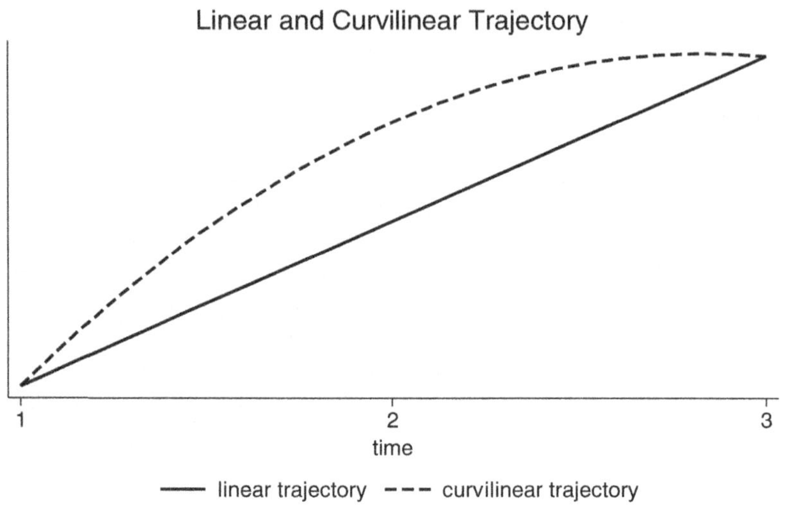

Fig. 5.1 Hypothetical example of a linear and curvilinear trajectory of the relationship of the dependent variable with time

and Bryk 2002; Singer and Willett 2003; Willett et al. 1998). When only two time points are available in the study, the best fitting, or most parsimonious, trajectory is by definition a straight line. One of the major advantages of estimating a trajectory with more than two time points is that it allows the researcher to get a better sense of the functional form of the relationship of the dependent variable with time. The functional form may be linear, implying a constant rate of change, or curvilinear, indicating that the rate of change may itself be changing over time (see Fig. 5.1).

For example, in the aforementioned hypothetical study of cigarette smoking, it might be informative to know whether smoking increases at a constant rate over time, or whether there is a particular time period, when youth drops out of school, during which cigarette use increases rapidly. In terms of intervention, discovering that growth in cigarette use is particularly rapid during a specific time period might offer the ability to develop a particularly effective intervention that was targeted at that particular developmental stage.

Further, as one develops the idea of a growth trajectory, it becomes apparent that estimating such trajectories allows one to distinguish the intercept, or starting point of a trajectory from the slope, or rate of change. At the end of the observation period in a longitudinal study, two different participants may have different outcomes on a particular measure. For example, in the study of smoking mentioned above, one study participant might smoke considerably more cigarettes than another participant. Such differences may occur either because of different starting initial levels of cigarette smoking (see Fig. 5.2), because of different rates of change in cigarette smoking (see Fig. 5.3), or because of a combination of both factors. Distinguishing among these possibilities would generally not be possible

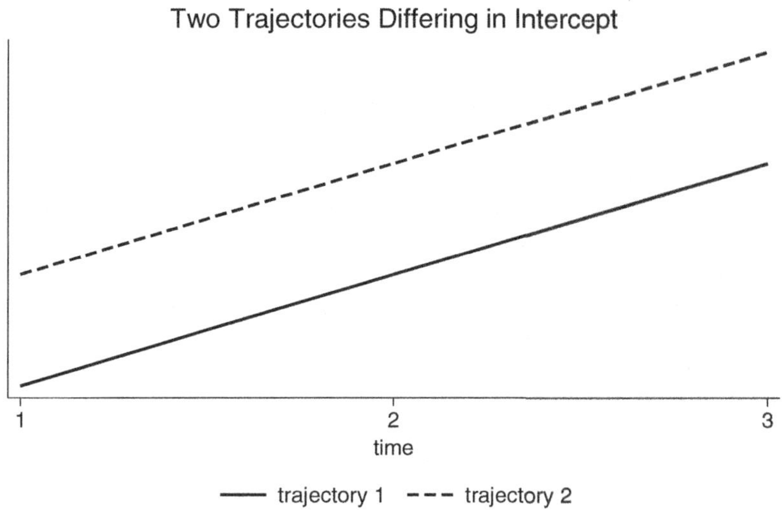

Fig. 5.2 Hypothetical illustration of smoking trajectories differing in initial starting levels (intercepts)

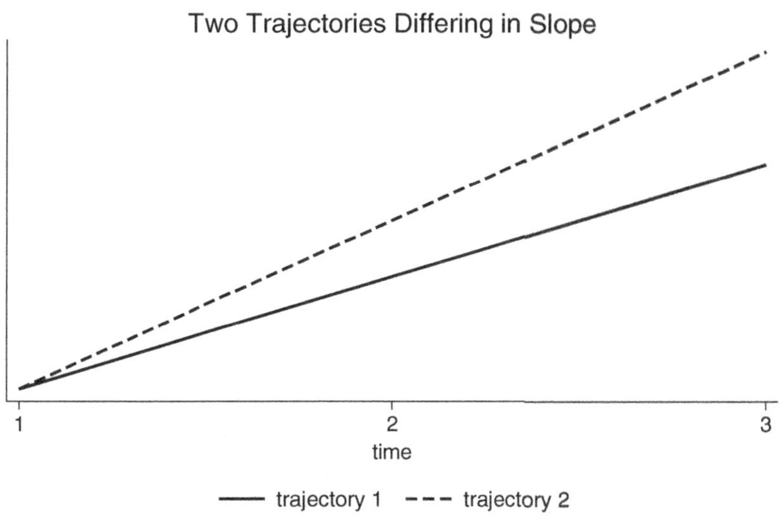

Fig. 5.3 Hypothetical illustration of smoking trajectories differing in rates of change (slopes)

in a cross-sectional study. Discerning whether differences at the end of the study are due to pre-existing differences at the beginning of the study (intercept differences) or are due to differential change over time (slope differences) might have implications for understanding the onset of cigarette smoking and thus for designing programs, policies and interventions.

Multilevel Modeling

Within the study of repeated measures, over the last several decades, a rapid growth has been seen in the area of statistical investigation known as multilevel modeling. Statistically, multilevel models can be complicated models. Our intent in this chapter is not to reprise the sometimes intricate notation of these models or to provide an in-depth review of the statistical concepts underlying these models. Thoroughgoing treatments of multilevel models are available elsewhere (Luke 2004; Rabe-Hesketh and Skrondal 2005; Raudenbush and Bryk 2002; Singer 1998; Singer and Willett 2003; Willett et al. 1998). Instead, we wish to lift up several conceptual insights and advantages offered by multilevel models, particularly in the context of longitudinal statistical analysis, which may not be apparent to the investigator newly contemplating the application of multilevel models to the area of substance use.

Statistically, in the longitudinal context, multilevel models estimate what can be called *person specific* growth trajectories. A hypothetical example of such person specific growth trajectories are illustrated in the "spaghetti plot" shown below where the thick bold line illustrates the overall average trajectory, while the thinner lines illustrate each person specific trajectory (see Fig. 5.4). One statistical motivation for multilevel models is that observations for the same individual may be correlated. More concretely, following our earlier example, cigarette smoking for a particular individual at a particular time is likely to have some relationship with later smoking by that individual: initial non-smokers may be more likely to be non-smokers, or light smokers later on; initial heavy smokers may be more likely to be heavy smokers later on. Statistically, this relationship of observations for each individual will often lead to underestimation of standard errors unless the model is adjusted for the use of repeated measures data.

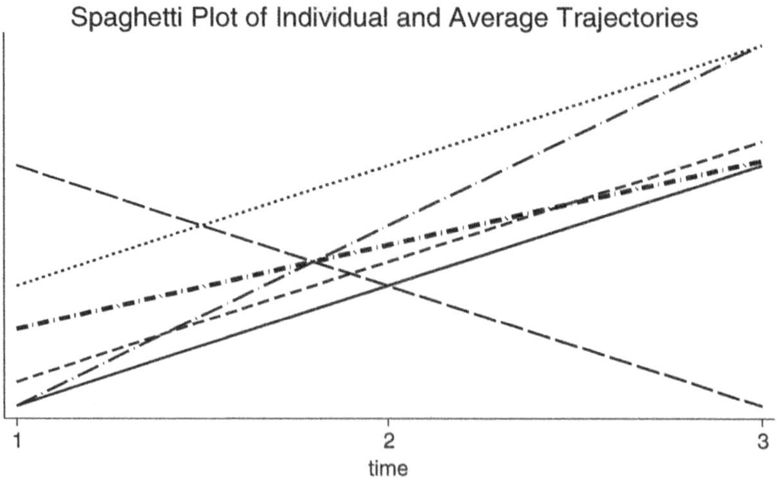

Fig. 5.4 Hypothetical example of person specific growth trajectories of cigarette smoking

Underestimation of standard errors would lead to artificially low p values and might thus cause the researcher to conclude that particular relationships were statistically significant, even when this was not appropriate. The researcher might thus make substantive conclusions about the phenomena under investigation that would not be supported by a correctly specified model.

For example, in applying a simpler ordinary least squares regression to the hypothetical study of cigarette smoking that we are using as an illustration, a researcher might conclude that dropping out of school was a statistically significant predictor of cigarette smoking. However, a multilevel model that adjusted standard errors for the clustering of observations inside individuals might conclude that the relationship of school dropout and cigarette smoking was not statistically significant. With repeated measures data, the more appropriate conclusion would be the one derived from the multilevel model. The multilevel model for repeated measures data adjusts for the correlated nature of repeated measures data and thus avoids such false attributions of statistical significance.

Multilevel models also allow for the researcher to separate between person and within person effects. Consider the hypothetical study of cigarette smoking that we have so far employed as an example. In this study, the amount of cigarette smoking may be observed at each time point for each individual person. Thinking about the variation in the number of cigarettes smoked by different people at different time points allows us to decompose this variation into two sources, within person variation and between person variation. Between person differences are difference in average levels of cigarette smoking. Within person differences are differences of each person's observation-specific level of cigarette smoking from the mean for that person. This gives rise to the simple heuristic equation:

$$\text{Total variation} = \text{between person variation} + \text{within person variation}$$

Within our hypothetical study of cigarette smoking, it might be substantively interesting to think about predictors of changes in day to day smoking as compared with predictors of differences between people in average levels of smoking.

Fixed Effects Models

Fixed effects models are a type of regression model suitable for use with repeated measures data. Fixed effects models are closely related to multilevel models although the relationship between these two classes of models is not commonly acknowledged and the two groups of models seem to have different disciplinary affiliations. We have discussed the difference between within person variation and between person variation. Multilevel models make use of both sources of variation. Fixed effects models only make use of within person change over time, and the between person variation is factored out.

The advantage of this approach is that fixed effects models are able to control for *all* time invariant (between person) quantities. In our hypothetical model of cigarette smoking, one might imagine a situation in which respondents came from different neighborhoods and characteristics of these neighborhoods were hypothesized to influence cigarette smoking. Now imagine that, unfortunately, such neighborhood characteristics were not observed, and therefore not available in the data. In a standard model, such as an ordinary least squares regression model, failure to include such unobserved variables could lead to bias in the regression coefficients. In a fixed effects regression model, estimation of model parameters would not be biased by the fact that neighborhood measures had not been included in the model. Indeed the fixed effects regression model would control for *any* omitted time invariant variable.

Despite this notable advantage of the fixed effects regression model, some cautions are in order. Fixed effects regression models control for all time invariant variables essentially by treating the effects of such variables as a nuisance parameter that is not directly estimated. One real disadvantage of this approach is that even when measures *are* available for time invariant variables, such as commonly is the case for characteristics like respondent sex and respondent race, such variables cannot be included in the fixed effects regression model (Grogan-Kaylor 2004). Indeed many software packages will automatically drop such time invariant quantities if the analyst attempts to include them in a fixed effects regression model. What this means is that fixed effects regression models for longitudinal data *cannot* provide estimates for variables like sex or race. This inability of the fixed effect model to provide parameter estimates for time invariant variables such as sex and race must be balanced against the ability of such models to provide less biased estimates for variables that change value over the course of time.

Also, the fact that fixed effects models make use only of variables that change over time means that fixed effects models do not exploit all of the variation in observed variables since some of that variation may be between individuals rather than within individuals over time (e.g. differences in average cigarette use versus changes in cigarette use). The fact that fixed effects models use only some of the available variation means that fixed effects models may have two disadvantages. First, fixed effects models are likely to be less "statistically efficient." In some cases they may not find statistically significant effects in cases where multilevel models would do so. Second, fixed effects models may be less appropriate in situations where more variation is between people than within people over time.

As an example, consider the hypothetical discussion of cigarette smoking that we have been using as an illustration. We have considered that dropping out of school might be associated with cigarette smoking. If more of the variation in school participation is within people over time rather than between people, a situation may arise in which multilevel models are able to detect a statistically significant relationship while fixed effect regression methods are unable to do so.

In any specific research project, the relative advantages and disadvantages of fixed effects regression and multilevel modeling must be balanced against each other with consideration for the specific research question and phenomena under investigation. In this section we reviewed a selected number of approaches to analyze longitudinal data, namely growth modeling, multilevel modeling, and fixed-effects models. Finally, one could prospectively investigate the association between dropping out of school and substance use while also examining the potential effects of individual, familial, and social, neighborhood moderators and mediators on this association.

We presented these models because they are intuitively 'easy' to understand conceptually and as such we find them to be considerably helpful to explain various aspects of longitudinal data analysis. Also, these are the approaches we are most intimately familiar with in our own work. However, several other approaches exist to analyze longitudinal data such as structural equation modeling (Byrne 2001), and latent trajectory modeling (Jones et al. 2001) that readers are encouraged to investigate as well.

Conclusion

As documented by prior research and summarized in the beginning of this chapter, our understanding of substance use among Hispanic populations youth relies considerably on cross-sectional studies. Although important, these studies only provide a snapshot of an individual's drug involvement. The changing demographic and socioeconomic trends for Hispanic populations raise serious concerns about what we know about their lives and the potential links to substance use, resulting in a serious knowledge gap to address substance abuse problems these populations experience. For example, scarce attention has been paid to distinguishing the effects of income and wealth on children's substance using behaviors in the general population, let alone among Hispanics. Wealth, defined as the assets possessed by a family, has been found to add significant explanatory power to models of other family and child outcomes (such as children's educational outcomes), over and above income or poverty, in other areas of the social scientific literature (Conley 1999). How variations in income and wealth uniquely and together impact a Hispanic youth likelihood of dropping out of school may serve explain the link between school drug out and substance use.

Interestingly, although researchers have begun to examine the relationship of community characteristics with children's substance use, the literature on neighborhood effects has understudied questions of non-random selection into neighborhoods. More concretely, when correlations between aggregated neighborhood measures and children's behaviors are found, do these relationships indicate that neighborhoods have a causal effect on families and children, or do these relationships merely reflect that families with different levels of socio-economic resources are selected into different kinds of neighborhoods. Relatedly, how do these factors have an impact on a Hispanic youth likelihood of dropping out of or completing high school? This is a particularly important issue with racial and ethnic minorities such as Hispanics and African Americans due to the overrepresentation of these populations among the poorer neighborhoods in the U.S. More attention to how different kinds of families are selected into different kinds of neighborhoods and how their lives evolve is warranted.

Longitudinal research is needed to understand differential substance use pathways Hispanic youth experience as a function of the interplay of their biological and temperamental make up with their parenting, familial, and cultural experiences. More longitudinal research is needed to understand how the association of parenting behaviors with children's substance use will be partially moderated by family income, family wealth, and parental education. In addition, more longitudinal research is needed to understand how the socioeconomic stress Hispanic families experience resulting from their acculturation and assimilation experiences, including experiences of discrimination due to their heritage, skin color, and language differences, impact Hispanic families and their children.

In sum, these are but a few examples of a large number of research questions that remain unanswered with Hispanics and that are ideally suited for investigation with longitudinal studies. Perhaps longitudinal studies will help us understand the link between dropping out of school and substance use and will shed light into the perplexing findings that at the youngest ages, Hispanic adolescents have a higher prevalence of substance use than all other groups (the exception being American Indian youth) but among adults without a high school education Hispanics are not overrepresented among substance users. Are there unique familial and community factors that serve to protect Hispanic youth who have dropped out of school? Longitudinal research should help shed light into this question.

References

Byrne, B. M. (2001). *Structural equation modeling with AMOS: Basic concepts, applications, and programming*. Mahwah: Lawrence Erlbaum Associates.

Conley, D. (1999). *Being African American, living in the red: Race, wealth, and social policy in America*. Berkeley: University of California Press.

Delva, J., Wallace, J. M., Bachman, J., O'Malley, P. M., Johnston, L. D., & Schulenberg, J. (2005). The epidemiology of alcohol, cigarettes, and illicit drugs among Mexican American, Puerto Rican, Cuban American, and other Latin American youths in the US: 1991–2002. *American Journal of Public Health, 95*, 696–702.

Delva, J., Lorenzo-Blanco, E., Horner, P., Hallman, S., Sanchez, N., Grogan-Kaylor, A., & Castillo, M. (2009, June). *Association between discrimination and substance use is not limited to the US: Preliminary evidence from the Santiago Longitudinal Study*. Poster presented at the 2009 NIDA international forum and the College on Problems of Drug Dependence (CPDD) conference, Reno, Nevada.

Federal Bureau of Investigation. (2009, November). *Uniform crime report hate crime statistics, 2007*. Unpublished report. Available at: http://www.fbi.gov/ucr/hc2007/downloadablepdfs/victims.pdf. Accessed on 6 Nov 2009.

Gee, G., Delva, J., & Takeuchi, D. T. (2007). Relationships between self-reported unfair treatment and prescription medication use, illicit drug use, and alcohol dependence among Filipino Americans. *American Journal of Public Health, 97*, 933–940.

Gil, A., & Vega, W. (2010). Alcohol, tobacco, and other drugs. In M. Aguirre-Molina, L. N. Borrell, & W. Vega (Eds.), *Health issues in Latino males* (pp. 99–122). New Brunswick: Rutgers University Press.

Grogan-Kaylor, A. (2004). The effect of corporal punishment on antisocial behavior in children. *Social Work Research, 28*(3), 153–164.

Humes, K. R., Jones, N. A., & Ramirez, R. R. (2011, March). Overview of race and hispanic origin: 2010. *2010 Census Briefs. US Census Bureau.* Available: http://www.census.gov/prod/cen2010/briefs/c2010br-02.pdf

Johnston, L. D., O'Malley, P. M., Bachman, J. G., & Schulenberg, J. E. (2009). *Monitoring the future national survey results on drug use, 1975–2008: Volume I, Secondary school students (NIH Publication No. 09–7402).* Bethesda: National Institute on Drug Abuse.

Jones, B., Nagin, D., & Roeder, K. (2001). A SAS procedure based on mixture models for estimating developmental trajectories. *Sociological Methods and Research, 29*(3), 374–393.

Kochar, R. (2004). *The wealth of hispanic households: 1996 to 2002.* Washington, DC: The Pew Hispanic Center. Unpublished report available at: http://pewhispanic.org/files/reports/34.pdf. Report accessed on 10 Jan 2009.

Kochar, R. (2005). *Latino labor report, 2004: More jobs for new immigrants but at lower wages.* Washington, DC: The Pew Hispanic Center. Unpublished report available at: http://pewhispanic.org/files/reports/45.pdf. Report accessed on 10 Jan 2009.

Kochar, R. (2008). *Latino labor report, 2008: Construction reverses job growth for latinos.* Washington, DC: The Pew Hispanic Center. Unpublished report available at: http://pewhispanic.org/files/reports/88.pdf. Report accessed on 10 Jan 2009.

Luke, D. A. (2004). *Multilevel modeling.* Thousand Oaks: Sage.

Pew Hispanic Center .(2009). *Statistical portrait of hispanics in the United States, 2007.* Author. Unpublished report available at: http://pewhispanic.org/factsheets/factsheet.php?FactsheetID=46. Report accessed on 10 Jan 2009.

Rabe-Hesketh, S., & Skrondal, A. (2005). *Multilevel and longitudinal modeling using STATA.* College Station: STATA Press.

Raudenbush, S. W., & Bryk, A. S. (2002). *Hierarchical linear models: Applications and data analysis methods.* Thousand Oaks: Sage Publications.

Singer, J. (1998). Using SAS PROC MIXED to fit multilevel models, hierarchical models, and individual growth models. *Journal of Educational and Behavioral Statistics, 24*(4), 323–355.

Singer, J. D., & Willett, J. B. (2003). *Applied longitudinal data analysis: Modeling change and event occurrence.* Oxford/New York: Oxford University Press.

United States Census Bureau. (2009, May 14). *Census Bureau estimates nearly half of children under age 5 are minorities. Estimates find nation's population growing older, more diverse* [Press release]. Unpublished report available at http://www.census.gov/Press-Release/www/releases/archives/population/013733.html. Accessed on 6 Nov 2009.

Vega, W., & Gil, A. (Eds.). (1998). *Drug use and ethnicity in early adolescence.* New York: Plenum Press.

Willett, J. B., Singer, J. D., & Martin, N. C. (1998). The design and analysis of longitudinal studies of development and psychopathology in context: Statistical models and methodological recommendations. *Development and Psychopathology, 10*(2), 395–426.

Chapter 6
The Stress of Discrimination: A Possible Influence on Drug Use Trajectory?

Clifford L. Broman

Introduction

This chapter discusses racial discrimination as a possible factor in African American – white differences in drug use trajectory. The central question addressed in this chapter is whether racial discrimination, at both the macro and the micro level, play a significant role in changing patterns of substance use over the life course for African Americans and whites. We specifically focus on the transition to young adulthood, and the growing adversity likely encountered by minority young adults. Two central arguments are made: First, we argue that the transition to adulthood is one that can be problematic for all adolescents. For some the road may be rocky, and for others, relatively smooth. Secondly, there is the added stress of dealing with the harsh world of racial disadvantage that confronts minority youth as they transition. This may be an added burden for minority youth. Both of these factors may combine to increase substance use over the young adulthood for African Americans.

We first briefly review the evidence on differences in substance use over the life course for African Americans and whites. Then we move into a discussion of the role of stress in substance use. From there, we review and suggest that race and race-related socioeconomic adversity may be factors of critical importance in these changing patterns over the life course.

C.L. Broman (✉)
Department of Sociology, Michigan State University, East Lansing, MI, USA
e-mail: broman@msu.edu

Y.F. Thomas, L.N. Price (eds.), *Drug Use Trajectories Among Minority Youth*,
DOI 10.1007/978-94-017-7491-8_6

Racial Differences in Substance Use Across the Life Course

Several studies have provided excellent epidemiological data with convergent findings regarding substance use by substance type, age and race (Johnston et al. 2000; SAMHSA 2007). The evidence is clear that almost all adolescents and young adults are significantly more likely than people over age 26 to be users of alcohol and illicit drugs (SAMHSA 2007). Among young adults, alcohol is the most widely used substance, followed by marijuana (SAMHSA 2007). Young adults are much more likely to be heavy users of alcohol, and to engage in heavy episodic drinking (SAMHSA 2007). Other illegal substances, such as heroin, cocaine and ecstasy typically have prevalence rates of less than six percent among adolescents and young adults (SAMHSA 2007).

Studies also consistently find differences in the lifetime and current prevalence of substance use among African American and White adolescents (Barnes et al. 1994; Cataeno and Clark 1998; Horton 2007; Kandel 1995; O'Malley and Johnston 2002; SAMHSA 2007; Wallace et al. 1995, 2002). For example, in The Monitoring the Future (MTF) data, among White adolescents, 26.0 % of 8th, 52.2 % of 10th, and 67.2 % of 12th graders reported having been drunk in the past 30 days, while for African American adolescents, the figures are 17.8 % of 8th, 29.5 % of 10th, and 40.6 % of 12th graders (Johnston et al. 2000). For marijuana/hashish, lifetime use is 20.0 % of 8th, 39.9 % of 10th, and 50.3 % of 12th grade White students, and 23.9 % of 8th, 37.3 % of 10th, and 45.1 % of 12th graders who are African American. Among Whites, 17.2 % of 8th, 25.8 % of 10th, and 32.1 % of 12th graders reported illicit drug use other than marijuana/hashish over their lifetime, whereas among African American adolescents, 7.9 % of 8th, 8.3 % of 10th, and 11.3 % of 12th graders reported lifetime use of illicit drugs (Johnston et al. 2000). Thus there is ample evidence that African American youth report lower rates of legal and illegal substance use in adolescence, and initiate substance use at older ages than their White counterparts (Keefe and Newcomb 1996; Strycker et al. 2003). However, it is worth noting that recent NSDUH data show patterns of illicit substance use among adolescents to be essentially the same for African American and white adolescents (SAMHSA 2007; Shih et al. 2010), a pattern that diverges from most other research. In young adulthood, however, ages 18–25, African Americans are less likely than whites to be users of illicit drugs and alcohol (SAMHSA 2007).

However, as people move into young adulthood these patterns begin to shift, and by the later 30s, African Americans are using substances at disproportionately higher rates than Whites (Muthen and Muthen 2000). In addition, African Americans are experiencing comparatively more severe substance-related consequences than are whites (Herd 1989, 1990, 1997; Mudar et al. 2002; Muthen and Muthen 2000; Nielson 1999; Wallace et al. 1999). For example, African Americans have higher rates of cirrhosis of the liver as a consequence of alcohol abuse. African Americans also suffer from higher rates of substance use-related interpersonal and neighborhood violence (Galvan and Caetano 2003; SAMSHA 2007; Sanders-Phillips 1996a, b). And though African American young adults are less likely than white young adults to use alcohol, in adulthood, African Americans are more likely than whites to develop alcohol dependence (Galea and Rudenstine 2005).

Combined with other forms of social and economic disparities that face minority groups, it is imperative that interpretations of national data of these trends occur within a context-sensitive framework. Therefore, it is important to examine which sociocultural and psychological factors influence the shift in patterns of substance use among African Americans.

Why Do These Rates Change?

During adolescence, the period when substance use initiation and experimentation are most likely to occur, there may be several dimensions that are especially protective for African American youth against substance use. However, during emerging adulthood, the period from ages 18 to 25, there are stressors that may be harmful for African American young adults. These factors may impact the racial cross-over in substance use.

One is the transition to adulthood that follows high school. This transition is one in which most individuals face new and more stressful experiences. This is a factor of importance for all young adults, independent of race and ethnicity. A second factor, which is much more significant for minority young adults, is subtle and overt race-related stressors, including racial discrimination in various life domains. Discrimination operates at both the system level and at the individual level. That is, because of systemic racism (Feagin 2000), African Americans are disadvantaged in several ways in the U.S. (Farley and Allen 1987; Feagin 2000; Feagin and Vera 1995). This means that the average African American young adult starts young adulthood at a disadvantage, when compared to the average white young adult (APSE Research Brief 2009). Individual-level discrimination also may take a toll in terms of experiencing subtle or overt discrimination, which is known to have a variety of detrimental sociopsychological consequences (Broudy et al. 2007; Hill et al. 2004; Satcher 2001; Seaton and Yip 2009; Wagner and Abbott 2007). Discrimination processes may exacerbate the difficulty of transition into adulthood. In sum, minority young adults likely face both the burden of transitioning to adulthood, as well as the increased stress from discrimination in various life domains. This added stressor may place African American young adults at increased risk for substance use. This may be of great importance in understanding the pattern of increased use of substances among African American adults during a life stage period, in the late twenties, when substance use among White adults decreases.

The Role of Stress

Stress has been shown to be of great importance in the use of substances. Therefore if the transition to adulthood is typically stressful, as is discrimination, then we might expect changes in substance use as adolescents' transition into young adulthood. This is what the literature shows. Stress exposure, some responses to stress, and strategies for coping with stress, are found to change significantly as adolescents move into young adulthood.

There have been many studies providing evidence that stress is associated with substance use among young adults. The research literature shows that stress-motivated substance use is fairly common (Bradizza et al. 1999; Carpenter and Hasin 1999; Colder and Chassin 1993; Cooper et al. 1992, 1995; Cox and Klinger 1988, 1990; Ham et al. 2007; Lemke et al. 2007; Lewis et al. 2008; McCreary and Sadava 2000; Perkins 1999; Read et al. 2003; Rutledge and Sher 2001). The basic finding is that young adults who experience higher levels of stress tend to use substances at greater levels and, correspondingly, have a greater number of substance-related problems (Colder and Chassin 1993; McCreary and Sadava 2000; Perkins 1999). Among the college attending population, during the post-collegiate years, alcohol use decreases, but the relationship between stress and alcohol use for the purpose of stress reduction increases (Perkins 1999). That is, even though levels of alcohol use are lower in the post-collegiate years, people are increasingly likely to use alcohol as a means for coping with stress. Among young adults who do not attend college, the rates of substance use are even higher as they age into their later 20s and 30s (Merline et al. 2004; White et al. 2005). Factors such as dropping out of high school, and unemployment in adulthood are critical in substance use (Merline et al. 2004; Muthen and Muthen 2000). Other studies suggest that substance use as a means of enhancing positive emotions or to cope with troubling emotions is common. Cooper et al. (1992) showed that there was a varying pattern of stress and alcohol use across race. African Americans with a greater number of negative life events had more problems with alcohol than was true for whites. Therefore the evidence is fairly strong that stress is a factor in substance use for both whites and African Americans.

The pattern of stress-motivated substance use also holds true for research on traumatic stressors. The evidence shows that traumatic stressors, such as childhood physical and sexual abuse, are generally related to substance use. When examined more closely by type of trauma and gender, nuances emerge. For example, while many studies report a link between childhood abuse and substance use (Bensley et al. 2000; Kilpatrick et al. 1997; Widom et al. 1995), some have found that childhood physical and sexual abuse are related to substance abuse for women only (Bensley et al. 2000; Widom et al. 1995; Langeland and Hartgers 1998). However other studies have found that sexual abuse is linked to problem substance use for both men and women (Galaif et al. 2001). Considering results from studies of both general and traumatic stressors, the evidence is fairly clear that stress is an important factor in substance use among young adults, and that the type and intensity of the stressor is an important factor in this stress and substance use relationship.

Stress and Life Transitions

One of the two important areas of increasing stress for young adults is the transition from adolescence. This is an important stressor for young adults, and we know that some navigate it easier than others. Nonetheless, the transition to adulthood that follows high school age and experiences is one in which most individuals face new and more stressful experiences (Arnett 2004; Bachman et al. 2002; Schulenberg et al. 2004). The transition to adulthood heralds significant life changes, beginning with greater emotional and financial independence from family members and other adults. This life stage period generally includes the start of college or a job/career, movement toward more serious romantic relationships and for some, the beginning of a new family. Arnett (2004) suggests that the time of emerging adulthood, ages 18–25, is a time for explorations, instability and possibilities. Young adults are exploring possibilities in love, work, and careers. Consistent with the notion of exploration, instability may characterize this period. Romantic relationships are typically frequent, may be short-lived, and this may be stressful. Career goals may change drastically, either because of learning of new possibilities, or the failure to thrive in one's initial career choice. Arnett (2005) shows that this period is also characterized by relatively frequent geographical moves. Young adults move from the home of origin, into dormitories, apartments, and other houses. Cohabitation without marriage is relatively frequent in this period as well, which may sometimes lead to a more permanent relationship, such as marriage.

These are all critical developmental tasks, for which there is no clear pathway to successful adulthood. Many young adults "muddle through". Schulenberg et al. (2004) discuss the transition to adulthood as a possible contributor to changes in mental health and related phenomena. There is a lack of structure and no clear normative guidelines for making this transition. As noted, the value of this lack of structure and clear guidelines is that it allows for individuality in charting a course to adulthood. For some individuals, this works well. But for others, it may not. For example, many young adults choose to attend college, a normative pathway, that contributes to greater learning and the likelihood of a more successful and satisfying career. At the same time, however, the transition to college is a stressor that places students at increased risk for alcohol abuse, both from college attendance, and the stress they bring with them to college and experiences in college (Lindsay 2006; Pritchard et al. 2007). Thus, this normatively positive step increases the likelihood of substance use.

At the same time this possibly stressful life transition is occurring for young adults, factors found to be protective against problematic substance use are also changing. Two of these protective factors are family and religion. Both begin to decline in influence as adolescents' transition into young adulthood.

As the evidence suggests, as adolescents move into young adulthood, both family influences and religious participation and commitment begin to decline. This may be an important factor in the increased likelihood of substance use among young adults. However, it may be especially problematic for African American young adults, given their greater religiousness (Wallace et al. 2003). Religion is a factor that has been found to be protective against alcohol use in adolescent, college, and adult community samples (Galen and Rogers 2004; Wills et al. 2003; Wallace et al. 2003, 2007; Wechsler et al. 1995). Religiosity has been assessed as dimensions of individual commitment and practice (Weaver et al. 2006; Wills et al. 2003), as well as a form of coping (Carver et al. 1989). As a coping mechanism, religiosity provides an explanatory framework buffering the impact of stressful life events and lowering rates of substance use over time (Wills et al. 2003). There is also evidence that religiosity provides a resource that buffers the influence of negative affect on substance abuse. Aspects of religious involvement may also reduce the likelihood of substance use-related coping. Attending religious services can also provide a natural network from which to derive social support, known to be protective against substance use. Individuals who belong to particular religious faiths or denominations may also adhere to specific prohibitions related to substance use (Galen and Rogers 2004), such as the prohibition against alcohol use in Islam.

Religion is also related to culture and ethnic identification, and in turn, differences in substance use among African Americans and Whites. On average, levels of religiosity are higher among African Americans than Whites (Brook et al. 1998; Brody et al. 1996; Chatters 2000; Wallace et al. 2003). The high endorsement of religious dimensions among African Americans may account for their reduced substance use during adolescence. In African American families, parental religiosity is protective for the family unit and is related to fewer conflicts and risky behaviors among the children (Brody et al. 1996). However, there is evidence that religious dimensions are differentially protective against substance use at various stages in development. Religiosity of parents and significant adults may be more influential on substance-related behaviors during childhood and adolescence (Smith 2003). But religious participation frequently decreases in late adolescence and early adulthood (Hunsberger 2002). It is during this period that young African American adults are at risk for increased substance use, which may indicate that they have lost some of the protection provided by religious participation. Therefore, while religion may exert its protective influence for African American adolescents, the decline of religious behavior in young adulthood may increasingly expose African American young adults to greater substance using behaviors.

Family influence may similarly decline as adolescents move into young adulthood. Because families provide a socializing agent for adolescents, parents and other significant adults may provide early protective influences with respect to both racial socialization and substance use behaviors (Hughes 2003; Hops et al. 1996). For example, the majority of White youth, as opposed to African American youth, report that their first drink was with parents (Strycker et al. 2003). This fact suggests greater parental supervision and control with regard to substance using behavior. As significant adults are protective against substance use (Caldwell et al. 2004; Brook et al. 1998; Maton and Zimmerman 1992; Schinke et al. 2006), the decline of this influence for all young adults leads to greater substance use. In the context of increased stressors as African American youth mature into young adults, the decline of parental influence, a normal process as youth become young adults, may also contribute to greater substance using behavior among specifically African American young adults, given the increased level of environmental stressors.

To understand this pattern, we argue that much more attention must be paid to the role of race and related phenomena. The life conditions of African American and white young adults may change greatly as they age out of adolescence, and this likely affects changes in the rates of substance use.

Racial Disadvantage in the Lives of African-Americans

Many scholars suggest that we cannot realistically expect to explain racial disparities in any health-related outcome without considering the effects of racism on health (Brown 2003; Krieger 2003; LaViest 2003; Williams and Collins 2001), and the critical impact of racial disadvantage that operates in the U.S. In a society such as the U.S., this is of great importance. This is because the context of daily life in the U.S. is a racialized one. The U.S. social system is one that is racialized. By this is meant that in the U.S., individuals are placed into racial categories that predict their placement within ideological, political, social, and economic systems (Bonilla-Silva 1997; Forman 2003; Wallace 1999). As a result, stereotypes and prejudices are enmeshed in the fabric of U.S. culture and social systems (see Broman et al. 2000, for a discussion; Feagin 2000). In such a society, discrimination operates at both the system and individual level. Racial discrimination is systemic (Feagin 2000). Since U.S. society was developed along racial lines, with a system of white supremacy (Fredrickson 1982), racism is structured into the foundation of American society. Most U.S. institutions were developed under this system of white supremacy, and systematically disenfranchised and degraded African Americans (and other racial and ethnic minorities) (Bonilla-Silva 1997; Feagin 2000; Feagin and Vera 1995). For example, Loewen (2005) extensively documents many systemic disadvantages faced by African Americans during the nadir of racism in the United States, a period following reconstruction, and continuing into the 1940s. During this period, there was systematic disenfranchisement of African Americans, many instances of African American property destruction, the formation to African American ghettos, and the consolidation of racial segregation. Many from the burgeoning African American middle class saw their homes and properties destroyed, and were forced to flee for their lives from various Northern and Southern towns into Northern urban ghettos. Of course, as African Americans moved north from the South during the World War II era, they were segregated into ghettos. This pattern of racial disadvantage persists today (Bonilla-Silva 1997; Broman et al. 2000; Feagin 2000; Feagin and Vera 1995). Thus, racial and ethnic minorities are disadvantaged by a system of institutionalized inequality that affects life chances and opportunities for success (Bonilla-Silva 1997; Feagin 2000).

The systemic nature of racial disadvantage that plays a role in the lives of African American adolescents and young adults can be seen in the neighborhoods and schools attended by a large proportion of these people. Systemic discrimination means that an individual is disadvantaged simply due to the fact that she or he grows up in a poor neighborhood and attends a poor school. Because this disadvantage is structured into the American social system, most African American and Latino children are born into disadvantage, which may cumulate over time. This is a burden to bear in addition to the burden of individual level discrimination, and the general stress of making the transition to adulthood, that all children face.

The Neighborhood Context

Neighborhoods are a key framework for considering how discrimination operates in the lives of poor and disadvantaged communities, because of the continuing and persistent residential segregation in the United States (Fuller et al. 2005; LaViest 2003; LaViest and Wallace 2000; Massey and Denton 1993; Williams and Collins 2001). In fact, levels of racial segregation are as great today as in the 1960s (Aguirre and Turner 2007). The effects operate both at the individual and the institutional level. Systematic racial segregation means that minorities live in the poorest neighborhoods, which are severely disadvantaged.

Racially segregated communities mean that racial and ethnic minorities are isolated from the more positive amenities of U.S. society, such as good schools, housing and jobs. Racial minorities (African Americans and Latinos) are usually the most affected by poor neighborhood quality (LaViest 2003; Williams and Collins 2001).

Segregation by race influences a variety of health outcomes (Diez-Roux et al. 2001; Fuller et al. 2005; Williams and Collins 2001). Research suggests that the negative effects of racial residential segregation on health are related to the differential characteristics of the neighborhoods where racial groups reside. For example, found that living in neighborhoods characterized by a poor quality built environment is associated with a greater likelihood of depression.

Why is residential segregation a negative influence on health? Recent conceptual work (Boardman 2004; Browning and Cagney 2003; Ellen et al. 2001; Robert 1999) has increased our understanding of the impact of neighborhood factors on various health outcomes. Four sets of neighborhood factors have been identified as critical in health outcomes. These are (1) neighborhood institutions and resources; (2) Physical stresses; (3) Social networks; and (4) undermining individual-level resources. Neighborhoods differ in access to various resources, including health resources. Extreme neighborhood isolation is associated with poor health outcomes because the neighborhood conditions that encourage such outcomes are present.

One key aspect of racial residential segregation is that poor African Americans are isolated from resources that may exist in the broader community (Subramanian et al. 2005). Poorer communities are less likely to have access to health care resources, social services, and transportation (Pickett and Pearl 2001; Polednak 1993). In addition, disadvantaged neighborhoods often lack green space and have lower access to nutritious fresh fruits and vegetables. There are also greater physical stresses in disadvantaged neighborhoods. Racially segregated and poor neighborhoods are often located in areas with poorer quality housing, which may contain lead-based paints and other environmental hazards. One example is the exposure to pesticides used to control pests in older and deteriorating homes. Other physical stresses concern environmental pollution. Hazardous industries are often located in poor neighborhoods which increase the risk of exposure to pollutants, such as volatile organic compounds in the water (Goldberg et al. 1995), and particulate matter in the air (Parker et al. 2005).

Racially segregated and poor neighborhoods may also be more likely to experience decrements from social networks. Residents of poorer neighborhoods may lack the social support that residents of more resource-rich neighborhoods have. Family and friends may be more exposed to stressors themselves, and this may lessen the socially supportive network available to people living in poorer communities (Boardman et al. 2001).

Living in racially segregated and poor neighborhoods also undermines individual-level resources. Residents of such neighborhoods may suffer from more psychological distress (Boardman et al. 2001; Cutrona et al. 2000; Ellen et al. 2001; Schulz et al. 2000) because of stressors in the environment, poverty, neighborhood hazards and lower social support. Therefore, the average resident of a poorly resourced neighborhood is not only more likely than those from richer neighborhoods to suffer personal decrements, but they are also more likely to be confronted with a network of suffering people. This may be a 'double-stressor'; that is, suffering more from watching friends and family suffer. This may be one reason for the finding of lower levels of collective efficacy in racially segregated and poor neighborhoods (Boardman and Robert 2000; Browning and Cagney 2003). When too many people in the immediate environment are under duress, one may feel unable to count on people in an emergency situation.

Therefore, youth from poorly resourced and distressed neighborhoods are more likely to be confronted with a greater number of stressors, themselves, and may be more likely to confront people under duress, and decrements in social support as they reach adulthood.

In addition, poor neighborhoods have a greater number of drug users and drugs (LaViest and Wallace 2000). Availability of illegal drugs is much greater in poor neighborhoods (Cataeno and Clark 1998; Crum et al. 1996; O'Malley and Johnston 2002). In the face of growing stressors in young adulthood, this greater exposure to drugs is hazardous for young African American adults.

One associated factor with living in poor neighborhoods is poor schools. African American and Latino children are significantly more likely to attend schools which are sub-standard (Aguirre and Turner 2007; Feagin 2000; Peske and Haycock 2006). This refers to schools that are lacking in basic amenities, as well as books, supplies and even more experienced teachers.

This results in schools that simply 'warehouse' children, and very little learning takes place. These schools are prone to violence. Some students carry knives, guns and other weapons. For example, Waters (1999) discusses this in detail in her study of African American immigrants. The schools attended in the African American neighborhoods of New York were described as shocking in the level of drugs, weapons and chaos. Teachers described their fear of being assaulted. Students would push teachers down stairs, threaten them, and throw things at them. Some of the teachers noted their fear of students with weapons. Students were similarly afraid. Some would carry weapons because of their fear of being assaulted. All of the schools have metal detectors, but students described how people sneak into back windows to bypass those metal detectors. This results in weapons possession inside of the school (Waters 1999). In addition, students would commonly walk to school through streets where drugs are openly sold. It is not an atmosphere conducive to learning.

Poor schools and neighborhoods are examples of how systemic racism is harmful to African American and other minority children. When schools are in chaos, the children who attend those schools are hard pressed to learn. Teachers are more concerned with safety and discipline than with teaching. Students may be reluctant to attend classes because of safety concerns. When poor resources of the schools are added to the mix, the schools become a daunting place where learning is a secondary concern.

We have briefly discussed some of macro-level aspects of racial discrimination that likely affect African American youth both in adolescence and young adulthood. Now we turn to a discussion of micro-level discrimination, or the personal experiences of race and ethnic based discrimination.

The Experience of Racial discrimination

Racial discrimination at the level of personal experience is commonly conceptualized as a pervasive traumatic stressor with harmful individual, institutional, and cultural manifestations (Jones 1997; Sanchez-Hucles 1998). Individual-level racial discrimination refers to a variety of negative comments or behaviors that result in an individual being treated less favorably on the basis of his/her race, color, ethnicity, or national origin; these behaviors reach legal status as discrimination when denial of opportunities is based on these characteristics (e.g., housing, employment) (CRA of 1964, Title VI, as amended 42 U.S.C. § 2000d; CRA of 1964, Title VII, as amended 42 U.S.C. § 2000e; Schneider et al. 2000).

Experiencing racial discrimination is a common occurrence for African Americans; anywhere from 60 to 90 % of African Americans report experiencing at least one instance of discrimination a year (Broman et al. 2000; Feagin 1991; Kessler et al. 1999; Pachter et al. 2010). African Americans are significantly more likely than whites to be the victims of racially discriminatory treatment (Broman et al. 2000; Feagin 1991; Kessler et al. 1999; Seaton et al. 2008). This effect is also found for African American youth (Berkel et al. 2009; Pachter et al. 2010; Seaton et al. 2008) and young adults (Taylor and Turner 2002). Significantly, parental protection from discrimination declines as adolescents move into young adulthood (Berkel et al. 2009). This means that the protective effect of parents, along with that of religious behavior, as was noted earlier, are in decline as African American adolescents become African American young adults.

Many have begun to investigate explanatory and causal models for how racism and discrimination contribute to observed health disparities (Broman 2007; Krieger 2003; Lambert et al. 2009; LaViest 2003; Williams and Collins 2001). A large body of research supports the idea that racial discrimination is associated with a host of negative physical and psychological sequelae (Broudy et al. 2007; Hill et al. 2004; Satcher 2001; Seaton and Yip 2009; Wagner and Abbott 2007) including increased stress (Barnes and Lightsey 2005; Utsey et al. 2000) and increased licit and illicit substance use (Broman 2007; Choi et al. 2006). Martin and colleagues (2003) found support for both indirect and direct effects of racial discrimination on substance use. For example, discrimination was associated with escapist thinking, which in turn was associated with increased substance use. Further, research suggests that anger regarding racial discrimination accounts for over 30 % of alcohol use among African American adolescents (Terrell et al. 2006). Studies investigating the impact of experiencing racial discrimination on substance use are continuing (this volume), and the evidence is consistent with prior research.

The personal experience of discrimination in schools also functions to harm achievement in school. Research shows that minority children perceive that their teachers prefer Asian American children to other minority children (Rosenbloom and Way 2004). This may be good for the academic achievement of Asian American children, but not other minority children. Other research shows that discrimination experiences of adolescents tend to decrease academic curiosity, persistence and grades in school (Neblett et al. 2006). Therefore to the extent that African American children are exposed to discrimination, achievement in school is harmed. Lower levels of academic control have also been associated with discriminatory experiences (Lambert et al. 2009). Experiencing discrimination also decreases school bonding among minority children (Dotterer et al. 2009). This may increase the likelihood that a child will drop out of school.

In addition to neighborhood decrements commonly experienced by African American (and some other minority children), poorer schools, and the academic difficulties experienced in school, create adolescents and young adults who are ill-equipped to compete for quality higher education and in the economic arena for good jobs. The consequence of this is increasing adversity for the young African American adult, as compared to the young white adult.

Increasing Socioeconomic Adversity for African Americans

One of the most important transitions made in young adulthood is moving from financial dependence on parents, to being financially self-sufficient. Many young adults report that financial independence is necessary for identification as an adult (Arnett 2004). Some research suggests that there may be important differences in this transition by race. First, African Americans are disproportionately vulnerable to unemployment (Broman et al. 2001), with rates twice that of whites and are more likely to experience prolonged unemployment, and greater economic hardship as a result of joblessness (Blank 2001; Williams and Collins 2001). Second, African Americans are more likely than whites to be underemployed; African Americans are overrepresented in unskilled occupations (Aguirre and Turner 2007), with low wages, poor working conditions and job instability (Blank 2001; Broman et al. 2001; Williams and Collins 2001). Lastly, African Americans are more vulnerable to job displacement due to recessions and economic changes (Blank 2001; Broman et al. 2001; Williams and Collins 2001; Wilson 1990). Therefore, the average African American young adult is at much greater risk of economic hardship and deprivation than is the average white adolescent (See Wallace and Muroff 2002).

As a result of the change from a manufacturing to a service economy, income wages over the past three decades have declined for African American Americans, particularly for African American men (Blank 2001; Williams and Collins 2001). These issues are exacerbated by residence in urban areas, where the majority of the African American population lives. Jobs are predominantly located in suburban areas, usually far from where the bulk of young African Americans live. As the economy has shifted from being manufacturing-based to service-based, it is important to note that a greater number of positions require a college degree, while those that do not, such as department store clerks (e.g. Wal-Mart, Target) are frequently located in suburban areas and provide substantially lower wages and benefits. Lastly, because downsizing and rehiring after layoffs are often decided by seniority, younger workers in these positions may have even greater job instability than older co-workers.

One consequence of the greater probability of under- and unemployment among African Americans has been a rise in the number of African American families experiencing economic hardship or poverty. African Americans are about three times more likely to be in poverty than are whites (Aguirre and Turner 2007; Blank 2001; Williams and Collins 2001), and this means that many African American children grow up poor. Thus, African American young adults (as well as African Americans of other ages) are more likely than their White counterparts to experience economic hardship. Research has shown that African American adults suffer poorer mental health as a consequence of economic hard times (Blank 2001; Broman et al. 2001; Williams and Collins 2001). This may be another way in which the African American young adult grows up in an environment where there is the 'double stressor' situation. Parents are stressed because of economic hardship, and adolescents may also feel the burden of this. As they grow into young adults, African American children who come from poor backgrounds face their own economic hardship, in addition to that of their parents, siblings and close friends. This seems to have implications for substance abuse.

Poverty has been linked to increased likelihood of alcohol and drug use problems for both men and women (Fothergill and Ensminger 2006; Jones-Webb et al. 1997b). Studies have found that socio-economic status is associated with substance problems among African American and white men (Galea et al. 2004; Jones-Webb et al. 1995). Together, these factors may further increase stress and substance use-related coping and disproportionately affect African American young adults. This may be one of the reasons that studies have found that neighborhood disadvantage is an important factor in substance use (Boardman et al. 2001; Crum et al. 1996; Fuller et al. 2005; Jones-Webb et al. 1995,1997a, b).

Therefore, in addition to the normative transition from adolescence to young adulthood, and the stress they may engender, African American individuals may also face additional subtle and overt race-related stressors (Wallace and Muroff 2002). This includes racial discrimination in various life domains (e.g., education, housing, work, finances) (Bonilla-Silva 1997; Brown 2003; Forman 2003; Krieger 2003; LaViest 2003; Wallace 1999; Williams and Collins 2001), which may exacerbate the difficulty of transitions in other domains (e.g., lower wages associated with workplace discrimination may make the transition to beginning a new family more stressful) (Blank 2001; Broman et al 2001). These new stressors may place individuals at increased risk for substance use (Bachman et al. 2002; Broman 2007; Cooper et al. 1992; Martin et al. 2003; Schuckit and Smith 2006; Schulenberg et al. 1996, 2004). Thus, this confluence of life circumstances and events may be of great importance in helping us to understand the trajectory of substance abuse over the 20s and 30s for African Americans and whites.

Conclusions

This chapter has raised the issue of discrimination at both the individual and system level as factors in the increased substance use of African Americans in early adulthood. Structural disadvantage faced by African Americans portends a much more negative future for the average African American young adult than is true for white young adults. This may be of great importance in the changing patterns of substance use as adolescence fades and adulthood comes.

This is a possibility that will need to be addressed by systematic research. There is great interest in the health impacts of discrimination, as witnessed by the many studies of the issue, but our conceptualization of the problem must also deal with the structural level. This is because it is this structural discrimination that has proven to be much less amenable to individual self-striving and life plans. Changing discrimination at the system level demands a different kind of effort than the eradication of individual level discrimination. Only this kind of effort may ultimately prove valuable in reducing health disparities in general, and disparities in substance use in particular.

References

Aguirre, A., Jr., & Turner, J. H. (2007). *American ethnicity*. New York: McGraw-Hill.

APSE Research Brief. (2009, July). *Vulnerable youth and the transition to adulthood*. U.S. Department of Health and Human Services, Office of the Assistant Secretary for Planning and Evaluation.

Arnett, J. J. (2004). *Emerging adulthood: The winding road from late teens through early twenties*. New York: Oxford University Press.

Arnett, J. J. (2005). The developmental context of substance use in emerging adulthood. *Journal of Drug Issues, 35*, 235–254.

Bachman, J. G., O'Malley, P. M., Schulenberg, J. E., Johnston, L. D., Bryant, A. L., & Merline, A. C. (2002). *The decline of substance use in young adulthood*. Mahwah: Lawrence Erlbaum.

Barnes, P. W., & Lightsey, O. R. (2005). Perceived racist discrimination, coping, stress, and life satisfaction. *Journal of Multicultural Counseling and Development, 33*, 48–61.

Barnes, G., Farrell, M., & Banerjee, S. (1994). Family influences on alcohol abuse and other problem behaviors among African American and White adolescents in a general population sample. *Journal of Research on Adolescence, 4*, 183–201.

Bensley, L. S., Van Eenwyk, J., & Simmons, K. W. (2000). Self-reported childhood sexual and physical abuse and adult HIV-risk behaviors and heavy drinking. *American Journal of Preventive Medicine, 18*, 151–158.

Berkel, C., Murry, V. M., Hurt, T. R., Chen, Y. F., Brody, G. H., Simons, R. L., Cutrona, C., & Gibbons, F. X. (2009). It takes a village: Protecting rural African American youth in the context of racism. *Journal of Youth & Adolescence, 38*, 175–188.

Blank, R. M. (2001). An overview of trends in social and economic well-being by race. In *America becoming*. Washington, DC: National Academies Press.

Boardman, J. (2004). Stress and physical health. *Social Science and Medicine, 58*, 2473–2483.

Boardman, J., & Robert, S. (2000). Neighborhood socioeconomic status and perceptions of self-efficacy. *Sociological Perspectives, 43*, 117–136.

Boardman, J., et al. (2001). Neighborhood disadvantage, stress, and drug use among adults. *Journal of Health and Social Behavior, 42*, 151–165.

Bonilla-Silva, E. (1997). Rethinking racism: Toward a structural interpretation. *American Sociological Review, 62*, 465–479.

Bradizza, C. M., Reifman, A., & Barnes, G. M. (1999). Social and coping reasons for drinking: Predicting alcohol misuse in adolescents. *Journal of Studies on Alcohol, 60*, 491–499.

Brody, G. H., Stoneman, Z., & Flor, D. (1996). Parental religiosity, family processes, and youth competence in rural two parent African American families. *Developmental Psychology, 32*, 696–706.

Broman, C. L. (2007). Perceived discrimination and alcohol use among African American and white college students. *Journal of Alcohol and Drug Education, 51*, 8–16.

Broman, C. L., Mavaddat, R., & Hsu, S. (2000). The experience and consequences of racial discrimination. *Journal of African American Psychology, 26*, 165–180.

Broman, C. L., et al. (2001). *Stress and distress among the unemployed*. New York: Kluwer Academic Plenum Publishers.

Brook, J. S., Balka, E. B., Brook, D. W., Win, P. T., & Gursen, M. D. (1998). Drug use among African Americans: Ethnic identity as a protective factor. *Psychological Reports, 83*, 1427–1446.

Broudy, R., Brondolo, E., Coakley, V., Brady, N., Cassells, A., Tobin, J. N., & Sweeney, M. (2007). Perceived ethnic discrimination in relation to daily moods and negative social interactions. *Journal of Behavioral Medicine, 30*, 31–43.

Brown, T. N. (2003). Critical race theory speaks to the sociology of mental health: Mental health problems produced by racial stratification. *Journal of Health and Social Behavior, 44*, 292–301.

Browning, C. R., & Cagney, K. A. (2003). Moving beyond poverty. *Journal of Health and Social Behavior, 44*, 552–571.

Caetano, R., & Clark, C. L. (1998). Trends in alcohol consumption patterns among Whites, African Americans and Hispanics: 1984 and 1995. *Journal of Studies on Alcohol, 59*, 659–668.

Caldwell, C. H., Sellers, R. M., Bernat, D. H., & Zimmerman, M. A. (2004). Racial identity, parental support, and alcohol use in a sample of academically at-risk African American high school students. *American Journal of Community Psychology, 34*, 71–82.

Carpenter, K., & Hasin, D. (1999). Drinking to cope with negative affect and DSM-IV alcohol use disorders. *Journal of Studies on Alcohol, 60*, 694–704.

Carver, C. S., Scheier, M. F., & Weintraub, J. K. (1989). Assessing coping strategies: A theoretically based approach. *Journal of Personality and Social Psychology, 56*, 267–283.

Chatters, L. M. (2000). Religion and health: Public health research and practice. *Annual Review of Public Health, 21*, 335–367.

Choi, Y., Harachi, T. W., Gillmore, M. R., & Catalano, R. F. (2006). Are multiracial adolescents at greater risk? *American Journal of Orthopsychiatry, 76*, 86–97.

Colder, C. R., & Chassin, L. (1993). The stress and negative affect model of adolescent alcohol use and the moderating effects of behavioral undercontrol. *Journal of Studies on Alcohol, 54*, 326–333.

Cooper, M. L., Russell, M., Skinner, J. B., Frone, M. R., & Mudar, P. (1992). Stress and alcohol use: Moderating effects of gender, coping and alcohol expectancies. *Journal of Abnormal Psychology, 101*, 139–152.

Cooper, M. L., Frone, M. R., Russell, M., & Mudar, P. (1995). Drinking to regulate positive and negative emotions: A motivational model of alcohol use. *Journal of Personality and Social Psychology, 69*(5), 990–1005.

Cox, W. M., & Klinger, E. (1988). A motivational model of alcohol use. *Journal of Abnormal Psychology, 97*, 168–180.

Cox, W. M., & Klinger, E. (1990). Incentive motivation, affective change, and alcohol use: A model. In W. M. Cox (Ed.), *Why people drink: Parameters of alcohol as reinforcer* (pp. 291–314). New York: Gardner Press, Inc.

Crum, R. M., Lillie-Blanton, M., & Anthony, J. C. (1996). Neighborhood environment and opportunity to use cocaine and other drugs in late childhood and early adolescence. *Drug and Alcohol Dependence, 43*, 155–161.

Cutrona, C. E., Russell, D. W., Hessling, R. M., Brown, P. A., & Murry, V. (2000). Direct and moderating effects of community context on the psychological well-being of African American women. *Journal of Personality and Social Psychology, 79*, 1088–1101.

Diez-Roux, A. V., et al. (2001). Neighborhood of residence and incidence of coronary heart disease. *New England Journal of Medicine, 345*, 99–106.

Dotterer, A. M., McHale, S., & Crouter, A. C. (2009). Sociocultural factors and school engagement among African American youth. *Applied Developmental Science, 13*, 61–73.

Ellen, I. G., et al. (2001). Neighborhood effects on health. *Journal of Urban Affairs, 23*, 391–408.

Farley, R., & Allen, W. R. (1987). *The color line and the quality of life in America*. New York: Russell Sage.

Feagin, J. R. (1991). The continuing significance of race: Anti-African American discrimination in public places. *American Sociological Review, 56*, 101–116.

Feagin, J. R. (2000). *Racist America*. New York: Routledge.

Feagin, J. R., & Vera, H. (1995). *White racism*. New York: Routledge.

Forman, T. A. (2003). The social psychological costs of racial segmentation in the workplace: A study of African Americans' well-being. *Journal of Health and Social Behavior, 44*, 332–352.

Fothergill, K. E., & Ensminger, M. E. (2006). Childhood and adolescent antecedents of drug and alcohol problems: A longitudinal study. *Drug and Alcohol Dependence, 82*, 61–76.

Fredrickson, G. M. (1982). *White supremacy*. New York: Oxford University Press.

Fuller, C. M., et al. (2005). Effects of race, neighborhood, and social network on age at initiation of injection drug use. *American Journal of Public Health, 95*, 689–695.

Galaif, E. R., Stein, J. A., Newcomb, M. D., & Bernstein, H. C. G. (2001). Gender differences in the prediction of problem alcohol use in adulthood: Exploring the influence of family factors and childhood maltreatment. *Journal of Studies on Alcohol, 62*, 332–340.

Galea, S., & Rudenstine, S. (2005). Challenges in understanding disparities in drug use and consequences. *Journal of Urban Health, 82*, iii5–iii12.

Galea, S., Nandi, A., & Vlahov, D. (2004). The social epidemiology of substance use. *Epidemiologic Reviews, 26*, 36–52.
Galen, L. W., & Rogers, W. M. (2004). Religiosity, alcohol expectancies, drinking motives and their interaction in the prediction of drinking among college students. *Journal of Studies on Alcohol and Drugs, 65*, 469–476.
Galvan, F. H., & Caetano, R. (2003). Alcohol use and related problems among ethnic minorities in the United States. *Alcohol Research & Health, 27*, 87–94.
Goldberg, M., Al-Homsi, N., Goulet, L., & Riberdy, H. (1995). Incidence of cancer among persons living near a municipal solid waste landfill in Montreal, Ouebec. *Archives of Environmental Health, 5*, 416–424.
Ham, L. S., Bonin, M., & Hope, D. A. (2007). The role of drinking motives in social anxiety and alcohol use. *Journal of Anxiety Disorders, 21*, 991–1003.
Herd, D. (1989). The epidemiology of drinking patterns and related problems among U.S. African Americans. In NIAAA research monograph 18, *Alcohol use among U.S. ethnic minorities*. Washington, DC: US Government Printing Office.
Herd, D. (1990). Subgroup differences in drinking patterns among African American and white men: Results from a National Survey. *Journal of Studies on Alcohol, 51*, 221–232.
Herd, D. (1997). Sex ratios of drinking patterns and problems among African Americans and whites: Results from a national survey. *Journal of Studies on Alcohol and Drugs, 58*, 75–83.
Hill, C. V., Neighbors, H., & Gayle, H. (2004). The relationship between racial discrimination and health for African American Americans: Measurement challenges and the realities of coping. *African American Research Perspectives, 10*, 89–98.
Hops, H., Duncan, T. E., Duncan, S. C., & Stoolmiller, M. (1996). Parent substance use as a predictor of adolescent use: A six-year lagged analysis. *Annals of Behavioral Medicine, 18*, 157–164.
Horton, E. G. (2007). Racial differences in the effects of age of onset on alcohol consumption and development of alcohol-related problems among males from mid-adolescence to young adulthood. *Journal of Ethnicity in Substance Abuse, 6*, 1–14.
Hughes, D. (2003). Correlates of African American and Latino parents' messages to children about ethnicity and race: A comparative study of racial socialization. *American Journal of Community Psychology, 31*, 15–33.
Hunsberger, B. (2002). A longitudinal study of religious doubts in high school and beyond. *Journal for the Scientific Study of Religion, 41*, 255–266.
Johnston, L. D., O'Malley, P. M., & Bachman, J. G. (2000). *The monitoring the future national results on adolescent drug use*. Bethesda: National Institute on Drug Abuse.
Jones, J. M. (1997). *Prejudice and racism* (2nd ed.). New York: McGraw-Hill.
Jones-Webb, R., Hsiao, C. Y., & Hannan, P. (1995). Relationships between socioeconomic status and drinking problems among African American and white men. *Alcoholism: Clinical and Experimental Research, 19*(3), 623–627.
Jones-Webb, R., Hsiao, C. Y., Hannan, P., & Caetano, R. (1997a). Predictors of increases in alcohol-related problems among African American and white adults: Results from the 1984 and 1992 National Alcohol Surveys. *American Journal of Drug and Alcohol Abuse, 23*, 281–299.
Jones-Webb, R., Snowden, L., Herd, D., Short, B., & Hannan, P. (1997b). Alcohol-related problems among African American, Hispanic and White men: The contribution of neighborhood poverty. *Journal of Studies on Alcohol, 58*, 539–545.
Kandel, D. B. (1995). Ethnic differences in drug use: Patterns and paradoxes. In G. Botvin, S. Schinke, & M. Orlandi (Eds.), *Drug abuse prevention with multiethnic youth*. Thousand Oaks: Sage.
Keefe, K., & Newcomb, M. D. (1996). Demographic and psychosocial risk for alcohol use: Ethnic differences. *Journal of Studies on Alcohol, 57*, 521–530.

Kessler, R. C., Mickelson, K. D., & Williams, D. R. (1999). The prevalence, distribution, and mental health correlates of perceived discrimination in the United States. *Journal of Health and Social Behavior, 40*, 208–230.
Kilpatrick, D. G., Aciemo, R., Resnick, H. S., Saunders, B. E., & Best, C. L. (1997). A 2-year longitudinal analysis of the relationships between violent assault and substance use in women. *Journal of Consulting and Clinical Psychology, 65*, 834–847.
Krieger, N. (2003). Does racism harm health? Did child abuse exist before 1962?-on explicit questions, critical science, and current controversies: An ecosocial perspective. *American Journal of Public Health, 93*, 194–199.
Lambert, S. F., et al. (2009). Perceptions of racism and depressive symptoms in African American adolescents. *Journal of Youth and Adolescence, 38*, 519–532.
Langeland, W., & Hartgers, C. (1998). Child sexual and physical abuse and alcoholism: A review. *Journal of Studies on Alcohol, 59*, 336–348.
LaViest, T. A. (2003). Racial segregation and longevity among African-Americans. *Health Services Research, 38*, 1719–1733.
LaViest, T. A., & Wallace, J. M., Jr. (2000). Health risks and the inequitable distribution of liquor stores in African-American neighborhoods. *Social Science & Medicine, 51*, 613–617.
Lemke, S., Brennan, P. L., Schutte, K. K., & Moos, R. H. (2007). Upward pressures on drinking: Exposure and reactivity in adulthood. *Journal of Studies on Alcohol and Drugs, 68*, 437–445.
Lewis, M. A., et al. (2008). Fitting in and feeling fine: Conformity and coping motives as mediators of the relationship between social anxiety and problematic drinking. *Psychology of Addictive Behaviors, 22*, 58–67.
Lindsay, V. (2006). Factors that predict freshmen college students' preference to drink alcohol. *Journal of Alcohol and Drug Education, 50*, 7–19.
Loewen, J. W. (2005). *Sundown towns: A hidden dimension of American racism*. New York: Touchstone.
Martin, J. K., Tuch, S. A., & Roman, P. M. (2003). Problem drinking patterns among African American: The impacts of reports of discrimination, perceptions of prejudice, and "risky" coping strategies. *Journal of Health and Social Behavior, 44*, 408–425.
Massey, D., & Denton, N. (1993). *American apartheid*. Cambridge, MA: Harvard University Press.
Maton, K. I., & Zimmerman, M. A. (1992). Psychosocial predictors of substance use among urban African American male adolescents. *Drugs and Society, 6*, 79–113.
McCreary, D. R., & Sadava, S. W. (2000). Stress, alcohol use and alcohol-related problems: The influence of negative and positive affect in two cohorts of young adults. *Journal of Studies on Alcohol, 61*, 466–474.
Merline, A. C., O'Malley, P. M., Schulenberg, J. E., Bachman, J. G., & Johnston, L. D. (2004). Substance use among adults 35 years of age: Prevalence, adulthood predictors, and impact of adolescent substance use. *American Journal of Public Health, 94*, 96–102.
Mudar, P., Kearns, J. N., & Leonard, K. E. (2002). The transition to marriage and changes in alcohol involvement among African American couples and white couples. *Journal of Studies on Alcohol, 63*, 568–576.
Muthen, B., & Muthen, L. K. (2000). The development of heavy drinking and alcohol-related problems from ages 18 to 37 in a U.S. National sample. *Journal of Studies on Alcohol, 61*, 290–300.
Neblett, E. W., Phillip, C. L., Cogburn, C. D., & Sellers, R. M. (2006). African American adolescent's discrimination experiences and academic achievement: Racial socialization as a cultural compensatory and protective factor. *Journal of African American Psychology, 32*, 199–218.
Nielsen, A. L. (1999). Testing Sampson and Laub's life course theory: Age, race/ethnicity and drunkenness. *Deviant Behavior, 20*, 129–151.
O'Malley, P. M., & Johnston, L. D. (2002). Epidemiology of alcohol and other drug use among American college students. *Journal of Studies on Alcohol, 14*, 23–39.

Pachter, L. M., et al. (2010). Perceived racism and discrimination in children and youths. *Health & Social Work, 35*, 61–70.
Parker, J. D., Woodruff, T. J., Basu, R., & Schoendorf, K. C. (2005). Air pollution and birth weight among term infants in California. *Pediatrics, 115*, 121–128.
Perkins, H. W. (1999). Stress-motivated drinking in collegiate and postcollegiate young adulthood: Life course and gender patterns. *Journal of Studies on Alcohol, 60*, 219–227.
Peske, H., & Haycock, K. (2006). *Teaching inequality: How poor and minority students are short-changed on teacher quality*. Washington, DC: The Education Trust.
Pickett, K. E., & Pearl, M. (2001). Multilevel analyses of neighborhood socioeconomic context and health outcomes: A critical review. *Journal of Epidemiology and Community Health, 55*, 111–122.
Polednak, A. P. (1993). Poverty, residential segregation, and African American/white mortality ratios in urban areas. *Journal of Health Care for the Poor and Underserved, 4*, 363–373.
Pritchard, M. E., Wilson, G. S., & Yamnitz, B. (2007). What predicts adjustment among college students? A longitudinal panel study. *Journal of American College Health, 56*, 15–21.
Read, J. P., Wood, M. D., Kahler, C. W., Maddock, J. E., & Palfai, T. P. (2003). Examining the role of drinking motives in college student alcohol use and problems. *Psychology of Addictive Behaviors, 17*, 13–23.
Robert, S. A. (1999). Socioeconomic position and health. *Annual Review of Sociology, 25*, 489–516.
Rosenbloom, S. R., & Way, N. (2004). Experiences of discrimination among African American, Asian American, and Latino adolescents in an urban high school. *Youth & Society, 35*, 420–451.
Rutledge, P. C., & Sher, K. J. (2001). Heavy drinking from the freshman year into early young adulthood: The roles of stress, tension-reduction drinking motives, gender and personality. *Journal of Studies on Alcohol, 62*, 457–466.
Sanchez-Hucles, J. (1998). Racism: Emotional abusiveness and psychological trauma for ethnic minorities. *Journal of Emotional Abuse, 1*, 69–87.
Sanders-Phillips, K. (1996a). Correlates of health promotion behaviors in low-income African American women and Latinas. *American Journal of Preventive Medicine, 12*, 450–458.
Sanders-Phillips, K. (1996b). The ecology of urban violence. *American Journal of Health Promotion, 10*, 308–317.
Satcher, D. (2001). *Mental health: Culture, race, and ethnicity.* A supplement to mental health: A report of the Surgeon General. Department of Health and Human Services. Rockville: Public Health Service.
Schinke, S., DiNoia, J., Schwinn, T., & Cole, K. (2006). Drug abuse risk and protective factors among African American, urban adolescent girls: A randomized trial of computer delivered, mother-daughter intervention. *Psychology of Addictive Behaviors, 20*, 496–500.
Schneider, K. T., Hitlan, R. T., & Radhakrishnan, R. (2000). An examination of the nature and correlates of ethnic harassment experiences in multiple contexts. *Journal of Applied Psychology, 85*, 3–12.
Schuckit, T. A., & Smith, T. L. (2006). An evaluation of the level of response to alcohol, externalizing symptoms, and depressive symptoms as predictors of alcoholism. *Journal of Studies on Alcohol and Drugs, 67*, 215–227.
Schulenberg, J. E., O'Malley, P. M., Bachman, J. G., Wadsworth, K. N., & Johnston, L. D. (1996). Getting drunk and growing up: Trajectories of frequent binge drinking during the transition to adulthood. *Journal of Studies on Alcohol and Drugs, 57*, 289–304.
Schulenberg, J. E., Sameroff, A. J., & Cicchetti, D. (2004). The transition to adulthood as a critical juncture in the course of psychopathology and mental health. *Development and Psychopathology, 16*, 799–806.
Schulz, A. J., et al. (2000). Unfair treatment, neighborhood effects, and mental health in the Detroit metropolitan area. *Journal of Health and Social Behavior, 41*, 14–32.

Seaton, E. K., & Yip, T. (2009). The name assigned to the document by the author. This field may also contain sub-titles, series names, and report numbers. School and neighborhood contexts, perceptions of racial discrimination, and psychological well-being among African American adolescents. *Journal of Youth and Adolescence, 38*, 153–163.

Seaton, E. K., Caldwell, C. H., Sellers, R. M., & Jackson, J. S. (2008). The prevalence of perceived discrimination among African American and Caribbean African American youth. *Developmental Psychology, 44*, 1288–1297.

Shih, R. A., et al. (2010). Racial/ethnic differences in adolescent substance use. *Journal of Studies on Alcohol and Drugs, 71*, 640–651.

Smith, C. (2003). Theorizing religious effects among American adolescents. *Journal of Scientific Study of Religion, 42*, 17–30.

Strycker, L. A., Duncan, S. C., & Pickering, M. A. (2003). The social context of alcohol initiation among African American and White youth. *Journal of Ethnicity and Substance Abuse, 2*, 35–42.

Subramanian, S. V., et al. (2005). Racial disparities in context: A multilevel analysis of neighborhood variations in poverty and excess mortality among African American populations in Massachusetts. *American Journal of Public Health, 95*, 60–265.

Substance Abuse and Mental Health Services Administration (SAMSHA). (2007). *Results from the 2006 National survey on drug use and health: National findings.* (Office of applied studies, NSDUH). Rockville.

Taylor, J., & Turner, R. J. (2002). Perceived discrimination, social stress, and depression in the transition to adulthood: Racial contrasts. *Social Psychological Quarterly, 65*, 213–225.

Terrell, F., Miller, A. R., Foster, K., & Watkins, C. E. (2006). Racial discrimination – Induced anger and alcohol use among African American adolescents. *Adolescence, 41*, 485–492.

Utsey, S. O., Ponterotto, J. G., Reynolds, A. L., & Cancelli, A. A. (2000). Racial discrimination, coping, life satisfaction, and self-esteem among African Americans. *Journal of Counseling and Development, 78*, 72–80.

Wagner, J., & Abbott, G. (2007). Depression and depression care in diabetes: Relationship to perceived discrimination in African Americans. *Diabetes Care, 30*, 364–366.

Wallace, J. (1999). Explaining race differences in adolescent and young adult drug use: The role of racialized social systems. *Drugs & Society, 14*, 21–36.

Wallace, J. M., & Muroff, J. R. (2002). Preventing substance abuse among African American children and youth: Race differences in risk factor exposure and vulnerability. *The Journal of Primary Prevention, 22*(3), 235–261.

Wallace, J. M., Bachman, J., O'Malley, P., & Johnston, L. (1995). Racial/ethnic differences in adolescent drug use. In G. Botvin, S. Schinke, & M. Orlandi (Eds.), *Drug abuse prevention with multiethnic youth.* Thousand Oaks: Sage.

Wallace, J. M., Forman, T. A., Guthrie, B. J., Bachman, J. G., O'Malley, P. M., & Johnston, L. D. (1999). The epidemiology of tobacco, alcohol, and other drug use among African American youth. *Journal of Studies on Alcohol, 60*, 800–809.

Wallace, J. M., Bachman, J. G., O'Malley, P. M., Johnston, L. D., Schulenberg, J. E., & Cooper, S. M. (2002). Tobacco, alcohol, and illicit drug use: Racial and ethnic differences among U.S. high school seniors, 1976–2000. *Public Health Reports, 117*, S67–S75.

Wallace, J. M., Brown, T. N., Bachman, J. G., & LaViest, T. A. (2003). The influence of race and religion on abstinence from alcohol, cigarettes and marijuana among adolescents. *Journal of Studies on Alcohol, 64*, 843–859.

Wallace, J. M., Yamaguchi, R., Bachman, J., O'Malley, P., et al. (2007). Religiosity and adolescent substance use. *Social Problems, 54*, 308–328.

Waters, M. C. (1999). *African American identities*. Cambridge, MA: Harvard University Press.

Weaver, A. J., Pargament, K. I., Flannelly, K. J., & Oppenheimer, J. E. (2006). Trends in the scientific study of religion, spirituality, and health. *Journal of Religion and Health, 45*, 208–214.

Wechsler, H., Davenport, A., Dowdall, G., Moeykens, B., & Castillo, S. (1995). Binge drinking in college. *Journal of the American Medical Association, 273*, 1903.

White, H. R., Labouvie, E. W., & Papadaratsakis, V. (2005). Changes in substance use during the transition to adulthood: A comparison of college students and their noncollege age peers. *Journal of Drug Issues, 35*(2), 281–306.
Widom, C. S., Ireland, T., & Glynn, P. J. (1995). Alcohol abuse in abused and neglected children followed up: Are they at increased risk? *Journal of Studies on Alcohol, 56*, 207–217.
Williams, D. R., & Collins, C. (2001). Racial residential segregation: A fundamental cause of racial disparities in health. *Public Health Reports, 116*, 404–416.
Wills, T. A., Yaeger, A. M., & Sandy, J. M. (2003). Buffering effect of religiosity for adolescent substance use. *Psychology of Addictive Behaviors, 17*, 24–31.
Wilson, W. J. (1990). *The truly disadvantaged.* Chicago: University of Chicago Press.

Chapter 7
Race/Ethnicity, Religiosity and Differences and Similarities in American Adolescents' Substance Use

John M. Wallace Jr., Patrick M. O'Malley Jr., Jerald G. Bachman, John E. Schulenberg, and Lloyd D. Johnston

Introduction

Despite recent declines in the use of some drugs, substance use remains widespread among American young people. In fact, by the time that they reach their senior year in high school, 72 % have used alcohol, 45 % have tried cigarettes and 47 % have used an illicit drug (Johnston et al. 2009). The use and abuse of drugs by American young people and adults remains a pressing problem that costs the nation more than half a million premature deaths each year and an estimated $467 billion in federal, state, and local government expenditures for lost productivity, healthcare, social welfare administration, criminal justice and other costs (Robert Wood Johnson Foundation 2001; National Center on Addiction and Substance Abuse 2009).

Past research suggests that although racial/ethnic differences in the prevalence of substance use are not large, especially among adults, African American, Hispanic and Native American populations experience disproportionately its adverse health, social and economic consequences (National Institute on Drug Abuse [NIDA] 2003). Regardless of race or ethnicity, however, most adult substance *abusers* initiated *use* during adolescence (Substance Abuse and Mental Health Services Administration 2009; NIDA 2003). In light of this fact, it is critical to identify those factors that might protect young people from initiating or abusing drugs (Department of Health and Human Services [DHS] 2001; Hawkins et al. 1992).

J.M. Wallace Jr. (✉)
University of Pittsburgh's School of Social Work, Pittsburgh, PA, USA
e-mail: johnw@pitt.edu

P.M. O'Malley Jr. • J.G. Bachman • L.D. Johnston
University of Michigan's Institute for Social Research, Ann Arbor, MI, USA

J.E. Schulenberg
Department of Psychology, Center for Human Growth and Development, University of Michigan's Institute for Social Research, Ann Arbor, MI, USA

Y.F. Thomas, L.N. Price (eds.), *Drug Use Trajectories Among Minority Youth*,
DOI 10.1007/978-94-017-7491-8_7

One potentially important protective factor that past research has identified is religion. Prior to the 1990s, religion did not receive a significant amount of attention as a potential protective factor against adolescent substance use (see Gorsuch 1988, 1995). Since that time, however, numerous studies have explored the issue (see Hill et al. 2009 for a review). In fact, the literature on the relationship between religion and substance use has grown so much that there are now a number of reviews of the primary research (see Cotton et al. 2006; Dew et al. 2008; Hill et al. 2009; National Center on Addiction and Substance Abuse (2001); Rew and Wong 2006; Wong et al. 2006). Summarizing their findings, one of these recent reviews concluded that, "all 61 articles addressing substance use found at least one correlation linking religious variables with less usage—thus, there is a preponderance of evidence linking greater religious involvement with lower levels of substance use among adolescents" (Dew et al. 2008:388).

Despite the conclusion that religion relates inversely to adolescent substance use, a number of important questions remain to be answered. One of the most important questions concerns the extent to which this broad conclusion generalizes to the millions of non-white young people who comprise America's increasingly racially and ethnically diverse population. Given that the vast majority of the extant research is based upon relatively small, often non-representative, racially homogenous (i.e., white) samples, the answer to this question remains largely unknown. Even past studies that have attempted to address this question are limited in that they have only compared African American and white youth (Wallace et al. 2003) or in one instance, African American, white and Hispanic youth (Wallace et al. 2007). According to another recent review of the literature on the relationship between religiosity and adolescent health behaviors like substance use, "no study investigated differences in R/S [religion/spirituality] between Asian Americans or Native Americans and other racial groups" (Rew and Wong 2006:388). Similarly, a review of studies on religion and health concludes that, "…research has neglected specific subpopulations, such as Hispanics, Asian Americans, Native Americans and groups of low socioeconomic status" (Williams and Sternthal 2007:49).

In an effort to begin to address this gap in the literature, the present study uses large, nationally representative samples of white, African American, Mexican American, Puerto Rican, Other Latin American, Asian American and Native American young people to examine empirically the relationship between religiosity and adolescent alcohol, cigarette and marijuana use. This paper replicates and extends our earlier analyses (i.e., Wallace et al. 2007) that examined the relationship between religiosity and substance use among white, African American and Hispanic youth. Consistent with the earlier study, two questions motivate the present investigation: First, "how religious are American youth?" and second, "to what extent does religiosity 'protect' African American, Hispanic (including Mexican American, Puerto Rican, Other Latin American), Asian American and Native American young people from drug use, as past research suggests that it protects white youth?"

Methods

Sample

The data for this investigation were drawn from the University of Michigan's Monitoring the Future study. The study design and methods are summarized briefly below; a detailed description is available elsewhere (see Johnston et al. 2009). Monitoring the Future uses a multi-stage sampling procedure to obtain nationally representative samples of 8th, 10th, and 12th graders from the 48 contiguous states.[1] Stage one is the selection of geographic region; stage two is the selection of specific schools—approximately 420 each year; and stage three is the selection of students within each school. This sampling strategy has been used to collect data annually from high school seniors since 1975 and from 8th and 10th graders since 1991. Sample weights were assigned to each student to take into account differential probabilities of selection. Students completed self-administered, machine-readable questionnaires during a normal class period, with a response rate of 88 %. Absence on the day of data collection was the primary reason that students were missed; it is estimated that less than one percent of students refused to complete the questionnaire. To simplify presentation, data are presented for only 10th graders. The findings for 8th and 12th graders are generally consistent with those for 10th graders.

In order to ensure sufficient numbers of cases to perform race-specific analyses, we combined 5 years of data (2005–2009). This strategy resulted in samples of 47,183 white, 9343 African-American, and 4529 Mexican American, 936 Puerto Rican, 3145 Other Latin American, 3197 Asian American, and 658 Native American 10th graders.

Dependent Variables

The substance-use measures ask about students' 30-day (any use in the last 30 days) alcohol and tobacco use and 12-month marijuana use. All three of the measures are dichotomized, with the response categories being 0 = no use and 1 = any use. The alcohol measure asks, "On how many occasions (if any) have you used alcohol during the last 30 days?" The cigarette measure asks, "How often have you smoked cigarettes in the past 30 days?" The marijuana measure asks, "On how many occasions (if any) have you used marijuana during the last 12 months?"

[1] Data are not included for the state of California for the years in question because of a state regulation requiring advance written parental consent for questions about religion. For this reason, we were unable to include the religiosity questions in the California administrations.

Independent Variables

Religion is a multidimensional construct that has attitudinal, behavioral, and organizational dimensions. In order to operationalize these three dimensions, we use measures of religious importance, attendance, and denominational affiliation, respectively. The specific wording of the religion measures and their associated response categories are as follows: Religious importance was measured by the question: How important is religion in your life? Possible responses ranged from "not important" (1) to "very important" (4). Religious attendance was measured by the question: How often do you attend religious services? Possible responses ranged from "never" (1) to "about once a week or more" (4). Denominational affiliation was measured by the following question: What is your religious preference? Guided by the classification schemes of past research (see Roof and Mckinney 1987; Smith 1990), respondents' denominational affiliation was coded into a four-category religious conservatism measure ranging from "no affiliation" (1) to "conservative" (4). The classification scheme of the affiliation measure is as follows: 1 = none; 2 = liberal (i.e., Episcopal, Presbyterian, United Church of Christ, Unitarian, Roman Catholic, Jewish); 3 = moderate (i.e., Disciples of Christ, Lutheran, Methodist, Eastern Orthodox); 4 = conservative (i.e., Baptist, Churches of Christ, Other Protestant, Other Religion, Latter Day Saints, Muslim/Moslem, Buddhist). The correlations between the religion measures are moderate to strong ($r = .60$ between attendance and importance, $r = .32$ between attendance and religious conservatism, and $r = .41$ between importance and conservatism).

Analysis Strategy

The analyses presented below proceed in four stages: first we display data on the epidemiology of alcohol, cigarette, and marijuana use separately by racial/ethnic subgroup (Table 7.1). Second, to answer the question, "How religious are American youth?" we present data on the distribution of the religious importance, attendance, and denomination measures, also separately by racial/ethnicity (Tables 7.2a, b and c). Third, to test the extent to which the inverse relationship between religiosity and substance use that research has identified among white youth also exists among African American, Mexican American, Puerto Rican, Other Latin American, Asian and Native American youth, we investigate the bivariate relationship between drug use and religiosity, by race/ethnicity (Tables 7.3a, b and c). Finally, we examine the results of multivariate logistic regression analyses, also run separately by racial/ethnic group, to determine the extent to which the bivariate relationships identified in Tables 7.3a, b and c hold when key socio-demographic factors that have been found to relate to both drug use and religion (e.g., gender, socio-economic status, region) are statistically controlled (Tables 7.4a, b and c).

Given the relatively large numbers of cases that we use in these analyses, many of the findings may reach traditional levels of "statistical" significance (i.e., $p < .05$) and yet be of little substantive significance. Recognizing this possibility, we treat as

Table 7.1 Use of alcohol, cigarettes and marijuana among 10th graders by race (2005–2009 weighted data combined)

			95 % Confidence limits	
Race/ethnicity	*N*	Percent	Upper limit	Lower limit
Alcohol				
White	*47,183*	34.4	33.2	35.7
Black	*9343*	20.9	19.6	22.1
Mexican	*4529*	36.8	34.9	38.8
Puerto Rican	*936*	34.7	30.3	39.1
Other Latin	*3145*	31.2	28.5	33.9
Asian	*3197*	15.6	13.1	18.2
Native American	*658*	38.7	34.7	42.7
Cigarettes				
White	*47,966*	15.7	14.3	17.0
Black	*9930*	7.1	6.0	8.3
Mexican	*4776*	12.6	10.3	15.0
Puerto Rican	*980*	11.4	8.1	14.6
Other Latin	*3330*	9.7	8.1	11.3
Asian	*3260*	6.3	5.0	7.6
Native American	*686*	23.9	18.7	29.1
Marijuana				
White	*47,988*	25.8	24.2	27.3
Black	*9901*	22.8	20.9	24.6
Mexican	*4758*	29.3	27.3	31.4
Puerto Rican	*982*	28.0	23.7	32.4
Other Latin	*3319*	23.1	20.4	25.8
Asian	*3262*	11.6	8.7	14.5
Native American	*681*	33.0	26.3	39.8

significant only those differences that equal or exceed $P<.01$, and we limit our discussion to those differences that we judge to be both statistically and substantively important.

Results

Alcohol, Tobacco, and Marijuana Use

Table 7.1 presents data on the epidemiology of 30-day alcohol use, 30-day tobacco use, and 12-month marijuana use among American 10th graders, separately by racial/ethnic subgroup. The table shows the percent of youth in each subgroup who have used the specific substance and the associated 95 % confidence intervals (to facilitate comparisons across subgroups). Nationally, between 16 % and 39 % of

Table 7.2a Religious importance among 10th graders by race (2005–2009 weighted data combined)

				95 % Confidence limits	
Race/ethnicity	*N*	Responses	Percent	Upper limit	Lower limit
White	*44,666*	Not important	20.1	17.2	23.0
		A little important	27.1	25.3	28.9
		Pretty important	28.1	26.7	29.5
		Very important	24.7	21.1	28.3
Black	*9256*	Not important	7.8	6.1	9.4
		A little important	15.6	13.3	17.9
		Pretty important	29.6	28.1	31.0
		Very important	47.0	43.4	50.6
Mexican	*2297*	Not important	9.5	7.5	11.6
		A little important	26.7	22.8	30.6
		Pretty important	36.7	34.1	39.2
		Very important	27.1	20.6	33.6
Puerto Rican	*942*	Not important	16.2	12.7	19.7
		A little important	27.7	22.9	32.5
		Pretty important	30.0	25.0	35.1
		Very important	26.1	23.2	29.0
Other Latin	*2427*	Not important	12.2	9.8	14.5
		A little important	25.2	21.9	28.5
		Pretty important	33.7	31.2	36.1
		Very important	28.9	25.0	32.8
Asian	*1979*	Not important	18.0	15.8	20.1
		A little important	22.0	19.7	24.3
		Pretty important	30.7	28.3	33.1
		Very important	29.4	26.8	31.9
Native American	*628*	Not important	25.4	19.2	31.5
		A little important	24.3	18.3	30.3
		Pretty important	25.8	21.0	30.5
		Very important	24.6	19.5	29.6

10th graders are current users of alcohol, between 6 % and 24 % currently smoke cigarettes and between 12 % and 33 % percent have used marijuana in the past year. Consistent with the findings of past research, alcohol, cigarette and marijuana use are, on average, highest among Native Americans, at an intermediate level among White youth and youth in the Hispanic subgroups, lower still among African Americans and lowest among Asian American youth.

Religiosity—Importance, Attendance and Affiliation

Tables 7.2a, b and c display the univariate distributions of the religious importance, attendance, and denominational affiliation measures, separately by race/ethnicity. Within each of the racial/ethnic groups, the data suggest that the majority of 10th

Table 7.2b Religious attendance among 10th graders by race (2005–2009 weighted data combined)

Race/ethnicity	*N*	Responses	Percent	95 % Confidence limits Upper limit	Lower limit
White	*44,649*	Never	19.5	17.3	21.8
		Rarely	28.6	27.5	29.6
		Once or twice a month	15.8	15.0	16.5
		About once a week or more	36.2	32.9	39.4
Black	*9257*	Never	11.0	8.9	13.0
		Rarely	28.6	26.7	30.5
		Once or twice a month	18.2	17.2	19.3
		About once a week or more	42.2	39.6	44.9
Mexican	*2300*	Never	13.4	10.1	16.7
		Rarely	32.9	30.7	35.2
		Once or twice a month	17.9	16.3	19.6
		About once a week or more	35.7	30.8	40.6
Puerto Rican	*937*	Never	20.5	17.6	23.4
		Rarely	35.7	30.5	40.8
		Once or twice a month	13.8	10.4	17.2
		About once a week or more	30.1	27.2	32.9
Other Latin	*2427*	Never	16.2	13.9	18.4
		Rarely	34.2	31.2	37.2
		Once or twice a month	17.7	15.3	20.0
		About once a week or more	31.9	27.3	36.5
Asian	*1982*	Never	18.3	15.0	21.7
		Rarely	27.8	25.9	29.7
		Once or twice a month	19.4	16.3	22.4
		About once a week or more	34.5	31.2	37.8
Native American	*625*	Never	26.1	19.6	32.5
		Rarely	30.7	26.9	34.6
		Once or twice a month	15.1	11.7	18.4
		About once a week or more	28.1	23.2	33.1

graders are at least somewhat religious and that a quarter or more can be considered very religious. For example, three quarters of African American 10th graders, nearly two thirds of Mexican American, Other Latin American and Asian American 10th graders, and half or more of Puerto Rican, White and Native American 10th graders say religion is a "pretty" or "very" important part of their life. More than 40 % of

Table 7.2c Religious affiliation among 10th graders by race (2005–2009 weighted data combined)

				95 % Confidence limits	
Race/ethnicity	*N*	Responses	Percent	Upper limit	Lower limit
White	*43,926*	None	18.5	16.4	20.7
		Liberal	19.6	15.6	23.5
		Moderate	30.1	26.4	33.9
		Conservative	31.8	25.2	38.3
Black	*9007*	None	11.8	10.0	13.7
		Liberal	6.2	4.6	7.8
		Moderate	21.0	19.0	23.0
		Conservative	61.0	58.2	63.7
Mexican	*2217*	None	13.0	9.4	16.6
		Liberal	3.8	2.7	4.8
		Moderate	57.9	54.2	61.7
		Conservative	25.3	23.6	27.1
Puerto Rican	*919*	None	18.4	15.4	21.4
		Liberal	4.5	2.0	6.9
		Moderate	44.7	39.9	49.5
		Conservative	32.5	28.4	36.5
Other Latin	*2359*	None	15.0	13.0	17.0
		Liberal	3.8	2.6	5.1
		Moderate	53.3	50.3	56.3
		Conservative	27.9	25.0	30.8
Asian	*1960*	None	18.2	15.0	21.4
		Liberal	5.9	4.2	7.7
		Moderate	17.2	14.5	19.9
		Conservative	58.7	54.4	63.0
Native American	*611*	None	25.3	18.9	31.7
		Liberal	10.5	7.7	13.2
		Moderate	23.1	18.3	27.8
		Conservative	41.2	31.9	50.5

10th graders indicate that they attend religious services at least once or twice a month, and three quarters or more indicate that they are affiliated with a religious denomination.

Comparing racial/ethnic differences in religiosity, the data in Tables 7.2a, b and c reveal that African American youth are most likely to say that religion is an important or very important part of their life (76 %), most likely to attend religious services monthly or more (60 %), and most likely to be affiliated with a conservative denomination (61 %). Among the other subgroups, the proportions who say that religion is an important or very important part of their life is roughly comparable among Mexican American, Other Latin American and Asian American 10th graders (approximately 60 %) and somewhat lower among White, Puerto Rican, and Native American 10th graders (around 50 %). Roughly half of White, Mexican American,

Table 7.3a Prevalence rates of 30-day alcohol and effect sizes (Gamma) among 10th grade students by religion and race/ethnicity (2005–2009) weighted data combined

	White	Black	Mexican	Puerto Rican	Other Latin	Asian	Native American
Religious importance							
Not important	42.1	28.8	43.8	36.3	42.8	26.3	51.1
A little important	41.1	25.1	44.0	37.7	36.0	18.4	37.4
Pretty important	35.2	22.6	40.7	31.7	32.9	17.1	35.4
Very important	20.1	17.4	25.3	30.5	22.1	12.9	29.7
Gamma	–0.26	–0.18	–0.22	–0.09	–0.23	–0.21	–0.23
N	*43,703*	*7883*	*2182*	*729*	*2148*	*1800*	*619*
Attendance							
Never	40.9	26.0	42.0	37.3	36.6	23.9	50.4
Rarely	40.6	23.7	43.0	36.6	35.9	16.9	34.0
Once or twice a month	39.0	22.3	38.9	31.6	35.0	21.0	41.1
About once a week or more	24.2	17.5	30.9	29.7	23.5	13.6	30.4
Gamma	–0.23	–0.15	–0.16	–0.11	–0.18	–0.16	–0.20
N	*43,694*	*7879*	*2185*	*724*	*2146*	*1800*	*616*
Affiliation							
None	40.2	24.2	44.2	27.9	37.5	22.1	48.2
Liberal	30.6	29.0	39.8	54.5	28.0	16.5	35.7
Moderate	35.6	20.2	38.0	34.8	33.9	21.5	34.0
Conservative	32.3	19.9	32.6	35.1	26.6	15.3	36.1
Gamma	–0.06	–0.09	–0.13	0.04	–0.13	–0.16	–0.14
N	*42,981*	*7667*	*2100*	*707*	*2085*	*1786*	*606*

Other Latin American, and Asian American 10th graders report that they attend religious services monthly or more, compared to approximately 40 % of Puerto Rican and Native American 10th graders. More than half of Asian American 10th graders are affiliated with a conservative religious denomination, followed by forty percent of American Indians, a third of Whites and Puerto Ricans, and a quarter of Mexican Americans and Other Latin Americans.

The Relationship Between Substance Use and Religiosity

The data in Tables 7.1, 7.2a, b and c reveal important racial/ethnic similarities and differences in substance use and in measures of religiosity. The next question we examine concerns the extent to which religiosity is inversely related to substance use among non-white youth (especially Asian American, Native American, and the Hispanic subgroups) as it has been found to be among white youth. Specifically, the data presented in Tables 7.3a, b and c show racial/ethnic differences in the prevalence of 30 day alcohol use (Table 7.3a), 30 day cigarette use (Table 7.3b) and

Table 7.3b Prevalence rates of 30-day cigarettes and effect sizes (Gamma) among 10th graders by religion and race/ethnicity (2005–2009) weighted data combined

	White	Black	Mexican	Cuban	Other Latin	Asian	Native American
Religious importance							
Not important	22.8	12.3	26.9	33.3	13.5	11.8	33.4
A little important	19.1	9.7	18.1	12.3	11.8	8.0	29.3
Pretty important	14.7	7.6	16.8	21.3	8.6	5.9	22.7
Very important	8.5	5.3	9.3	13.2	8.0	5.8	13.5
Gamma	−0.28	−0.24	−0.26	−0.26	−0.16	−0.21	−0.29
N	*44,432*	*8348*	*2305*	*204*	*2282*	*1835*	*643*
Attendance							
Never	23.3	11.3	21.1	34.7	14.5	9.2	29.2
Rarely	19.2	10.1	19.3	22.7	10.9	9.4	25.8
Once or twice a month	15.5	5.5	17.4	9.8	8.0	6.2	28.8
About once a week or more	9.7	5.1	10.5	9.8	7.4	5.5	17.7
Gamma	−0.29	−0.27	−0.23	−0.46	−0.21	−0.18	−0.15
N	*44,421*	*8347*	*2307*	*204*	*2280*	*1837*	*640*
Affiliation							
None	23.5	10.0	23.8	37.3	15.0	9.5	28.0
Liberal	11.4	12.6	16.0	40.0	7.5	7.6	26.1
Moderate	13.1	6.7	14.0	6.1	8.8	4.9	25.5
Conservative	17.3	6.3	16.2	26.4	9.9	7.4	22.5
Gamma	−0.05	−0.17	−0.09	−0.15	−0.09	−0.04	−0.09
N	*43,687*	*8120*	*2225*	*196*	*2211*	*1820*	*630*

annual marijuana use (Table 7.3c) by the importance that American 10th graders ascribe to religion, by the frequency with which they attend religious services, and by their denominational affiliation. The tables also show the strength of the bivariate associations (i.e., gamma coefficients) between the religion measures and the substance use measures. (Because the substance use and religion measures are ordinally scaled, the gamma coefficient was selected as the appropriate measure of association.)

Consistent with the findings of past research on white youth, the data in Tables 7.3a, b and c reveal weak to moderate inverse relationships (gammas = −.09 to −.36) between the substance use measures and religious importance and attendance, across the racial/ethnic subgroups. Similarly, there are inverse, albeit smaller, relationships (gammas = −. 03 to −.21) between denominational affiliation and the substance use measures. In general, within each racial/ethnic group, young people for whom religion is more important, who attend religious services more frequently and who are affiliated with a religious denomination are less likely to drink, smoke cigarettes, or use marijuana than their less religiously committed counterparts.

Although the strength of the inverse relationships between the religion measures and the substance use measures are only modest, it should be noted that many of the

Table 7.3c Prevalence rates of annual marijuana and effect sizes (Gamma) among 10th graders by religion and race/ethnicity (2005–2009) weighted data combined

	White	Black	Mexican	Puerto Rican	Other Latin	Asian	Native American
Religious importance							
Not important	38.3	30.4	42.8	33.2	31.8	19.5	42.2
A little important	31.4	29.7	35.6	35.7	27.3	14.5	33.5
Pretty important	23.1	23.3	30.0	23.8	22.5	12.9	32.1
Very important	12.1	18.5	21.6	19.0	13.6	10.6	20.1
Gamma	–0.36	–0.20	–0.23	–0.24	–0.27	–0.18	–0.26
N	*44,435*	*8334*	*2296*	*768*	*2277*	*1846*	*640*
Attendance							
Never	37.5	26.0	39.7	39.8	30.5	19.5	38.3
Rarely	31.2	27.0	35.0	31.3	24.4	14.1	31.0
Once or twice a month	25.5	24.1	30.3	21.1	21.5	12.4	39.7
About once a week or more	15.1	18.2	22.9	17.1	16.2	11.3	23.7
Gamma	–0.34	–0.16	–0.22	–0.33	–0.21	–0.17	–0.15
N	*44,429*	*8330*	*2299*	*736*	*2276*	*1847*	*637*
Affiliation							
None	37.6	27.7	39.9	32.1	27.0	15.9	41.1
Liberal	20.3	25.3	31.2	49.5	19.7	10.4	30.9
Moderate	23.0	21.9	27.9	22.9	22.2	17.0	32.7
Conservative	24.4	21.6	30.0	27.9	20.9	12.4	25.3
Gamma	–0.12	–0.08	–0.08	–0.07	–0.07	–0.10	–0.21
N	*43,697*	*8109*	*2218*	*746*	*2210*	*1830*	*627*

proportional differences are substantial. For example, within each racial/ethnic subgroup (except Puerto Ricans), the prevalence of 30 day alcohol use, 30 cigarette use, and annual marijuana use is typically 1.5–2 times lower among the most religious youth (i.e., those for whom religion is very important and who attend religious services weekly) and those who are least religious (see Tables 7.3a, b and c). In general, the conservatism of the religious denomination with which students are affiliated is not strongly correlated with substance use among any of the racial/ethnic subgroups.

Tables 7.4a, b and c present the results of logistic regression analyses in which the substance use measures were regressed on the three religion measures while controlling for gender, family structure, parental education (a proxy for socio-economic status), urbanicity, and region. The purpose of these analyses was to ascertain the extent to which the bivariate relationships between religiosity and substance use would hold when important socio-demographic factors were statistically controlled. In order to facilitate interpretation, the logistic regression coefficients for the relationship between the independent variables and the specific substance use measure are presented as odds ratios. For each of the independent variables one

Table 7.4a Odds ratios for 30-day alcohol by race controlling sociodemographic factors and religiosity, for 10th graders, 2005–2009, data combined

	White	Black	Mexican	Puerto Rican	Other Latin	Asian	Native American
Gender							
Female	1.06	1.00	1.02	1.16	0.96	0.99	0.75
Male	–	–	–	–	–	–	–
Family structure							
0 Parents	1.48	1.06	1.12	1.27	0.96	1.53	1.40
1 Parent	1.27	0.97	1.25	1.20	1.11	1.57	1.51
2 Parents	–	–	–	–	–	–	–
Parental education							
Low (1.0–3.0)	1.17	1.02	0.88	0.75	0.86	0.93	1.08
Medium (3.5–4.0)	1.11	1.04	1.02	1.17	0.98	1.08	0.93
High (4.5–6.0)	–	–	–	–	–	–	–
Population density							
Non MSA	1.04	1.47	1.28	1.31	1.64	0.78	0.76
Other MSA	0.98	1.13	1.14	0.90	1.20	1.21	1.10
Large MSA	–	–	–	–	–	–	–
Region							
Northeast	1.30	0.78	1.01	2.00	1.64	0.57	1.35
Midwest	1.17	0.81	0.81	3.91	1.01	0.71	2.13
South	1.46	0.75	0.83	2.45	1.28	0.81	1.74
West	–	–	–	–	–	–	–
Religious importance							
Not important	2.58	1.63	2.28	1.93	2.40	2.13	1.75
A little important	2.27	1.41	2.24	1.29	1.59	1.28	1.30
Pretty important	1.92	1.34	2.08	1.02	1.57	1.08	1.20
Very important	–	–	–	–	–	–	–
Attendance							
Never	1.37	1.26	1.18	1.52	1.33	1.40	1.51
Rarely	1.47	1.28	1.30	1.50	1.43	1.24	0.99
Once or twice a month	1.59	1.26	1.09	1.16	1.47	1.58	1.30
About once a week or more	–	–	–	–	–	–	–
Affiliation							
None	0.90	0.94	1.15	0.50	1.13	0.93	1.11
Liberal	0.94	1.51	1.45	2.36	1.21	1.10	1.07
Moderate	1.11	1.04	1.17	0.83	1.32	1.68	1.07
Conservative	–	–	–	–	–	–	–
N	*41,298*	*6999*	*1876*	*643*	*1853*	*1583*	*540*
Model Chi Square	2136^{c}	103^{c}	79.9^{c}	34.4^{a}	77.3^{c}	45.7^{c}	36.2^{b}

$^{a}p<.05$, $^{b}p<.01$, $^{c}p<.001$

Table 7.4b Odds ratios for 30-day cigarettes by race controlling sociodemographic factors and religiosity, for 10th graders, 2005–2009, data combined

	White	Black	Mexican	Puerto Rican	Other Latin	Asian	Native American
Gender							
Female	1.18	0.62	1.03	1.02	0.92	0.77	0.73
Male	–	–	–	–	–	–	–
Family structure							
0 Parents	2.28	2.26	2.67	2.25	2.06	2.52	2.89
1 Parent	1.58	1.10	1.47	1.08	1.23	2.16	1.47
2 Parents	–	–	–	–	–	–	–
Parental education							
Low (1.0–3.0)	1.70	1.37	0.65	1.25	0.62	1.18	2.08
Medium (3.5–4.0)	1.38	1.06	0.87	1.42	0.62	1.33	1.26
High (4.5–6.0)	–	–	–	–	–	–	–
Population density							
Non MSA	1.28	2.37	1.64	0.96	1.96	0.75	0.94
Other MSA	1.02	1.13	1.08	0.81	1.26	1.25	0.85
Large MSA	–	–	–	–	–	–	–
Region							
Northeast	1.21	0.72	1.30	0.62	1.01	0.41	1.56
Midwest	1.62	0.91	1.10	0.97	2.37	0.55	4.16
South	1.88	0.73	1.11	1.06	1.93	0.59	2.26
West	–	–	–	–	–	–	–
Religious importance							
Not important	2.50	1.44	2.45	1.32	1.20	2.18	3.72
A little important	2.16	1.39	1.72	1.83	1.32	1.16	2.76
Pretty important	1.78	1.32	1.84	1.35	1.09	0.91	1.83
Very important	–	–	–	–	–	–	–
Attendance							
Never	1.55	1.72	1.43	1.88	2.13	1.04	1.12
Rarely	1.44	1.76	1.74	1.38	1.67	1.52	1.28
Once or twice a month	1.37	0.96	1.67	1.10	1.18	0.94	1.76
About once a week or more	–	–	–	–	–	–	–
Affiliation							
None	0.93	0.94	1.15	0.75	1.14	0.66	0.60
Liberal	0.71	1.68	0.92	4.20	0.79	1.25	1.06
Moderate	0.81	1.10	0.80	0.72	0.93	0.74	1.15
Conservative	–	–	–	–	–	–	–
N	*41,949*	*7390*	*1973*	*671*	*1956*	*1608*	*560*
Model Chi square	2350[b]	232[b]	92.4[b]	29.6	69.5[b]	4.5[a]	59.9[b]

[a] $p < .01$
[b] $p < .001$

Table 7.4c Odds ratios for annual marijuana by race controlling sociodemographic factors and religiosity, for 10th graders, 2005–2009, data combined

	White	Black	Mexican	Puerto Rican	Other Latin	Asian	Native American
Gender							
Female	0.94	0.69	0.84	0.73	0.95	0.74	0.53
Male	–	–	–	–	–	–	–
Family structure							
0 Parents	2.23	1.84	2.69	2.15	2.58	0.82	2.81
1 Parent	1.58	1.24	1.53	1.61	1.33	2.58	1.91
2 Parents	–	–	–	–	–	–	–
Parental education							
Low (1.0–3.0)	1.29	1.13	1.09	2.03	0.84	1.19	1.07
Medium (3.5–4.0)	1.13	1.18	1.11	1.58	1.08	1.03	0.92
High (4.5–6.0)	–	–	–	–	–	–	–
Population density							
Non MSA	0.87	0.90	1.29	0.94	1.35	0.67	1.16
Other MSA	0.98	0.95	1.41	0.86	1.57	1.05	1.51
Large MSA	–	–	–	–	–	–	–
Region							
Northeast	1.05	0.50	0.95	0.36	0.80	0.37	1.24
Midwest	0.95	0.67	0.81	0.49	0.88	0.54	1.76
South	1.18	0.64	1.00	0.74	0.99	0.74	1.16
West	–	–	–	–	–	–	–
Religious importance							
Not important	3.08	1.65	2.00	2.67	2.63	1.80	2.36
A little important	2.53	1.63	1.58	3.65	2.13	1.29	2.11
Pretty important	1.95	1.28	1.52	2.30	1.95	0.97	2.03
Very important	–	–	–	–	–	–	–
Attendance							
Never	1.65	1.06	1.64	2.50	1.76	1.66	1.16
Rarely	1.58	1.37	1.57	1.43	1.41	1.12	1.15
Once or twice a month	1.49	1.27	1.28	1.38	1.23	0.94	2.01
About once a week or more	–	–	–	–	–	–	–
Affiliation							
None	0.96	1.02	1.13	0.70	0.79	0.66	1.38
Liberal	0.82	1.15	1.06	3.19	0.98	0.88	1.20
Moderate	0.90	0.98	0.83	0.60	0.98	1.62	1.33
Conservative	–	–	–	–	–	–	–
N	*41,967*	*7381*	*1959*	*673*	*1959*	*1624*	*559*
Model Chi square	3051[a]	226[a]	116[a]	86.2[a]	97.9[a]	78.5[a]	59.5[a]

[a]$p < .001$

of the categories is omitted. Thus the odds ratios presented in the tables indicate the relationship between the specific category of the independent variable, *relative to* the omitted category, when the other independent variables are controlled. An odds ratio of 1 indicates that the likelihood of a 10th grader in that category using the substance in question is no different than that of 10th grader in the omitted category. An odds ratio greater than 1 indicates an increased chance that students in that category will use the substance, and an odds ratio less than 1 indicates that there is a decreased chance that students in that category will use the substance.

The data presented in Tables 7.4a, b and c generally suggest that even after key demographic variables are controlled, religious importance, and to a lesser extent, religious attendance continue to relate inversely to drug use, not only among white and African American youth, but also among the various Hispanic populations, and among Asian Americans and Native Americans. This conclusion is derived from the fact that the vast majority of the coefficients for the relationship between these two religion measures and alcohol, cigarette and marijuana use exceed 1. This means that the odds of using alcohol, cigarettes or marijuana are, on average, greater among young people who are not highly religious (i.e., for whom religion is not very important and who do not attend religious services once a week or more). In general, denominational differences in alcohol, cigarette, and marijuana use are relatively small and inconsistent across the racial/ethnic subgroups and across the three substances.

Discussion

The primary purposes of the present study were to describe the prevalence of religiosity—including importance, attendance and denominational affiliation, among a racially and ethnically diverse, nationally representative sample of American adolescents; and to determine to what extent religiosity "protects" African American, Mexican American, Puerto Rican, Other Latin American, Asian American and Native American young people from drug use, as past research suggests that it protects white youth.

Consistent with the findings of our prior research, and that of others, we found that substantial numbers of American young people, regardless of their racial/ethnic identification indicate that religion is at least a "pretty" important part of their life, that they attend religious services with some regularity (once a month or more) and that the majority of them claim a religious affiliation.

Also consistent with the findings of previous research, based on predominantly white samples, we found statistically and substantively significant inverse bivariate relationships between religious importance and substance use and between religious attendance and substance use for African American, Mexican American, Puerto Rican, Other Latin American, Asian American, Native American and white young people. Further, we found that, on average, these relationships continued to exist, even after we controlled for important socio-demographic factors such as family structure, parent education, region and urbanicity.

Despite the consistency of the findings with past research, the study has a number of important limitations. First, the cross-sectional design of this study precludes any firm conclusions about causality. Second, because the data are drawn from samples of students, the findings do not generalize to populations who are not in school (e.g., dropouts, incarcerated youth). Third, although the existing research does not confirm the hypothesis, there is a chance that students who are more highly religious might underreport their substance use because of social desirability. Fourth, despite our ability to disaggregate the Hispanic population, there is significant diversity among the subgroups of young people who identify themselves as "white" "African American" "Asian" and "Native American." Our current measures of race/ethnicity do not allow us to examine these important within group distinctions.

Despite these and its other limitations the present study adds to our collective knowledge about the relationship between race/ethnicity, religiosity and adolescent substance use and provides some direction for future research. One important task for future research includes the need to better understand and more rigorously examine the potential mechanisms by which religiosity may influence adolescents' substance use. Past research suggests that potential pathways through which religion may influence adolescent substance use include establishing moral directives or rules of self-control (e.g., abstinence from substance use), providing opportunities to acquire learned competencies (e.g., drug prevention instruction), and providing social (e.g., abstaining peer and role models) and organizational ties (e.g., support groups) (see Smith 2003; Regnerus 2003 and Wallace et al. 2007, for a fuller discussion). Another related task for future research is to explore the extent to which these potential mechanisms are similar or differ across racial/ethnic groups. In light of the consistency of the findings, across the racial/ethnic groups, that religion may "protect" adolescents against substance use, an additional task for both researchers and practitioners is to consider the ways in which religiosity and communities of faith might be engaged to assist the nation in the fight against adolescent substance abuse and its sequelae.

Acknowledgements The Monitoring the Future Study was funded by grant R01 DA01411 from the National Institute on Drug Abuse. We would like to thank Tim Perry for assistance with data analysis and preparation of tables.

References

Cotton, S., Zebracki, K., Rosenthal, S. L., Tsevat, J., & Drotar, D. (2006). Religion/spirituality and adolescent health outcomes: A review. *Journal of Adolescent Health, 38*, 472–480.

Department of Health and Human Services. (2001). *Science-based substance abuse prevention: A guide* (DHHS publication number (SMA) 01-3505). Washington, DC: UD Government Printing Office.

Dew, R. E., Daniel, S. S., Armstrong, T. D., Goldston, D. B., Triplett, M. F., & Koenig, H. G. (2008). Religion/spirituality and adolescent psychiatric symptoms: A review. *Child Psychiatry and Human Development, 39*, 381–398.

Gorsuch, R. L. (1988). Psychology of religion. *Annual Review of Psychology, 39*, 201–221.

Gorsuch, R. L. (1995). Religious aspects of substance abuse and recovery. *Journal of Social Issues, 51*, 65–83.

Hawkins, J. D., Catalano, R. F., & Miller, J. Y. (1992). Risk and protective factors for alcohol and other drug problems in adolescence and early adulthood: Implications for substance abuse prevention. *Psychological Bulletin, 112*, 64–105.

Hill, T. D., Burdette, A. M., Weiss, M. L., & Chitwood, D. D. (2009). In C. G. Leukfeld, T. P. Gollata, & M. Staton-Tindall (Eds.), *Adolescent substance abuse: Evidence based approaches to prevention and treatment*. New York: Springer.

Johnston, L. D., O'Malley, P. M., Bachman, J. G., & Schulenberg, J. E. (2009). *Monitoring the future national survey results on drug use, 1975–2008: Volume I, Secondary school students* (NIH publication no. 09-7402). Bethesda: National Institute on Drug Abuse.

National Center on Addiction and Substance Abuse. (2001). *So help me God: Substance abuse, religion and spirituality*. New York: Columbia University.

National Center on Addiction and Substance Abuse. (2009). *Shoveling up II: The impact of substance abuse on federal, state and local budgets*. New York: Columbia University.

National Institute on Drug Abuse. (2003). *Drug use among racial and ethnic minorities* (NIH publication no. 03-3888). Washington, DC: US Government Printing Office.

Regnerus, M. (2003). Religion and positive adolescent outcomes: A review of research and theory. *Review of Religious Research, 44*, 394–413.

Rew, L., & Wong, Y. J. (2006). A systematic review of associations among religiosity/spirituality and adolescent health attitudes and behaviors. *Journal of Adolescent Health, 38*, 433–442.

Robert Wood Johnson Foundation. (2001). *Substance abuse: The nation's number one health problem*. Princeton: Authors.

Roof, W. C., & McKinney, W. (1987). *American mainline religion: Its changing shape and future*. New Brunswick: Rutgers University Press.

Smith, T. W. (1990). Classifying protestant denominations. *Review of Religious Research, 31*, 225–245.

Smith, C. (2003). Theorizing religious effects among American adolescents. *Journal for the Scientific Study of Religion, 42*, 17–30.

Substance Abuse Mental Health Services Administration. (2009). *Results from the 2008 national survey on drug use and health: National findings* (Office of applied studies, NSDUH series H-36, HHS publication no. SMA 09-4434). Rockville: Department of Health and Human Services, Substance Abuse and Mental Health Services Administration, Office of Applied Studies.

Wallace, J. M., Jr., Brown, T. N., Bachman, J. G., & LaVeist, T. A. (2003). The influence of race and religion on abstinence from alcohol, cigarettes and marijuana among adolescents. *Journal of Studies on Alcohol, 64*, 843–848.

Wallace, J. M., Delva, J., Jr., O'Malley, P. M., Bachman, J. G., Schulenberg, J. E., Johnston, L. D., & Stewart, C. (2007). Race/ethnicity, religiosity and adolescent alcohol, cigarette and marijuana use. *Social Work in Public Health, 23*, 193–213.

Williams, D. R., & Sternthal, M. J. (2007). Spirituality, religion and health: Evidence and research directions. *Medical Journal of Australia, 186*, S47–S50.

Wong, Y. J., Rew, L., & Slaikeu, K. D. (2006). A systematic review of recent research on adolescent religiosity/spirituality and mental health. *Issues in Mental Health Nursing, 27*, 161–183.

Chapter 8
Racial Differences in Substance Use: Using Longitudinal Data to Fill Gaps in Knowledge

Helene Raskin White, Rolf Loeber, and Tammy Chung

Racial differences in substance use have been described in numerous studies and publications. It has generally been found that White, compared to African-American, adolescents are more likely to drink alcohol, smoke cigarettes, and use hard drugs (Johnston et al. 2011; Lee et al. 2010; White et al. 2007a; White et al. 2004). Data on racial differences in adolescent marijuana use have been less consistent with some studies showing higher rates among Whites (e.g., Tragesser et al. 2007; Wallace et al. 2003b), others showing higher rates among African Americans (Lee and Abdel-Ghany 2004; Lee et al. 2010; White et al. 2007a), and still others showing that racial differences depend on age and gender (Wallace et al. 2003a; Johnston et al. 2009). Whereas African Americans, compared to Whites, report lower rates of most types of drug use in adolescence and young adulthood, the former report higher rates of use of some drugs in middle adulthood (Kandel 1991) and, in adulthood, African-American and White subgroups have similar rates of substance use disorders (Substance Abuse and Mental Health Services Administration 2010). Thus, racial differences in substance use depend on developmental stage as well as type of substance examined.

Despite the fact that several studies have addressed racial differences in substance use, there still exist gaps in understanding the development of substance use among African Americans. In this chapter, we focus on three of these gaps. The first has to do with the inadequate attention to heterogeneity among substance users and substances used, which includes: (1) differences across drugs consumed, (2) differences across contexts, such as where people live and when they were born, and

H.R. White (✉)
Center of Alcohol Studies and Sociology Department, Rutgers University, New Brunswick, NJ, USA
e-mail: hewhite@rci.rutgers.edu

R. Loeber • T. Chung
Department of Psychiatry, University of Pittsburgh, Pittsburgh, PA, USA

Y.F. Thomas, L.N. Price (eds.), *Drug Use Trajectories Among Minority Youth*, DOI 10.1007/978-94-017-7491-8_8

(3) intraindividual differences, such as gender and attitudes. The second gap has been the failure to adequately explain *why* there are differences in the age of onset, prevalence, and frequency of substance use for African Americans, compared to Whites. The third gap is the lack of research that examines how racial differences in substance use relate to other problems behaviors, such as violence. A major reason for these gaps is the paucity of long-term studies of African Americans that follow them from childhood to adulthood with short time gaps, and permit the observation of developmental processes, such as escalation and cessation. In this chapter, we use data from two prospective longitudinal studies to address these issues, the Pittsburgh Youth Study (PYS) and the Pittsburgh Girls Study (PGS). These two studies provide a unique opportunity to delineate trajectories of substance use among African Americans because both are general population samples and are more than half African-American with the remainder being primarily White.

First, we examine trajectories of different types of substance use among White and African-American young men from early adolescence into early adulthood in the PYS. We then examine different types of heterogeneity that may affect racial differences, by examining cohort differences in the PYS and gender differences between the PYS and PGS. Next we summarize data from two previous PYS studies and emerging findings from the PGS on alcohol and tobacco use. The first PYS study examined racial differences in developmental stages of substance use and movement from initiation to regular use. The second attempted to explain late onset smoking among African Americans. PGS data have been used to examine racial differences in girls' alcohol expectancies and alcohol use, as well as racial differences in girls' expectancies for and use of cigarettes. Finally, we examine racial differences in the association between substance use and violence.

Method

Samples

Pittsburgh Youth Study Most of the data presented in this chapter come from the PYS (Loeber et al. 2008). The PYS is a prospective, longitudinal study of the development of delinquency, substance use, and mental health problems. In 1987–1988, random samples of first, fourth, and seventh grade boys enrolled in the City of Pittsburgh public schools were selected. Approximately 850 boys in each grade (85 % of the target sample) were screened. Families were paid for their participation, and informed written consent was obtained from both the participants and their legal guardians. The 15 % nonparticipation rate did not result in sample bias, at least in regard to achievement test results and racial distribution, which were the only two variables that could be compared from school records (Loeber et al. 2008).

About 500 boys in each grade (the 30 % who scored highest on a risk assessment for later antisocial behavior and another 30 % randomly selected from the remainder) were selected for the first follow up 6 months later; therefore, approximately half of each cohort was selected for potential high risk and half was not. There was no significant difference in the percent high risk by race.

The present analyses focused on the youngest cohort born primarily in 1980–1981 (from approximately age 6 through age 19; $N = 503$) and oldest cohort born primarily in 1973–1975 (from approximately age 12 through age 25; $N = 506$). After the first follow up, members of the youngest cohort were subsequently followed up at 6-month intervals for six additional assessments and then at yearly intervals for a total of 14 years and members of the oldest cohort were followed at 6-month intervals for five additional assessments and then at yearly intervals for a total of 14 years. For these analyses we combined the early 6-month assessments to form annual assessments across all ages. Data were coded by age at assessment rather than wave. Attrition has remained relatively low and the completion rate has averaged above 90 % across 14 years of data collection (for more information about recruitment and retention see Stouthamer-Loeber and Van Kammen 1995). Primary caretakers were interviewed also until the boys were approximately 18 years old and teachers provided data until approximate age 16. A little more than half (55 %) of the sample is African-American, and almost all the rest (41 %) is White, reflecting the racial composition of the Pittsburgh Public Schools when the study began. The population of other ethnic/racial minority groups in Pittsburgh is very low. For this chapter we eliminated the other and mixed group (4 %). Over 90 % of the boys lived with their natural mother at the initial assessment and over one third of the boys' families received public assistance or food stamps. (For greater detail on design and sample see Loeber et al. 2008.) Respondents were interviewed separately in-person. All study procedures were approved by the University of Pittsburgh Institutional Review Board. Families were compensated for their participation. Although the PYS is a high-risk sample, it is possible to weight the data back to the original population of all public school students in Pittsburgh, which we have done for most of these analyses.

Pittsburgh Girls Study The PGS involves a household sample of four cohorts of girls, ages 5–8 at the first assessment, and their primary caretaker, who have been followed annually according to an accelerated longitudinal design (Hipwell et al. 2002). As with the PYS, we wanted to study a high-risk sample and, thus, recruitment of participants followed an enumeration of 103,238 households in the City of Pittsburgh. Based on the 1990 Census data on poverty, the lower third of neighborhoods were labeled as "disadvantaged" (i.e., >25 % of families were known to be living in poverty) and the upper two thirds were labeled "advantaged." Every disadvantaged neighborhood and 50 % of the advantaged neighborhoods were enumerated over a 1-year period, which elicited 3241 5- to 8-year-old girls representing 83.7 % of the girls identified by the subsequent 2000 Census. No differences were revealed in the rates of success in identifying girls according to the type of neighborhood. Of 2876 girls who were age eligible and who could also be subsequently located, 2451 (85.2 %) agreed to participate in the longitudinal study (see Hipwell et al. 2002 and Keenan et al. 2010 for further details). Separate in-home interviews for both the girl and caretaker were conducted annually by trained interviewers using a laptop computer. All study procedures were approved by the University of Pittsburgh Institutional Review Board. Families were compensated for their participation.

For the present analyses, we used data collected from the two oldest cohorts, who were ages 7 and 8 at the initial assessment in 2000–2001 (n=611 and 622, respectively). Like the PYS, the two oldest cohorts are predominantly African-American (52.4 %) and White (41.8 %), with a small proportion (5.8 %) of girls representing other race/ethnic groups (e.g., bi-racial). At the initial assessment, the majority of caretakers was female; most caretakers (57 %) were cohabiting with a spouse or domestic partner; about half (51 %) completed 12 years of education or less; and 35 % of households were receiving public assistance (Hipwell et al. 2005).

The oldest PGS cohort, which provided prevalence data (Figs. 8.7 and 8.8), has been followed annually through age 15. Retention over 8 years of data collection has been high, with an average participation rate above 90 %. For prevalence analyses conducted in the oldest cohort, we weighted the data to compensate for the over-sampling of girls from low socioeconomic (SES) neighborhoods, and excluded the small number of girls representing other race/ethnic groups.

Measures

The data for this chapter come primarily from the young men's and women's self reports of their own substance use. Although self-report data on substance use has been validated in national studies (Oetting and Beauvais 1990; Winters et al. 1991), it should be noted that some research suggests that African Americans, compared to Whites, are more likely to under-report use of drugs, such as cigarettes (Bauman and Ennett 1994; Fisher et al. 2008). Thus, racial differences should be interpreted with caution.

In the PYS, at each wave, participants were asked whether in the past year (or 6 months), they used (a) tobacco (cigarettes, pipes, or chewing tobacco); (b) drank alcohol (beer, wine, or hard liquor); (c) used marijuana or hashish; and (d) used hallucinogens, cocaine, crack, heroin, PCP, and tranquilizers, barbiturates, codeine, amphetamines, and other prescription medications for non medical reasons. These latter ten substances were combined to reflect hard drug use. These items were use to create *annual* and *lifetime prevalence* rates. For alcohol, if the caretaker knew about their use or the reason for drinking was a "special occasion or religious ritual" or with adults at dinner, then use was not counted. At screening the respondents were asked if they had ever used each substance and, if yes, at what age they first used. Subsequent to that, they were asked about use in the last year (or last 6 months in the first few assessments for the PYS). The first time the respondent reported using a substance was coded as their *age of onset*. (If onset was reported prior to age 5, it was coded as age 5.) Respondents also reported on the number of times they used each substance in the last year. These variables provide information on *frequency* of use. For alcohol, frequency is the sum of beer, wine, and hard liquor. Frequency was truncated at 365 times in a year.

In the PGS, at ages 7–10, respondents were asked if they had ever taken a sip of beer, wine or hard liquor without their parents' permission, whereas at ages 11–15 they were asked if they ever tried beer, wine, or hard liquor. A "yes" at any age for any beverage was used to determine *annual prevalence of alcohol use*. In addition, respondents reported on whether they had tried marijuana at each age, which was used to create the *annual prevalence of marijuana use* variable. Girls also reported on whether they had smoked a cigarette or used tobacco in the past year.

In the PGS, we also measured alcohol and tobacco expectancies, which are beliefs about the possible outcomes of substance use (e.g., feel relaxed, feel happy). Substance use expectancies may play an important role as an early cognitive precursor of substance use, and have been found to predict substance use (review: Jones et al. 2001). Analyses focused on positive, rather than negative, substance use expectancies because positive expectancies consistently predict substance use in youth (Jones et al. 2001). Alcohol expectancies were assessed at ages 7–10 in the second oldest PGS cohort using the Children's Expectancy Questionnaire-Revised (Dunn and Goldman 1996). Tobacco expectancies were assessed in the oldest PGS cohort at ages 11–14 using items adapted from Hornik et al. (2002).

In this chapter, we use two different measures of violence in the PYS. *All-source serious violence prevalence* (Loeber et al. 2008) is a construct that combines reported serious violence (based on reports from the youth, primary caretaker, and teacher) and serious violence conviction abstracted from official court records. Serious violence (all-source) is considered prevalent if either source is positive. Examples of serious reported violence are: forcible robbery, attacking with intent to injure, sexual coercion, and rape. Examples of serious violence based on official records are: robbery, homicide, rape, aggravated assault, involuntary deviate sexual intercourse, aggravated indecent assault, or spousal sexual assault. *Self-reported violence* is the youth's self-report of engaging in serious violence each year, including forcible robbery, attacking with intent to injure, sexual coercion, and rape.

Results

Differences by Drug Types

As the chapter by Kandel in this volume indicates, in national studies, racial differences in drug use depend on the type of drug examined. We replicated those findings in the PYS sample. Figure 8.1 shows the annual prevalence of alcohol use for African Americans and Whites in the oldest cohort. At most ages, especially from middle adolescence through the early 1920s, Whites were more likely to drink than African Americans. At age 21, the alcohol prevalence rate peaked for Whites at 91 %, whereas the peak for African Americans only reached 74.6 % at age 22.

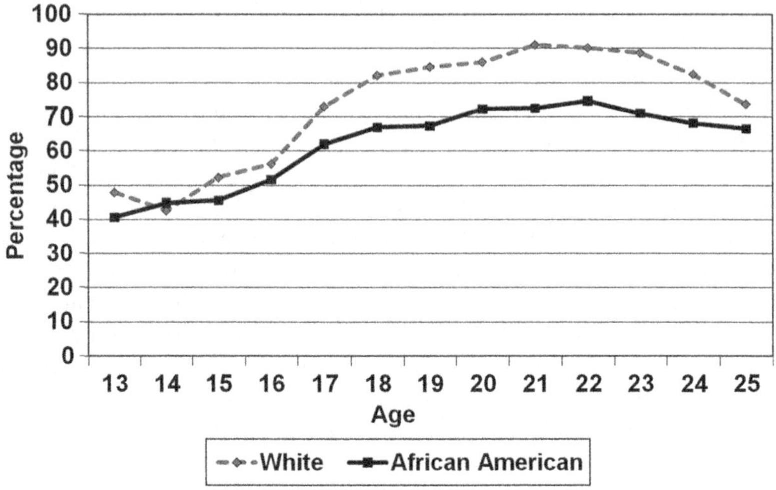

Fig. 8.1 Annual prevalence of alcohol use in the PYS oldest cohort by race (weighted)

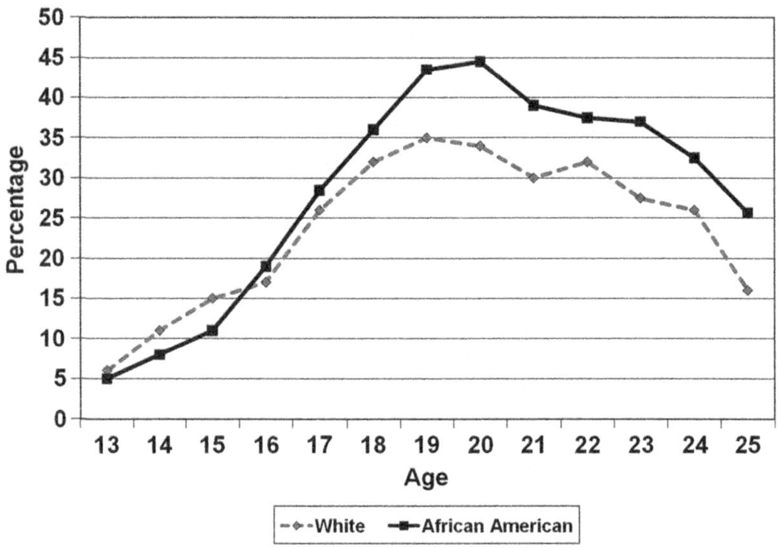

Fig. 8.2 Annual prevalence of Marijuana use in the PYS oldest cohort by race (weighted)

Figure 8.2 shows annual prevalence rates of marijuana use for African Americans and Whites in the oldest cohort. Whites reported slightly higher prevalence rates in early adolescence but there was a cross-over effect in middle adolescence so that from age 16 on African Americans were increasingly more likely to smoke marijuana than Whites. The highest annual prevalence rate for Whites was 34.8 % at age 19, whereas the highest rate for African Americans was 44.5 % at age 20.

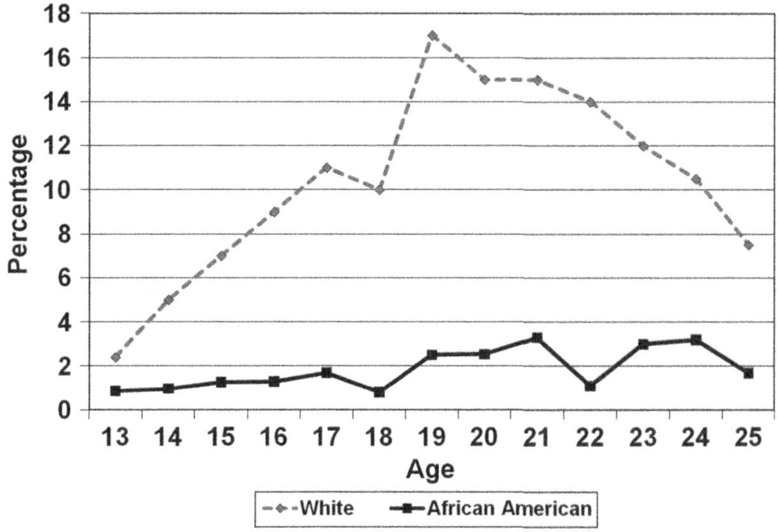

Fig. 8.3 Annual prevalence of hard drug use in the PYS oldest cohort by race (weighted)

In contrast to the findings for marijuana, Whites reported considerably higher rates of hard drug use (which included hallucinogens, cocaine, crack, heroin, PCP, and non-medical use of prescription drugs) than African Americans (Fig. 8.3). This difference extended from early adolescence through emerging adulthood. The peak in hard drug use for Whites was 16.6 % at age 19 compared to a peak of 3.3 % at age 21 for African Americans.

Cohort Differences

One must consider cohort differences when examining trajectories of substance use for Whites and African Americans. Golub and Johnson (1999) identified three different drug use generations using data from the Arrestee Drug Abuse Monitoring (ADAM) system: the "heroin generation" born between 1945 and 1959, the "cocaine/crack generation" born between 1955 and 1969, and the "blunts (marijuana cigars) generation" born since 1970. Both cohorts in the PYS sample were born after 1970 and theoretically represent the "blunts generation." Even though the two PYS cohorts were born approximately only 6 years apart, they reported very different patterns of drug use.

Figure 8.4 shows the percent who first used marijuana each year for both cohorts and both races. For African Americans in the oldest cohort there was a large increase in the percent of users between age 16 and age 17, with a peak at age 19, and then a drop off of new users after age 20. In contrast, for African Americans in the youngest cohort, the largest increase in use occurred between age 12 and age 13, with a

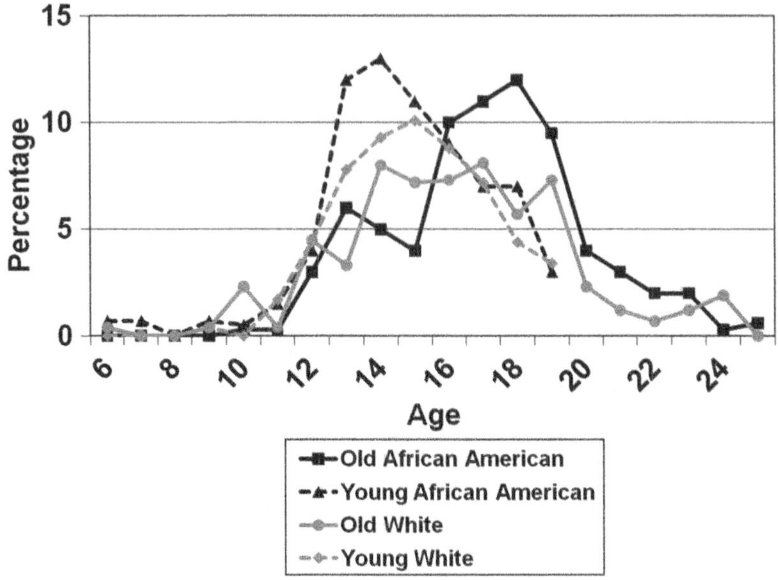

Fig. 8.4 Percentage of Whites and African Americans who began using marijuana each year in the PYS oldest and youngest cohorts (weighted)

peak at age 15 and a large drop off after that. Note, however, that these peaks occurred around the same years (early 1990s) for both cohorts suggesting that, in the case of marijuana initiation, historical influences were stronger than age influences for African Americans. The differences for the Whites were not apparent with both cohorts displaying similar patterns through age 19 years. Thus, historical differences for Whites were not as strong as those for African Americans.

The cohort effect is also apparent when one compares racial differences in annual prevalence of marijuana use from age 13 through age 19 for both cohorts (Fig. 8.5). In the oldest cohort, there was a slight cross-over effect at age 16, when African Americans surpassed Whites in marijuana use and this difference increased over time. By age 19, 62.7 % of the African Americans and 54.7 % of the Whites in the oldest cohort had used marijuana (not shown). For the youngest cohort, there was no cross-over effect and, in fact, from age 13 on, annual prevalence of marijuana use was greater for African Americans than Whites. By age 19, 70.5 % of the African Americans and 57.4 % of the Whites had used marijuana (not shown). Overall, in both cohorts annual prevalence rates were higher for African Americans than Whites and these differences appeared to increase with age. For both races, the younger cohort, compared to the older cohort, reported higher prevalence rates during early and middle adolescence.

In contrast, for alcohol use in both cohorts (except at age 14 for the older cohort), Whites reported higher annual prevalence rates than African Americans (Fig. 8.6). In early to middle adolescence, these differences were more pronounced in the

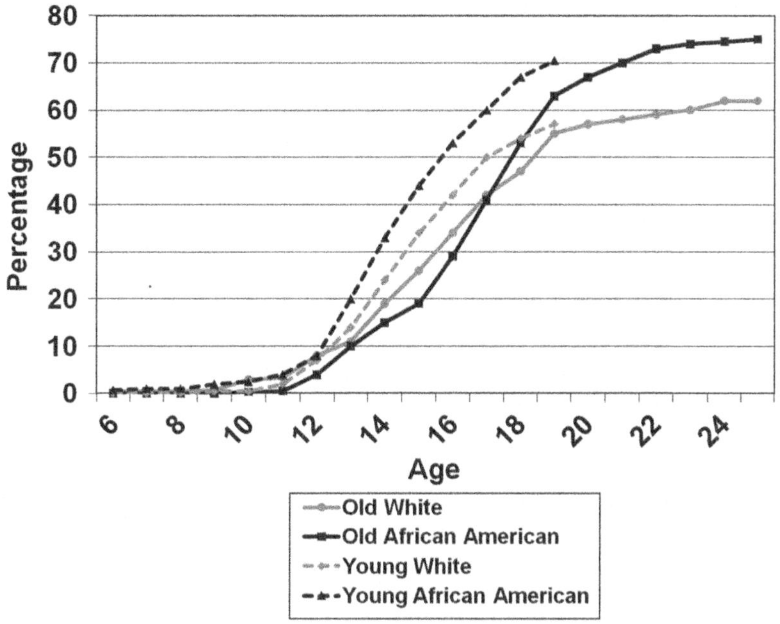

Fig. 8.5 Lifetime prevalence of marijuana use in the PYS oldest and youngest cohorts by race (weighted)

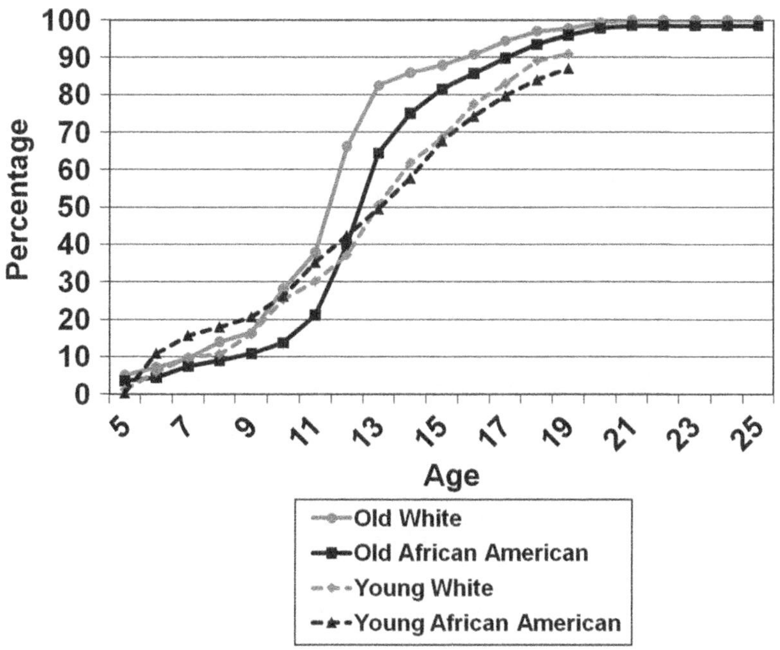

Fig. 8.6 Lifetime prevalence of alcohol use in the PYS oldest and youngest cohorts by race (weighted)

younger than older cohort. For both races, the older cohort reported higher rates of alcohol use than the younger cohort. By age 19, 97.7 % of the Whites in the oldest cohort, 96.0 % of the African Americans in the oldest cohort, 90.9 % of Whites in the youngest cohort, and 87.0 % of African Americans in the youngest cohort had initiated alcohol use (not shown).

Gender Differences

Few longitudinal studies have examined whether racial differences in substance use are similar for young men and young women. We compared African-American and White young men and young women in terms of annual prevalence rates for the youngest PYS cohort from age 8 through age 15 to those for the oldest cohort of the PGS sample from age 8 through age 15. Due to the difference in the measures and the years of data collection (i.e., 20 years separate the data collected in PYS and PGS cohorts), caution should be used when comparing these two samples.

As discussed above, African-American males generally reported higher rates of marijuana use than White males, especially from age 13 on (Fig. 8.7). In contrast, for young females, racial differences in use were not as apparent. For both races, rates were higher for males than females.

For alcohol use, the picture was more complex (Fig. 8.8). At most ages, White males reported higher annual prevalence than African-American males. African-

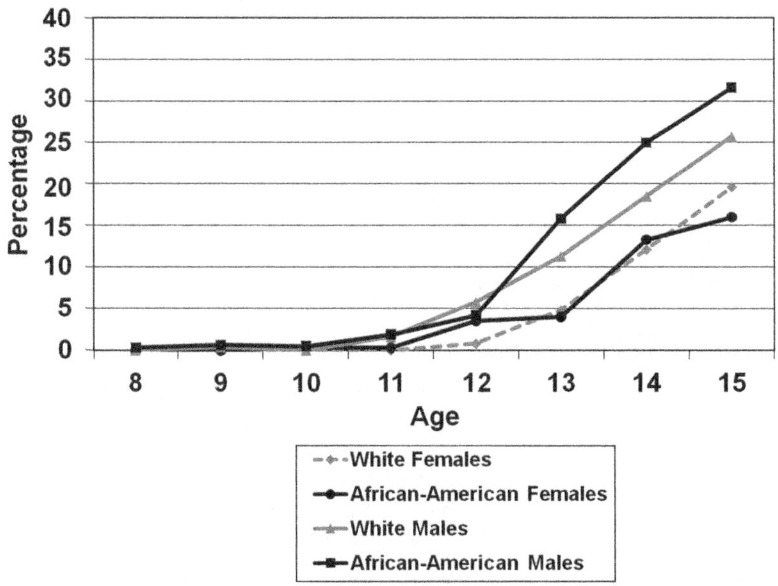

Fig. 8.7 Annual prevalence of marijuana use among females in the PGS oldest cohort and males in the PYS youngest cohort by race (weighted)

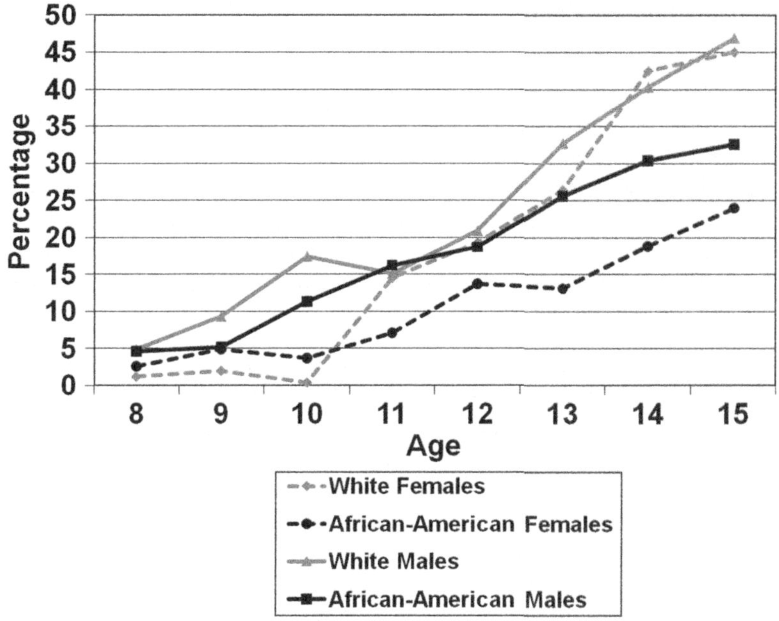

Fig. 8.8 Annual prevalence of alcohol use among females in the PGS oldest cohort and males in the PYS youngest cohort by race (weighted)

American, compared to White, females reported higher rates of alcohol use in childhood, but then increasingly lower rates throughout adolescence. Throughout childhood and adolescence, African-American males were more likely to drink than African-American females. Although White males were also more likely to drink than White females during childhood, by age 11 females caught up to males. In fact, at ages 14 and 15 White females were more likely to drink than African-American males. Thus, gender variations in racial differences depend on age and substance examined.

Thus far we have demonstrated that individual and contextual factors influence racial differences in substance use. In the PYS sample, racial differences in marijuana depended on the ages examined, whereas differences were more consistent across all ages for alcohol. Racial differences in marijuana use were more pronounced for the males than females, although historical changes need to be considered given that the boys in the PYS youngest cohort were initially interviewed 20 years earlier than the girls in the PGS sample. Gender differences in alcohol use were more pronounced for African Americans than Whites. Other sources of heterogeneity that should be considered when studying racial differences include: residential and SES differences, individual difference factors (family history, temperament, etc.), acceptability (norms) and availability of substances, differences in physiological reactions, and differences in subjective experiences. In addition, one should consider heterogeneity within racial/ethnic groups.

Explaining Racial Differences in Substance Use

As stated earlier, although several studies have examined similarities and differences in risk factors for substance use among adolescents (e.g., Skinner et al. 2009; Wallace and Muroff 2002; White et al. 2006), few studies have explained racial differences in substance use onset and patterns of use. One explanation for racial differences in substance use could be differential physiological reactions to substances, which could affect addiction liability. Furthermore, differences in the subjective experiences of different drugs could also affect addiction liability (Ridenour et al. 2006).

Racial Differences in Sequencing and Progression of Drug Use If there are racial differences in the addiction potential of different drugs, one would expect to see differences in the percent of initiators who become regular users. In a previous study using the PYS data we examined the percent of initiators of alcohol, cigarettes, marijuana, and hard drugs who became regular users (White et al. 2007a). The first year that the participant indicated that he had used the substance became the age of onset for initiation of that substance. Regular use was defined across each drug class as daily tobacco use, weekly alcohol use, weekly marijuana use, and use of hard drugs (all other illicit drugs other than marijuana) at a frequency of three or more times in the past year. The first year that the participant indicated that he had met the criterion for regular use became the age of onset of regular use for that substance.

For both races, the typical developmental sequence for substance use initiation and regular use was alcohol and/or tobacco, then marijuana, and then hard drugs. However, African Americans were more likely to deviate from this sequence than were Whites. A large minority of African-American users either began their substance use initiation with marijuana or hard drugs rather than alcohol or tobacco or first tried hard drugs before trying marijuana (Fig. 8.9).

We compared the percentage of initiators who became regular users and found only one significant racial difference (Fig. 8.10). More White than African-American alcohol initiators became regular (weekly) users. The percent of initiators who became regular (3+ times per year) users of hard drugs was also higher for Whites than African Americans but was not statistically significant probably because of the small number of African Americans who had used hard drugs in the sample. When we looked at the lag time from initiation to regular use for those who became regular users, we found the same pattern for Whites and African Americans (Fig. 8.11). The longest lag time was for alcohol, whereas the shortest was for hard drugs. There were no significant differences in the lag time from initiation to regular use for African Americans and Whites for any substance. Therefore, these data did not suggest greater "addiction" liability for one race compared to another (for greater detail, see White et al. 2007a). Nevertheless, we did not measure abuse or dependence and we did not follow these young men past age 19. As we follow these young men into adulthood, we should be able to shed more light on this issue.

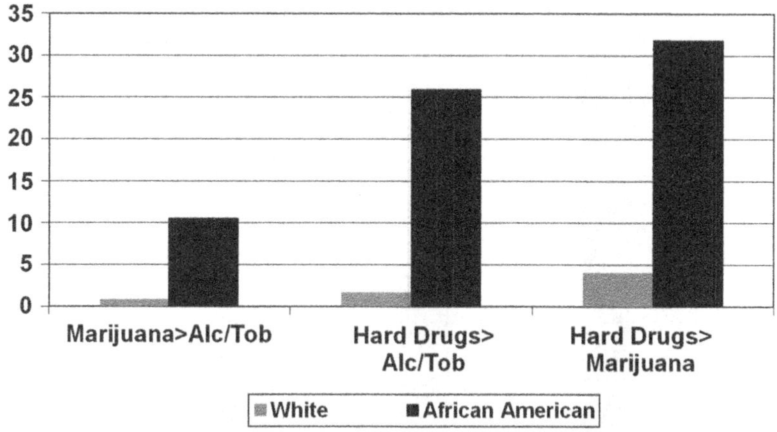

Fig. 8.9 Percentage not following the typical sequence in initiation by race in the PYS youngest cohort (weighted) (Note: Adapted from White et al. 2007. *Alc* alcohol, *Tob* tobacco)

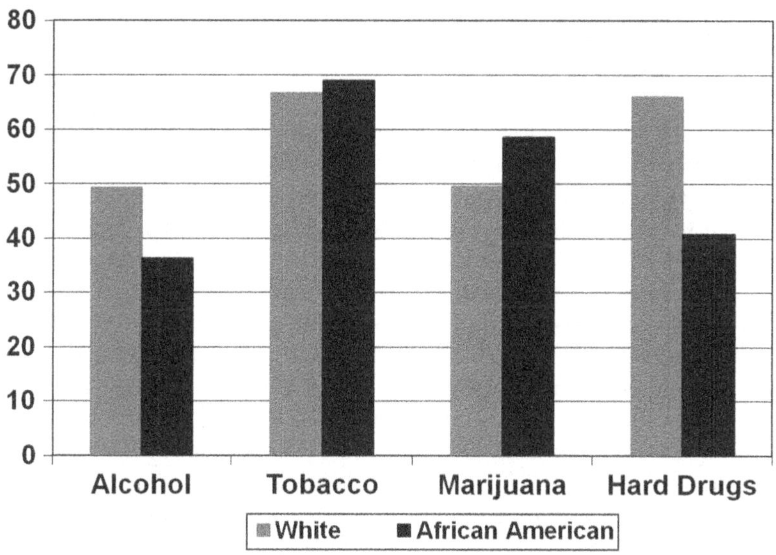

Fig. 8.10 Percentage of initiators who became regular users by age 19 in the PYS youngest cohort (weighted) (Note: Adapted from White et al. 2007a)

Explaining Racial Differences in the Onset of Cigarette Smoking In another previous study with the PYS sample (White et al. 2007b), we attempted to address why African Americans start smoking cigarettes much later than Whites, which is a fairly robust finding across studies (e.g., Greisler and Kandel 1998; White et al. 2004). Specifically we wanted to identify adolescent risk factors for late-onset smoking among African-American young men. We divided the African Americans

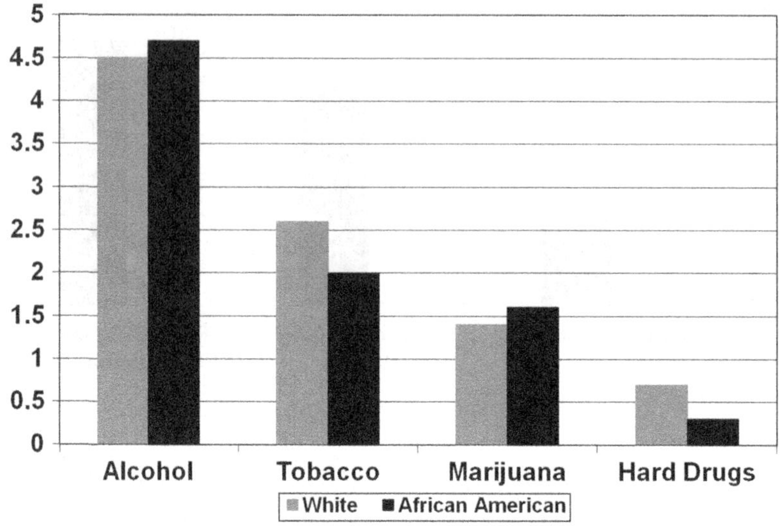

Fig. 8.11 Lag time from initiation to regular use in the PYS youngest cohort by race (weighted) (Note: Adapted from White et al. 2007a)

in the oldest PYS cohort into: (1) *Nonsmokers:* those who never tried a cigarette (N=99); (2) *Late-onset Smokers:* those who smoked at least twice in any year between the ages of 19 and 25 and began smoking at age 17 or older (N=60); and (3) *Early-onset Smokers:* those who smoked at least twice in any year between the ages of 19 and 25 and began smoking by age 16 (N=61). We examined a large number of known psychological, behavioral, and environmental predictors of smoking. Many previously identified risk factors for cigarette use measured in adolescence (at age 16) were able to distinguish early-onset smokers from non-smokers and late-onset smokers. Based on bivariate analyses, at age 16 early-onset smokers, compared to non-smokers and late-onset smokers, were more likely to report positive attitudes towards substance use, lower academic achievement, higher levels of delinquency, truancy, alcohol use, marijuana use, and peer substance use, to be less involved in their families, and to have poorer relationships with their parents. In multivariate analyses, only truancy and peer substance use differentiated early-onset smokers from non-smokers and late-onset smokers. None of these risk factors at age 16, however, was able to reliably distinguish those who would begin smoking later from those who would remain non-smokers. Therefore, we attempted to identify factors in later adolescence that might help explain late-onset smoking. We measured similar psychological, behavioral, and environmental risk factors in late adolescence (ages 17–19) as we had measured at age 16 and also examined measures of current (ages 17–19) life circumstances including highest grade completed, age left home, being on welfare, attending college, working full-time, amount of spending money, having a driver's license, and being incarcerated. Multivariate analyses indicated that only higher marijuana use between ages 17–19 and lower educational attainment by age 19 differentiated late-onset smokers from non-smokers. Thus, most of the risk factors and life circumstance variables that were examined could not explain late-onset smoking among African Americans (for greater detail, see White et al. 2007b). There are other life situation changes (e.g., joining the military, living with a smoker), which we did not measure, that may precipitate the onset of smoking in late adolescence and emerging adulthood and future research should explore these transitional variables. In addition, we did not measure contextual factors (except neighborhood quality), which need to be explored during the transition from late adolescence to young adulthood. We also did not measure racial identity and racial discrimination, which may be related to the onset of cigarette smoking among African Americans. For example, feelings of discrimination or stress may intensify as African-American youths leave home and attempt to enter the work force, which could account for their later onset of smoking (see chapters by Broman and Wallace in this volume). More research is clearly needed to identify factors that explain the later onset of substance use, especially cigarette smoking, among African Americans.

Expectancies as an Explanation for Racial Differences in Substance Use In the PGS, we examined race differences in positive alcohol and tobacco expectancies as cognitive precursors of use during childhood and early adolescence. In the second oldest PGS cohort, positive alcohol expectancies tended to increase over ages 7–10 among White girls, possibly signaling their increasing readiness to engage in alcohol use (Hipwell et al. 2005; Chung et al. 2008). A similar increase in positive alcohol expectancies was not observed for African American girls. Analyses of racial differences in alcohol use in the oldest PGS cohort at ages 11–15 (Loeber et al. 2010) provide indirect support for White girls' increasing readiness to engage in alcohol use during early adolescence; White, compared to African-American, girls were more likely to report alcohol use at age 11, and showed more rapid increase in alcohol use through age 15.

We also examined positive tobacco expectancies in the oldest PGS cohort as a predictor of tobacco use in early adolescence (Chung et al. 2010). At ages 11–14, African- American, relative to White, girls initially had more positive tobacco expectancies, and less rapid change in positive expectancies. White girls tended to show an increase in positive tobacco expectancies over time, generally in parallel with an increase in smoking prevalence, as would be predicted by expectancy theory. Racial differences in smoking prevalence emerged at age 14, with greater past year smoking prevalence among White (17 %), compared to African-American (8 %), girls.

We observed some racial differences in risk factors for positive alcohol expectancies, and early alcohol and tobacco use in PGS. For example, among White, but not African-American, girls in the second oldest PGS cohort, greater physical aggression was associated with more positive initial alcohol expectancies in childhood (Chung et al. 2008). Other analyses in the oldest PGS cohort indicated that, among White girls, conduct problems prospectively predicted alcohol use at ages 11–13, whereas this prospective association was observed at ages 13–14 for African-American girls (Loeber et al. 2010). Taken together, these results indicate the role of early externalizing behaviors as a risk factor for early alcohol use, particularly for White girls, and highlight racial differences in the developmental emergence of risk. In addition, we observed some racial differences in the presence of risk factors commonly associated with tobacco use. Despite experiencing more risk factors for smoking (e.g., greater socio-economic disadvantage, more depression), African-American, compared to White, girls reported lower smoking prevalence during early adolescence (Chung et al. 2010). African-American girls' apparent resistance to risks commonly associated with early tobacco use warrants further investigation to identify protective factors that buffer African-American girls' risk for using certain substances, particularly under adverse conditions.

Associations Between Substance Use and Violence

As stated earlier, more research is needed to examine the associations between substance use and other problem behaviors and how these associations differ by race. One problem behavior that has often been linked to substance use is violence (White et al. 2009). Several race-related issues are important when reviewing the evidence for the association between substance use and violence: (1) Violence is more common among African-American, compared to White, youth (e.g., Loeber et al. 2005, 2008; Snyder and Sickmund 2006); (2) As reviewed earlier in this chapter, alcohol use tends to be more common among White, compared to African-American, youth during adolescence and early adulthood; and (3) Drug dealing is more common among African-American than White, youth (Johnson et al. 1990). These findings elicit two important questions, which are addressed below: (1) Does substance use play a different role in violence by White, compared to African-American, youth? (2) Is violence associated with drug dealing more common among African-American than White young men?

Racial Differences in Violence Earlier in this chapter, we presented data showing that African-American young men, compared to White young men, reported *lower* prevalence of alcohol and hard drug use. However, African-American young men, compared to White young men, reported *higher* marijuana use after age 15. These findings should be evaluated in the light of developmental changes in violence. Studies have consistently found that violence increases with age and peaks in late adolescence and early adulthood (Elliott 1994; Loeber et al. 2008). Figure 8.12 shows racial differences in the annual prevalence of serious violence (all source violence measure, which combines youth, caretaker and teacher report with official records, weighted; see Measures section above) in the PYS oldest cohort. African-American, compared to White, young men report a higher annual prevalence of violence between ages 13 and 25. The magnitude of the difference is most distinct between 18 and 23 when the prevalence of violence by African-American young

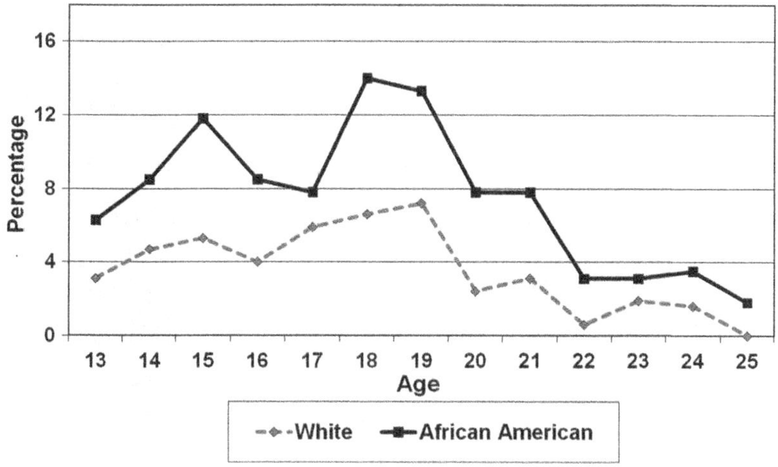

Fig. 8.12 Prevalence of violence (combined report and official record data) in the PYS oldest sample by race (weighted)

men is at least twice as high as Whites (except at age 19). Seen from a developmental perspective, on average African-American young men outgrow violence at a slower pace than do White young men.

It can be argued that what matters is whether race itself predicts violence. This important question, however, should be examined in the context of other factors that are known to predict violence. Therefore, the key question is whether race predicts violence once other factors are taken into account. These other factors should represent factors in the social environment such as the family, peer group, school, and neighborhood, as well as individual factors. Loeber et al. (2008) showed that African-American status significantly predicted all source violence (unweighted) in seven out of seven bivariate analyses for violence across different age periods in the youngest and oldest PYS cohorts. However, once other factors were taken into account, African-American status no longer contributed to the prediction of violence in seven out of seven regression analyses. Thus, there is overwhelming evidence in the PYS that the association between African-American status and violence can be accounted for by the presence of several other, non-race factors that are associated with race.

Racial Differences in the Relationship Between Intoxication and Violence Although previous studies have shown that substance use predicts violence, a key question is whether this association differs by race. Whereas patterns of substance use, especially alcohol use, have been shown to be associated with violence, it can be argued that it is not regular use, but intoxication by either alcohol or drugs that is particularly associated with violence (White et al. 2009). Also, an argument can be made that given that the prevalence of alcohol and hard drug use tend to be lower for African-American, compared to White, young men, the probability of intoxication and associated vio-

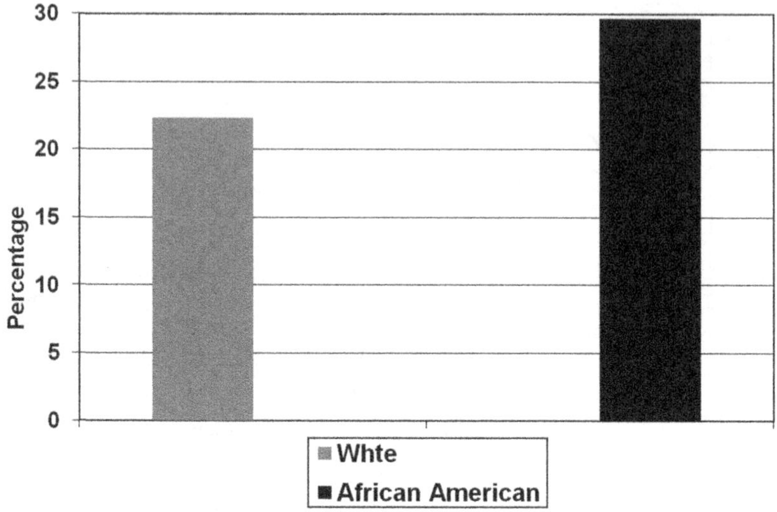

Fig. 8.13 Percentage of self-reported violent offenders in the PYS sample who were intoxicated during a violent offense ages 13–19 by race (weighted)

lence will be lower for African-American than White young men. In other words, the question we addressed is whether the association between intoxication and violence is the same for each racial group. To answer this question, we used self-reports of intoxication proximal to the violent act. Those youths who self-reported that they committed a violent offense in any year were then asked if they committed their most serious offense that year under the influence of alcohol and/or drugs.

Figure 8.13 shows the percentage of violent (those who self-reported any serious violent offense between ages 13–19) young men in the PYS sample who reported being intoxicated when committing a serious violent offense at any age from 13 to 19 years. (These data were weighted.) Although the percentage of young men intoxicated while committing violence was slightly higher for African-American than White young men, the difference was not statistically significant (chi square = 1.21, $df = 1$, $p = .27$). Thus, although violence was more common among African-American, compared to White, young men, intoxication during violence was equally common within both racial groups.

Racial Differences in the Relationship Between Dealing and Violence The next questions concerned drug dealing. Given that both drug dealing and violence have been found to be common among African-American young men, we examined whether African-American young men who dealt drugs were more violent than Whites who dealt drugs and whether the findings held for dealing marijuana and dealing hard drugs.

First, in support of prior research, we found that, in both PYS cohorts combined, self-reported commission of a violent offense between ages 13–19 (weighted) was

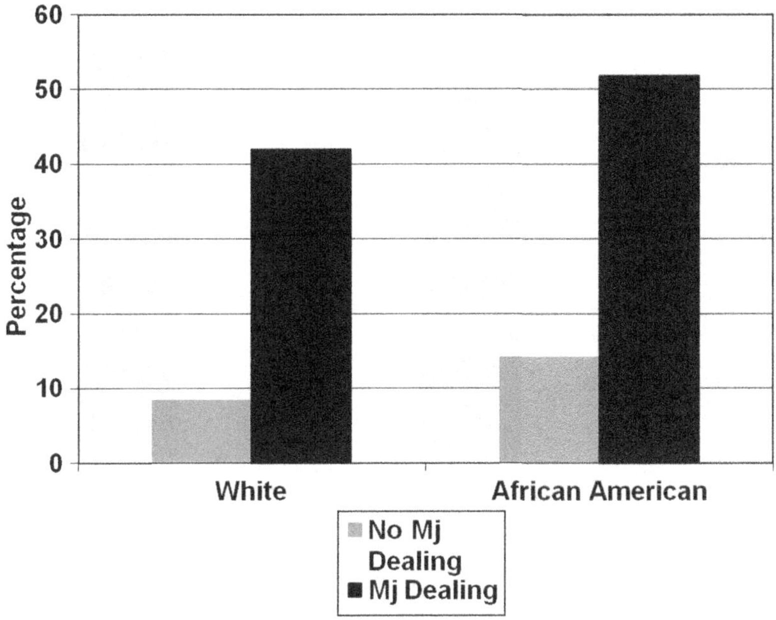

Fig. 8.14 Prevalence of self-reported violence in the PYS sample among those who did and did not deal marijuana (Mj) ages 13–19 by race (weighted)

significantly (chi square = 15.28, $df = 1$, $p < .001$) more prevalent among African Americans (26.4 %) than Whites (15.9 %). In addition, between ages 13–19 in both PYS cohorts combined, marijuana dealing was significantly (chi square = 12.01, $df = 1$, $p < .001$) more prevalent among African Americans than Whites (32.5 % vs. 22.4 %, respectively), as was hard drug dealing (29.5 % vs. 7.2 %, respectively; chi square = 74.10, $df = 1$, $p < .001$).

Next, we conducted weighted hierarchical logistic regression analyses using both PYS cohorts combined to determine whether race, dealing marijuana at any time between ages 13 and 19, and their interaction predicted committing a serious violent offense (self-reported) at any age between ages 13 and 19. Figure 8.14 shows the prevalence (weighted) of violence for those who dealt and did not deal marijuana among both Whites and African Americans. The logistic regression analyses indicated that marijuana dealers were 7.0 times ($p < .001$) more likely to report violence than those who did not deal marijuana and African Americans were 1.6 times ($p < .01$) more likely to commit violence than Whites. However, the interaction between marijuana dealing and race was not statistically significant ($p = .59$). Thus, even though marijuana dealing and violence were more common among African-American than White young men, the association between violence and marijuana dealing was similar across races. In other words, marijuana dealers, compared to those who did not deal marijuana, were more prone to be violent, but the association was not different for each race.

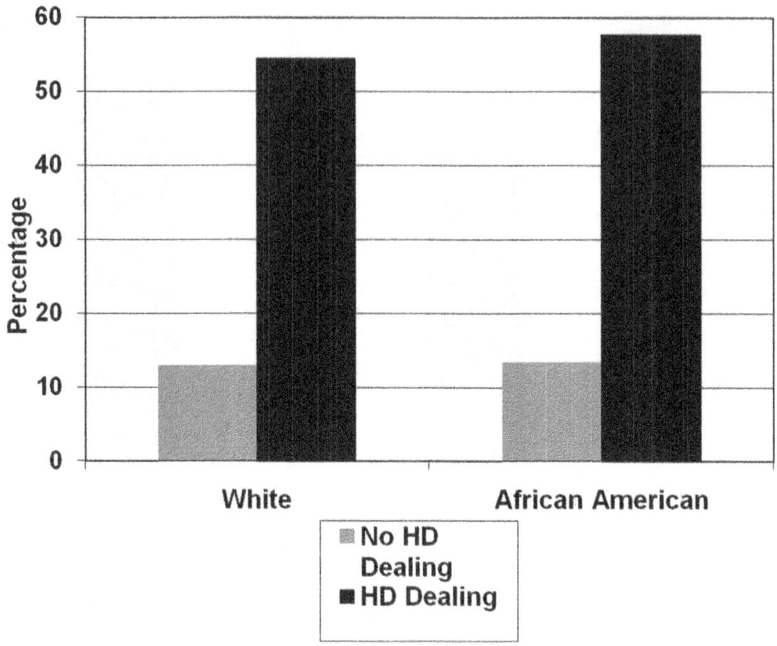

Fig. 8.15 Prevalence of self-reported violence in the PYS sample among those who did and did not deal hard drugs (HD) ages 13–19 by race (weighted) (Adapted from White et al. 2007)

Figure 8.15 shows the same information for dealing hard drugs. Dealing hard drugs was a significant predictor of violence (Odds Ratio = 8.7, $p < .001$), but, with dealing in the model, race (being African-American) was not a significant predictor (Odds Ratio = 1.1, $p = .76$). In addition, the interaction between race and hard drug dealing was not statistically significant ($p = .83$). Thus, the two races did not differ in the association between hard drug dealing and violence.

In summary, although drug dealing was more common among African-American than White young men, the association between dealing both marijuana and hard drugs and violence was the same across racial groups. Thus, both types of drug dealing are related to the commission of violence to the same degree for both races.

Discussion

Summary and Limitations

As stated earlier, there are several cautions that need to be considered when evaluating the findings reported in this chapter. First, our samples came from one city in the United States and were limited to specific cohorts who grew up during unique drug eras. Thus, the findings may not generalize to other populations and historical periods. Furthermore, the substance use measures were based on self reports from participants and were not validated against biochemical measures. In addition, for most comparisons, we discussed racial differences in substance use trends without performing any statistical analyses to determine if the observed differences were significant.

Within our limited samples, we found that there are racial differences in substance use but they vary by type of drug examined and characteristics of respondents. For example, whereas in the PYS sample Whites generally reported higher rates of alcohol and hard drug use than African Americans, for marijuana there was a cross-over effect in adolescence. In the youngest PYS cohort this cross-over occurred in early adolescence when African Americans surpassed Whites, whereas in the oldest PYS cohort, the cross-over occurred in middle adolescence. Previously (White et al. 2008) we had compared annual prevalence rates at age 18 for the PYS cohorts with those for high school seniors at the same time (1998 for the youngest cohort and 1992 for the oldest cohort) using data from the Monitoring the Future (MTF) Study (Johnston et al. 2001). Despite our over-representation of African Americans, the annual prevalence rates for marijuana (39 % PYS vs. 38 % MTF) and alcohol (79 % PYS vs. 81 % MTF) in the youngest cohort were virtually identical to the national rates; the annual prevalence rates for marijuana were somewhat higher in the PYS oldest cohort (37 %) compared to the MTF sample (22 %), whereas rates for alcohol were the same (74 % PYS vs. 73 % MTF). Using the Woodlawn cohort, an inner-city sample of African-American boys and girls born 15–20 years before the PYS sample, Ensminger and colleagues (in this volume, Chap. 4) compared their rates of adolescent substance use to those in the MTF at a comparable time. The Woodlawn rates were higher for all substances, especially compared to African Americans in the MTF. Historical changes probably account for the larger discrepancies for the Woodlawn data, compared to the PYS data. From the late 1960 and through the 1970s, both legal and illegal drug use became more normative among all youth regardless of race or community (Johnston et al. 2001).

Among males in the youngest PYS cohort, more African Americans used marijuana than Whites and more Whites used alcohol than African Americans. Race differences for females in the PGS were less consistent. Across the races, males reported higher use of both alcohol and marijuana than females. When analyses were conducted by race, we found greater discrepancies between African-American males and females than between White males and females. Although males of both races reported greater annual prevalence of marijuana use during adolescence than females, the gap was greater for African-American boys and girls than for White boys and girls. Further, while there appeared to be much higher prevalence rates for alcohol use among African-American boys than girls during adolescence, gender differences for White boys and girls were negligible. Consistent with our findings, Ensminger et al. (2002) found higher rates of all substance use during adolescence for males compared to females among African Americans in the Woodlawn Study. In young adulthood, Ensminger et al. (2002) did not find gender differences in alcohol or tobacco use, whereas illegal drug use was still higher among males than females in the Woodlawn study (see chapter by Ensminger et al. in this volume, Chap. 4). Overall, our findings and theirs support our original contention that research on racial differences in substance use needs to consider heterogeneity among users and types of drugs.

In our previous research, White et al. (2007a) found that African Americans, compared to Whites, were less likely to follow the typical sequence of substance use, and many African-American drug users initiated marijuana or hard drugs prior to trying alcohol and tobacco. Nevertheless, there did not appear to be any difference between the races in terms of addiction liability through late adolescence. In the Woodlawn study, Doherty et al. (2007) also examined the sequence of drug use onset. They found 84 % followed the typical sequence starting either with alcohol *or* tobacco followed by marijuana. However, for most youth, marijuana onset preceded tobacco onset. Thus, in two different cohorts of African-Americans, separated by about 20 years, marijuana was often used prior to legal substances, especially tobacco.

Emerging findings from the PGS identified some racial differences in cognitive precursors, such as positive expectancies, and risk factors for substance use. Specifically, White girls generally showed an increase in positive expectancies through early adolescence, which was generally followed by increases in alcohol and tobacco use. In contrast, African-American girls tended to have relatively high initial positive expectancies, and to show little change in positive expectancies over time. These racial differences in the developmental trajectories of positive expectancies begin to explain some of the observed racial differences in substance use prevalence and should be studied further. In addition, African-American girls' greater experience of risks commonly associated with tobacco use, and apparent resistance to these risks, highlight the importance of identifying protective factors that reduce African-American girls' risk for early tobacco use.

Among African-American males in the PYS, White et al. (2007b) found that a sizeable percentage of emerging adult cigarette smokers began smoking after age 16. Although the researchers were able to identify risk and protective factors in adolescence that could differentiate early-onset smokers from later-onset and non-onset smokers (i.e., peer drug use and truancy in the multivariate analyses), they could not predict which African-American nonsmokers would go on to become smokers in late adolescence and emerging adulthood. Even at ages 17–19 very few risk and protective factors differentiated nonsmokers from late-onset smokers. Using the Woodlawn data, Juon et al. (2002) also found that the majority of current cigarette smokers in young adulthood had first initiated after age 17. More research is needed to examine *why* there are differences in initiation and patterns of cigarette use for African Americans and Whites.

We found that more African-American, compared to White, males in the PYS engaged in drug dealing and violence. Despite these race differences, however, intoxication during violence occurred at a rate that was similar for African Americans and Whites and violence related to drug dealing occurred at a rate that was similar across the two racial groups. Thus, the associations of substance intoxication and dealing with violence did not differ by race.

Future Research

Little is known about the maturation process out of substance use for African Americans. Studies generally indicate that youth peak in their heavy drinking and drug use in their early 1920s and then mature out in their late 20s and early 30s (Bachman et al. 1997; White et al. 2005). Nevertheless, most of this research has been conducted on Whites and especially on college students. One, therefore, needs to follow African Americans regularly from early adolescence into adulthood to understand more about escalation and cessation. We are currently collecting data on the PYS sample in young adulthood (ages 28–34). When these data become available, we will be in a better position to understand developmental trajectories of substance use for African Americans. Doherty and colleagues (2007) demonstrated that current drug use in adulthood persisted longer for the Woodlawn sample, compared to national samples, suggesting that African Americans may not mature out of substance use, especially illegal drug use, as quickly as Whites (see also Ensminger et al. 1997). It is important to determine if these findings hold up in more recent cohorts.

There are several additional issues that we hope to address in our future research that we have not focused on in this chapter. Specifically, we plan to focus on what is going on during the transition from emerging adulthood to young adulthood to explain why some individuals mature out of substance use and others do not. Furthermore, we plan to determine whether the predictors of maturation (e.g., marriage, parenthood, career) are similar for Whites and African Americans. While in this chapter we conducted preliminary analyses of racial differences in the associations between substance use and violence, we will continue to examine these associations as well as the associations between trajectories of drug use and other problem behaviors (e.g., crime, mental health) and whether these associations differ by race. Finally, future research needs to identify intra-racial factors that affect trajectories of substance use and abuse.

Implications for Prevention

The PYS and PGS studies indicate clear racial differences in levels and shapes of developmental trajectories of, and risk factors for, substance use (e.g., development of substance use expectancies), as well as some commonalities in the associations of substance use with other risk behaviors, such as violence. Other research also has identified differences between African Americans and Whites in exposure and vulnerability to risk factors for substance use in (Wallace and Muroff 2002). Importantly, PYS data suggest that race differences may reflect differences in the social environment (e.g., neighborhood conditions, exposure to violence, poverty) that are often correlated with race (Loeber et al. 2008). Given some differences, as well as similarities, in risk and protection for African-American and White youth, a key question is how to tailor preventive interventions to be maximally effective for subgroups of youth, since risk may differ across certain subgroups (e.g., race, gender) for specific substances, timing of initiation to use, and factors associated with substance use (Castro et al. 2004).

With regard to differences in the timing and type of substances used among African-American and White youth, PYS data indicate that, in early adolescence, prevention efforts could target alcohol and tobacco among White youth, but that it might make more sense to target marijuana use, rather than alcohol or cigarette use, for African Americans (White et al. 2007a). Based on PGS data, in early adolescence, White girls were more likely to report substance use than African-American girls, suggesting the potential utility of targeting reduction of risk in White girls, and reinforcing protective factors among African-American girls to reduce early substance use. In both the PYS and PGS, we found that African Americans appear to be protected, at least during adolescence, from smoking cigarettes. Nevertheless, African-American, compared to White, adult smokers experience greater morbidity and mortality related to smoking (Juon et al. 2002). Thus, tobacco prevention programs delivered to African-American youth could focus on addressing risks specific to middle to late adolescence, the period during which risk for initiation and escalation of tobacco use increases in this subgroup. Overall, our studies indicate that there are different ages that may be optimal to intervene for different substances and these ages depend on race and gender.

Extending prevention implications into emerging adulthood, PYS data in emerging adulthood indicate less hard drug use among African-American, compared to White, males. Longer-term results from the Woodlawn study suggest that African Americans are more likely to initiate illegal drug use in adulthood, and less likely to desist from use (see the chapter by Ensminger et al. in this volume, Chap. 4). In addition, the Woodlawn study found higher rates of dependence among African-American illegal drug users, compared to White illegal drug users. We are in the process of collecting data in the PYS during adulthood to identify changes in risk and protection in adulthood that contribute to increased risk for illegal drug use during this developmental period. Preventive substance use interventions that continue into adulthood, and which target illicit drug use, may reduce risk for substance-related harms among African-American individuals.

Our research with the PGS suggests that risk factors, such as substance use expectancies and co-occurring psychopathology, may differ developmentally in their prediction of African-American and White substance use. Specifically, racial differences in developmental trajectories of positive alcohol and tobacco expectancies were correlated with trajectories of use, suggesting the possibility that expectancies represent a potentially malleable mechanism associated with risk for substance use. Racial differences in the factors that help to shape girls' expectancies, such as parental attitudes toward substance use and parenting behaviors, need to be identified to develop culturally-tailored interventions that reduce risk for substance use. For example, Doherty et al. (2007) showed the importance of parental rule setting and family cohesion for urban African-American girls. Thus, prevention efforts that aim to amplify and support protective factors that buffer risk in specific subgroups are needed.

We also identified common associations between substance use and violence across race in the PYS. Specifically, some of the risk factors for violence were similar for African Americans and Whites, and substance use and drug dealing related similarly to violence for both races. Thus, these findings suggest that, once young males are identified for violence prevention programs, focusing on similar risk factors for violence may work for both races. Furthermore, data from both the PYS and PGS suggest that early conduct problems and aggression are risk factors for substance use. These findings are consistent with those from the Woodlawn study (Ensminger et al. 2002), as well as numerous other studies (for a review see White and Gorman 2000). There is robust support for the utility of screening children for externalizing behaviors and providing targeted interventions for high-risk youth to reduce these behaviors early in development. Doing so may help to reduce or delay the onset of substance use.

In determining whether different intervention approaches are needed for specific youth subgroups, it is important to recognize that race/ethnicity is often associated with socio-ecological and socio-economic factors that may explain observed differences by race/ethnicity, and that even within racial/ethnic subgroups heterogeneity exists (e.g., age, gender, neighborhood conditions). One community-based approach to intervention (i.e., Communities that Care; Hawkins et al. 2009) uses epidemiological data to develop a community-specific profile of elevated risk and reduced protective factors in order to select empirically-based interventions (school-based, family-based, individual-level) that are matched to local socio-ecological and socio-economic resources and needs, and to specific developmental periods (e.g., ages 10–14). Within this broader context of matching on community-level risk and resources, interventions can be tailored to accommodate the core values and beliefs of specific subgroups. For example, ethnic/racial identity has been found to play a protective role in reducing risk for substance use, particularly among African-American youth (Brook and Pahl 2005). Effective prevention and intervention to reduce substance use needs to recognize the role of common, as well as subgroup-specific risk and protective factors, that operate across development.

Acknowledgements This research was supported, in part, by grants from the National Institute on Drug Abuse (P20 DA17552; R01 DA411018; R01 DA012237), the National Institute of Alcohol Abuse and Alcoholism (ARRA R01 AA016798; AA014357; AA017128), the National Institute of Mental Health (P30 MH079920; R01 MH050778; R01 MH73941; R01 MH056630), the Office of Juvenile Justice and Delinquency Prevention (96-MU-FX-0012; OJJDP 2005-JK-FX-0001; Department of Health, Commonwealth of the State of Pennsylvania, the FISA Foundation, and the Falk Fund. Points of view in this document are those of the authors and do not necessarily represent the official position or policies of the U.S. Department of Justice. We thank Rebecca Stallings for her assistance in preparing the data files.

References

Bachman, J. G., Wadsworth, K. N., O'Malley, P. M., Johnston, L. D., & Schulenberg, J. E. (1997). *Smoking, drinking, and drug use in young adulthood: The impacts of new freedoms and new responsibilities*. Mahwah: Lawrence Erlbaum Associates.

Bauman, K. E., & Ennett, S. E. (1994). Tobacco use by African American and white adolescents: The validity of self-reports. *American Journal of Public Health, 84*, 394–398.

Brook, J., & Pahl, K. (2005). The protective role of ethnic and racial identity and aspects of Africentric orientation against drug use among African American young adults. *Journal of Genetic Psychology, 166*, 329–345.

Castro, F. G., Barrera, M., & Martinez, C. R. (2004). The cultural adaptation of prevention interventions: Resolving tensions between fidelity and fit. *Prevention Science, 5*, 41–45.

Chung, T., Hipwell, A. E., Loeber, R., White, H. R., & Stouthamer-Loeber, M. (2008). Ethnic differences in positive alcohol expectancies during childhood: The Pittsburgh Girls Study. *Alcoholism, Clinical and Experimental Research, 32*, 966–974.

Chung, T., White, H. R., Hipwell, A. E., Stepp, S. D., & Loeber, R. (2010). A parallel process model of the development of positive smoking expectancies and smoking behavior during early adolescence in Caucasian and African-American girls. *Addictive Behaviors, 35*, 647–650.

Doherty, E. E., Green, K. M., Resinger, H. S., & Ensminger, M. E. (2007). Long-term patterns of drug use among an urban African-American cohort: The role of gender and family. *Journal of Urban Health, 85*, 250–267.

Dunn, M., & Goldman, M. (1996). Empirical modeling of an alcohol expectancy memory network in elementary school children as a function of grade. *Experimental and Clinical Psychopharmacology, 4*, 209–217.

Elliott, D. S. (1994). Serious violent offenders: Onset, developmental course, and termination – The American Society of Criminology 1993 Presidential Address. *Criminology, 32*, 1–21.

Ensminger, M. E., Anthony, J. C., & McCord, J. (1997). The inner city and drug use: Initial findings from an epidemiological studies. *Drug and Alcohol Dependence, 48*, 175–184.

Ensminger, M. E., Juon, H. S., & Fothergill, K. E. (2002). Childhood and adolescent antecedents of substance use in adulthood. *Addiction, 97*, 833–844.

Fisher, M. A., Taylor, G. W., Shelton, B. J., & Debanne, S. (2008). Age and race/ethnicity-gender predictors of denying smoking, United States. *Journal of Health Care for the Poor and Underserved, 19*, 75–89.

Golub, A. L., & Johnson, B. D. (1999). Cohort changes in illegal drug use among arrestees in Manhattan: From the heroin injection generation to the blunts generation. *Substance Use and Misuse, 34*, 1733–1763.

Griesler, P. C., & Kandel, D. B. (1998). Ethnic differences in correlates of adolescent cigarette smoking. *Journal of Adolescent Health, 23*, 167–180.

Hawkins, J. D., Oesterle, S., Brown, E. C., Arthur, M. W., Abbott, R. D., Fagan, A. A., & Catalano, R. F. (2009). Results of a type 2 translational research trial to prevent adolescent drug use and delinquency. *Archives of Pediatric and Adolescent Medicine, 163*, 789–798.

Hipwell, A. E., Loeber, R., Southamer-Loeber, M., Keenan, K., White, H. R., & Kroneman, L. (2002). Characteristics of girls with early onset disruptive and antisocial behavior. *Criminal Behavior and Mental Health, 12*, 99–118.

Hipwell, A. E., White, H. R., Loeber, R., Stouthamer-Loeber, M., Chung, T., & Sembower, M. A. (2005). Young girls' expectancies about the effects of alcohol, future intentions and pattern of use. *Journal of Studies on Alcohol, 66*, 630–639.

Hornik, R., Maklan, D., Cadell, D., Prado, A., Barmada, C., Jacobsohn, L., Orwin, R., Sridharan, S., Zador, P., Southwell, B., Zanutto, E., Baskin, R., Chu, A., Morin, C., Taylor, K., & Steel, D. (2002). *Evaluation of the national youth anti-drug media campaign: Fourth semi-annual report of findings*. Rockville: Westat.

Johnson, B. D., Williams, T., Dei, K. A., & Sanabria, H. (1990). Drug use in the inner city: Impact on hard-drug users and the community. In M. Tonry & J. Q. Wilson (Eds.), Drugs and crime. *Crime and justice, a review of research*, vol. 13 (pp. 9–68). Chicago: Chicago University Press.

Johnston, L. D., O'Malley, P. M., & Bachman, J. G. (2001). *Monitoring the future: National survey results on drug use*. Rockville: National Institute on Drug Abuse.

Johnston, L. D., O'Malley, P. M., Bachman, J. G., & Schulenberg, J. E. (2009). *Demographic subgroup trends for various licit and illicit drugs, 1975–2008* (Monitoring the future occasional paper no. 71). Ann Arbor: Institute for Social Research. Available: http://www.monitoringthefuture.org/

Johnston, L. D., O'Malley, P. M., Bachman, J. G., & Schulenberg, J. E. (2011). *Monitoring the future national results on adolescent drug use: Overview of key findings, 2010*. Ann Arbor: Institute for Social Research, The University of Michigan.

Jones, B. T., Corbin, W., & Fromme, K. (2001). A review of expectancy theory and alcohol consumption. *Addiction, 96*, 57–72.

Juon, H. S., Ensminger, M. E., & Sydnor, K. D. (2002). A longitudinal study of developmental trajectories to young adult cigarette smoking. *Drug and Alcohol Dependence, 66*, 303–314.

Kandel, D. B. (1991). The social demography of drug use. *Milbank Quarterly, 69*, 365–414.

Keenan, K., Hipwell, A. E., Chung, T., Stepp, S., Stouthamer-Loeber, M., Loeber, R., & McTigue, K. (2010). The Pittsburgh girls study: Overview and initial findings. *Journal of Clinical Child and Adolescent Psychology, 39*, 506–521.

Lee, Y. G., & Abdel-Ghany, M. (2004). American youth consumption of licit and illicit substances. *International Journal of Consumer Studies, 28*, 454–465.

Lee, C., Mun, E. Y., White, H. R., & Simon, P. (2010). Substance use trajectories of African American and white young men from adolescence to emerging adulthood: A two-part growth curve analysis. *Journal of Ethnicity in Substance Abuse, 9*, 301–319.

Loeber, R., Pardini, D., Homish, D. L., Wei, E. H., Crawford, A. M., Farrington, D. P., Stouthamer-Loeber, M., Creemers, J., Koehler, S. A., & Rosenfeld, R. (2005). The prediction of violence and homicide in young men. *Journal of Consulting and Clinical Psychology, 73*, 1074–1088.

Loeber, R., Farrington, D., Stouthamer-Loeber, M., & White, H. R. (2008). *Violence and serious theft: Developmental course and origins from childhood to adulthood*. New York: Routledge Press.

Loeber, R., Stepp, S. D., Chung, T., Hipwell, A. E., & White, H. R. (2010). Time-varying associations between conduct problems and alcohol use in adolescent girls: The moderating role of race. *Journal of Studies on Alcohol and Drugs, 71*, 544–553.

Oetting, E. R., & Beauvais, F. (1990). Adolescent drug rise: Findings of national and local surveys. *Journal of Consulting and Clinical Psychology, 58*, 385–394.

Ridenour, T. A., Lanza, S. T., Donny, E. C., & Clark, D. B. (2006). Different lengths of time for progressions in adolescent substance involvement. *Addictive Behaviors, 31*, 962–983.

Skinner, M., Haggerty, K., & Catalano, R. (2009). Parental and peer influences on teen smoking: Are White and African American families different? *Nicotine and Tobacco Research, 11*, 558–563.

Snyder, H. N., & Sickmund, M. (2006). *Juvenile offenders and victims: 2006 national report*. Washington, DC: U. S. Department of Justice, Office of Justice Programs, Office of Juvenile Justice and Delinquency Prevention.

Stouthamer-Loeber, M., & Van Kammen, W. B. (1995). *Data collection and management: A practical guide*. Newbury Park: Sage.

Substance Abuse and Mental Health Services Administration. (2010). *Results from the 2009 National Survey on Drug Use and Health: Volume I. Summary of National Findings* (Office of applied studies, NSDUH series H-38A, HHS publication no. SMA 10- 4586 findings). Rockville.

Tragesser, S. L., Beauvais, F., Swaim, R. C., Edwards, R. W., & Oetting, E. R. (2007). Parental monitoring, peer drug involvement, and marijuana use across three ethnicities. *Journal of Cross-Cultural Psychology, 38*, 670–694.

Wallace, J. M., Jr., & Muroff, J. R. (2002). Preventing substance abuse among African American children and youth: Race differences in risk factor exposure and vulnerability. *Journal of Primary Prevention, 22*, 235–261.

Wallace, J. M., Bachman, J. G., Jr., O'Malley, P. M., Schulenberg, J. E., Cooper, S. M., & Johnston, L. D. (2003a). Gender and ethnic differences in smoking, drinking, and illicit drug use among American 8th, 10th, and 12th grader students, 1976–2000. *Addiction, 98*, 225–234.

Wallace, J. M., Jr., Brown, T. N., Bachman, J. G., & Laveist, T. A. (2003b). The influence of race and religion on abstinence from alcohol, cigarettes and marijuana among adolescents. *Journal of Studies on Alcohol, 64*, 843–848.

White, H. R., & Gorman, D. M. (2000). Dynamics of the drug-crime relationship. In G. LaFree (Ed.), *Criminal justice 2000* (The nature of crime: Continuity and change, Vol. I, pp. 151–218). Washington, DC: U.S. Department of Justice.

White, H. R., Nagin, D., Replogle, E., & Stouthamer-Loeber, M. (2004). Racial differences in trajectories of cigarette use. *Drug and Alcohol Dependence, 76*, 219–227.

White, H. R., Labouvie, E. W., & Papadaratsakis, V. (2005). Changes in substance use during the transition to adulthood: A comparison of college students and their noncollege age peers. *Journal of Drug Issues, 35*, 281–305.

White, H. R., Metzger, L., Stouthamer-Loeber, M., Violette, N., & Nagin, D. (2006). Racial differences in developmental risk and protective factors for cigarette smoking. In E. V. Metrosa (Ed.), *Racial and ethnic disparities in health and health care* (pp. 101–130). New York: Nova.

White, H. R., Jarrett, N., Valencia, E. Y., Loeber, R., & Wei, E. (2007a). Stages and sequences of initiation and regular substance use in a longitudinal cohort of African-American and White male adolescents. *Journal of Studies on Alcohol and Drugs, 68*, 173–181.

White, H. R., Violette, N. M., Metzger, L., & Stouthamer-Loeber, M. (2007b). Adolescent risk factors for late-onset smoking among African-American young men. *Nicotine & Tobacco Research, 9*, 153–161.

White, H. R., Loeber, R., & Farrington, D. P. (2008). Substance use, dealing, gang membership, and gun carrying and their predictive associations with serious violence and serious theft. In R. Loeber, D. Farrington, M. Stouthamer-Loeber, & H. R. White (Eds.), *Violence and serious theft: Developmental course and origins from childhood to adulthood* (pp. 137–166). New York: Routledge.

White, H. R., Jackson, K., & Loeber, R. (2009). Developmental sequences and comorbidity of substance use and violence. In M. Krohn, A. Lizotte, & G. P. Hall (Eds.), *Handbook of deviance and crime* (pp. 433–468). New York: Springer Publications.

Winters, K. C., Stinchfield, R. D., Henly, G. A., & Schwartz, R. H. (1991). Validity of adolescent self-reports of alcohol and other drug involvement. *The International Journal of the Addictions, 25*, 1379–1395.

Chapter 9
Predictors of Growth Trajectories of Substance Use from 9th to 11th Grade Among Hispanic Adolescents in Southern California

Jennifer B. Unger, Lourdes Baezconde-Garbanati, Anamara Ritt-Olson, Daniel W. Soto, and Chih-Ping Chou

Hispanic adolescents in the United States experience a wide range of cultural challenges including acculturative stress (Romero et al. 2007), discrimination from members of the majority culture and other cultures (Pérez et al. 2008), family conflict resulting from parent-child acculturation discrepancies (Szapocznik et al. 1978; Unger et al. 2009), and the challenges of finding a balance between adopting the cultural norms and values of the United States and preserving those of the culture of origin (Balcazar et al. 1995; Gil et al. 1994; Szapocznik et al. 1989). Most Hispanic adolescents learn to navigate these challenges and become well-adjusted, productive members of society. Unfortunately, some turn to maladaptive behaviors such as substance use. Because Hispanics are the most rapidly-growing foreign-born group in the United States (Ramirez and de la Cruz 2002), it is important to gain a more complete understanding of their risk and protective factors for problem behaviors including substance use.

Culture-related risk and protective factors for substance use among Hispanic adolescents are numerous and complex. As described below, the risk factors identified in previous studies include identification with U.S. cultural norms and values, discrimination, and acculturative stress. (For this study, US cultural norms and values are defined of those of the majority culture. Nevertheless we recognize that some youth may be acculturating to a mixture of cultures if they live in multicultural environments.) Protective factors include maintenance of connections with the culture of origin and a strong ethnic identity.

J.B. Unger (✉) • L. Baezconde-Garbanati • A. Ritt-Olson • D.W. Soto • C.-P. Chou
Keck School of Medicine, University of Southern California, Los Angeles, CA, USA
e-mail: jenniferbethunger@gmail.com; unger@usc.edu

Y.F. Thomas, L.N. Price (eds.), *Drug Use Trajectories Among Minority Youth*,
DOI 10.1007/978-94-017-7491-8_9

Culture-Related Risk Factors for Substance Use

Acculturation to the United States

Early acculturation theories have proposed that adolescents living in diverse cultural contexts can choose to identify with a new culture, with the culture of origin, with both cultures, or with neither culture (Berry 1980). In today's multicultural environments such as Los Angeles, the acculturation may be to a mixture of cultures, especially among youth who are raised in multicultural communities or have parents from different cultural backgrounds. Multiple studies have found that Hispanic adolescents who identify primarily with the U.S. culture and not with their cultures of origin are at increased risk of tobacco, alcohol, and other drug use (Brook et al. 1998; De La Rosa 2002; Epstein et al. 2001; Love et al. 2006; McQueen et al. 2003; Unger et al. 2000). Other studies have found that a bicultural orientation (identifying with both cultures) is protective against substance use and emotional health problems (Bacallao and Smokowski 2005; Carvajal et al. 2002).

Discrimination

Ethnic/racial discrimination can include overt acts such as name-calling, violence, harassment by police, or discourteous treatment by store clerks, or more subtle acts such as speaking in a manner that implies that a person is uneducated, unintelligent, or untrustworthy (Williams et al. 2008). Discrimination has been associated with smoking (Guthrie et al. 2002), aggressive behavior (Smokowski and Bacallao 2006), depression (Greene et al. 2006; Romero and Roberts 2003; Szalacha et al. 2003), and physical health problems (Williams et al. 2008) across diverse ethnic and racial groups.

Acculturative Stress

Acculturative stress includes the psychological impact of experiences related to the immigration process, being an immigrant, or being a member of a minority group. These experiences include instrumental stressors (e.g., poverty, language barriers, lack of access to healthcare, unsafe neighborhoods, poor housing, unemployment or underemployment), social stressors (e.g., loss of social networks, loss of social status, loss of family support, and intergenerational conflict), and societal stressors (e.g., stigma, political events) (Caplan 2007). Among adolescents, acculturative stress has been associated with an increased risk of substance use (Romero et al. 2007; Vega and Gil 1998).

Culture-Related Protective Factors

Ethnic Identity

Ethnic identity includes knowledge of one's membership in an ethnic group, perceptions of the value and emotional significance of that membership, feelings of belonging and commitment to the ethnic group, sharing values and attitudes with the group, and understanding and practicing the norms and customs of the group (Phinney 1990). Several studies have found that a strong ethnic identity, including strong ethnic affiliation, attachment, and pride, is protective against substance use among Hispanic youth (Brook et al. 2007b; Love et al. 2006; Marsiglia et al. 2004).

Maintenance of Hispanic Cultural Orientation

In Berry's (1980) bidimensional model of acculturation, individuals can maintain strong identification with the culture of origin, regardless of the degree to which they adopt aspects of the new culture. Maintenance of Hispanic cultural orientation may lead to better relationships between adolescents and their parents and may protect adolescents from peer influences to engage in risky behavior. Among adolescents, maintenance of Hispanic cultural orientation has been associated with a reduced risk of substance use (Allen et al. 2008; Love et al. 2006; Szapocznik et al. 2007; Unger et al. 2009).

Cultural Values

Several traditional Hispanic cultural values are relevant to adolescent substance use. *Familism/Familismo* includes the desire to maintain strong family ties, the expectation that the family will be the primary source of instrumental and emotional support, the feeling of loyalty to the family, and the commitment to the family over individual needs and desires (Negy and Woods 1992; Staples and Mirandé 1980). Familism might cause adolescents to abstain from substance use to prevent their family members from experiencing negative consequences (e.g., problems with law enforcement, exposure to secondhand smoke, or a tarnished family reputation), and/or because their obligations to care for family members are incompatible with a substance-using lifestyle. *Respeto* addresses the maintenance of harmonious interpersonal relationships through respect for self and others. This includes verbal and nonverbal rules of respect such as politely greeting elders, not challenging an elder's point of view, and not interrupting conversations between adults (Valdés 1996). Respeto might make adolescents more likely to obey their parents' rules about substance use. *Fatalism* is the belief that one's destiny is not under one's personal control, but instead is controlled by God or fate (Neff and Hoppe 1993). Adolescents with an internal locus of control may abstain from substance use to protect their health, whereas adolescents with fatalistic beliefs may experiment with substance use freely because they believe that their risk of disease is controlled by fate, rather than by their own actions.

Trajectories of Substance Use Among Hispanic Adolescents

Several previous studies have examined trajectories in substance use among Hispanic adolescents. For cigarettes, alcohol, and illicit drugs, Hispanics typically show a pattern of earlier initiation compared with Whites, African Americans, and Asians (Ellickson et al. 2004; Szapocznik et al. 2007). For example, in the Monitoring The Future Surveys, Hispanic 8th and 10th graders report the highest lifetime, annual, and 30-day prevalence rates of alcohol, cigarette, and illicit drug use of any racial/ethnic groups studied (Johnston et al. 2009). Later in adolescence and young adulthood, Whites catch up and surpass Hispanics in substance use. By 12th grade, Whites typically have a higher prevalence of substance use than Hispanics do (Johnston et al. 2009).

Some researchers have attributed this reversal in ranking between Hispanics and Whites to the higher school dropout rates among Hispanics. If more substance-using Hispanics drop out of school before 12th grade, fewer substance-using Hispanics remain in school to complete school-based surveys. However, non-school-based studies such as the National Household Survey on Drug Use and Health (SAMHSA 2008) and longitudinal cohort studies that followed adolescents regardless of their dropout status (Ellickson et al. 2004) have also demonstrated this pattern.

Of course, not all Hispanic adolescents follow the same substance use trajectory. There is considerable individual variation in age of first experimentation and progression from experimentation to more frequent use. Similar to studies of other racial/ethnic groups (e.g., Colder et al. 2002; Jackson et al. 2008; Windle and Wiesner 2004), studies of Hispanic adolescents have identified distinct trajectory groups, including nonusers, early initiators who quit or decrease their use, late initiators, and early initiators who continue using (Brook et al. 2006a). In general, Hispanic adolescents who begin using substances early have the greatest risk for substance use problems later in adulthood (Brook et al. 2006b), and these early initiators also show the most problematic patterns of risk factors such as mental health problems, peer and parental use, and family conflict (Bray et al. 2001, 2003; Brook et al. 2006a).

More research is needed to understand the risk and protective factors for substance use throughout the process of adolescent development. Specifically, it is important to understand which factors influence substance use early in adolescence, and which factors influence the growth trajectories of substance use throughout adolescence and young adulthood.

This article examines the predictors of growth trajectories of substance use with data from a longitudinal study of substance use among Hispanic adolescents in Southern California. This study surveyed Hispanic adolescents at three timepoints: 9th, 10th, and 11th grade. This made it possible to examine the effects of hypothesized predictor variables on two parameters of growth in substance use: the *initial status* (an estimate of the respondent's level of drug use in 9th grade), and the *slope* (an estimate of the rate of increase in drug use from 9th to 11th grade). This analysis provides a more complete representation of each individual's drug use pattern than could be obtained from a single measure taken at one point in time.

Method

School Recruitment

Project RED (Reteniendo y Entendiendo Diversidad para Salud) is a longitudinal study of acculturation patterns and substance use among Hispanic/Latino adolescents in Southern California. The respondents in this study were students attending seven high schools in the Los Angeles area who completed surveys in 9th, 10th, and 11th grade. Because this is a study of Hispanic adolescents, schools were approached and invited to participate if they contained at least 70 % Hispanic students, as indicated by data from the California Board of Education, and were not participating in other studies or interventions designed to address variables of interest in this study. Efforts were also made to obtain a sample of schools with a wide range of socioeconomic characteristics. The median annual household incomes in the ZIP codes served by the schools ranged from $29,000 to $73,000, according to 2000 U.S. Census data. Approval was obtained from the school principals and/or district superintendents, according to their established procedures.

Student Recruitment

The 9th, 10th, and 11th grade surveys were conducted in the Fall of 2005, 2006, and 2007, respectively. In 2005, all 9th-grade students in the school were invited to participate in the survey. Trained research assistants visited the students' classrooms, explained the study, and distributed consent forms for the students to take home for their parents to sign. If students did not return the consent forms, the research assistants telephoned their parents to ask for verbal parental consent. Students with written or verbal parental consent were allowed to participate. Although the students were minors and could not give legal consent, we also gave them the opportunity to assent or decline to participate, as a way of involving them in the decision-making process. This procedure was approved by the university's Institutional Review Board.

Across the 7 schools, 3218 students were invited to participate. Of those, 2420 (75 %) provided parental consent and student assent. Of those, 2222 (92 %)

completed the survey in 9th grade. Of the 2222 students who completed the 9th grade survey, 1773 (80 %) also completed surveys in 10th and 11th grade, 182 (8 %) completed a survey in 10th grade but not in 11th grade, 50 (2 %) completed a survey in 11th grade but not in 10th grade, and 217 (10 %) were lost to attrition before the 10th grade survey.

Of the 2222 students who completed the 9th grade survey, 1963 (88 %) self-identified as Hispanic or Latino or reported a Latin American country of origin. Of those 1963 Hispanic/Latino students, 1668 (85 %) provided complete data on all variables of interest on the 9th grade survey. These 1668 students are included in this analysis. Because growth curve analyses can handle missing data, all students with complete 9th grade data could be included in the analysis, regardless of whether they participated in the 10th or 11th grade surveys.

Survey Procedure

On the day of the survey, the data collectors distributed surveys to all students who had provided parental consent and student assent. Using a standardized script, they reminded the students that their responses were confidential and that they could skip any questions they did not want to answer. The classroom teachers were present during survey administration, but the data collectors instructed them not to participate in the survey process to ensure that they would not inadvertently see the students' responses. To help students with low literacy skills, the data collectors also read the entire survey aloud during the class period so the students could follow along.

The data collectors returned to the schools when the students were in 10th and 11th grade. Students who could be located in the same schools (and students who had transferred to another school participating in the study) completed follow-up surveys in their classrooms, using the same procedure used in 9th grade. Extensive tracking procedures were used to locate the students who had transferred schools. For the 9th grade survey, students filled out a Student Information Sheet with contact information such as their home addresses, home phone numbers, cell phone numbers, parents' cell phone numbers, email addresses, and addresses and phone numbers of a relative or close family friend who would know their whereabouts if they moved. School personnel also provided forwarding information if available. Data collectors telephoned the missing students in the evenings and surveyed them by telephone.

Measures

Surveys were available in English and Spanish. To create the Spanish translations, we first looked for the translated items that were published or recommended by the scales' authors. If none were available, one translator translated the items from

English to Spanish, and then the translation was checked by a translation team including bilingual researchers of Mexican, Salvadoran, and Argentinean descent. This procedure was used to ensure that the Spanish translation reflected the idioms that are used among Mexican-Americans and other Hispanic/Latinos living in Southern California. Although English and Spanish versions were available, only 17 students (0.8 %) chose to complete the survey in Spanish. The survey assessed substance use, acculturation, family and peer characteristics, psychological variables, and demographic characteristics.

Adolescents' Acculturation

Adolescents responded to 12 items from the Revised Acculturation Rating Scale for Mexican Americans (ARSMA-II; Cuellar et al. 1995b): 7 from the Anglo orientation subscale and 5 from the Hispanic orientation subscale. These 12 items were selected based on a pilot study in a similar school (Unger, unpublished), in which these items had the highest factor loadings on their respective scales. Recently, a shorter version of the ARSMA-II was validated among adolescents (Bauman 2005). Unfortunately, this scale had not yet been published at the time when this survey was conducted. Of the 12 items that we selected for our scale, 10 were also included in Bauman's scale; Bauman included 2 items that we did not include (enjoying reading in Spanish and enjoying English movies), and we included 2 items that Bauman did not include (enjoying English music and enjoying reading in English). The remaining 10 items were identical across the two scales (enjoying Spanish language TV, enjoying speaking Spanish, enjoying Spanish movies, speaking Spanish, thinking in Spanish, speaking English, writing letters in English, associating with Anglos, thinking in English, and having Anglo friends). The scores on each subscale (U.S. Orientation and Hispanic Orientation) were rescaled so that they ranged from 0 = lowest to 1 = highest. The wording of the response options was not changed. The Cronbach's alphas were .77 for U.S. Orientation and .88 for Hispanic Orientation.

Ethnic Identity Development

Ethnic identity development was assessed with the Multigroup Ethnic Identity Measure (Phinney 1992). This 12-item scale has shown good reliability among adolescents and adults of diverse ethnic backgrounds, with Cronbach's alphas typically above .80. In our sample, the Cronbach's alpha was .88. Examples of items include, "I have spent time trying to find out more about my ethnic group, such as its history, traditions, and customs" and "I have a clear sense of my ethnic background and what it means for me."

Discrimination

Discrimination was assessed with a 10-item measure of the respondents' perceptions of instances of discrimination that they personally experience (Guyll et al. 2001). Examples of items include, "You are treated with less respect than other people" and "People act as if they're better than you." Responses were rated on a four-point scale from "never" to "often." Because this scale can apply to various types of discrimination (e.g., based on race/ethnicity, gender, age, sexual orientation, physical handicaps, etc.), the scale was preceded by the following text: "Sometimes people feel that they are treated differently because of their ethnic or cultural background. How do people treat you?" The Cronbach's alpha was 0.87.

Acculturative Stress

Acculturative stress was assessed with the following items (adapted from Gil et al. 2000 and validated by Booker et al. 2008): "Do you ever feel uncomfortable when you have to choose between doing things like Americans or non-Americans?" "Do you ever have problems with your family because you like to do things the American way?" and "Do you ever get upset with your parents because they don't understand the American lifestyle?" Responses were rated on a four-point scale from "never" to "very often." The Cronbach's alpha was .68.

Cultural Values

Measures of cultural values were adapted from previous studies of adolescents in Southern California (Unger et al. 2006) and other studies of Hispanics (Cuellar et al. 1995a). *Familism* was assessed with four items (Cronbach's alpha=.77): "If one of my relatives needed a place to stay for a few months, my family would let them stay with us," "I expect my relatives to help me when I need them," "When a family makes an important decision, they should talk about it with their close relatives," and "If anyone in my family needed help, we would all be there to help them." *Respeto* was assessed with four items (Cronbach's alpha=.89): "I will take care of my parents when they are old," "It is important to honor my parents," "It is important to respect my parents," and "I want to be a good person so that people know that my parents raised me right." *Fatalism* was assessed with four items (Cronbach's alpha=.72): "It's more important to enjoy life now than to plan for the future," "People can't really do much to change what happens in life. You just have to accept things," "I live for today because I don't know what will happen in the future," and "I don't plan ahead because most things in life are a matter of luck." All cultural values items were rated on a 4 point scale ranging from "definitely no" to "definitely yes."

Substance Use

Past-month cigarette, alcohol, and marijuana use were the outcome measures. Respondents were asked, "During the past 30 days, on how many days did you smoke cigarettes?" "During the past 30 days, on how many days did you have at least one drink of alcohol?" "In the last 30 days, how many times have you used marijuana (grass, pot, weed)?" The questions about the number of days in the past month were rated on a 7-point scale from "0 days" to "all 30 days." The question about the number of times the respondent used marijuana was rated on a 6-point scale from "0 times" to "40 or more times." All three substance use variables were log-transformed because their distributions were skewed.

Demographic Characteristics

Demographic characteristics included age, gender, generation in the United States, number of rooms per person in the respondent's home (an indicator of socioeconomic status; Myers and Choi 1992), friends' substance use, and parents' substance use. Generation in the U.S. was assessed with three items: "In what country were you born?" "Where was your mother born?" and "Where was your father born?" Response options were "United States" or "Other." The "Other" option included a line for the respondent to write the name of the country. Respondents were classified as 1th generation if the student and parents were born outside the U.S., as 2th generation if the student was born in U.S. but both parents were born outside the U.S., and as 3th generation if the student and at least one parent were born in the U.S.

Friends' substance use and parents' substance use were included as covariates because they have been significantly associated with adolescent substance use in numerous studies (Petraitis et al. 1998). To assess friends' substance use, we asked respondents to think about their 5 best friends in their class (e.g., the 9th grade at their school). Respondents were then asked whether each friend had ever tried cigarettes, alcohol, or marijuana. Some respondents named fewer than 5 friends. Therefore, the friends' cigarette use score was the proportion of named friends who had ever tried cigarettes (e.g., if the respondent listed 4 friends and 3 of them had tried cigarettes, the friends' cigarette use score would be .75). Similarly, the friends' alcohol and marijuana use scores were the proportion of named friends who had ever tried alcohol or marijuana, respectively. Parents' substance use was assessed with the following 3 questions: "Think of the two adults that you spend the most time with. How many of them smoke cigarettes every day or most days?" "Think of the two adults that you spend the most time with. How many of them drink alcohol at least once per week?" and "Think of the two adults that you spend the most time with. How many of them smoke marijuana?" Response options were "None of them," "1 of them," and "2 of them."

Data Analysis

Individual growth curve models were used to examine the effects of the hypothesized predictor variables on substance use trajectories from 9th to 11th grade, using the MIXED procedure in SAS. Individual growth curve models can be used to examine the unique trajectories of individuals with repeated measures data (Raudenbush and Bryk 2002; Singer 1998). This method overcomes some of the limitations of traditional repeated measures ANOVA techniques. Repeated measures ANOVA requires complete data from all respondents at all timepoints, whereas individual growth curve models can include individuals with missing data. Individual growth curve models also make it possible to examine individual trajectories of change, including variation in initial status (intercept) and rate of change (slope), instead of assuming that all individuals follow the same trajectory.

The growth curve model investigated Level 1 (within person) change in substance use over time and Level 2 (between person) effects of the predictors on the intercept and slope of substance use over time. The full maximum likelihood model in SAS PROC MIXED was used. Cigarette, alcohol, and marijuana use were evaluated in separate models. In each model, the dependent variable was the log of the number of days in the past month that the respondent reported using the substance.

The growth curve analysis proceeded in two steps. First, we evaluated an unconditional growth model, which models the intercept and slope of growth in substance use, but does not include any covariates. The fit of this model was evaluated with the −2 log likelihood statistic. Next, the hypothesized covariates were added to the model. The covariates were evaluated both as predictors of the intercept (main effects) and predictors of the slope (covariate X time interactions). The −2 log likelihood statistic of this model was compared with that of the unconditional growth model to determine whether the addition of the covariates significantly improved the model fit. An improvement in model fit was indicated by a statistically significant decrease in the −2 log likelihood statistic, which follows a chi-square distribution. If the model with the covariates represented a significant improvement over the unconditional growth model, the significance of each covariate was evaluated by examining the beta coefficients and their associated p-values.

We attempted to fit a three-level model (timepoints nested within individuals nested within schools), but the models failed to converge. Preliminary analyses indicated that the intraclass correlations of students nested within schools were very small (.0006 for smoking, .0082 for alcohol, and .0033 for marijuana), indicating that intraclass correlations had minimal influence on the model. Furthermore, the parameter estimates from the three-level model with convergence problems were nearly identical to those from the two-level model that converged successfully. Therefore, this article reports the results of the simpler two-level model of timepoints nested within individuals.

Results

Demographic Characteristics of Respondents

As shown in Table 9.1, the students' mean age was 14.0 years and the sample was approximately half male and half female. Because students self-reported their race/ethnicity with a "check all that apply" question, some selected multiple categories. Some of the Hispanic/Latino youth also self-reported as White (5 %), African-American (1 %), American Indian (1 %), Asian (1 %), or Pacific Islander (1 %). Their countries of origin included Mexico (84 %), the United States (29 %), El Salvador (9 %), Guatemala (6 %), and Honduras (1 %) (respondents could select more than one country of origin). Over half of the students (62 %) were second-generation (student born in the U.S. but neither parent born in the U.S.). Figure 9.1 shows the lifetime and past-month substance use prevalence in 9th, 10th, and 11th grade. In 9th grade, nearly half (47 %) of the students had tried alcohol; 24 % had tried cigarettes and 10 % had tried marijuana. By 11th grade, 68 % had tried alcohol, 39 % had tried cigarettes, and 17 % had tried marijuana.

Table 9.1 Self-reported baseline demographic characteristics and substance use of Hispanic/Latino students

	N	%
Age (years)		
12–13	123	8 %
14	1338	85 %
15–16	112	7 %
Missing	2	0 %
Gender		
Female	839	53 %
Male	728	46 %
Missing	8	1 %
Other race/ethnicity (in addition to Hispanic/Latino)		
White	82	5 %
African-American	21	1 %
Asian	14	1 %
Pacific Islander	13	1 %
American Indian	19	1 %
Generation in the United States		
1 (Student and parents born outside U.S.)	214	14 %
2 (Student born in U.S., both parents born outside U.S.)	978	62 %
3 (Student and at least one parent born in U.S.)	364	23 %
Other/missing	19	1 %

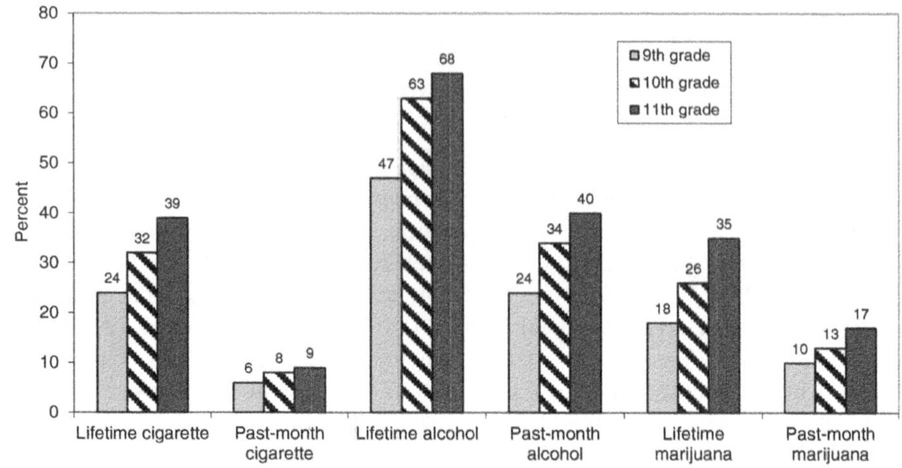

Fig. 9.1 Lifetime and past-month substance use prevalence

Growth Curve Models of Associations between Cultural Variables and Substance Use

The growth curve models provide estimates for the effects of the predictor variables on the intercept (the respondent's initial level of substance use at 9th grade) and on the slope (the respondent's growth trajectory of substance use from 9th to 11th grade). For cigarettes, alcohol, and marijuana, the model with covariates represented a significant improvement in model fit over the unconditional growth model without covariates. Therefore, the parameter estimates within these models were examined.

Variables Associated with Initial Level of Substance Use

The standardized parameter estimates and their p-values are shown in Table 9.2. For cigarettes, the variables significantly associated with initial status (cigarette use in 9th grade) were perceived discrimination, low respeto, fatalism, friends' cigarette smoking, and parents' cigarette smoking. For alcohol, the variables significantly associated with initial status were acculturative stress, perceived discrimination, low respeto, friends' alcohol use, and parents' alcohol use. For marijuana, the variables significantly associated with initial status were perceived discrimination, low respeto, fatalism, friends' marijuana use, and parents' marijuana use.

Table 9.2 Predictors of growth in substance use from 9th to 11th grade

	Cigarettes		Alcohol		Marijuana	
	STB	p-value	STB	p-value	STB	p-value
Intercept	−.012	.6073	**−.126**	.0001	**.085**	.0001
Time	.029	.0864	**.160**	.0001	**−.069**	.0001
Age	.014	.5383	−.025	.2174	−.031	.1519
Female (vs. male)]	.032	.1762	.002	.9236	−.020	.3703
Rooms per person	.019	.3982	.014	.4826	.011	.6150
U.S. Acculturation	.009	.7053	.002	.9318	.014	.5551
Hispanic acculturation	−.004	.8531	.021	.3090	−.030	.1747
Ethnic identity	−.030	.2199	.012	.5759	.029	.2007
Acculturative stress	.003	.8970	**.048**	.0212	−.001	.9527
Perceived discrimination	**.092**	.0002	**.087**	.0001	**.070**	.0026
Familism	.019	.4627	.022	.3393	.025	.3189
Respeto	**−.162**	.0001	**−.123**	.0001	**−.132**	.0001
Fatalism	**.063**	.0072	.034	.0987	**.077**	.0004
Friends' substance use	**.255**	.0001	**.332**	.0001	**.509**	.0001
Parents' substance use	**.085**	.0003	**.116**	.0001	**.192**	.0001
Age X time	−.010	.5862	−.010	.5513	.020	.1837
Female X time	**−.057**	.0014	**−.034**	.0493	**−.048**	.0015
Rooms per person X time	−.015	.3717	.003	.8597	−.018	.2161
U.S. Acculturation X time	−.026	.1547	−.009	.6238	−.009	.5852
Hispanic acculturation X time	−.026	.1462	**−.036**	.0443	−.022	.1587
Ethnic identity X time	.017	.3581	−.003	.8860	.003	.8245
Acculturative stress X time	.032	.0729	.008	.6437	.004	.7879
Perceived discrimination X time	−.020	.2836	**−.038**	.0364	−.013	.4313
Familism X time	−.016	.4155	−.010	.6235	−.021	.2120
Respeto X time	**.080**	.0001	.016	.4336	**.055**	.0020
Fatalism X time	−.008	.6679	**.036**	.0344	−.017	.2724
Friends' substance use X time	**−.037**	.0359	**−.064**	.0002	**−.155**	.0001
Parents' substance use X time	−.009	.6057	**−.034**	.0446	**−.087**	.0001

Variables Associated with Growth in Substance Use

The covariate X time interaction terms in Table X indicate the associations between the covariates and growth in substance use from 9th to 11th grade. For cigarettes, the variables significantly associated with the slope (change in cigarette smoking from 9th to 11th grade) were male gender, respeto, and low friends' use. For alcohol, the variables significantly associated with the slope were male gender, low Hispanic acculturation, low perceived discrimination, fatalism, and low friends' and parents' alcohol use. For marijuana, the variables significantly associated with the slope were male gender, respeto, and low friends' and parents' use.

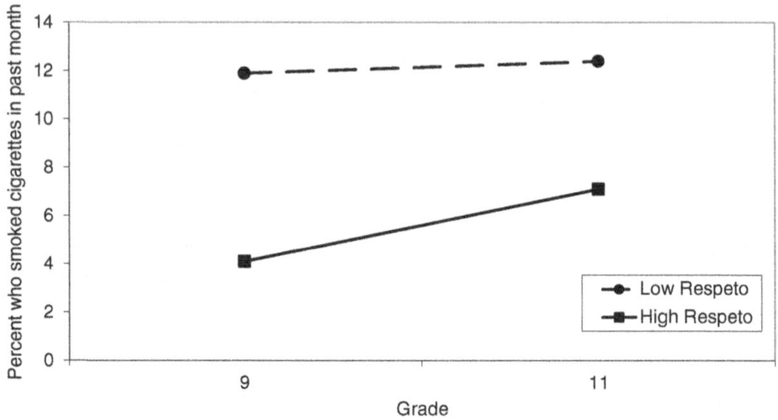

Fig. 9.2 Prevalence of past-month alcohol use among students with high vs. low levels of respeto

Some of the predictor variables had opposite effects on the intercept and the slope. For example, high levels of respeto were associated with less initial substance use (intercept) but steeper growth in substance use over time (slope). A closer examination of these patterns revealed that adolescents above the median level of respeto were protected from substance use before 9th grade, but their substance use increased from 9th to 11th grade. However, their substance use remained lower than that of their peers with low respeto scores, who initiated substance use earlier and increased more slowly, but still maintained a higher level of substance use at each timepoint. This pattern is shown in Fig. 9.2. Similar patterns were observed for friends' and parents' use.

Discussion

This study identified some of the predictors of growth trajectories of substance use among Hispanic adolescents in Southern California. Consistent with previous studies (Guthrie et al. 2002; Romero et al. 2007), this study identified discrimination and acculturative stress as risk factors for substance use among Hispanic adolescents, in addition to well-documented risk factors such as friends' use and parents' use. We also found that cultural values were associated with substance use; respeto was a protective factor, and fatalism was a risk factor.

Respeto had interesting associations with both the intercept and slope of substance use. The negative associations between respeto and the intercept indicate that adolescents with higher levels of respeto are less likely to be using cigarettes, alcohol, and marijuana by 9th grade. However, these adolescents are at higher risk for escalating their cigarette and marijuana use during high school. One explanation for this finding is that the Hispanic cultural value of respeto may delay experimentation prior to 9th grade, but the adolescents who have not yet tried cigarettes or marijuana prior to 9th grade are still at risk of experimenting later, even if they have high levels of respeto. However, it is important to keep in mind that regardless of their steeper trajectory of substance use, the students with high respeto scores had lower levels of substance use at every timepoint compared with the students with low respeto scores. Therefore, it appears that respeto was protective against substance use throughout the study period, although it may have been more protective in 9th grade than in 11th grade.

The Hispanic cultural value of fatalism was a risk factor for substance use; it was positively associated with the intercept of cigarette and marijuana use, and with the slope of alcohol use. Many researchers have made broad assumptions that traditional cultural values are protective. Indeed, most Hispanic cultural values including familism and respeto have been shown to be protective against substance use in previous studies. However, fatalism may have the opposite effect. Adolescents with fatalistic beliefs may perceive that there is no reason to avoid substance use, because their health and well-being are controlled by external forces. Research among other ethnic groups has shown that traits similar to fatalism such as lack of future orientation (Haegerich and Tolan 2008) or external health locus of control (Eiser et al. 1989) are also risk factors for substance use.

Most previous studies of the development of substance use among Hispanic adolescents have assessed substance use as an outcome at a single point of time, rather than as a growth trajectory. The results of this analysis indicate that risk and protective factors might exert their influences at different developmental stages. For example, perceived discrimination, friends' substance use, and parents' substance use influenced the intercept of the substance use trajectory, indicating that their effects likely occurred prior to 9th grade. In other words, adolescents with friends or parents who use substances, and/or who experience high levels of discrimination, may begin using substances before 9th grade and arrive in high school with an elevated risk of substance use. Conversely, other variables such as male gender do not influence initial level of substance use in 9th grade, but are associated with a steeper growth in substance use during high school.

Limitations and Directions for Further Research

These results are based on adolescents' self-reports of their substance use, which may have been underreported. However, the respondents were assured that their surveys were completely confidential, and previous studies have found adolescents' self-reports of substance use to be quite accurate under confidential survey conditions (Harrison and Hughes 1997).

As in many longitudinal studies, higher-risk students (e.g., those who had already experimented with substance use by 9th grade) were more likely to be lost to attrition. These results might not generalize to the students who were not followed up successfully. Intensive attempts were made to locate all participants and survey them at school or by telephone, resulting in an 88 % follow-up rate, but some attrition is inevitable in longitudinal research. One of the strengths of growth curve modeling is that we were able to include data from participants who were absent at one or more assessments. However, the estimates of the growth curve parameters may be less precise among these individuals. Additional research is needed to understand the acculturation and family-related stresses and substance use behavior of students who drop out of school or change schools and/or residences frequently.

Conclusion

These findings suggest that cultural phenomena such as discrimination, acculturative stress, and cultural values can influence the trajectories of substance use among Hispanic adolescents. It may be possible to increase the effectiveness of prevention interventions by addressing these cultural issues. For example, substance use prevention curricula could help adolescents develop more effective coping skills to reduce the harmful impact of discrimination, or they could encourage adolescents to explore the facets of their cultural values that are consistent with healthy lifestyles. Culturally relevant substance use prevention programs have the potential to slow the trajectories of drug use among Hispanics and reduce health disparities.

Acknowledgments This research was supported by the National Institute on Drug Abuse (grant # DA016310). The authors thank Alicia Mayaute, Melissa Esmero, and David Danneker for their assistance. Please address reprint requests to Jennifer B. Unger, Ph.D., Keck School of Medicine, University of Southern California, 2001 N. Soto St., SSB 302P, Los Angeles, CA 90089, Phone: 324-442-8234, .

References

Allen, M. L., Elliott, M. N., Fuligni, A. J., Morales, L. S., Hambarsoomian, K., & Schuster, M. A. (2008). The relationship between Spanish language use and substance use behaviors among Latino youth: A social network approach. *Journal of Adolescent Health, 43*, 372–379.

Bacallao, M. L., & Smokowski, P. R. (2005). "Entre dos mundos" (Between Two Worlds): Bicultural skills training with Latino immigrant families. *Journal of Primary Prevention, 26*, 485–509.

Balcazar, H., Castro, F., & Krull, J. L. (1995). Cancer risk reduction in Mexican American women, The role of acculturation, education and health risk factors. *Health Education Quarterly, 22*, 61–84.

Bauman, S. (2005). The reliability and validity of the brief acculturation rating scale for Mexican Americans-II for children and adolescents. *Hispanic Journal of Behavioral Sciences, 27*, 426–441.

Berry, J. W. (1980). Acculturation as varieties of adaptation. In A. Padilla (Ed.), *Acculturation: Theory, models, and findings* (pp. 9–25). Boulder: Westview.

Booker, C. L., Unger, J. B., Azen, S. P., Baezconde-Garbanati, L., Lickel, B., & Johnson, C. A. (2008). A longitudinal analysis of stressful life events, smoking behaviors, and gender differences in a multicultural sample of adolescents. *Substance Use & Misuse, 43*, 1521–1543.

Bray, J. H., Adams, G. J., Getz, J. G., & Baer, P. E. (2001). Developmental, family, and ethnic influences on adolescent alcohol usage: A growth curve approach. *Journal of Family Psychology, 15*, 301–314.

Bray, J. H., Adams, G. J., Getz, J. G., & McQueen, A. (2003). Individuation, peers, and adolescent alcohol use: A latent growth analysis. *Journal of Consulting and Clinical Psychology, 71*, 553–564.

Brook, J. S., Whiteman, M., Balka, E. B., Win, P. T., & Gursen, M. D. (1998). Drug use among Puerto Ricans: Ethnic identity as a protective factor. *Hispanic Journal of Behavioral Sciences, 20*, 241–254.

Brook, J. S., Ning, Y., & Brook, D. W. (2006a). Personality risk factors associated with trajectories of tobacco use. *American Journal on Addictions, 15*, 426–433.

Brook, J. S., Pahl, K., & Ning, Y. (2006b). Peer and parental influences on longitudinal trajectories of smoking among African Americans and Puerto Ricans. *Nicotine & Tobacco Research, 8*(5, Oct), 639–651.

Brook, J. S., Balka, E. B., Ning, Y., & Brook, D. W. (2007a). Trajectories of cigarette smoking among African Americans and Puerto Ricans from adolescence to young adulthood: Associations with dependence on alcohol and illegal drugs. *American Journal on Addictions, 16*, 195–201.

Brook, J. S., Duan, T., Brook, D. W., & Ning, Y. (2007b). Pathways to nicotine dependence in African American and Puerto Rican young adults. *American Journal of Addictions, 16*, 450–456.

Caplan, S. (2007). Latinos, acculturation, and acculturative stress: A dimensional concept analysis. *Policy, Politics & Nursing Practice, 8*, 93–106.

Carvajal, S. C., Hanson, C. E., Romero, A. J., & Coyle, K. K. (2002). Behavioural risk factors and protective factors in adolescents: A comparison of Latinos and non-Latino whites. *Ethnicity and Health, 7*, 181–193.

Colder, C. R., Campbell, R. T., Ruel, E., Richardson, J. L., & Flay, B. R. (2002). A finite mixture model of growth trajectories of adolescent alcohol use: Predictors and consequences. *Journal of Consulting and Clinical Psychology, 70*, 976–985.

Cuellar, I., Arnold, B., & González, G. (1995a). Cognitive referents of acculturation: Assessment of cultural constructs in Mexican American families. *Journal of Community Psychology, 23*, 339–356.

Cuellar, I., Arnold, B., & Maldonado, R. (1995b). The acculturation rating scale for Mexican Americans-II: A revision of the original ARSMA scale. *Hispanic Journal of Behavioral Sciences, 17*, 275–304.

De La Rosa, M. (2002). Acculturation and Latino adolescents' substance use: A research agenda for the future. *Substance Use & Misuse, 37*, 429–456.

Eiser, J. R., Eiser, C., Gammage, P., & Morgan, M. (1989). Health locus of control and health beliefs in relation to adolescent smoking. *British Journal of Addiction, 84*, 1059–1065.

Ellickson, P. L., Orlando, M., Tucker, J. S., & Klein, D. J. (2004). From adolescence to young adulthood: Racial/ethnic disparities in smoking. *American Journal of Public Health, 94*, 293–299.

Epstein, J. A., Botvin, G. J., & Diaz, T. (2001). Linguistic acculturation associated with higher marijuana and polydrug use among Hispanic adolescents. *Substance Use & Misuse, 36*, 477–499.

Gil, A. G., Vega, W. A., & Dimas, J. M. (1994). Acculturative stress and personal adjustment among Hispanic adolescent boys. *Journal of Community Psychology, 22*, 43–54.

Gil, A. G., Wagner, E. F., & Vega, W. A. (2000). Acculturation, familism, and alcohol use among Latino adolescent males: Longitudinal relations. *Journal of Community Psychology, 28*, 443–458.

Greene, M. L., Way, N., & Pahl, K. (2006). Trajectories of perceived adult and peer discrimination among African American, Latino, and Asian American adolescents: Patterns and psychological correlates. *Developmental Psychology, 42*, 218–236.
Guthrie, B. J., Young, A. M., Williams, D. R., Boyd, C. J., & Kintner, E. K. (2002). African American girls' smoking habits and day-to-day experiences with racial discrimination. *Nursing Research, 51*, 183–190.
Guyll, K. A., Matthews, K. A., & Bromberger, J. T. (2001). Discrimination and unfair treatment: Relationship to cardiovascular reactivity among African American and European American women. *Health Psychology, 20*, 315–325.
Haegerich, T. M., & Tolan, P. H. (2008). Core competencies and the prevention of adolescent substance use. *New Directions in Child and Adolescent Development, 122*, 47–60.
Harrison, L., & Hughes, A. (Eds.). (1997). *The validity of self-reported drug use: Improving the accuracy of survey estimates* (NIDA research monograph 167). Rockville: U.S. Department of Health and Human Services.
Jackson, K. M., Sher, K. J., & Schulenberg, J. E. (2008). Conjoint developmental trajectories of young adult substance use. *Alcoholism: Clinical and Experimental Research, 32*, 723–737.
Johnston, L. D., O'Malley, P. M., Bachman, J. G., & Schulenberg, J. E. (2009). *Monitoring the future national survey results on drug use, 1975-2008. Volume I: Secondary school students* (NIH Publication No. 09-7402). Bethesda: National Institute on Drug Abuse.
Love, A. S., Yin, Z., Codina, E., & Zapata, J. T. (2006). Ethnic identity and risky health behaviors in school-age Mexican-American children. *Psychological Reports, 98*, 735–744.
Marsiglia, F. F., Kulis, S., Hecht, M. L., & Sills, S. (2004). Ethnicity and ethnic identity as predictors of drug norms and drug use among preadolescents in the US Southwest. *Substance Use & Misuse, 39*, 1061–1094.
McQueen, A., Getz, J. G., & Bray, J. H. (2003). Acculturation, substance use, and deviant behavior: Examining separation and family conflict as mediators. *Child Development, 74*, 1737–1750.
Myers, D., & Choi, S. Y. (1992). Growth in overcrowded housing: A comparison of the states. *Applied Demography, 7*, 1–4.
Neff, J. A., & Hoppe, S. K. (1993). Race/ethnicity, acculturation, and psychological distress: Fatalism and religiosity as cultural resources. *Journal of Community Psychology, 21*, 3–20.
Negy, C., & Woods, D. (1992). The importance of acculturation in understanding research with Hispanic- Americans. *Hispanic Journal of Behavioral Sciences, 14*, 224–247.
Pérez, D. J., Fortuna, L., & Alegría, M. (2008). Prevalence and correlates of everyday discrimination among U.S. Latinos. *Journal of Community Psychology, 36*, 421–433.
Petraitis, J., Flay, B. R., Miller, T. Q., Torpy, E. J., & Greiner, B. (1998). Illicit substance use among adolescents: A matrix of prospective predictors. *Substance Use & Misuse, 33*, 2561–2604.
Phinney, J. (1990). Ethnic identity in adolescents and adults: Review of research. *Psychological Bulletin, 108*, 499–514.
Phinney, J. (1992). The multigroup ethnic identity measure: A new scale for use with adolescents and young adults from diverse groups. *Journal of Adolescent Research, 7*, 156–176.
Ramirez, R. R., & de la Cruz, G. P. (2002). *The Hispanic population in the United States: March 2002* (Current population reports, P20-545). Washington, DC: U.S. Census Bureau.
Raudenbush, S. W., & Bryk, A. S. (2002). *Hierarchical linear models: Applications and data analysis methods* (2nd ed.). Thousand Oaks: Sage.
Romero, A. J., & Roberts, R. E. (2003). Stress within a bicultural context for adolescents of Mexican descent. *Cultural Diversity and Ethnic Minority Psychology, 9*, 171–184.
Romero, A. J., Martinez, D., & Carvajal, S. C. (2007). Bicultural stress and adolescent risk behaviors in a community sample of Latinos and non-Latino European Americans. *Ethnicity and Health, 12*, 443–463.
Singer, J. D. (1998). Using SAS PROC MIXED to fit multilevel models, hierarchical models, and individual growth models. *Journal of Educational and Behavioral Statistics, 24*, 323–355.

Smokowski, P. R., & Bacallao, M. L. (2006). Acculturation and aggression in Latino adolescents: A structural model focusing on cultural risk factors and assets. *Journal of Abnormal Child Psychology, 34*, 657–671.

Staples, R., & Mirandé, A. (1980). Racial and cultural variations among American families: A decennial review of the literature on minority families. *Journal of Marriage and the Family, 42*, 887–903.

Substance Abuse and Mental Health Services Administration, Office of Applied Studies. (2008). *Results from the 2007 national survey on drug use and health: National findings* (NSDUH Series H-34, DHHS Publication No. SMA 08-4343). Rockville: Substance Abuse and Mental Health Services Administration, Office of Applied Studies.

Szalacha, L. A., Erkut, S., García Coll, C., Alarcón, O., Fields, J. P., & Ceder, I. (2003). Discrimination and Puerto Rican children's and adolescents' mental health. *Cultural Diversity and Ethnic Minority Psychology, 9*, 141–155.

Szapocznik, J., Scopetta, M. A., Kurtines, W. M., & Aranalde, M. A. (1978). Theory and measurement of acculturation. *Interamerican Journal of Psychology, 12*, 113–130.

Szapocznik, J., Santisteban, D., Rio, A., Perez-Vidal, A., Santisteban, D., & Kurines, W. M. (1989). Family effectiveness training: An intervention to prevent drug abuse and problem behaviors in Hispanic adolescents. *Hispanic Journal of Behavioral Sciences, 11*, 4–27.

Szapocznik, J., Prado, G., Burlew, A. K., Williams, R. A., & Santisteban, D. A. (2007). Drug abuse in African American and Hispanic adolescents: Culture, development, and behavior. *Annual Review of Clinical Psychology, 3*, 77–105.

Unger, J. B., Cruz, T. B., Rohrbach, L. A., Ribisl, K. M., Baezconde-Garbanati, L., Chen, X., Trinidad, D. R., & Johnson, C. A. (2000). English language use as a risk factor for smoking initiation among Hispanic and Asian American adolescents: Evidence for mediation by tobacco-related beliefs and social norms. *Health Psychology, 19*, 403–410.

Unger, J. B., Shakib, S., Gallaher, P., Ritt-Olson, A., Mouttapa, M., Palmer, P. H., & Johnson, C. A. (2006). Cultural/interpersonal values and smoking in an ethnically diverse sample of Southern California adolescents. *Journal of Cultural Diversity, 13*, 55–63.

Unger, J. B., Ritt-Olson, A., Soto, D. W., & Baezconde-Garbanati, L. (2009). Parent–child acculturation discrepancies as a risk factor for substance use among Hispanic adolescents in Southern California. *Journal of Immigrant and Minority Health, 11*, 149–157.

Valdés, G. (1996). *Con Respeto: Bridging the distances between culturally diverse families and schools. An ethnographic portrait.* New York: Teachers College Press.

Vega, W. A., & Gil, A. G. (1998). *Drug use and ethnicity in early adolescence*. New York: Plenum Press.

Williams, D. R., Neighbors, H. W., & Jackson, J. S. (2008). Racial/ethnic discrimination and health: Findings from community studies. *American Journal of Public Health, 98*, S29–S37.

Windle, M., & Wiesner, M. (2004). Trajectories of marijuana use from adolescence to young adulthood: Predictors and outcomes. *Development and Psychopathology, 16*, 1007–1027.

Chapter 10
Predicting Change in Substance Use and Substance Use Cognitions of Mexican Origin Youth during the Transition from Childhood to Early Adolescence

Rand D. Conger, Gary D. Stockdale, Hairong Song, Richard W. Robins, and Keith F. Widaman

Substance use trajectories typically are flat during childhood when they reflect essentially little or no use of substances. For a small number of children they begin to escalate when the children transition into early adolescence. This early initiation of alcohol, tobacco or other drug (ATOD) use during late childhood or early adolescence can have long-term adverse consequences, including increased risk for academic failure, impaired social relations, emotional problems, antisocial behaviors, physical illness, precocious and risky sex, and the development of substance use disorders involving abuse and dependence (Conger and Rueter 1996; Dawes et al. 2000; Glantz and Leshner 2000; Guo et al. 2001; Hops et al. 2000; Jessor et al. 1995; Rebhun 1998; Staton et al. 1999). Given this constellation of negative outcomes, there has been increasing scientific attention directed to studying early risk for ATOD use and mechanisms that increase resilience to such risk (Conger 1997; Conger and Rueter 1996; Glantz and Leshner 2000). Moreover, these risk processes are of special interest in specific population subgroups, including children and adolescents of Mexican origin. The study of risk and resilience for ATOD use has received insufficient attention in this ethnic group, yet demographic trends indicate that by the middle of this century Latinos will increase from 12 % in 2000 to approximately 25 % of the U.S. population (Cauce and Domenech Rodriguez 2002).

R.D. Conger (✉)
Human Ecology, University of California – Davis, Davis, CA, USA
e-mail: rdconger@ucdavis.edu

G.D. Stockdale • R.W. Robins • K.F. Widaman
Psychology Department, University of California – Davis, Davis, CA, USA

H. Song
University of Oklahoma, Norman, OK, USA

Y.F. Thomas, L.N. Price (eds.), *Drug Use Trajectories Among Minority Youth*,
DOI 10.1007/978-94-017-7491-8_10

The focus of the current report involves data from the California Families Project (CFP), a study of Mexican origin children and their families living in California. The majority of Latinos or Hispanics in the U.S. are of Mexican descent, representing 77 % of the Latino population in California (U.S. Census Bureau 2000). California serves as a bellwether for national trends inasmuch as Mexican Americans currently constitute 25 % of the State's population. For the CFP, we define Mexican Americans as people of Mexican origin, including both recent immigrants to the U.S. as well as citizens who were born in this country. Consistent with the theme of this volume, the CFP is concerned with trajectories of ATOD use from middle or late childhood, beginning in fifth grade, through the adolescent years. Because the study is still in its early stages, in this report we focus on the period from fifth to seventh grade, a time during which involvement with substances is expected to grow from essentially zero to initial thoughts about use or actual experimentation with ATODs. As noted, this early initiation is often especially problematic for later risk trajectories to more serious involvement with substances.

An important issue addressed in the CFP concerns the identification of developmental processes associated with risk and resilience. To a significant degree, risk for ATOD use and abuse emerges over time as part of a developmental process beginning during childhood, well before the period when actual involvement with ATODs typically occurs (Cicchetti 1999; Conger 1997; Dawes et al. 2000; Glantz and Leshner 2000; Tarter and Vanyukov 1994). In addition to its theoretical importance, this perspective has enormous practical significance, inasmuch as better understanding of these early risk processes could lead to the development of more effective interventions designed to prevent the initial onset of ATOD use, an outcome much preferred to dealing with the problem after it has occurred (Cicchetti 1999; Sloboda 1999). Moreover, a developmental perspective considers different trajectories of ATOD use, some of which may lead to long-term disability while others may reflect a temporary pathway of normal adolescent experimentation (e.g., Schulenberg et al. 2001).

Despite these strengths of the developmental approach, only a few prospective, longitudinal studies have empirically evaluated hypothesized early predictors of risk or resilience that are expected to be present by middle to late childhood (e.g., Anthony 1993; Brook et al. 1998; Chilcoat et al. 1995; Dawes et al. 2000; Ensminger et al. 2002; Glantz and Leshner 2000; Guo et al. 2001; Kellam et al. 1983). Moreover, those developmental studies beginning in childhood that do exist typically have not addressed either the Latino experience in general or the Mexican American experience in particular (see Burton and Jarrett 2000; McLoyd et al. 2000). The primary goal of the CFP is to help address these important gaps in previous research. The project draws on earlier research and theory in pursuit of this goal.

The Conceptual Model

As shown in Fig. 10.1, the conceptual model for the present report is especially concerned with the role of the family in youth development and how general family context and parent functioning affect the risk for involvement with ATODS during the critical transition from late childhood (fifth grade) to early adolescence (seventh grade). Because seventh grade is still a relatively early point in development for significant levels of substance use or abuse, our conceptual model relates both to actual substance use and to expressions of intention or willingness to use in the future. Recent research has shown that these types of cognitions during childhood and adolescence are robust predictors of later ATOD use (Andrews et al. 2008). Thus, when we use the term "involvement" with substances, it refers to actual consumption and also to expressed intentions or willingness to initiate substance use in the near future.

Both theory (e.g., Brook et al. 2001; Catalano and Hawkins 2000; Conger 1997; Glantz and Leshner 2000) and empirical findings from basic and prevention research underscore the importance of family processes in increasing or decreasing risk for ATOD use and abuse (e.g., Brook et al. 1998; Chassin et al. 2004; Ensminger et al. 2002; Guo et al. 2002; Hops et al. 2000; Sloboda 1999). Family influences appear to be especially important in deterring both the early onset of ATOD use as well as association with friends who use substances, two of the primary predictors of persistent ATOD use and later substance disorders (e.g., Brook et al. 1998; 2001; Conger 1997; Conger and Rueter 1996; Glantz and Leshner 2000; Guo et al. 2001; Hops et al. 2000; Melby et al. 1993). Also important, family processes appear to be particularly significant with regard to ATOD use and problem behavior by Mexican American youth and by children in poor families (e.g., Bray et al. 2001; Cauce and Domenech Rodriguez 2002; Ellickson and Morton 1999; Swaim et al. 1998).

Consistent with these observations, the model that guides the present analyses (Fig. 10.1) derives from earlier research on family economic stress (e.g., Conger and Conger 2002) and from current research and theory on contextual and cultural influences on the development of ATOD use (e.g., Chassin et al. 2004; Glantz and Leshner 2000; Gonzales et al. 2002; Vega and Gil 1999). The original family stress model traced the specific processes through which economic hardship and other life stressors affect family relationships and the development of children and adolescents (e.g., Conger and Conger 2002; Conger et al. 2002). The inclusion of economic pressure as an exogenous variable in Fig. 10.1 reflects this component of the study. The economic aspect of family stress processes is especially salient for families of Mexican origin inasmuch as a too common situation for individuals of Mexican heritage in the U.S. is to live with extremely low incomes or below the official poverty line (approximately 30 % in 1996 compared to about 12 % for the nation as a whole; Baca Zinn and Wells 2000).

The conceptual framework guiding the CFP extends this earlier model by delineating specific, stressful conditions to which many Mexican American parents and children may be exposed, such as living in dangerous and disadvantaged neighborhoods, experiencing ethnic discrimination, and living in a single-parent household (Fig. 10.1). Consistent with findings from earlier research, the model proposes that economic pressure, such as the inability to meet basic material needs involving food, clothing, housing, and medical care will increase parent emotional distress (Conger and Conger 2002). Economic hardship also increases the likelihood that a family will be forced to reside in a high risk living environment with a history "...of violence, poor educational systems, few ladders of opportunity, and where

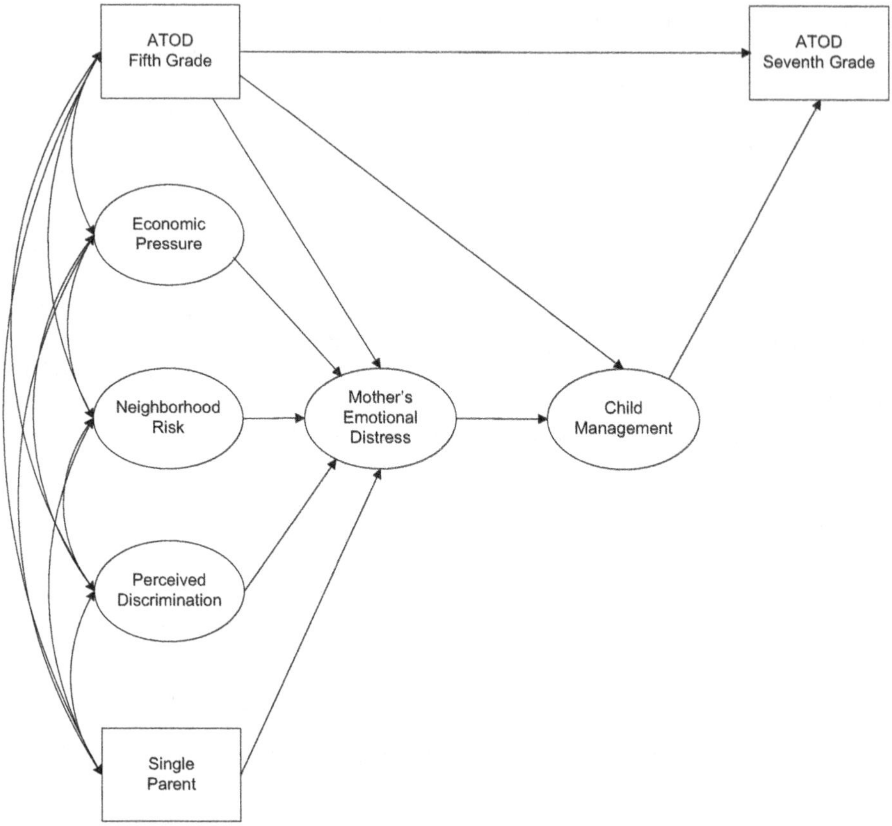

Fig. 10.1 Conceptual alcohol use and intent to use model

intensive drug distribution and sales, gang activity, and other forms of criminal deviance are rampant" (Vega and Gil 1999, p. 62). Thus, neighborhood risk is included as an additional marker of family stress in Fig. 10.1.

The model in Fig. 10.1 also incorporates stress related to occupying a minority role in the dominant culture through the experience of actual or perceived discrimination (Portes and Rumbaut 2001; Vega et al. 1998). Vega and his colleagues found that perceived discrimination was associated with depressed mood in Mexican American adults (Finch et al. 2000). Similarly, discrimination related to minority status has been related to emotional distress for Latinos in general (e.g., Salgado de Snyder 1987; Amaro et al. 1987). The conceptual framework also recognizes the importance of family structure in the family stress process. For that reason we include being a single parent, typically single-mother status, in the set of predictor variables in the model. Previous research demonstrates that individuals in single mother families are particularly susceptible to a variety of risks ranging from economic hardship to poor psychological functioning (Brown and Moran 1997; Murry et al. 2001). Consistent with these earlier findings, we expect that single parents in the CFP will be at greater risk for emotional distress than those from 2-parent families.

The model in Fig. 10.1 predicts that these four markers of stressful life experiences or conditions – economic pressure, living in a risky neighborhood environment, experiencing discrimination because of one's race or ethnicity, and raising children as a single parent – will increase parent emotional distress, resulting in depressed mood and increased anxiety. Earlier research supports these predictions for economic stress (Conger 1995; Conger and Donnellan 2007; Conger et al. 1999; Evans and English 2002; Yeung et al. 2002), for stressful events and conditions in general (Conger et al. 1995; Ge et al. 1994), and for minority, including Mexican American, as well as majority families (Brody et al. 2002; Conger et al. 2002; Mistry et al. 2002; Parke et al. 2004). Consistent with the family stress model, we predict that stress-related disruptions in caregivers' emotional functioning will negatively affect their parenting practices (e.g., Conger et al. 2002; Conger and Donnellan 2007; Cummings et al. 2001; Fincham 1998; Gonzales et al. 2000; Harold and Conger 1997). In the present analyses, we are especially concerned with the positive management of child behaviors through practices such as explanation, positive reinforcement of desired behaviors, and monitoring of children's activities.

Figure 10.1 indicates that child management practices are expected to predict ATOD use in the seventh grade. Specifically, we expect that effective management practices will reduce risk for the onset and continuation of use during early adolescence. Note also that the influence of earlier family stress and parent symptoms of distress are expected to be associated with later child ATOD use only indirectly through child management. These predictions are consistent with the family stress model and findings from previous research (Conger and Donnellan 2007). Notice also, however, that child ATOD use during fifth grade is expected to predict both parent distress and child management. In the case of parenting, earlier research has shown that the behavioral problems of children can have a disruptive influence on parents' management practices (e.g., Conger and Ge 1999; Kim et al. 2001; Patterson et al. 1992; Stewart et al. 2000); thus, we include this path in the model. Moreover, earlier child behavioral problems involving substances are expected to act as an additional source of stress in a parent's life and increase the risk of emotional distress consistent with the expected impact of other family stresses. Finally, the model proposes that there will be continuity in ATOD use from the fifth to seventh grades. In the following analyses we test these predictions from the conceptual model (Fig. 10.1).

After testing predictions from the model outlined in Fig. 10.1, we turn to evaluating the effect of two culture-related variables, religion and familism, on the course of ATOD use and cognitions from the fifth to seventh grades. Religion and commitment to family are traditional priorities in Latino families that are expected to reduce risk for child maladjustment (Cauce and Domenech-Rodriguez 2002). Here we are interested in both the direct effects of religion and familism measured at the fifth grade on ATOD use or cognitions at the seventh grade and also on the moderating effects of these variables. Direct effects of a culture-related variable on ATOD use or cognitions at seventh grade are expected to be negative suggesting resilience in

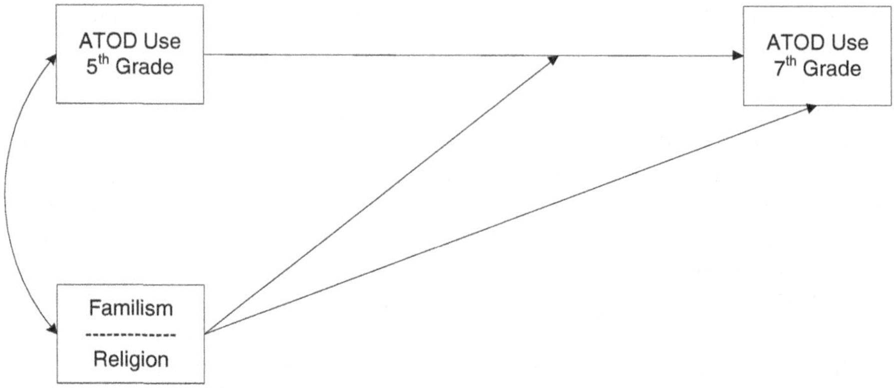

Fig. 10.2 Conceptual interaction model

terms of a compensatory or promotive effect. Furthermore, an interaction (product term) of the culture-related variable with the problem behavior at an earlier time is also expected to produce a negative path to the problem behavior at a later time, yielding a buffering or protective effect also indicative of resilience (Luther et al. 2000; Masten 2001). That is, we expect that higher levels of familism or religious beliefs in fifth grade will reduce the risk that ATOD involvements during fifth grade will continue or increase into seventh grade. The conceptual interaction model for these analyses is shown in Fig. 10.2.

Methods

Participants and Procedures

The participant sample consisted of 317 Mexican-origin families with a non-handicapped, normally-functioning child who attended the fifth grade in a public or Catholic school within one of two cities in northern California. Children were drawn at random from the student rosters for the school districts of these two cities. Families of these children were then recruited by telephone or, for cases in which there was no listed phone number, by a recruiter who went to their home. For the present sample, 68.6 % of the eligible families agreed to participate. All family members were of Mexican origin as determined by their ancestry and their self-identification as being of Mexican heritage. First, second, and third generation children of Mexican origin were eligible for the study. Also, the focal child had to be living with his or her biological mother. Either two-parent (82 % of the sample) or single-parent (18 % of the sample) families were eligible to participate. In two-parent families, the father had to be the child's biological father. For the present analyses, we include only the focal child and the mother in order to maximize sample size. This procedure also makes the findings more consistent across one and two parent families.

Trained research staff interviewed the participants (child, mother, and, in most cases, father) in their homes using laptop computers. Interviewers were all bilingual and most were of Mexican heritage. They received 2 weeks of training and continuing supervision in the field by the interviewer coordinator. Continuing checks were made to assure that interviewers complied with a standardized set of interviewing procedures. The data used for this report come from computer-assisted interviews that were programmed using the BLAISE 4.8.1 software program (Statistics Netherlands 2009). During fifth grade and then again during seventh grade, interviewers visited the families on two separate occasions usually within a 1-week period. Each visit lasted between 2 and 3 h and each participant was interviewed separately by one of the two interviewers. Every effort was made to assure that the interviews were completed independently so that other family members could not hear the questions or answers for mother, father, or focal child. At both waves of data collection for the fifth and seventh grades, the mother provided demographic information about the family and household members. Interviews were conducted in Spanish or English based on the preference of each participant.

Measures

All measures were available in both Spanish and English for parents and children to complete. The English versions of all measures were translated into Spanish by native Spanish speakers who were members of the project research staff. The translations were from English to Spanish and then an independent group of bilingual staff members back translated the measures from Spanish to English to assure that the original meaning of each item was maintained. All variables were measured during the fifth grade and ATOD use and intent were also measured during the seventh grade.

Economic Pressure To measure economic pressure, mothers completed three subscales of a self-report inventory developed by Conger and colleagues (Conger et al. 1991; Conger and Elder 1994). The first subscale comprises two items tapping whether the mother felt that she could not "make ends meet" during the past 3 months. The two items asked whether there was enough income to pay monthly bills and if there was any money left over at the end of the month. Higher scores on this scale indicate a higher level of not being able to make ends meet. The second subscale measured whether the family could meet its basic material needs related to clothing, a car, a home, furniture and household appliances, food, and medical services. The responses used a 4-point Likert-type scale (1 = *not at all true*, 4 = *very true*). The responses were reverse coded such that higher scores indicate a higher degree of unmet material needs. The third subscale queried whether the family had made significant financial cutbacks in many areas, such as food and utilities, because of economic hardship. These responses were dichotomously scored (1 = yes, 0 = no). These three subscales (alpha = 0.69) each were used as a separate indicator of a latent variable estimating economic pressure.

Perceived Discrimination This instrument measured the mother's experiences with and perceptions of discrimination or prejudices against Mexicans/Mexican-Americans in the workplace, neighborhood, and schools (Johnston and Delgado 2004). Participants responded to a 4-point Likert-type scale (1 = *almost never or never*, 4 = *almost always or always*. This measure contains one scale assessing overall perceived discrimination. The items in this measure tap feeling less respected and being insulted because of the respondent's ethnicity. To obtain higher quality of indicators for the latent construct of discrimination, we randomly assigned each of the 17 items into one of four subscales, or parcels, yielding three 4-item parcels and one 5-item parcel. The alpha for these four parcels was 0.86. Recent research (e.g., Kishton and Widaman 1994; Little et al. 2002) supports parceling items to create indictors for latent variables because it decreases the effect of measurement errors and produces indicators with high reliability, which in turn better defines the latent variable. A high score on the construct indicates higher levels of discrimination.

Neighborhood Risk The target children and their mothers completed two scales assessing the degree of environmental risk existing in their neighborhoods. One scale contains 10 items adapted from existing scales by Aneshensel and Sucoff (1996) and Bowen and Chapman (1996) indicating how often criminal events (e.g., violent crimes, theft, gang fights, and public uses of alcohol and drugs) occur in the neighborhood. Responses were made on a 4-point Likert-type scale (1 = *almost never or never*; 4 = *almost always or always*). Mother and child reports were then averaged to form a composite measure of neighborhood criminal events. Another questionnaire elicits personal evaluations of the attractiveness of the neighborhood (Lansing and Marans 1969). The item response was also on a 4-point Likert-type scale (1 = *not at all true*; 4 = *very true*). All the items in this scale were reverse coded; correspondingly, higher scores indicated lower neighborhood quality. Sample items tap the following neighborhood characteristics: an attractive place to live, safe, good neighbors, etc. We averaged across mother and child reports on each scale to create two indicators for a latent factor of neighborhood risk. A high score indicates that there is high risk for criminal activities and low environmental quality in the neighborhood.

Mother's Emotional Distress To measure mother's emotional distress, a short form of the Center for Epidemiologic Studies Depression Scale (CES-D; Radloff 1977) and the Mini Mood and Anxiety Symptom Questionnaire (MASQ; Clark and Watson 1995) were completed. The short form of the CES-D was developed by Cole et al. (2004). It contains 10 items designed to measure depressive symptomatology in the general population. The response format is a 4-point Likert scale (1 = *almost never or never*; 4 = *almost always or always*). The MASQ includes 90 items related to symptoms of anxiety and mood disorders. The Mini-MASQ is a shortened version that includes 26 items (Casillas and Clark 2001). In this study, the response categories were modified from the original 5-point to a 4-point Likert-type scale (1 = *not at all*; 4 = *very much*). The alphas were 0.77 and 0.92 for the CES-D and MASQ, respectively. These two scales were used to create a two-indicator factor of mother's emotional distress.

Mother's Child Management The child's mother answered a questionnaire on parenting practices using measures from the Iowa Youth and Families Project (Conger and Elder 1994). The scores from the subscales for Inductive Reasoning (IR) and Positive Reinforcement (PR) were used for the analysis in this study. Both subscales use a 4-point Likert-type response format (1 = *almost never or never*; 4 = *almost always or always*). The IR subscale measures the degree to which parents care about their children's opinions and feelings, or they explain decisions on family matters that involve the child. The PR subscale (3 items) measures how often the mother positively reinforces the child's good behaviors in situations such as earning good grades, doing chores, and participating in sports at school. The mother also completed a questionnaire reporting the degree to which she monitored the child in daily life (Small and Kerns 1993; Small and Luster 1994). Items from this scale asked whether or not the mother knew what was going on in most aspects of the child's life, including school work, work and play activities. The monitoring scale and the two subscales for parenting practices were used to create a three-scale factor. The alpha for the three scales was 0.75.

Single Parent The family was coded as single (1) or non-single (0). Single indicated the mother was living without a romantic partner whereas non-single implied she was either married or living with a male partner in a continuing romantic relationship. For two-parent families, during fifth grade this was always the focal child's biological father.

ATOD Cognitions These items were assessed during both the fifth grade and the seventh grade and asked the focal child about his or her intention to use tobacco, alcohol, or illegal drugs in the next year (from the date of the interview). The measure also asked if the child would be willing to participate with a group of friends who were using tobacco, alcohol, or other drugs (Gibbons et al. 1998). Nine items compose this scale with response options from 1 to 4. These nine items (alpha of 0.91) were averaged for each participant to create a single, manifest scale score named ATOD cognitions in the analyses.

ATOD Use These nine items asked about actual substance use by the target child (Elliott et al. 1985). All substances are queried, from tobacco to hard liquor and other illegal drugs. Two of the items tap binge drinking by asking whether four or more drinks of some type of alcohol were consumed at a single sitting. The items are rated on a 1–5 scale where 1 indicates 'never' and 5 'almost every day.' The responses for each participant for all nine items were averaged to yield a single, manifest scale score with an alpha of 0.78.

Religion The target child completed the Religion Scale which consists of seven items from the Mexican American Acculturation/Enculturation Scale (MAAS:

developed by N.A. Gonzales, G.P. Knight, & D. Saenz; Arizona State University). All seven items indicate the importance of God and the strength He provides to the individual. The items are responded to on a scale from 1 to 4, where 1 indicates 'Not at all' and 4 indicates 'Very much.' A representative question is, 'It is important to thank God every day for all one has.' The alpha for the seven items was 0.75.

Familism The target child completed the Familism Scale which consists of 16 items also from the MAAS. All 16 items tap the importance and primacy of the family, that it is up to each family member to make sacrifices if need be for the good of the family. All items are responded to on a scale from 1 to 4, where 1 indicates 'Not at all' and 4, 'Very much.' A representative question is, 'A person should always think about his/her family when making important decisions.' The alpha for the 16 items was 0.82.

Data Analysis

The first sets of analyses based on the model shown in Fig. 10.1 were designed to evaluate predictions from the overall conceptual model. The outcomes of interest measured during seventh grade included ATOD use and intent to use. With the ATOD variables controlled during fifth grade, paths predicting ATOD involvement during seventh grade can be thought of as predicting relative change in these behaviors. This model was evaluated by first regressing the ATOD variable at seventh grade on all the fifth grade variables and then deleting paths that were not statistically significant and arriving at the optimum model based on chi-square difference tests for nested models and fit statistics. A nested model is obtained by deleting one or more paths from a given model (i.e. forcing the path coefficient to 0) while leaving all other paths unchanged and adding no new paths.

Figure 10.2 illustrates the secondary sets of analyses in which familism and religion as well as their interactions with the two ATOD variables measured at the fifth grade were predictors of the two ATOD variables measured at the seventh grade. The path from the familism/religion variable that touches with an arrowhead the path between the two ATOD variables measured across grades indicates the effect of the interaction. All models were fit to the data using Mplus 5.21 (Muthén and Muthén 2007).

For our analyses, we used the chi-square statistic associated with full information maximum likelihood (FIML) estimation as an overall fit statistic. We used FIML estimation to account for missing data which are present in all longitudinal studies. FIML estimation uses all available information in the moment matrix thereby providing values on variables where data are missing at random (Enders 2001). Previous research shows that FIML provides a more accurate estimate of model parameters than ad hoc procedures such as listwise or pairwise deletion. Several standard fit indices were used to evaluate competing, or nested, models, all of which are based on the chi-square statistic. Because the chi-square value is sensitive to sample size, three alternative fit indexes also are reported. The root mean square error of approximation (RMSEA) indicates close fit when values are less than .05 and reasonable fit when values are between .05 and .08 (Browne and Cudeck 1993). The other two indexes are the Tucker-Lewis index (TLI; Tucker and Lewis 1973) and the comparative fit index (CFI; Bentler 1990), both indicating acceptable fit for values above .90.

Results

Descriptive Statistics

As noted earlier, even though the transition from fifth to seventh grades represents the transition from late childhood to early adolescence, seventh grade is still an early point in development to see significant substance use or abuse. For example, the Monitoring the Future Study has shown that real escalation in substance use typically begins around the eighth or ninth grades (Johnston et al. 2000). For that reason, we examined not only actual substance use but also intent or willingness to use substances inasmuch as these cognitions are good predictors of later involvement with ATODs (Andrews et al. 2008). The data from the current study are consistent with the expectation that some change should be observed in relation to substance involvement, especially for substance use cognitions. Table 10.1 provides the means, standard deviations, and minimum and maximum scores for the variables examined in this report. Instructive is the finding that ATOD cognitions increase from a mean of .013 in fifth grade to a mean of .039 in the seventh grade. Moreover, the maximum score on this variable increases from .89 to 1.78 across the two grades. These results indicate that most children in the study are not

Table 10.1 Manifest means, standard deviations, minimum, and maximum values of constructs used in SEM analyses

Variable	*N*	Mean	Standard deviation	Minimum	Maximum
Child management	315	3.170	0.477	1.57	4.00
Mother's emotional distress	306	1.742	0.442	1.00	3.44
Perceived discrimination	312	1.595	0.359	1.00	3.47
Economic pressure	312	2.520	0.527	1.25	3.88
Neighborhood risk	310	1.711	0.352	1.00	2.94
Familism	290	3.611	0.314	2.63	4.00
Religion	289	3.467	0.472	1.43	4.00
ATOD cognitions fifth grade	306	0.013	0.074	0.00	0.89
ATOD cognitions seventh grade	262	0.039	0.156	0.00	1.78
ATOD use fifth grade	303	1.027	0.148	1.00	2.78
ATOD use seventh grade	262	1.048	0.165	1.00	2.67

Table 10.2 Standardized factor loading estimates and standard errors for indicators of the latent variables used in structural equation models

	ATOD Cognitions		ATOD Use	
Factor Indicator	Estimate	*SE*	Estimate	*SE*
	Economic pressure			
Ends meet	0.778	0.036	0.778	0.036
Material needs	0.770	0.038	0.771	0.038
Cutbacks	0.699	0.040	0.698	0.040
	Neighborhood risk			
Criminal events	0.886	0.066	0.941	0.075
Neighborhood quality	0.719	0.059	0.677	0.062
	Perceived discrimination			
Parcel 1	0.800	0.026	0.810	0.026
Parcel 2	0.825	0.025	0.823	0.025
Parcel 3	0.832	0.024	0.832	0.024
Parcel 4	0.675	0.036	0.674	0.036
	Mother's emotional distress			
CES-D	0.877	0.049	0.854	0.048
MASQ	0.810	0.048	0.832	0.047
	Child management			
Inductive reasoning	0.796	0.042	0.787	0.041
Positive reinforcement	0.689	0.043	0.692	0.043
Child monitoring	0.668	0.044	0.674	0.044

contemplating involvement with substances, but there is an increase in the degree to which respondents express interest in using substances from fifth to seventh grades. This result is consistent with the expectation of a trajectory of increasing risk for ATOD involvement during the transition from childhood to early adolescence. Actual substance use increased at an even slower but positive rate.

Prior to examining associations among the study variables, we examined the factor loadings of indicator variables on study constructs to assure the validity of the measurement strategy. The findings in Table 10.2 demonstrate that the indicator variables met acceptable expectations. For example, as described earlier four parcels were generated as indicators of perceived discrimination. The factor loadings for this latent variable were all positive and statistically significant and ranged from .674 (parcel 4) in the ATOD use model to .832 for parcel 3 in both the ATOD use and cognitions models. The loadings for the other constructs in the model were of similar magnitude and statistically significant. These findings provide strong support for the measurement approach used for tests of the conceptual framework.

Table 10.3 provides the correlations among the constructs employed in later structural equation models (SEMs) used to test the conceptual framework. The correlations demonstrate that three of the exogenous predictor variables (i.e., discrimination, neighborhood risk, and economic pressure) in the conceptual model

Table 10.3 Correlations among variables and factors used in SEM path analysis

Variable/Factor	1	2	3	4	5	6	7	8	9
1. Perceived discrimination									
2. Neighborhood risk	0.374*								
3. Economic pressure	0.163*	0.339*							
4. Mother's emotional distress	0.397*	0.249*	0.357*						
5. Child management	0.008	–0.231*	–0.179*	–0.141*					
6. Single mother	0.115	–0.006	0.060	0.132*	0.088				
7. ATOD use fifth grade	0.088	0.166*	0.049	–0.068	–0.207*	–0.043			
8. ATOD use seventh grade	0.019	0.155*	0.044	0.032	–0.246*	–0.038	0.558*		
9. ATOD cognitions fifth grade	–0.009	0.149*	0.086	0.034	–0.152*	0.104	0.500*	0.499*	
10. ATOD cognitions seventh grade	0.011	0.146*	0.134*	–0.003	–0.208*	0.015	0.517*	0.672*	0.423*

*Indicates correlation is significant $p<.05$, z-test, 2-tail

are positively and significantly related to one another (e.g., $r = .339$, $p < .05$ for economic pressure and neighborhood risk). Unexpectedly, none of these predictors is correlated with single-parent status. In addition, all four of the exogenous variables are positively and significantly related to maternal emotional distress, as expected. Moreover, both neighborhood risk and economic pressure predict child management and some dimensions of ATOD involvement. If these associations are not statistically significant in the test of the full model, that will be good evidence for the meditational processes proposed by the conceptual framework. Also important, ATOD cognitions in the fifth grade predict ATOD use in the seventh grade ($r = .499$, $p < .05$), consistent with our expectation that these dimensions of intention and willingness provide an important pathway to later use. As predicted by the conceptual model, good child management skills were negatively related to later ATOD involvement and, finally, both substance use and substance cognitions demonstrated stability over time. These promising associations among the constructs in the model suggested that more formal tests of the conceptual framework are worthwhile.

Structural Equation Models

Figure 10.3 provides the results for the model test predicting to ATOD cognitions. In the figure an asterisk indicates a significant standardized path coefficient estimate at $p < .05$ (1-tail). Because the model predicts the direction of effects, we used a 1-tail test. Path coefficients represented by a dotted line were not statistically significant. The fit of the final ATOD cognitions model was acceptable with a χ^2 $(135, N = 317) = 165.86$, $p < .001$; CFI = 0.962; TLI = .950; and RMSEA = 0.044.

Consistent with the correlations, willingness and intention to use substances (cognitions) demonstrated a moderate degree of stability from fifth to seventh grade ($\beta = .41$, $p < .05$). As expected, economic pressure and perceived discrimination were associated with greater maternal emotional distress, but neither neighborhood risk nor single parent status directly predicted distress with all exogenous variables in the model. Economic problems and experiences of discrimination were the stressors most strongly related to emotional difficulties for these mothers. As expected, mother's emotional distress was associated with a lower level of child management skills. The focal child's ATOD cognitions in fifth grade were also significantly and negatively related to child management. In fact, the indirect pathway from cognitions in fifth grade to cognitions in seventh grade through child management was statistically significant. Finally, child management was negatively related to ATOD cognitions in the seventh grade. This result suggests that good child management skills can blunt the expected upward trajectory in intentions or willingness to use substances during the transition from childhood to adolescence. Moreover, the absence of direct effects from earlier predictors in the model to seventh grade ATOD cognitions provides support for the mediating process proposed in the conceptual model (Fig. 10.1).

Figure 10.4 provides the findings for the model related to substance use in the fifth and seventh grades. The coefficients and indications of statistical significance

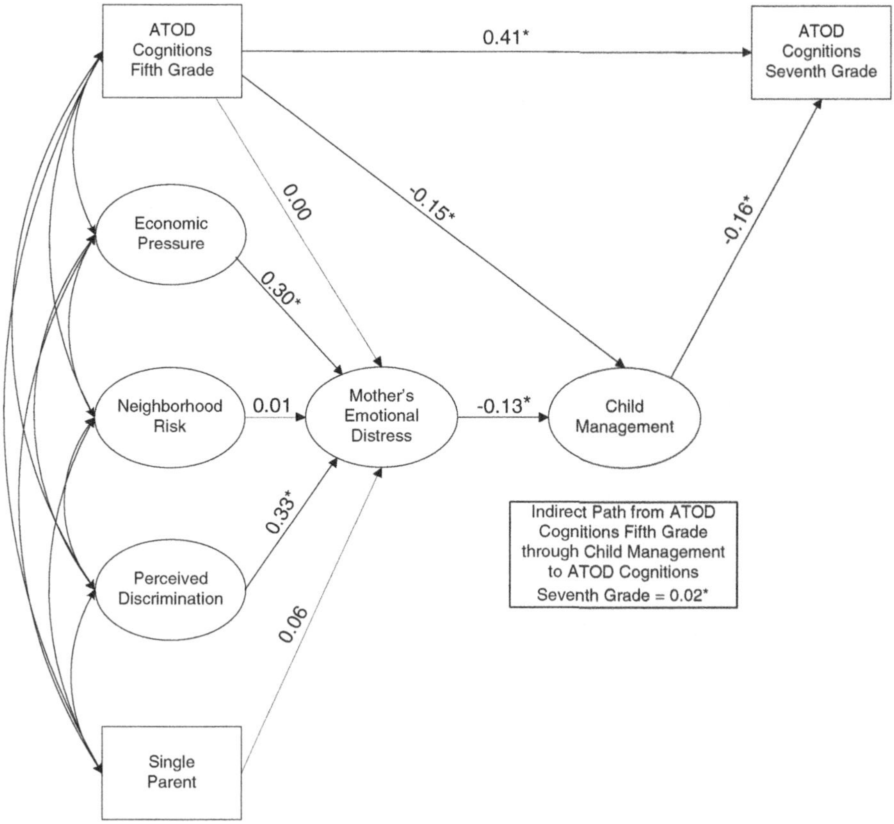

Fig. 10.3 Child alcohol, tobacco and other drugs intended use

in the figure can be interpreted in the same way as in Fig. 10.3. The fit of the final ATOD use model was also acceptable with a χ^2 (135, N=317)=167.40, p<.001; CFI=0.962; TLI=.951; and RMSEA=0.044. As before, both economic pressure and perceived discrimination predicted emotional distress; however, ATOD use during fifth grade also predicted mother's distress. This finding suggests that actual use rather than simply thinking about using substances is more likely to generate distress upon the part of the parent. In addition, distress predicted management which had a negative association with use in the seventh grade. And as earlier, ATOD involvement in fifth grade was negatively associated with skillful child management and had an indirect relationship with use in seventh grade through mother's managerial behaviors. Finally, there was significant stability in use from fifth to seventh grades (β=.54). These results are reasonably consistent with predictions from the conceptual model (Fig. 10.1).

Our second sets of analyses address the role of culturally relevant priorities in the development of ATOD involvements. These analyses are based on the model shown

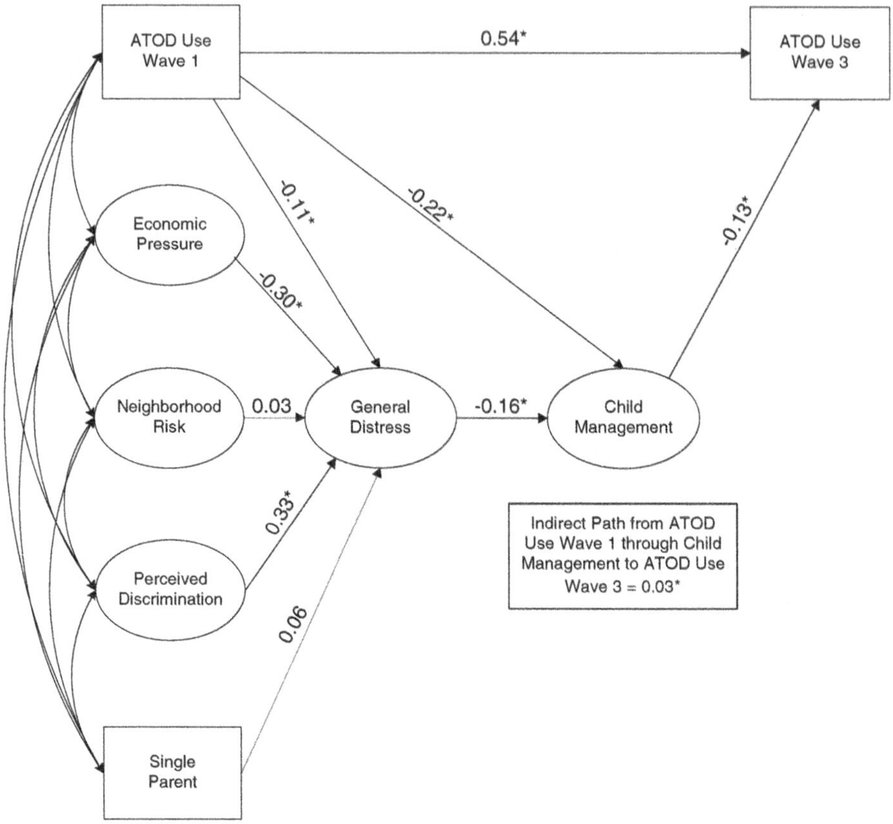

Fig. 10.4 Child alcohol, tobacco and other drugs use

in Fig. 10.2. Correlations among the variables are shown in Table 10.4. The correlations indicate that religion during fifth grade was negatively and significantly ($p < .05$) related to ATOD cognitions (−.182) and use (−.184) at seventh grade, consistent with the notion of a promotive or compensatory effect demonstrating resilience. These values indicate small effects, yet relations were in the expected direction. Contrary to expectations, familism during fifth grade was not a significant predictor of the seventh grade ATOD variables. Turning to the interactions, the correlations between the interaction terms and the ATOD variables were all significant and in the expected directions. For example, the interaction between ATOD cognitions and familism during fifth grade was highly predictive of ATOD cognitions during seventh grade ($r = -.552$), suggesting a buffering effect in the sense that high familism reduces continuity or stability in ATOD cognitions over time. All of the predicted interactions terms produced similar correlations consistent with expectations.

Figures 10.5 (ATOD Cognitions) and 10.6 (ATOD Use) show the results of the path analyses for familism and religion. The coefficients representing the direct

Table 10.4 Correlations among variables used in ethnicity analyses

Variable/interaction	1	2	3	4	5	6	7	8	9
1. Familism 5th grade									
2. Religion 5th grade	0.573*								
3. ATOD cognitions 5th grade	–0.084	–0.190*							
4. ATOD Cog 5th X Fam 5th	–0.111	0.065	–0.700*						
5. ATOD Cog 5th X Rel 5th	0.012	0.204*	–0.641*	0.685*					
6. ATOD cognitions 7th grade	–0.124	–0.182*	0.457*	–0.552*	–0.607*				
7. ATOD use 5th grade	–0.017	–0.159*	0.502*	–0.449*	–0.698*	0.550*			
8. ATOD use 5th X Fam 5th	–0.126*	0.145*	–0.440*	0.475*	0.740*	–0.558*	–0.452*		
9. ATOD use 5th X Rel 5th	0.031	0.235*	–0.564*	0.606*	0.966*	–0.633*	–0.632*	0.828	
10. ATOD use 7th grade	–0.099	–0.184*	0.530*	–0.717*	–0.678	0.687*	0.591*	–0.584*	–0.699*

Note: X indicates a product variable, or interaction term
*Indicates correlation is significant $p<.05$, t-test, 2-tail

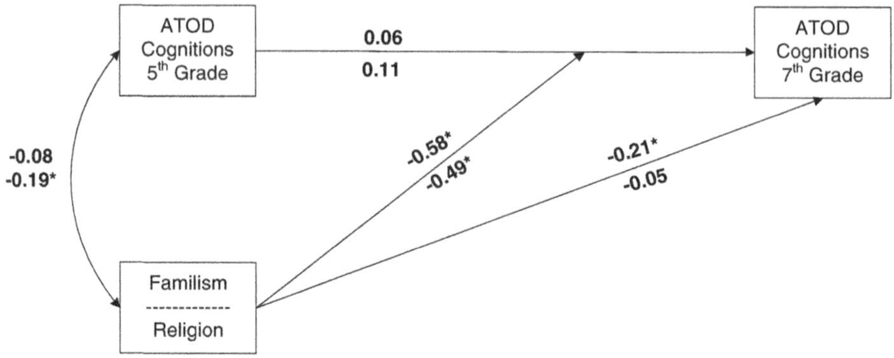

Fig. 10.5 Effect of familism and religion on child alcohol, tobacco, and other drugs intended use. *Note.* Familism estimated standardized path coefficients displayed above the directed paths, religion estimates displayed below the directed paths

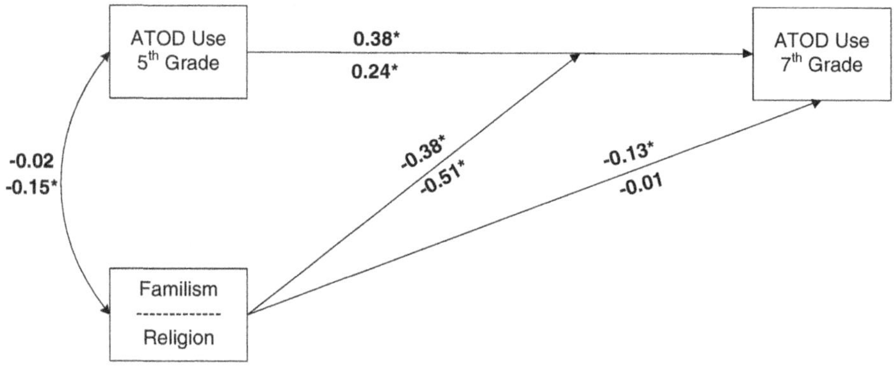

Fig. 10.6 Effect of familism and religion on child alcohol, tobacco, and other drugs use. *Note.* Familism estimated standardized path coefficients displayed above the directed paths, religion estimates displayed below the directed paths

effects from the religion and familism variables to the ATOD variables at seventh grade cannot be accurately interpreted because they were estimated in the presence of an interaction. For that reason, we focus on the results involving the interaction effects. Returning to the interactions illustrated in Figs. 10.5 and 10.6, the paths shown in these figures are standardized values with familism estimates shown above the path and religion estimates shown below the path. All variables were mean centered for these analyses and the model fit was perfect due to saturated models; that is, all possible paths were estimated. As expected, all of the paths representing interaction effects were negative and statistically significant (e.g., −.58 from familism to the path from ATOD cognitions in fifth to seventh grade).

Substantive interpretation of models with included interactions from the path values alone is difficult, however. Given that, the path values were re-interpreted by calculating the effect at −1, 0, +1 and +2 standard deviations from the mean (of 0) for each ATOD variable measured at the 5th grade for −1 (Low) and +1 (High) standard deviations from the mean (of 0) for familism and religion. These results are shown in Figs. 10.7, 10.8, 10.9, and 10.10. The scales for both the vertical and horizontal axes are standard deviation units. In all four combinations (e.g. Familism on ATOD Cognitions), those individuals above the mean on the ATOD variable at the fifth grade will have markedly lower predicted use on that ATOD variable at the seventh grade if they are high (1 *SD* above the mean) on either religion or familism. Conversely, if they are low (1 *SD* below the mean) on either religion or familism and above the mean on the ATOD variable at the fifth grade, the predicted use of that ATOD variable at the seventh grade will decrease. These results are quite consistent with the idea that familism or religious beliefs reduce continuity in ATOD use or

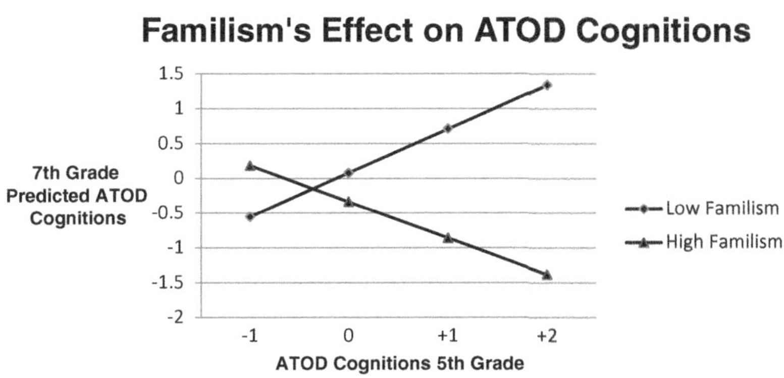

Fig. 10.7 The interaction between familism and 5th grade ATOD cognitions predicting 7th grade ATOD cognitions

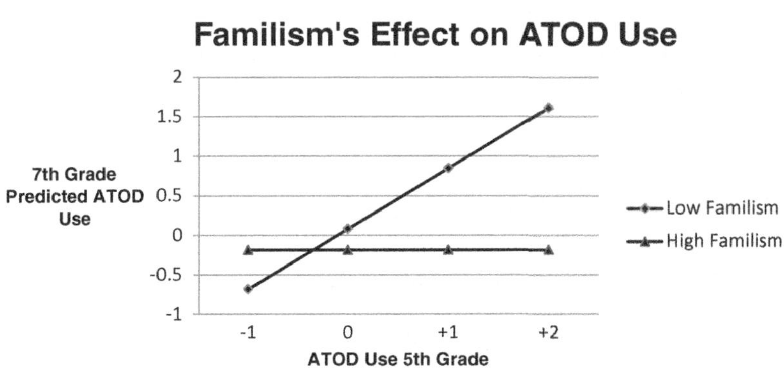

Fig. 10.8 The interaction between familism and 5th grade ATOD use predicting 7th grade ATOD use. *Note.* X- and Y-axes are in standard deviation units. Low and High refer to 1 standard deviation below and above the mean on Familism

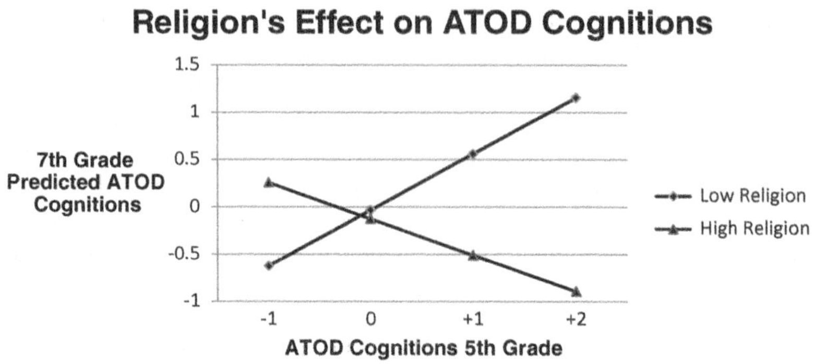

Fig. 10.9 The interaction between religion and 5th grade ATOD cognitions predicting 7th grade ATOD cognitions

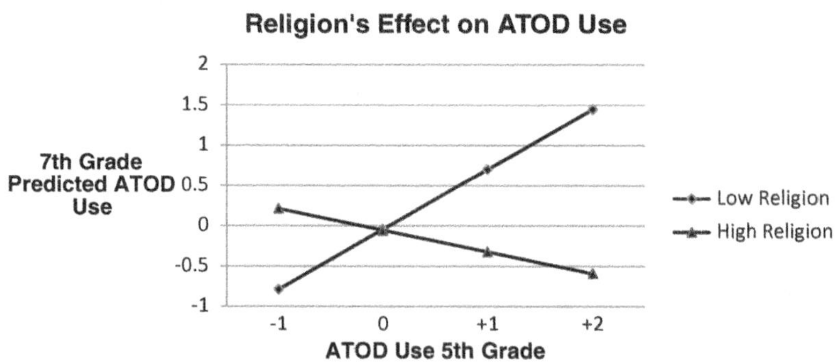

Fig. 10.10 The interaction between religion and 5th grade ATOD use predicting 7th grade ATOD use. *Note.* X- and Y-axes are in standard deviation units. Low and High refer to 1 standard deviation below and above the mean on Religion

cognitions from fifth to seventh grade, indicating that they may be a source of resilience in terms of blunting the negative effects of early involvement with substances (Figs. 10.7, 10.8, 10.9 and 10.10).

Discussion and Final Conclusions

The edited volume in which this chapter appears is concerned with trajectories of substance use by minority youth. The very first rise in these trajectories typically occurs during the transition from childhood to early adolescence. As noted at the beginning of this report, for those youth who experience an early escalation of ATOD use, these trajectories can increase quickly and lead to a multitude of problems including later substance abuse and dependence. Findings of this type have led researchers to focus increasingly both on the early initiation of ATOD use and on a developmental perspective. That is, if scientists can produce greater understanding of the antecedents of these risky trajectories, and of the circumstances that blunt their occurrence, then more effective prevention programs can be developed to reduce ATOD problems for specific youth and for the community in general. The present report took this approach in an attempt to add to our knowledge of the processes through which Mexican origin children and families in the U.S. experience risk for and resilience to substance use at an early point in development.

We drew on previous research and theory to propose a model that enumerated a number of stressors likely to be experienced by a large proportion of these families and children (Fig. 10.1). Consistent with general demographic trends, we expected that economic pressure would play an important role in the lives of these families and would be joined by related stressors involving the experience of racial or ethnic discrimination, living in risky and dangerous neighborhoods, or life as a single mother. We expected that these difficulties would exacerbate maternal emotional distress, as has been shown in previous research (Conger and Donnellan 2007). Earlier research has shown that, when mothers become depressed or anxious, these emotional difficulties tend to impair their ability to parent effectively. That is, a parent distracted by their own emotional problems is likely less able to concentrate on the oftentimes difficult tasks involved in raising a competent child, such as attending to the activities they are engaged in, explaining the reasons for rules and expectations and providing positive consequences for desired behaviors. However, when these parenting practices are maintained even in the face of the types of stressors studied here, we expect that they will operate as a source of resilience to the onset of ATOD use during the transition from childhood to early adolescence. The results of the present analyses were largely consistent with these expectations.

The findings showed that two of the stressors were especially damaging to the emotional health of these mothers. Economic pressure and the experience of discrimination were negatively related to maternal emotional distress in the final models. At the zero-order correlation level, neighborhood risk and single parent status also predicted emotional distress, but these main effects were not statistically significant with the other stressors in the model. Apparently, it is the economic difficulties they face and the discrimination they experience that are particularly troubling for Mexican American mothers. And as expected, mothers who were more anxious and depressed functioned less capably in terms of managing the development of their children. In turn, these emotional problems exacerbated increased trajectories of ATOD use or intentions to use through reduced parenting competence. Earlier child ATOD involvement also had a detrimental impact on effective parenting, making it even harder for the parent to engage in behaviors that would blunt early involvement during the fifth grade from continuing or escalating by the seventh grade.

However, all is not bleak. When mothers continued to engage in effective parenting practices despite the stress in their lives, children were less likely to increase their use or thoughts of use of substances. Thus, effective management practices by a parent can be thought of as a source of resilience in the face of adversity. Moreover, a child's subscription to traditional cultural priorities involving religion and familism appears to create a buffering or protective effect that reduces the likelihood that use or thoughts of use during late childhood will increase these problems during the early adolescent years. Simply put, these findings suggest that both specific maternal behaviors and traditional cultural values may serve as important sources of resilience in preventing early onset ATOD behaviors and cognitions from escalating into increased substance problems over time.

So what are the primary implications of these findings for theory, for preventive interventions, and for the design of future research? Theoretically, the results replicate earlier research with a variety of ethnicities and show that economic pressure has the same negative developmental influences on children of Mexican origin as for children in other population subgroups. This vulnerability to economic problems is especially problematic for families that are too often faced with difficult economic circumstances. In addition, financial adversities are compounded by the experience of ethnic discrimination which had a similar deleterious relationship with maternal emotional functioning. Interestingly, living in a risky neighborhood or being a single parent did not appear to be remarkably stressful above and beyond the economic and discriminatory problems associated with these situations. Especially important, to the degree that parents were able to maintain effective parenting practices, they apparently can reduce the likelihood that ATOD involvement will increase during the transition from childhood to early adolescence.

From a practical standpoint, these findings suggest that prevention programs aimed at supporting effective parenting practices should reduce the escalation of ATOD trajectories during this period of development, as also suggested by other research (e.g., Spoth et al. 2000). In addition, however, these results suggest that preventive interventions should focus not only on how the parent treats the child, but also on the parent's emotional capacity for engaging in effective parenting behaviors. That is, in addition to teaching parents how to more effectively raise a child, preventive interventions should increasingly address the degree to which the parent is psychologically capable of benefitting from such instruction. It may be the case that clinical interventions designed to reduce parental emotional distress may produce major benefits in terms of improving the effectiveness of educational programs aimed at improving parenting practices. Also important will be programs aimed at directly improving child behaviors that create greater emotional and parenting stress for their caregivers. For Mexican American families, supporting traditional values like commitments to family and religious beliefs may also play an important role in the prevention process. These findings also suggest that social and educational policies focused on reducing the experience of discrimination and economic hardship may also play an important role in fostering more effective parenting practices. Simply put, these results shine the spotlight on the contexts that affect parenting behavior, including the earlier characteristics of the child. Interventions at all of these points in the process should have a beneficial impact in terms of promoting the managerial capacities of the parent and the resilience of the child.

In terms of future research, these early findings from the CFP provide significant support for the importance of the developmental approach to understanding trajectories of risk or resilience for ATOD use. Also important is the inclusion of the environmental contexts in which these developmental processes occur. For minority families and children, it seems clear that experiences of discrimination add considerably to the difficulties with which they must contend in day to day life, including greater risk for economic hardship and marginality. Future research needs to improve understanding of how these contexts operate in the lives of minority families and their children. How can the capacity to cope with such challenges be improved and how can communities act to reduce the level of stress that minority families experience? These are key questions to be addressed in future studies. For California, they are particularly salient in terms of youth of Mexican origin inasmuch as they will become the majority of the population of the state within the next few decades. To ignore the issue of how to prevent developmental problems such as substance use and abuse and how to promote competent and successful development within this ethnic group will imperil the health of the State in terms of a productive and well-functioning citizenry by mid-century.

Acknowledgement This research was supported by a grant from the National Institute on Drug Abuse (DA017902) with supplementary support from the National Institute on Alcohol Abuse and Alcoholism. Correspondence regarding this manuscript should be addressed to Rand Conger in the Department of Human Ecology, University of California-Davis, One Shields Avenue, Davis, CA 95616 or to rdconger@ucdavis.edu.

References

Amaro, H., Russo, N. F., & Johnson, J. (1987). Family and work predictors of psychological well-being among Hispanic women professionals. *Psychology of Women Quarterly, 11*, 505–521.

Anashensel, C. S., & Sucoff, C. A. (1996). The neighborhood context of adolescent mental health. *Journal of Health and Social Behavior, 37*, 293–310.

Andrews, J. A., Hampson, S. E., Barckley, M., Gerrard, M., & Gibbons, F. X. (2008). The effect of early cognitions on cigarette and alcohol use during adolescence. *Psychology of Addictive Behaviors, 22*, 96–106.

Anthony, J. C. (1993). The scope of epidemiologic research on drug use: A rationale of change. In M. G. Montero & J. A. Inciardi (Eds.), *Proceedings of the Brazil-United States binational meeting on drug abuse research* (pp. 213–223), Sao Paulo, Brazil, May 1992. Wilmington: University of Delaware

Baca Zinn, M., & Wells, B. (2000). Diversity within Latino families: New lessons for family social science. In D. H. Demo & K. R. Allen (Eds.), *Handbook of family diversity* (pp. 252–273). London: Oxford University Press.

Bentler, P. M. (1990). Comparative fit indexes in structural models. *Psychological Bulletin, 107*, 238–246.

Bowen, G. L., & Chapman, M. V. (1996). Poverty, neighborhood danger, social support, and the individual adaptation among at-risk youth in urban areas. *Journal of Family Issues, 17*, 641–666.

Bray, J. H., Adams, G., Getz, J. G., & Baer, P. E. (2001). Developmental, family, and ethnic influences on adolescent alcohol usage: A growth curve approach. *Journal of Family Psychology, 15*, 301–314.

Brody, G. H., McBride Murry, V., Kim, S., & Brown, A. C. (2002). Longitudinal pathways to competence and psychological adjustment among African American children living in rural single-parent households. *Child Development, 73*, 1505–1516.

Brook, J. S., Whiteman, M., Balka, E. B., Win, P. T., & Gursen, M. D. (1998). Drug use among Puerto Ricans: Ethnic identity as a protective factor. *Hispanic Journal of Behavioral Sciences, 20*, 241–254.

Brook, J. S., Brook, D. W., De La Rosa, M., Whiteman, M., Johnson, E., & Montoya, I. (2001). Adolescent illegal drug use: The impact of personality, family, and environmental factors. *Journal of Behavioral Medicine, 24*, 183–203.

Brown, G. W., & Moran, P. M. (1997). Single mothers, poverty and depression. *Psychological Medicine, 27*, 21–33.

Browne, M. W., & Cudeck, R. (1993). Alternative ways of assessing model fit. In K. A. Bollen & J. S. Long (Eds.), *Testing structural equation models* (pp. 136–162). Newbury Park: Sage.

Burton, L. M., & Jarrett, R. L. (2000). In the mix, yet on the margins: The place of families in urban neighborhood and child development research. *Journal of Marriage & the Family, 62*, 1114–1135.

Casillas, A., & Clark, L. A. (2001). *The mini-mood and anxiety symptom questionnaire (Mini-MASQ)*. Poster presented at the 72nd Annual meeting of the Midwestern Psychological Association, Chicago, IL.

Catalano, R. F., & Hawkins, J. D. (2000). The social development model: A theory of antisocial behavior. In J. D. Hawkings (Ed.), *Delinquency and crime* (pp. 149–196). Cambridge: Cambridge University Press.

Cauce, A. M., & Domenech Rodriguez, M. (2002). Latino families: Myths and realities. In J. Contreras (Ed.), *Latino youth and families facing the future*. New York: Basic Books.

Chassin, L., Husson, A., Barrera, M., Molina, B. S. G., Trim, R., & Ritter, J. (2004). Adolescent substance use. In R. M. Lerner & L. Steinberg (Eds.), *Handbook of adolescent psychology* (pp. 665–696). New York: Wiley.

Chilcoat, H. D., Dishion, T. J., & Anthony, J. C. (1995). Parent monitoring and the incidence of drug sampling in urban elementary school children. *American Journal of Epidemiology, 141*, 25–31.

Cicchetti, D. (1999). A developmental psychopathology perspective on drug abuse. In M. D. Glantz & C. R. Hartel (Eds.), *Drug abuse: Origins and intervention* (pp. 97–118). Washington, DC: American Psychological Association.

Clark, L. A., & Watson, D. (1995). *The mini mood and anxiety symptom questionnaire (Mini-MASQ)*. Unpublished manuscript, University of Iowa, Iowa City.

Cole, J. C., Rabin, A. S., Smith, T. L., & Kaufman, A. S. (2004). Development and validation of a Rasch-derived CES-D short form. *Psychological Assessment, 16*, 360–372.

Conger, R. D. (1995). Unemployment. In D. Levinson (Ed.), *Encyclopedia of marriage and the family* (pp. 731–735). New York: Macmillan.

Conger, R. D. (1997). The social context of substance abuse: A developmental perspective. In E. B. Robertson, Z. Sloboda, G. M. Boyd, L. Beatty, & N. J. Kozel (Eds.), *Rural substance abuse: State of knowledge and issues* (Research Monograph No. 168, NIH Publication No. 97-4177, pp. 6–36). Washington, DC: National Institute on Drug Abuse.

Conger, R. D., & Conger, K. J. (2002). Resilience in Midwestern families: Selected findings from the first decade of a prospective, longitudinal study. *Journal of Marriage and Family, 64*, 361–373.

Conger, R. D., & Elder, G. H., Jr. (1994). *Families in troubled times: Adapting to change in rural America*. Hawthorne: Aldine de Gruyter.

Conger, R. D., & Ge, X. (1999). Conflict and cohesion in parent-adolescent relations: Changes in emotional expression from early to mid-adolescence. In M. Cox & J. Brooks-Gunn (Eds.), *Conflict and cohesion in families: Causes and consequences* (pp. 185–206). Mahwah: Erlbaum.

Conger, R. D., & Rueter, M. A. (1996). Siblings, parents, and peers: A longitudinal study of social influences in adolescent risk for alcohol use and abuse. In G. H. Brody (Ed.), *Advances in applied developmental psychology* (Sibling relationships: Their causes and consequences, Vol. 10, pp. 1–30). Westport: Ablex.

Conger, R. D., Lorenz, F. O., Elder, G. H., Jr., & Melby, J. N. (1991). A process model of family economic pressure and early adolescent alcohol use. *Journal of Early Adolescence, 4*, 430–449.

Conger, R. D., Patterson, G. R., & Ge, X. (1995). It takes two to replicate: A mediational model for the impact of parents' stress on adolescent adjustment. *Child Development, 66*, 80–97.

Conger, R. D., Conger, K. J., Matthews, L. S., & Elder, G. H., Jr. (1999). Pathways of economic influence on adolescent adjustment. *American Journal of Community Psychology, 27*, 519–541.

Conger, R. D., Ebert-Wallace, L., Sun, Y., Simons, R. L., McLoyd, V. C., & Brody, G. H. (2002). Economic pressure in African American families: A replication and extension of the family stress model. *Developmental Psychology, 38*, 179–193.

Cummings, E. M., Goeke-Morey, M. C., & Dukewich, T. L. (2001). The study of relations between marital conflict and child adjustment: Challenges and new directions for methodology. In J. H. Grych & F. D. Fincham (Eds.), *Interparental conflict and child development: Theory, research, and applications* (pp. 39–63). New York: Cambridge University Press.

Dawes, M. A., Antelman, S. M., Vanyukov, M. M., Giancola, P., Tarter, R. E., Susman, E. J., Mezzich, A., & Clark, D. B. (2000). Developmental sources of variation in liability to adolescent substance use disorders. *Drug & Alcohol Dependence, 61*, 3–14.

Ellickson, P. L., & Morton, S. C. (1999). Identifying adolescent at risk for hard drug use: Racial/ethnic variations. *Journal of Adolescent Health, 25*, 382–395.

Elliott, D. S., Huizinga, D., & Ageton, S. S. (1985). *Explaining delinquency and drug use*. Beverly Hills: Sage.

Enders, C. K. (2001). A primer on maximum likelihood algorithms available for use with missing data. *Structural Equation Modeling, 8*, 128–141.

Ensminger, M. S., Juon, H. S., & Fothergill, K. E. (2002). Childhood and adolescent antecedents of substance use in adulthood. *Addiction, 97*, 833–844.

Evans, G. W., & English, K. (2002). The environment of poverty: Multiple stressor exposure, psychophysiological stress, and socioemotional adjustment. *Child Development, 73*, 1238–1249.

Finch, B. K., Kolody, B., & Vega, W. A. (2000). Perceived discrimination and depression among Mexican-Origin adults in California. *Journal of Health and Social Behavior, 41*, 295–313.

Fincham, F. D. (1998). Child development and marital relations. *Child Development, 69*, 543–574.

Ge, X., Lorenz, F. O., Conger, R. D., & Elder, G. H. (1994). Trajectories of stressful life events and depressive symptoms during adolescence. *Developmental Psychology, 30*, 467–483.

Gibbons, F. X., Gerrard, M., Blanton, H., & Russell, D. W. (1998). Reasoned action and social reaction: Willingness and intention as independent predictors of health risk. *Journal of Personality and Social Psychology, 74*, 1164–1181.

Glantz, M. D., & Leshner, A. I. (2000). Drug abuse and developmental psychopathology. *Development & Psychopathology, 12*, 795–814.

Gonzales, N. A., Pitts, S. C., Hill, N. E., & Roosa, M. W. (2000). A mediational model of the impact of interparental conflict on child adjustment in a multiethnic, low-income sample. *Journal of Family Psychology, 14*, 365–379.

Gonzales, N. A., Knight, G. P., Morgan-Lopez, A., Saenz, D., & Sirolli, A. (2002). Acculturation and the mental health of Latino youths: An integration and critique of the literature. In J. Contreras, A. Neal-Barnett, & K. Kerns (Eds.), *Latino children in the United States: Current research and future directions*. Westport: Praeger Publishers.

Guo, J., Hawkins, J. D., Hill, K. G., & Abbott, R. D. (2001). Childhood and adolescent predictors of alcohol abuse and dependence in young adulthood. *Journal of Studies on Alcohol, 62*, 754–762.

Guo, J., Hill, K. G., Hawkins, J. D., Catalano, R. F., & Abbott, R. D. (2002). A developmental analysis of sociodemographic, family, and peer effects on adolescent illicit drug initiation. *Journal of the American Academy of Child and Adolescent Psychiatry, 41*, 838–845.

Harold, G. T., & Conger, R. D. (1997). Marital conflict and adolescent distress: The role of adolescent awareness. *Child Development, 68*, 333–350.

Hops, H., Andrews, J. A., Duncan, S. C., Duncan, T. E., & Tildesley, E. (2000). Adolescent drug use development: A social interactional and contextual perspective. In A. J. Sameroff, M. Lewis, & S. M. Miller (Eds.), *Handbook of developmental psychopathology* (pp. 589–606). New York: Plenum Publishers.

Jessor, R., Van Den Bos, J., Vanderryn, J., Costa, F. M., & Turbin, M. S. (1995). Protective factors in adolescent problem behavior: Moderator effects and developmental change. *Developmental Psychology, 31*, 923–933.

Johnston, K. E., & Delgado, M. Y. (2004). *Mexican American adolescents' experiences with ethnic discrimination*. Poster presentation presented at the Biennial conference of the society for research on adolescence, Baltimore, Maryland.

Johnston, L. D., O'Malley, P. M., & Bachman, J. G. (2000). *The Monitoring the Future national results on adolescent drug use: Overview of key findings, 1999*. Bethesda: National Institute on Drug Abuse.

Kellam, S. G., Brown, C. H., Rubin, B. R., & Ensminger, M. E. (1983). Paths leading to teenage psychiatric symptoms and substance use: Developmental epidemiological studies in Woodlawn. In S. B. Guze, F. J. Earls, & J. E. Barrett (Eds.), *Childhood psychopathology and development* (pp. 17–51). New York: Raven.

Kim, K. J., Conger, R. D., Lorenz, F. O., & Elder, G. H., Jr. (2001). Parent-adolescent reciprocity in negative affect and its relation to early adult social development. *Developmental Psychology, 37*, 775–790.

Kishton, J. M., & Widaman, K. F. (1994). Unidimensional versus domain representative parceling of questionnaire items: An empirical example. *Educational and Psychological Measurement, 54*, 757–765.

Lansing, J. B., & Marans, R. W. (1969). Evaluation of neighborhood quality. *Journal of the American Planning Association, 35*, 195.

Little, T. D., Cunningham, W. A., Shahar, G., & Widaman, K. F. (2002). To parcel or not to parcel: Exploring the question, weighing the merits. *Structural Equation Modeling, 9*, 151–173.

Luthar, S. S., Cicchetti, D., & Becker, B. (2000). The construct of resilience: A critical evaluation and guidelines for future work. *Child Development, 71*, 543–562.

Masten, A. S. (2001). Ordinary magic: Resilience processes in development. *American Psychologist, 56*, 227–238.

McLoyd, V. C., Cauce, A. M., Takeuchi, D., & Wilson, L. (2000). Marital processes and parental socialization in families of color: A decade review of research. *Journal of Marriage & the Family, 62*, 1070–1093.

Melby, J. N., Conger, R. D., Conger, K. J., & Lorenz, F. O. (1993). Effects of parental behavior on tobacco use by young male adolescents. *Journal of Marriage & the Family, 55*, 439–454.

Mistry, R. S., Vandewater, E. A., Huston, A. C., & McLoyd, V. C. (2002). Economic well-being and children's social adjustment: The role of family process in an ethnically diverse low-income sample. *Child Development, 73*, 935–951.

Murry, V. M., Bynum, M. S., Brody, G. H., Willert, A., & Stephens, D. (2001). African American single mothers and children in context: A review of studies on risk and resilience. *Clinical Child and Family Psychology Review, 4*, 133–155.

Muthén, L. K., & Muthén, B. O. (2007). *Mplus user's guide* (5th ed.). Los Angeles: Muthén & Muthén.

Netherlands, S. (2009). *Blaise for windows (Version 4.81) [Computer Software]*. Heerlen: Blaise Department.

Parke, R. D., Coltrane, S., Duffy, S., Buriel, R., Dennis, J., Powers, J., et al. (2004). Economic stress, parenting, and child adjustment in Mexican American and European American families. *Child Development, 75*, 1632–1656.

Patterson, G. R., Crosby, L., & Vuchinich, S. (1992). Predicting risk for early police arrest. *Journal of Quantitative Criminology, 8*, 335–355.
Portes, A., & Rumbaut, R. G. (2001). *Legacies: The story of the immigrant second generation*. Berkeley: University of California Press.
Radloff, L. S. (1977). The CES-D Scale: A self-report depression scale for research in the general population. *Applied Psychological Measurement, 1*, 385–401.
Rebhun, L. A. (1998). Substance use among immigrants to the United States. In L. Sana (Ed.), *Handbook of immigrant health* (pp. 493–519). New York: Plenum Press.
Salgado de Snyder, V. N. (1987). Factors associated with acculturative stress and depressive symptomatology among married Mexican immigrant women. *Psychology of Women Quarterly, 11*, 475–488.
Schulenberg, J., Maggs, J. L., Steinman, K. J., & Zucker, R. A. (2001). Developmental matters: Taking the long view on substance abuse etiology and intervention during adolescence. In P. M. Monti, S. M. Colby, & T. A. O'Leary (Eds.), *Adolescents, alcohol, and substance abuse: Reaching teens through brief interventions* (pp. 19–57). New York: The Guilford Press.
Sloboda, Z. (1999). The prevention of drug abuse: Interrupting the paths. In M. D. Glantz & C. R. Hartel (Eds.), *Drug abuse: Origins & interventions* (pp. 223–242). Washington, DC: American Psychological Association.
Small, S. A., & Kerns, D. (1993). Unwanted sexual activity among peers during early and middle adolescence: Incidence and risk factors. *Journal of Marriage and the Family, 55*, 941–952.
Small, S. A., & Luster, T. (1994). Adolescent sexual activity: An ecological, risk-factor approach. *Journal of Marriage and the Family, 56*, 181–192.
Spoth, R. L., Redmond, C., & Shin, C. (2000). Reducing adolescents' aggressive and hostile behaviors: Randomized trial effects of a brief family intervention 4 years past baseline. *Archives of Pediatric and Adolescent Medicine, 154*, 1248–1257.
Staton, M., Leukefeld, C., Logan, T. K., Zimmerman, R., Lynam, D., Milich, R., Martin, C., McClanahan, K., & Clayton, R. (1999). Risky sex behavior and substance use among young adults. *Health and Social Work, 24*, 147–155.
Stewart, E. A., Simons, R. L., & Conger, R. D. (2000). The effects of delinquency and legal sanctions on parenting behaviors. In G. L. Fox & M. L. Benson (Eds.), *Families, crime and criminal justice* (1st ed., Vol. 2, pp. 257–279). New York: Elsevier Science Inc.
Swaim, R. C., Bates, S. C., & Chavez, E. L. (1998). Structural equation socialization model of substance use among Mexican-American and White non-Hispanic school dropouts. *Journal of Adolescent Health, 23*, 128–138.
Tarter, R. E., & Vanyukov, M. (1994). Alcoholism: A developmental disorder. *Journal of Consulting & Clinical Psychology, 62*, 1096–1107.
Tucker, L. R., & Lewis, C. (1973). A reliability coefficient for maximum likelihood factor analysis. *Psychometrika, 38*, 1–10.
U.S. Census Bureau. (2000). Census of population and housing, profiles of general demographic characteristics. No. 36. Cities with 250,000 or more inhabitants in 2000-Hispanic and non-Hispanic groups: 2000.
Vega, W. A., & Gil, A. G. (1999). A model for explaining drug use behavior among Hispanic adolescents. *Drugs & Society, 14*, 57–74.
Vega, W. A., Kolodny, B., Aguilar-Gaxiola, S., Alderete, E., Catalano, R., & Caraveo-Anduaga, J. (1998). Lifetime prevalence of DSM-III-R psychiatric disorders among urban and rural Mexican Americans in California. *Archives of General Psychiatry, 55*, 771–778.
Yeung, W. J., Linver, M. R., & Brooks-Gunn, J. (2002). How money matters for young children's development: Parental investment and family processes. *Child Development, 73*, 1861–1880.

Chapter 11
Prospecting Prejudice: An Examination of the Long-Term Effects of Perceived Racial Discrimination on the Health Behavior and Health Status of African Americans

Frederick X. Gibbons, Michelle L. Stock, Ross E. O'Hara, and Meg Gerrard

There are many reasons why adolescents experiment with drugs and other substances, ranging from impulse and curiosity to chance and opportunity. The fact that it is multiply determined and unstable—adolescents often try a substance once or twice and then refrain for months or even years – makes initiation of use very hard to predict. That is less the case with escalation of use. A number of factors, individually and collectively, have been shown to effectively predict trajectories of use among adolescents over time. Foremost among these factors are peer use and peer pressure, substance availability, and motive – negative affect or stress. Adolescents who affiliate with peers who are using show steeper increases in use (Curran et al. 1997; Wills and Cleary 1999); the same is true for adolescents under stress (Tate et al. 2007).

The strength of these two relations highlights the paradox that researchers examining *racial differences* in use have addressed for years, and that is the fact that African American adolescents generally report less substance use than White adolescents (Bolland et al. 2007; Watt and Rogers 2007; White et al. 2004) even though they are much more likely to be raised in difficult or high-risk environments that provide more opportunity to use substances, and more reason to use—i.e., more stress (e.g., stress associated with living in low SES environments; Ardelt and Eccles 2001). Add to the mix the fact that African American adolescents experience additional stress due to racial discrimination (Sellers et al. 2003) and the paradox becomes more pronounced.

F.X. Gibbons (✉) • M. Gerrard
Department of Psychology, University of Connecticut, Storrs, CT, USA
e-mail: Rick.gibbons@uconn.edu

M.L. Stock
Department of Psychology, The George Washington University, Washington, DC, USA

R.E. O'Hara
Department of Psychiatry, University of Connecticut Health Center, Farmington, CT, USA

Y.F. Thomas, L.N. Price (eds.), *Drug Use Trajectories Among Minority Youth*,
DOI 10.1007/978-94-017-7491-8_11

This chapter describes our research and that of others that has examined the trajectories of substance use among African American adolescents. Specifically, our focus is on the role that perceived racial discrimination (PRD) plays in determining trajectories of drug use. This line of research comprises both field/survey studies and laboratory experiments, and includes factors that mediate the PRD→use relation and factors that moderate the relation—both risk and protective. Finally, we also discuss implications of this research for use in interventions and preventive-interventions.

Figures 11.1 and 11.2 are based on self-reports of drinking in a representative national sample of 6522 adolescents—66 % White, 19 % Hispanic, and 11 %

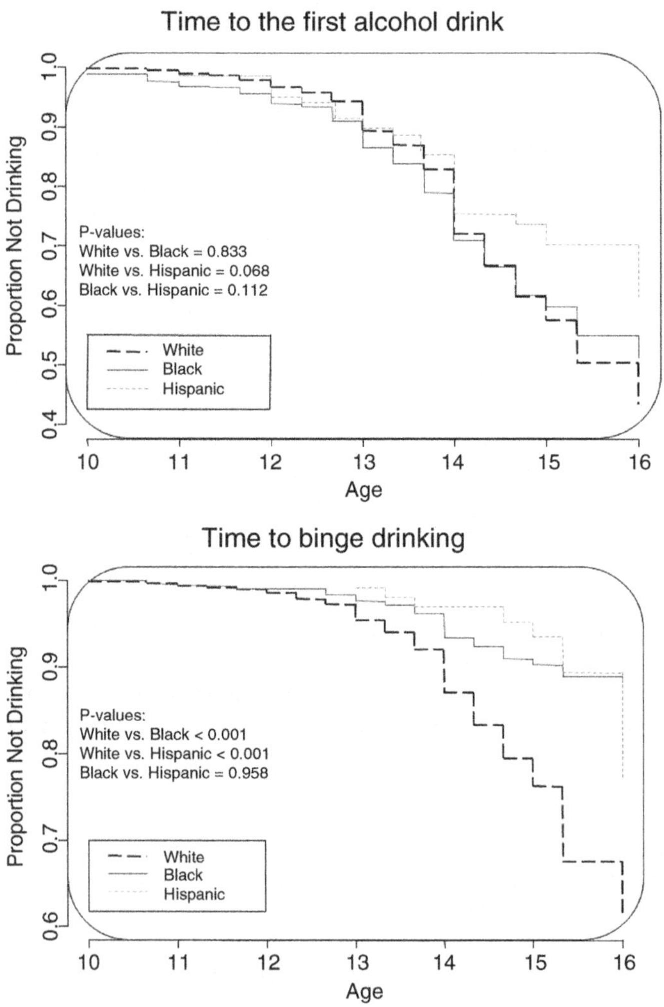

Fig. 11.1 Time to first drink and first binge episode by race/ethnic group in a national sample of adolescents 10–14 at Wave 1 (Note: adopted with permission from Tanski et al. (2010))

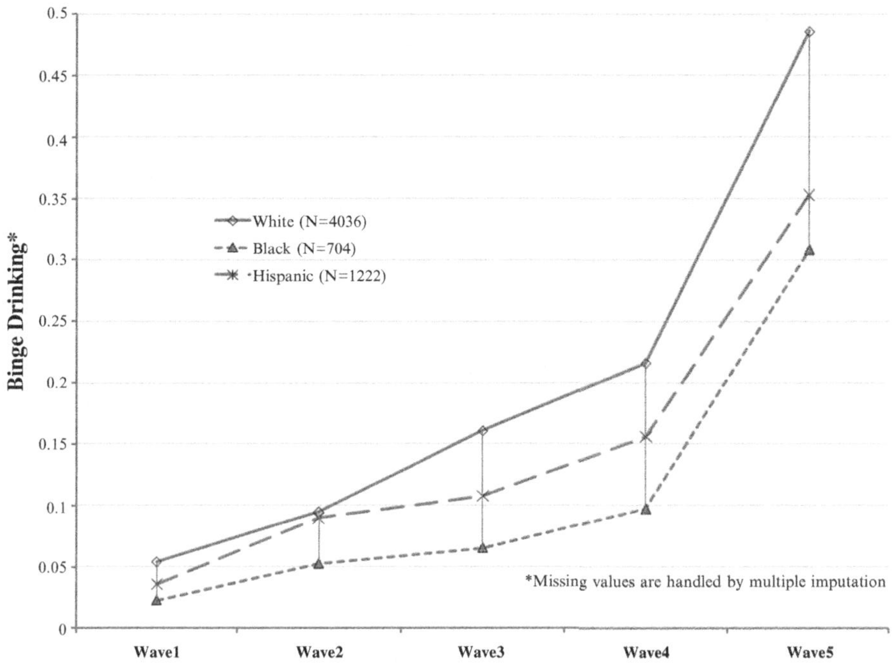

Fig. 11.2 Mean proportion reporting binge drinking by wave in a national sample of adolescents

African American, average age of 12 (range 10–14) at Wave 1 (W1) and 15 (range 13–17) at W5 (see Sargent et al. 2006; Tanski et al. 2010). The first figure is a survival analysis reflecting time to first drink and then to initial binge drinking episode. The second is percentage reporting drinking across five waves in the same sample. Two things are most notable about these figures. First is the consistency in age of initiation across the three racial/ethnic groups—the percentages reporting initiation of drinking for each of the three groups was essentially the same; there were no significant differences across the three groups: 3 %, 10 %, and 50 % for ages 10, 13, and 16, respectively. Second is the divergence in escalation (to binge drinking) after that time. This pattern of less use by African Americans relative to other racial/ethnic groups has proven consistent across studies and across time.

Importantly, however, these trajectories vary, considerably, by age. The pronounced deficit in use by African American adolescents continues into early adulthood. But some time in the mid to late 20s, the differences diminish, and then appear to reverse, a phenomenon referred to as a racial cross-over effect (Geronimus et al. 1993; Kandel 1995). Evidence of this can be seen in a variety of indicators (see Kandel & Thomas, this volume), including surveys of substance use (Pampel 2008; Schulenberg et al. 2005), especially drug use and problematic drinking (Watt 2008); arrests and clinic admissions (Levine et al. 2010); and statistics on drug-related mortality (Vega and Gil 1998).

Several factors have been proposed as contributors to this cross-over effect (see chapters by Kandel & Thomas; Ensminger et al.; and Broman, this volume). Those include a decline in parental influence in adulthood (Simons et al. 2003), delay of marriage and high divorce rates (marriage is protective against use; Arnett 2007; Kandel et al. this volume), and stressors, such as high unemployment (Kaiser Family Foundation 2006) and incarceration (Pettit and Western 2004). Our research has focused on one of those factors, stress: We have found evidence of a strong prospective relation—across periods of up to 11 or 12 years-- between PRD and substance use. This relation exists for the adolescents who are the focus of our research, and for their parents, and it maintains when controlling for a number of covariates that are reliable predictors of use, including some that have been suggested by others as possible alternatives to PRD as predictors of the racial cross-over (e.g., SES, negative life events, marital status). Most of this research comes from the Family and Community Health Study (FACHS).

FACHS

The FACHS is a longitudinal study of psychosocial and other factors related to the physical and mental health of African Americans. Six waves of data have been collected so far. The study began with a sample of 889 African American families, each with a target adolescent in 5th grade (*M* age of 10.5 at W1, 24.5 at W6) and a primary caregiver (parent; 92 % of whom were mothers of the adolescents; their *M* ages were 37 at W1 and 51 at W6). Half the families also had a secondary caregiver participating in the study, and 1/3 had an older sibling participating. At W4, we began collecting data from the person nominated by the adolescent as his/her best friend, and at W5, we added the adolescent's romantic partner (for additional description of the sample and it's recruitment, see Brody et al. 2001; Cutrona et al. 2000; Simons et al. 2005).

Retention rates have been very good; the W6 adolescent sample (completed in the fall of 2011) was 78 % of the W1 sample. In most cases, the interviews have been done in the homes of the participants, using Computer-Assisted Personal Interview (CAPI) and/or Audio Computer-Assisted Self-interview (ACASI). Both of these techniques promote validity and reliability of responding, especially when the questions involve personal information and/or illegal activities, such as risky sexual behavior or illegal drug use (Tourangeau and Smith 1996). The instrument includes two diagnostics to assess pathology: The NIMH Diagnostic Interview Schedule for Children (DISC IV; Shaffer et al. 1993), for the adolescents; and the University of Michigan Composite International Diagnostic Interview (UM-CIDI; Kessler 1991) for the parents (both of these generate clinical diagnoses according to the definitions and criteria of the DSM-IIIR and ICD-10 Systems). The interview also includes a variety of measures of individual differences, behaviors, and psychosocial factors related to physical and mental health.

We have used a modified version of Landrine and Klonoff's (1996) Schedule of Racist Events as a measure of PRD.[1] The scale assesses lifetime PRD that is thought to be attributable to one's race (e.g., "How often have you been treated unfairly because you were African American?"). The scale has been shown to be a reliable predictor of reactions to discrimination (see Pascoe and Richman 2009).

Levels of Use

Reports of heavy drinking among the FACHS adolescents were similar to those in the Sargent et al. national data set, and both levels were comparable to national norms for this age range; e.g., drinking rates from W1 to W4 were: 2.3 %, 2.7 %, 13.7 %, 25.2 % vs. national norms of 0.9 %, 2.1 %, 14.7 %, and 35.3 %–again, suggesting no differences in early experimentation, but then slower escalation for African Americans. The same was true for their reports of illicit drug use, with marijuana clearly being the preferred substance. Escalation in problematic use was more or less linear, with noticeable increases occurring around age 14, corresponding with the transition from middle school to high school (Johnston et al. 2008; SAMHSA 2008, 2015). By W5 (*M* age 22), 58 % reported some lifetime substance use, compared with national figures of 59 % for 21-year olds (SAMHSA 2008, 2015). In general, reports of use by the FACHS adolescents were within the range that is typical for their age levels. A different pattern emerged among the parents, however.

Focusing on problematic use (defined as more than occasional illicit drug use–i.e., lifetime use >5 times) and drinking problems (rather than social drinking), at age 37 (W1), 19.7 % reported one or the other or both (15.1 % drug use; 9.3 % drinking problems, with being under the influence in dangerous situations [driving] the most commonly reported problem). These figures are somewhat above national norms for this age group, but still well within the "normal" range. Of interest, however, is the trend in the later waves. Whereas the typical pattern is for use and abuse to decline from the late 30s into the late 40s (SAMHSA 2008, 2015), the opposite pattern can be seen in these data: reports of problematic use increased at nearly every wave (from 19.7 % at W1 to 33 % at W4 and then 35 % at W5).[2] This pattern is unusual in women of this age (SAMHSA 2015). We will return to this issue later.

[1] Modifications included simplifying the language somewhat so it could be understood by 5th graders, and replacing some of the items concerning workplace discrimination with more general, community-based discrimination experiences.

[2] Two issues about these analyses should be kept in mind. First, these are reports of lifetime use, so they will typically go up over time (though this amount of increase is unusual). Second, these figures are most likely (significant) underestimates for the overall sample. That is because in FACHS, like other longitudinal studies of health behavior, those who are engaging in more risky actions are more likely to attrite from the sample. In this case, the women who attrited from the study reported about three times as much problematic use at W1 as did the women who participated in all five waves. If we impute missing data across the last four waves, the percentages reporting problematic use at W5 exceed 40 %.

What Is Protective?

Although the focus of our research, and in fact, most of the research in this area, has been on risk, it is worth highlighting some of the protective factors in our data set before examining factors that have been positively associated with use. For the adolescents, each of the following constructs was negatively correlated with use synchronously and prospectively: (a) religiosity (cf. Wills et al. 2003), (b) parenting style—one that includes warmth and support along with monitoring and involvement (Cleveland et al. 2005; Pires and Jenkins 2007; see below); and, perhaps most important, (c) academic orientation and academic performance (Windle and Wiesner 2004). Of course, the *absence* of these protective factors usually constitutes a risk factor, but none of them is as strong a predictor (in our research) as is PRD.

The Effects of Perceived Racial Discrimination on Use Prospective Relations

The first study in this series (Gibbons et al. 2004a) examined the relations between parents' and adolescents' reports of PRD at W1 and their substance use reported at the same wave and then again 2 years later. Reports of discrimination were moderate to high for adolescents and their parents: approximately 90 % of each group reported some discriminatory experiences although the percentage reporting a lot of discrimination was, of course, a fair amount higher for the adults than for the children (20 % vs. 8 %). There was also similarity across family members in terms of the relations between reports of PRD and substance use. First, for the parents, the W1 measure was correlated with W1 use (tobacco, alcohol, and illicit drugs), and it predicted use 2 years later, controlling for the relation at W1—i.e., PRD predicted *change* in use over the 2 year period. In fact, PRD had the strongest zero-order correlation with W2 use of any measure we had, including a variety of constructs normally predictive of use (financial and relationship stress, social support, religiosity, optimism, etc.). A similar pattern emerged for the adolescents: W1 PRD predicted use (alcohol, smoking, and marijuana) at W2. This was initiation for the adolescents, since there was very little use by them at W1. Of particular interest (and relevance to this chapter), were the effects on the measures of what we call distress (depression and anxiety). For both the parents and their children, W1 PRD was associated with W1 distress ($ps < .01$), and it predicted increases in distress from W1 to W2 (both $ps < .01$).

To examine the relation between long-term change in PRD and change in use, we conducted a latent growth curve structural equation model (SEM) with FACHS adolescents (average age 10.5–18.5) that examined the relations between the initial level and slope (i.e., growth across all five waves) of PRD and level and slope of marijuana use from W3 to W5 (Gibbons et al. 2013). Covariates included targets' risk-taking, religiosity, negative life events, and their assessment of neighborhood risk, all reported at W1, along with the parents' self-reports of their substance use

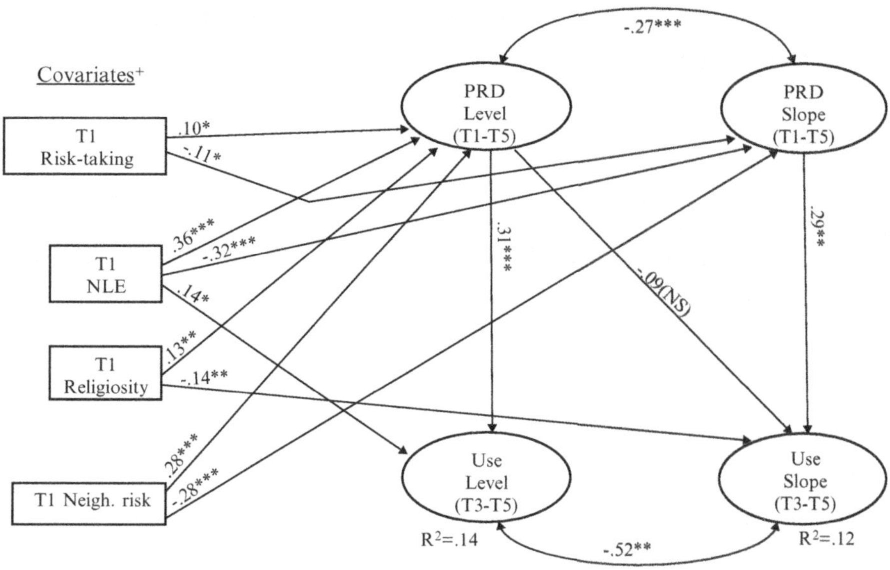

Fig. 11.3 A latent growth curve model of level and slope of PRD and substance use across five waves of FACHS (+Only significant paths from covariates are presented. *p<.05, **p<.01, ***p<.001. *PRD* perceived racial discrimination, *Use* combination of lifetime and past 12 months marijuana use, *NLE*: negative life event, *Neigh. risk*: neighborhood risk)

and their SES level.[3] Results of this analysis are presented in Fig. 11.3. Regarding the covariates, W1 negative life events was a risk factor and religiosity was protective. Level of PRD was a strong predictor of level of use, suggesting that early experience with PRD, as well as cumulative experience, is an important factor (see discussion below). Finally, and most interesting, the slope (trajectory) of PRD from W1 to W5 is a significant predictor of the trajectory (slope) of drug use.

Summary

The relation between PRD and substance use is robust and consistent: Discrimination is associated with increases in use, and changes in discrimination predict changes in use over time. Although many researchers have discussed these associations in the past, our analyses add a prospective element, across multiple waves.

[3]Although not always mentioned in the text, all of the studies involving either the Sargent et al. (2006) or FACHS data sets included a number of covariates. In FACHS, this typically consisted of contextual factors, such as neighborhood risk and SES; individual difference measures, such as sensation-seeking; and familial factors, such as parents' and sometimes older siblings' health risk behaviors.

Mediation of Discrimination Effects: Cognitions, Behavior, and Personality

In order to more fully understand the nature of the often-observed link between discrimination and substance use—and therefore both predict and possibly reduce the strength of the relation—it is essential to examine factors that mediate the association between the two. We have conducted a series of studies that have explored cognitive, behavioral, dispositional, and especially affective reactions to discrimination among FACHS adolescents as mediators of the relation. These analyses are described below.

Cognitions

Aversive experiences, such as those associated with racial discrimination, are likely to have an impact on the attitudes of those who must tolerate them, and these changes in cognitions can also affect vulnerability to substance use. To examine this impact, we have drawn on our model of adolescent health risk behavior, the Prototype/Willingness (Prototype) Model (Gibbons and Gerrard 1997; Gibbons et al. 2003, 2015). The prototype model focuses on the cognitions that mediate the relations between environmental factors (social, familial, contextual) and risky health behaviors, such as substance use or unprotected sex. Primary among these, as the name of the model implies, are the prototype or *image* that the adolescent has of the *type* of person his/her age who engages in the behavior, and the adolescent's behavioral willingness (BW) to engage in the behavior. BW is defined as an openness to *risk opportunity*—what the adolescent would be willing to do under certain circumstances. The more favorable the image, the more willing the adolescent is to do the behavior (Gerrard et al. 2005; for more detail on the model, see Gerrard et al. 1999, 2008; Gibbons et al. 2009). Typically, more favorable user images are reported by adolescents who live in high-risk neighborhoods where substances are more accessible (Gibbons et al. 2004b). The same is true for adolescents whose friends and/or older siblings use (Blanton et al. 1997; Pomery et al. 2005); those who have seen a lot of substance use in movies (Dal Cin et al. 2009; Gibbons et al. 2010b) – *and* those who have experienced higher amounts of discrimination (Gibbons et al. 2004a).

Behavior: Affiliation

Having a favorable image of others who engage in risky behaviors is important because it can lead to more affiliation with "deviant" peers (Gibbons et al. 2003); this relation also can be seen in the responses from the FACHS adolescents. Even though their self-reports of use were very low at W1 (<3 % said they were using), they did report higher percentages of use among their friends: 12 % said they had friends who used more than occasionally at W1. The correlation between W1 PRD and these reports of friends' use were strong ($r = .35$, $p < .001$); and W1 PRD predicted an *increase* in reports of friends' use, much as it had predicted increases in favorable images of those who use. Two sets of subsequent cross-lag analyses provided further information about the nature of these relations. Experience with discrimination was a stronger predictor of friends' use than vice-versa; and own use was a stronger predictor of friends' use than the opposite (i.e., selection more so than socialization; Schulenberg et al. 1999). In short, early experiences with discrimination appeared to encourage these adolescents to seek out the companionship of peers who were engaging in deviant behaviors, such as early substance use. Moreover, this relation was indirect, being mediated by distress (anxiety and depression; see below) and risk cognitions (images of users). But it was also direct; so, it wasn't just the fact that they were seeking solace and support from others to help them cope with the negative affect. This PRD → affiliation relation has held up across all waves of FACHS and appears to be one of the most important and perhaps dangerous reactions to discriminatory experiences.

Behavior: Conduct Disorder (CD)

Further evidence of the impact that PRD has on acceptance of deviance can be seen among the FACHS adolescents in the relation between early experience with discrimination and behavioral problems and pathology (as measured by the DISC-IV; Shaffer et al. 1993). These problems, in turn, also put adolescents at greater risk for substance use. At W1, 4.3 % of the adolescents had a diagnosis of CD (i.e., three or more delinquent behaviors, such as vandalism, physical abuse, or arson, within the past year); that figure was up to 6.9 % by W2. Both figures are above the national norm of 2.1% for ages 8–15 (Merikangas et al. 2010). CD diagnoses were correlated with W1 PRD ($r = .34$). And the two tended to covary over time: Brody et al. (2006a) conducted multi-group latent growth models on these data and found that the slope of reports of PRD predicted the slope of conduct problems, both of which increased over time. Cross-lag analyses indicated that discrimination was a stronger predictor of conduct problems than vice-versa (these trends have maintained in later waves of data), suggesting that experience with discrimination is an antecedent rather than a consequence of the behavior problems.

Subsequent analyses among the adolescents (Gibbons et al. 2007) extended the PRD → substance use prediction out to W3 (5 years after W1, at age 15 or 16) and found a significant relation with baseline PRD. In addition, these analyses provided evidence of cumulative risk: the combination of W1 CD diagnoses and high W1 PRD (top 25 % of the distribution) while rare–only about 2.5 % of the sample had both at this age– was associated with very high rates of use: 57 % of this group was using drugs at W3. More generally, these analyses provided evidence in support of a "critical period" hypothesis. First, early conduct problems are more diagnostic of later substance use and other types of serious delinquency than are later-onset problems (see Moffitt's theory of antisocial behavior, Moffit 1993). Second, although the stability of PRD across waves in the FACHS is high (e.g., 5-wave $\alpha = .77$), analyses have consistently shown that early experience with PRD tends to have a disproportionate impact on the adolescents' subsequent behavior relative to later experiences. For example, the correlation of W5 marijuana use (*M* age 21.5) with early cumulative discrimination (W1 + W2; age 10–13) is actually somewhat higher than the correlation with W4 discrimination, which was assessed at age 18 or 19; $rs = .11, p = .01$ vs. .08, $p = .06$). In short, early experiences with discrimination among African American children are associated with greater risk for conduct problems (Borrell et al. 2006), as well as early substance use. Moreover, the combination of conduct problems and experiences with discrimination before the age of 12 or 13 appears to put the child at much higher risk for later substance use problems and perhaps abuse.

Deviant Behavior: The Combination of CD and Affiliation

New analyses (for this chapter) adding affiliation into this SEM provided additional information about the extent to which behavior changes mediate the PRD/substance use relations. Early discrimination was prospectively associated with both CD problems and affiliation with deviant (i.e., using) peers at W2 (both $ps < .001$). These latter two constructs were, of course, related to one another ($p < .001$), and both of them, in turn, were associated with BW to use at W3, and then to W4 use. The indirect paths through CD and friends' use were both significant ($ps < .01$).

Summary

The evidence suggests that a common reaction among African American adolescents to discriminatory experiences, most of which involve interactions with Whites, is to reject conventional (majority) values (cf. Jessor and Jessor 1997), while increasing acceptance of deviance and deviant behavior. This can be seen in changes in both cognitions (attitudes) and behavior (acting out, seeking friends who are using, and substance use). Ironically, this behavior is actually more common among White adolescents, who tend to have more pro-risk cognitions (e.g., more favorable risk images) and are more likely to use substances (Gibbons et al. 2010b). In other words, African American adolescents who have experienced discrimination are likely to reject White societal values, and as a result, end up behaving more like White adolescents.

Mediation of Discrimination Effects: Negative Affect

Internalizing

At a more fundamental level, numerous studies have documented another PRD relation that appears to be self-evident: the association between discrimination and negative affect. In these studies, negative affect has included both *internalizing* reactions, such as anxiety and depression (Bynum et al. 2007; Hwang and Goto 2008), and *externalizing* reactions, such as anger and hostility (Dubois et al. 2002; for a general review, see Pascoe and Richman 2009). The vast majority of these studies have been cross-sectional, but the relations are strong and consistent enough, that few would argue with a claim of causality. In a series of longitudinal studies, we have examined different types of negative affective reactions to PRD as possible mediating mechanisms responsible for the eventual increase in substance use. The first of these studies (Gibbons et al. 2004a) looked at depression and anxiety—again, labeled distress—as mediators of the prospective relation (W1 to W2) with tobacco, drugs, and alcohol (combined) for both the FACHS adolescents and their parents. For the parents, there was a direct path from W1 discrimination to W2 use, controlling for W1 use, meaning W1 PRD predicted change in use; and the anticipated indirect path through distress was significant as well (both $ps < .01$). The direct effect was not significant for the adolescents, but the indirect effect was substantial ($t = 4.03$, $p < .0001$).

As strong as these relations between PRD and distress are, however, a review of the literature led us to conclude that distress was not the critical factor. For example, Simons et al. (2006) found that perceived discrimination was associated with violent delinquency among African American boys, and that anger, rather than depression or anxiety, mediated this effect. Similar results were reported by Terrell et al. (2006). Other studies with non-African American populations have also found stronger relations between use and anger as opposed to depression or anxiety: Hockey et al. (2000) reported that state and trait anxiety and depression were not consistently related to risk behavior, and Curry and Youngblade (2006) found that reports of anger were more strongly correlated with risk behavior than were reports of depression (cf. Fite et al. 2006). In the affect literature, correlational and experimental evidence suggests that risk-taking is related to anger (Lerner and Keltner 2001), whereas risk avoidance is related to sadness and anxiety, as well as fear (Michael and Ben-Zur 2007; Rydell et al. 2008). Finally, studies in social cognition have shown that anger (and not sadness) prompts heuristic processing (Bodenhausen et al. 2003; Moons and Mackie 2007), and heuristic processing has been linked with riskier behavior (Reyna and Farley 2006; Wang 2006).

Externalizing

In fact, when we redid the analyses from the first paper (i.e., Gibbons et al. 2004a), extending prediction out to W3 use, and including distress, but also adding constructs for hostility for the parents (from the UM-CIDI) and anger for the adolescents, a different pattern emerged (Gibbons et al. 2010a). First, among the parents, there was very high stability in reports of use (the W1 to W3 stability path was $\beta = .73$, $p < .001$), and hostility was strongly correlated with W1 PRD ($p < .001$). Nonetheless, W1 PRD predicted an increase in hostility from W1 to W2, as did W1 use. More important, the path from W1 PRD to W2 distress (i.e., change in distress) was no longer significant. The same was true for the path from W2 distress to W3 use, which had been significant, but was replaced by the W2 hostility to W3 use path ($p < .001$). As expected, the indirect path from W1 discrimination to use through (W1 and W2) hostility was significant ($p = .005$). Thus, as expected, hostility and not distress mediated the relation.

For the adolescents, both distress and anger were correlated with W1 PRD (both *ps* $< .001$), but W1 distress was not associated with any other construct. In contrast, W1 anger was directly associated with both W2 BW and with use at W3 – 5 years later. In addition, the path from discrimination to W3 use through anger and then BW, and the overall indirect effect of W1 discrimination on use at W3 were both significant ($p = .001$). In short, the pattern of results was very consistent across the two family members:

PRD exacerbated distress and hostility/anger, but only the latter was related to changes in use, which means the impact of PRD on use proceeded through anger and not distress (other health effects of discrimination- related distress are discussed below).

Summary

PRD was strongly linked with substance use within these families. In fact, just as the parents' W1 discrimination had the strongest zero-order correlation with their W2 use of any measure of stress (reported in Gibbons et al. 2004a), in Gibbons et al. (2010a), the correlation with W3 use was stronger for parents' W1 PRD ($r = .24$, $p < .001$) than any other risk measure in either W1 *or* W2, including neighborhood risk; negative life events; low SES; and financial, relationship, and job stress (all 12 *rs* at W1 and W2 $< .15$). These longitudinal relations are strong, and in many instances, they maintain controlling for synchronous relations that existed at earlier waves, indicating that PRD does predict change in affect or stress as well as health-impairing behaviors, such as substance use. Nonetheless, although the data are prospective, they are correlational—as is the case for the vast majority of studies in this area—which reduces our ability to draw conclusions about causality. To address this limitation, we have conducted a series of experimental lab studies in which we have simulated discrimination in the lab and then assessed affect and BW to use.

PRD and Affect: Lab Studies

Envisioning Discrimination

Two paradigms have been used in these lab studies. In Gibbons et al. (2010a, Study 2), 116 FACHS participants (*M* age 19) were brought into the lab individually. Given the focus of the study, we oversampled participants who reported in an earlier wave of FACHS that they were using drugs and/or drinking regularly. These individuals were asked to envision themselves in one of three situations at work: non-stressful; stressful, but not related to discrimination; or a discrimination experience (similar paradigms have been used by King 2005; Yoo and Lee 2008). After doing so, they responded to a 15-item mood adjective checklist that contained a mix of anxiety, depression, and anger items (the anger items were: bitter, aggressive, hostile, and angry), plus filler items; and then they indicated how willing they were to use drugs (there were filler tasks to reduce the apparent connection between the discrimination task and the drug items). This was followed by two sets of implicit measures intended to tap into drug cognition accessibility. First, participants were shown a string of words on the computer and were asked to say the first word that came to their mind. Some of the words were double entendre, and could be related to substance use (e.g., pitcher, pot), whereas the filler words were not related to substance use (e.g., boot, change). Then, later in the session, participants were asked to envision another work-related scenario. This one was positive, rather than stressful, and it was followed by an open-ended question: They were told to imagine they were at a party that some friends had arranged for them to celebrate a success at work, and then asked to describe their vision of the party in some detail—who was there, what were they and the other people at the party doing, etc. The word generation task (implicit measure) and the open-ended responses were coded for mention of substance use.

Results in the two nondiscrimination conditions were very similar and both were different from the discrimination condition, so they were combined, simplifying the design into a 2×2 (User vs. Nonuser x Discrimination vs. No discrimination). Looking first at the (imagined) party scenario, the anticipated Condition×Use interaction was significant, as reports of substance use at the party were highest for participants in the discrimination condition who had a history of prior substance use. The same interaction was significant on the word generation measure, with the same pattern: the most drug-related words came from the users in the Discrimination Condition.

A similar pattern emerged on the affect measures, but the results were more complex. First, relative to the nondiscrimination condition, the discrimination scenario elevated anger more than either anxiety or depression. Second, drug BW correlated (significantly) more highly with the anger index than it did with either the anxiety or depression indexes (differences in the correlations: both $ps < .005$). These experimental results suggest, once again, that discrimination does have more of an effect on anger than on either anxiety or depression, and that anger is more of an instigator to substance use than is either of the other two affective responses.

The mediational analyses were consistent with these conclusions: a bootstrapping procedure (Preacher et al. 2007) indicated that anger and not distress was a significant mediator of the relation between manipulated discrimination and drug BW. In fact, the Baron and Kenny (1986) mediation method suggested that about half of the effect of discrimination on BW was mediated by anger.

Cyberball

Williams and his colleagues have developed a very useful technique for examining the effects of social exclusion on a variety of outcomes, including affect, cognitions, and health status (Williams and Jarvis 2006). The technique involves a simulated game of catch on the computer in which participants, each alone in their own cubicle, believe they are playing with two or three other participants. Actually, the other "players" are created by the computer (typically, participants see bogus videos, photos, or read descriptive information of the other players). The computer is programmed so that any given participant(s) comes to perceive that the other players are excluding him/her from the game.[4] This type of social exclusion leads to high levels of self-reported distress and activation in the region of the brain associated with pain detection (Eisenberger et al. 2003). It is also linked to increases in anger and anti-social/aggressive behaviors, as well as a lowering of self-esteem, feelings of belonging, and control (see Williams 2007, for review).

We used a variation of this technique in which we had 199 young African American adults (*M* age = 21.5; from the Washington, D.C. area) play Cyberball with what they thought were three White persons of the same gender and age (Stock et al. 2011). Half the participants were in the exclusion condition, half in the non-exclusion (control) condition. After playing the game, participants' perceptions of exclusion in the game were assessed, along with their mood states and their drug and alcohol BW. As expected, excluded participants reported that they felt excluded, and most of them attributed it to their race—thus, the Cyberball exclusion paradigm appears to be an effective way of manipulating PRD. Also as expected, excluded participants reported feeling more angry than did non-excluded participants and they reported more anger than any other type of affect. Finally, excluded participants with a history of past use reported more substance use BW, on the BW questions as well as on another party projective measure (same type as before). Once again, this increased BW was linked to the elevated anger; the mediation was significant ($p = .04$ on a Sobel mediation test).

[4] We are also conducting studies that involve over inclusion (Stock et al. 2013a)—i.e., the participant receives the ball from the other "players" noticeably more than his/her share—to determine if this kind of noncontingent success may actually be stressful for both African American and White participants and, therefore, lead to more substance use BW (cf. Berglas and Jones 1978).

Summary

Results across several studies using very different methodologies, and samples ranging in age from early adolescence to late 40s, suggest the following: (a) Although discrimination is associated with increases in anxiety and depression (Pascoe and Richman 2009), its impact on anger is more pronounced; (b) Externalizing reactions to stress, such as anger and hostility, are more likely to promote drug use and general problematic use than are internalizing reactions, such as anxiety and depression; and (c) As a result of both sets of relations, anger appears to be the primary affective response that mediates the relation that has been consistently found between discrimination and substance use.

Discrimination and Self-control

One final factor that we have been examining as a mediator of the discrimination → substance use relation is self-control (Gibbons et al. 2012a). These analyses are based, in part, on a series of laboratory studies showing that interracial interactions between African Americans and Whites can lead to a temporary depletion in self-regulatory capacity (i.e., self-control) for members of both groups (Bair and Steele 2010; Richeson and Shelton 2007). Using W1 to W4 data from FACHS, we performed Latent Growth Curve analyses of the relation over time between PRD and self-control (assessed with a modified version of Kendall and Wilcox's 1979, scale). As expected, these analyses indicated that PRD and self-control did covary over time (Gibbons et al. 2012a Study 1). Specifically, increases in discrimination from age 10 to age 18 were associated with increases in anger and decreases in self-control; these changes in self-control, in turn, predicted changes (increases) in substance use. Once again, these results were conceptually replicated using experimental manipulations of PRD in a lab study (Gibbons et al. 2012a, Study 2).

Physical Health vs. Health Behavior: Effects of PRD among the FACHS Parents

Gibbons et al. (2004a), (2010a) and (2012a) all indicated that PRD was associated with an exacerbation of distress (anxiety and depression) as well as anger and hostility. However, the later studies demonstrated that only anger/hostility mediated the link between PRD and use. As suggested earlier, this tendency is consistent with literature linking externalizing behavior with substance use (Curry and Youngblade 2006; Terrell et al. 2006); but it also begs the question: What are the health consequences of these internalizing reactions to PRD? In fact, although externalizing appears to be more of a predictor of substance use and abuse, many studies have provided evidence of an association between internalizing and somatization or *physical* health problems. Bardone et al. (1998), for example, reported that aggression and hostility predicted risky sexual behavior, as well as drug and alcohol use, but not health problems; conversely, anxiety predicted medical problems (not use) and depression predicted medical problems. Laukkanen and colleagues (2002) also found that externalizing behavior was associated with health risk behavior, whereas internalizing was associated with poor health.

By the same token, PRD has also been linked with a number of physical health problems. Pascoe and Richman (2009) conducted a meta-analysis of the literature examining PRD's effects on both physical and mental health outcomes. They found a total of 110 studies that included measures of PRD and mental health, which produced a weighted average correlation of -.20 (95 % confidence interval [CI] = −.22, −.17) between PRD and the various mental health outcomes. Indicators of physical health identified included self-reports and/or diagnoses of specific ailments, ranging from nausea and headaches to hypertension and diabetes, as well as self-reports of overall health status. There were fewer studies of PRD and physical health (40); but the results were similar. The average correlation between PRD and physical health status was − .13 [95 % CI = −.16, −.10].

The correlations between PRD and mental vs. physical health were not significantly different, but Pascoe and Richman suggest that the relation with mental health may be somewhat stronger. Others have reached the same conclusion, pointing out that the effect of discrimination on physical health may be strong, but it takes longer for discrimination to produce physical health problems than to produce distress or hostility (Paradies 2006; Pavalko et al. 2003). The primary reason, it is suggested, is that the former relation is mediated by the latter (i.e., discrimination → distress → morbidity), and, therefore, may not appear in cross-sectional analyses. The evidence supports this. As stated above, numerous studies have documented synchronous and prospective relations between discrimination and distress. Moreover, a consistent series of studies has shown that distress is associated with morbidity, including self-reports of overall health status (Koopmans and Lamers 2005), as well as actual diagnoses or self-reported health problems (Robles et al. 2005; Wyatt et al. 2003).

The Pascoe and Richman meta-analysis and similar reviews of the PRD literature (Paradies 2006; Williams et al. 2003) led us to the following "differential mediation" hypothesis: (a) PRD is associated with increases in anxiety and depression as well as hostility / anger; and it is associated with both physical and mental health problems; (b) the relation between PRD and physical health problems is mediated by internalizing responses (distress), whereas the relation between PRD and substance use is mediated by externalizing responses (anger). These hypotheses were examined in the first four waves of data with the FACHS parents (average age 37 to 45; Gibbons et al. 2014). We included measures of PRD at W1 and W2 and of distress and hostility at W1 and W3. In addition, measures of problematic use (drinking problems and drug use) were included at W1 and W4. Finally, there was an index of morbidity, comprising: self-reports of physician-diagnosed health problems (such as asthma, high blood pressure, and cancer; cf. Williams et al. 1999); health functioning (i.e., the extent to which health status and/or pain interfered with daily activities, such as work or climbing stairs; Ware et al. 1996) and a measure of overall perceived health status. These multiple measures allowed us to examine the impact of change in PRD on change in the respective mediators (distress and hostility) and then change in both substance use and health status. Covariates included a number of factors shown to be related to both use and PRD, including SES, financial stress, as well as age and negative life events.

At W1, 17 % of the women reported their health was either fair or poor and 29 % reported high amounts of PRD (i.e., multiple experiences). By W4, the percentage reporting fair/poor health had risen to 21 % (40 % of the sample reported three or more diagnosed health problems, and 1/3 reported that their health interfered with daily activities); percentages reporting more than occasional drug use and/or some drinking problems increased from 24 to 35 % . Stability of all constructs across time was high (all $ps < .001$).

Nonetheless, the overall pattern was consistent with expectations based on our previous work. First, all four constructs—distress, hostility, PRD, and problematic use—were correlated at W1. Also, W1 PRD was significantly correlated with all W4 indicators of health problems and the measure of problematic use. Most important, *change* in PRD from W1 to W2 predicted changes in both distress and hostility from W1 to W3, and these changes, in turn, predicted a change (decline) in health status and an increase in problematic use, respectively (for both indirect effects), supporting the hypothesis of differential pathways to health problems and health behavior through internalizing and externalizing reactions.

Summary

Three results from this series of studies are noteworthy. First, PRD is strongly related, prospectively, to health-impairing behavior in African American adults as well as African American adolescents, and this relation is mediated by externalizing reactions to PRD more so than internalizing reactions. Second, internalizing reactions to PRD appear to predict worse health status. Given that our sample was only 45 at W4, and so had not yet experienced a lot of health problems, there is reason to expect this relation will get stronger over time. We will continue to explore these two pathways to health problems in subsequent waves of FACHS. Third, although W3 distress and hostility were modestly correlated ($r = .12$, $p < .001$), health problems and problematic use at W4 were not related at all ($r = -.04$, $p > .10$). One interpretation of these relations is that the women who had experienced a lot of racial discrimination responded in *either* an internalizing *or* an externalizing manner; these different reactions had different health consequences.

Moderation: Risk Factors

There are a number of factors – familial, personality, social, contextual – that increase risk for use among African American adolescents and young adults. Our focus, however, has been on those factors that moderate the effects of discrimination—i.e., constructs that may increase the strength of the relation between PRD and vulnerability for substance use and abuse.

Substance Use as Coping

Perhaps the most common explanation offered for the discrimination → substance use relation is that victims of PRD use alcohol and/or drugs as a way of coping with the stress produced by the discrimination (Borrell et al. 2006; Clark 2004; Guthrie et al. 2002). We examined this hypothesis in several studies (Gerrard et al. 2012). In the first of these studies, we took two items from the COPE (Carver et al. 1989) to assess reactions to stress ("I drink alcohol or take drugs, in order to…[think about it less]… [feel better]"). We then used these items as a moderator of the relation between early PRD and escalation of substance use over time among FACHS adolescents (Gerrard et al. 2012, Study 3). These analyses revealed that the participants who experienced early cumulative discrimination (W1-W3) and reported using substances as a coping mechanism had a significantly greater increase in substance use than did other participants ($p < .007$).

In order to examine this moderation more closely, we included coping scales in the two experimental studies mentioned above. In the first lab study (Gerrard et al. 2012, Study 2), regression analyses revealed that drug BW was predicted by discrimination condition (as mentioned before) and also use as coping; but more important, the interaction was significant. Specifically, the discrimination scenario led to higher reports of BW primarily among those who reported using drugs or alcohol to help them cope with stressful events. Importantly, this interaction emerged controlling for previous use, suggesting that it was use-as-coping rather than general (or social) use that was related to the discrimination reactions. In fact, more than 90 % of those in the discrimination condition, who endorsed using substances as a coping mechanism reported some drug BW, as opposed to fewer than 40 % of those who did not indicate they employed this coping style.

The second experimental study used a similar 4-item coping scale and the Cyberball manipulation mentioned above to promote perceptions of discrimination in non-FACHS African Americans (Stock et al. 2013a). The same Coping × Discrimination interaction emerged: Users in the Discrimination condition who reported higher levels of substance use as coping reported more substance use BW than did users in the control condition or those who reported lower levels of use as a coping mechanism.

Integration Level

As mentioned earlier, White adolescents smoke, drink alcohol, and use drugs more than do African American adolescents. Moreover, both White *and* African American adolescents in primarily White neighborhoods typically begin using substances at earlier ages and use more throughout adolescence than do those in more segregated contexts (e.g., Chen and Killeya-Jones 2006; Johnston et al. 2003). Integrated environments also appear to reduce the deterrent effects of ethnicity on cigarette use among African American and Hispanic youth (Johnson and Hoffman 2000). To examine these relations further, we surveyed 203 African American young adults in the Washington, DC metro area (Stock et al. 2013b). We asked them to report: what percentage of their peers, neighborhood populace, and co-workers was African American and White; their PRD, and their BW to get drunk in the future; and we included an assimilation scale (Sellers et al. 1997) to examine the extent to which the young adults stress an American (mainstream) identity. First, PRD was associated with more integrated environments (cf. Seaton and Yip 2009). Regression analyses, controlling for age and gender, revealed that these more integrated environments, along with PRD, and higher levels of assimilation to the "White" mainstream were all associated with greater levels of BW to drink a lot. Thus, it appears that general contact with Whites versus other African Americans is a risk factor for PRD and vulnerability to use, perhaps in order to fit in with the norms and/or reduce the likelihood of being discriminated against (Smalls et al. 2007).

Genetics

We have also examined a factor that moderates the PRD → use relation, but appears to do so in both a risk and buffering manner, and that is genetic architecture. Salivary DNA was collected from the FACHS adolescents and then we looked at two candidate genes—the serotonin transporter (5HTTLPR) and the dopamine receptor (DRD4)–as moderators of the effects of stressors, including PRD, on risk cognitions (e.g., prototypes, risk perceptions) and risk behavior (alcohol, drug, and tobacco use; Gibbons et al. 2012b). What we found was that adolescents with risk alleles on both genes were most likely to respond to high levels of PRD with risky cognitions and risky behavior (e.g., more drug use). We also found, however, that adolescents with these same genes were also the least likely to report risky cognitions and risky behavior if they reported relatively little PRD. In short, these adolescents were actually more "sensitive" to their environments (Ellis and Boyce 2011), responding poorly if they had experienced high levels of discrimination-based stress, but responding well if they had experienced relatively little discrimination. Examination of such Gene x Environment interactions can tell us a lot about how stressors, like PRD and other environmental factors, influence the substance use of African American and other adolescents, and will be a focus of future FACHS analyses.

Moderation: Protective Factors, Part 1: Parenting

Clearly, PRD puts African American adolescents at risk for substance use problems. However, most of them do not manifest such problems, and actually appear to be less at risk, overall, than are White adolescents. Obviously, some important factors are mitigating the effects that PRD can have. We have looked at several such factors suggested by the literature, including parenting style, and what could generally be termed "racial connection" – both racial socialization, and racial identity.

Supportive parenting has been associated with less adolescent substance use in a number of studies (see Caldwell et al. 2006; Pires and Jenkins 2007), with children of many different racial/ethnic groups (Brody et al. 2002; Wills and Cleary 1996). FACHS includes several measures of parenting style (Conger et al. 1992; Thornberry et al. 1989). Examination of these measures has identified a cluster of parenting behaviors that, in combination, appear to be particularly effective at reducing adolescent risk behavior; those include: provision of warmth and support, consistent application of discipline, high levels of communication, and most important, monitoring of the child's activities and whereabouts. SEMs using these four measures—each reported by the parent and/or the child – as indicators of a latent effective parenting construct have consistently shown that this kind of parenting is an important protective factor. In Gibbons et al. (2010a, Study 1), regression analyses were conducted in which the FACHS adolescents' W1 anger and W2 BW were regressed on the relevant predictors: parenting, adolescent and parent discrimination, and the adolescent Discrimination x Parenting interaction. The anger regression revealed main effects of both adolescents' and parents' discrimination (all positive; both $ps < .005$), and a negative effect for parenting ($p = .02$), as well as the anticipated adolescent Discrimination × Parenting interaction. The interaction pattern reflected buffering: there was less effect of discrimination on anger for those adolescents whose parents used supportive parenting (cf. Simons et al. 2006). The same buffering pattern emerged for W2 BW: Parenting again predicted negatively ($p < .001$), and the Discrimination × Parenting interaction was significant. When the same analysis was run with W3 use as a criterion, W1 discrimination predicted ($p < .004$), and the interaction was only marginal ($p = .06$), but it also reflected buffering. Thus, parenting style buffered against the effects of discrimination on anger, and BW, as well as use. Regressions predicting distress did not produce evidence of buffering, however, as the interaction was not significant ($p > .20$).

Anger

The same pattern could be seen in the lab study in Gibbons et al. (2010a, Study 2). An ANOVA was first conducted that included the parenting measure, as well as the Parenting × Discrimination (buffering) interaction term, and also type of mood measure (anger vs. distress) as a within-subjects factor. The anticipated Condition × Parenting × Mood measure interaction was significant, reflecting the fact that reports of anger were higher in the Discrimination Condition, and that was especially true for those whose parents were less supportive—the mean of that cell was significantly higher than all other means. Because of the results on the mood ANOVA, the regression was conducted only on anger. Prior use, Parenting, and Condition were included as predictors, along with the 2-way and the 3-way interaction terms. The anticipated 3-way interaction was significant, as the most anger was reported by the users in the Discrimination condition whose parents were less supportive. When the user and nonuser groups were analyzed separately, there was a marginal effect of Condition among the nonusers, but no other significant effects. In contrast, the Condition effect was significant among the users, as was the Parent × Condition interaction.

Behavioral Willingness

The same regression was conducted on the drug BW measure, and it produced very similar results. The 3-way (Condition × Use × Parenting) interaction was significant, and it followed the familiar pattern: The most BW was reported by users in the Discrimination Condition whose parents did not use supportive parenting practices. Looking at users and nonusers separately – again, there were no significant effects among the nonusers. Among the users, the Parenting effect was marginal ($p = .08$), and the Condition effect was significant. More important, the Parenting × Condition interaction was significant ($p < .001$), and it had the same pattern.

Summary

A parenting style that involves warmth and support, and especially monitoring of the child's whereabouts and activities has been shown to be protective against substance use and other types of risky behavior in adolescents. The same kind of buffering effect can be seen in terms of the relation between PRD and substance vulnerability (BW) and use. It appears that this buffering is indirect: effective parenting dampens the hostile/angry reaction to discrimination that is a typical response; and, less anger in response to the discrimination means less substance use.

Moderation: Protective Factors, Part 2: Racial Socialization and Racial Identity (RI)

Racial Socialization

Racial socialization involves communicating with children about their racial/ethnic background. More specifically, it includes interactions and communications between parents and their children that address such issues as African American culture and history, how African Americans should feel about their racial/ethnic heritage, awareness of discrimination, and ways of responding to race-related experiences (Stevenson et al. 2002). It has been positively linked with self-esteem (Murry et al. 2009; Wills et al. 2007), academic achievement (Neblett et al. 2006), and general resiliency in young African American adults (Brown 2008); and it has been shown to buffer the effects of PRD on mental health (Fischer and Shaw 1999). We found evidence of buffering vis a vis PRD and use in FACHS as well, using a scale developed by Hughes and Johnson (2001). Regression analyses indicated that the PRD to use relation was significantly weaker for those adolescents who indicated their parents had provided high amounts of racial socialization.

RI: Main Effects

Racial Identity is a complex construct, which includes several factors: a commitment and sense of belonging to one's racial group, active involvement with the group, and attributing significance and qualitative meaning to group membership (Phinney 1990; Sellers et al. 1998). Studies have demonstrated that minority youth who have high levels of RI or Africentric values are more likely to have negative attitudes toward substances, tend to initiate use at later ages, and are less likely to use substances than are those with low levels of RI (Brook et al. 1998; Corneille and Belgrave 2007; Holley et al. 2006; Pugh and Bry 2007). We found similar protective relations in FACHS, using Phinney's (1992) RI subscales, which focus on affirmation and belonging, and participation in activities with other African Americans ("racial group behaviors").

In this new study (Stock et al. 2013b, Study 1), we found a 3-way interaction among Condition Discrimination (yes or no), Previous Use, and RI, as the highest drug BW was reported in the Discrimination Condition among substance users with low levels of RI ($p < .02$). The same pattern was found on the projective party-scenario measure ($p < .04$). Related to the above analyses on substance use and integration, additional analyses revealed that both of these interactions were significantly stronger among young adults who lived in more integrated (i.e., mostly White) environments, where higher levels of perceived discrimination were reported. Additionally, an SEM with the FACHS adolescents revealed that RI at W1 was also more protective against substance-related cognitions among those living in mostly

White environments. For these adolescents, RI was associated with lower levels of affiliation with using peers, less BW, and then less substance use 5 years later (Stock et al. 2011).

Discrimination and RI Affirmation: An Experimental Study

The survey study suggested that self- reported RI is protective against discrimination experiences. However, the analyses were correlational, which presents the same interpretational problems mentioned earlier. Although RI tends to be relatively stable for young adults, its salience can be influenced by the social situation (Shelton and Sellers 2000), and it has been shown to be mutable—e.g., influenced by interventions (Cherry et al. 1998). To further examine possible causality, we designed an experimental study in which we manipulated RI via a self- affirmation writing task (Stock et al. 2013b, Study 2). Specifically, we asked half the participants in the Cyberball study to "...think about what it means to you to be an African American. For example, how is being connected to other members of your racial group important to who you are and how you feel about yourself?" Participants in the control condition wrote about what they did over the past 24 h. For both substance use BW and the projective party-scenario measure, the RI by Discrimination interactions (controlling for gender, self-reported RI, and self-esteem) indicated that the highest substance-use risk cognitions were among those who were excluded and did not engage in RI affirmation. However, these interactions were qualified by the RI × Discrimination × Previous Use interaction ($p < .01$), revealing a similar pattern to the first study: the RI × Discrimination effect was strongest among participants with higher levels of past substance use ($ps < .01$). Thus, once again, RI was protective against the potentially deleterious effects of discrimination on substance use cognitions, especially among those at highest risk.

Summary

These studies illustrate the important role that both racial socialization and RI can play in buffering young adults against substance use vulnerability when faced with multiple forms of discriminatory-based experiences. These findings are consistent with previous research showing that as connection to their ethnic group increases, perceived discrimination has less of an impact on the mental health of African Americans. However, these are the first lab-based studies to demonstrate the positive impact of RI, both self-reported and via RI affirmation, when faced with a racial discrimination scenario. For both studies, using two different methods of manipulating racial discrimination, the young African American adults who either had lower levels of RI or did not engage in RI affirmation reported the highest levels of BW to use drugs and the highest projective (scenario-based) substance use. In contrast, those with high levels of RI and little to no previous substance use reported the lowest levels of BW to use drugs in the future, and in the envisioned scenario. In short, individual differences in RI and racial socialization can help explain within-group variability among African American adolescents in reactions to discrimination, and also explain the tendency for African American adolescents to use substances less than Whites, in spite of the discrimination that they experience on a regular basis.

Implications for Interventions and Prevention

Many of the studies reviewed in this chapter support the often heard suggestion that substance use interventions for African American youth should focus on strengthening the protective factors that are most relevant to African American culture (Coard et al. 2004; Whaley and McQueen 2004). Specifically, intervention researchers have called for interventions that employ Africentric educational programs that increase ethnic identity (Belgrave et al. 2000; Cherry et al. 1998). One such program was developed by FACHS researchers using the Prototype model and also Brody and Murry's research on the protective aspects of regulated home environments (Brody et al. 2004). This intervention, called the Strong African American Families (SAAF) Program, is a dual-focus, family-centered intervention that takes advantage of ethnic identification by making salient the fact that African American adolescents typically engage in less substance use than do White adolescents.

The program has two central components that are consistent with our research on the importance of parenting as a buffer against African American youths' substance use (Brody et al. 2004), and the role of favorable risk images (prototypes) in promoting substance use (Gibbons and Gerrard 1997). It focuses on: (a) *parenting style:* promoting monitoring/involvement in the child's life, and communication about alcohol (e.g., setting clear expectations); and (b) promoting the negative aspects of children's perceptions of the *prototypical drinker* in order to decrease their BW to drink (Gerrard et al. 2002). This program has proven to be very effective in reducing the escalation of alcohol consumption among African American children ages 10–13 (Brody et al. 2006b; Gerrard et al. 2006). Specifically, the parenting component of the program was successful in increasing effective parenting, and this change resulted in a decrease in the adolescents' intentions to drink, and, ultimately, in less alcohol consumption. The prototype component of the program was successful at increasing the adolescents' negative perceptions of drinkers, which decreased their BW to drink, and ultimately reduced their alcohol consumption. In addition, this intervention demonstrated the utility of using a dual-process focus in interventions that target both parenting and images (Gerrard et al. 2006).[5]

[5]The contention here is based on the Prototype/Willingness model, which suggests that adolescent health risk behavior is both planful (and therefore involves deliberative, analytic reasoning or processing) *and* reactive, which means it also involves more image-based, heuristic processing (for a

Are African Americans Aware of Differences in Use?

The perspective taken in SAAF and similar Africentric interventions depends to some extent on African American adolescents believing that their racial group uses substances *less* than European Americans do. To check on this, we asked members of the FACHS sample which racial group they thought uses illegal drugs more. Thirty percent said it was Whites; only 5 % thought African Americans used more (the rest thought both groups used about the same amount). Adolescents with more positive racial identity were more likely to believe that African Americans use drugs less. A different sample of African American and White university students ($n = 164$; 60 % female, 55 % African American) produced a very different pattern of responses when questioned about *problematic use*, however. Both the African American and the White students thought African Americans were more likely to become substance (drug) abusers; 27.5 % of African Americans believed their own group was more likely to abuse substances compared to only 16.5 % who believed Whites were more likely. These data suggest that African American young adults have a reasonably accurate perception of trajectories of use among African Americans vs. Whites—*including* some awareness of the racial cross-over. In subsequent studies, we will examine whether these latter perceptions are associated with an increase in problematic use—as a kind of self-fulfilling prophecy (Buchanan and Hughes 2009; Madon et al. 2003), or actually have the opposite effect—due to inoculation ("forewarned is forearmed").

Summary

The utility of intervention and preventive-intervention programs that include elements of racial/ethnic identity is indicated by research suggesting that: (a) African American adolescents and young adults realize that African Americans generally tend to use substances less than Whites do; (b) measures of RI moderate the PRD to use relation in survey and in experimental/lab studies; and most important (c) these Africentric interventions appear to work well (Brody et al. 2006b; Gerrard et al. 2006).

---

discussion of dual processing models, see Chaiken and Trope 1999; Sherman et al. 2014). Therefore, an intervention or preventive-intervention that targets both types of social information processing—and both pathways in the Prototype model—should be more effective than one that targets just analytic processing (which is typical for intervention programs).

Conclusions

A series of studies that includes experimental/lab, field/survey, and intervention methods with African Americans ranging in age from 10 or 11 up to 48 and older provides some new information about an old relationship, or at least one that researchers have been speculating about for many years: The link between PRD and health-impairing behaviors. Numerous cross-sectional studies have documented a correlation between self-reported PRD and substance use. What the current set of studies shows is that experiences with racial discrimination do *predict* increases in drug use and problem drinking, that the two (PRD and problematic use) tend to *covary*—as discrimination increases, so does substance use – and that there are a number of identifiable factors that *mediate and moderate* this covariation. There are two primary mediators of this relation.

The first is a particular form of negative affect – anger and hostility; the second is a reduction in self-control. Understandably, PRD increases negative affect, in general, including both distress (anxiety and depression) and anger, but it is much more the latter reaction that leads to substance use and abuse. Associated with this increase in anger and hostility is a corresponding decrease in self-control. The stress and frustration associated with discrimination can deplete self-regulatory resources and capability, and that can be seen after brief interracial interactions as well as after multiple discriminatory interactions over time. Finally, PRD also leads to an increase in acceptance of deviance (in terms of attitudes and peer affiliation) among African American adolescents, as well as an increase in conduct problems; in combination, the conduct problems and affiliation with risky peers puts some of them at high risk for substance problems.

At the same time, a number of protective factors have been identified which serve to buffer African American adolescents against the negative affective response to PRD that can lead to abuse. These include effective parenting (e.g., monitoring of the child and his/her activities), parenting that includes elements of racial socialization, and, related to that, having a strong sense of racial identity. Interventions that incorporate efforts at promoting racial identity in adolescents and facilitating racial socialization in their parents appear to be effective, in part, because they can "take advantage" of the fact that being an African American child in the US means actually using substances less than children of other racial or ethnic groups, a fact of which older African American adolescents appear to be aware. Nonetheless, there is some reason for concern in this research, as well. Evidence from FACHS and other studies suggests that African American adults may be at higher risk for substance abuse problems.

Although it is not clear exactly when or why this "racial cross-over" effect occurs, there is reason to suspect that it reflects the cumulative effects of years of experience with racial discrimination (and the erosion of self-control that comes with it). These experiences undoubtedly have an effect on physical health status, as well as health-impairing behaviors, such as substance use. Determining why this is the case, and also if the long-term impact of PRD is a primary factor contributing to health disparities in the U.S.—as many scientists and policy makers have suggested (Mays et al. 2007; Williams and Mohammed 2009; Williams et al. 2003)—is a critical question for the next series of studies in FACHS and other research programs.

References

Ardelt, M., & Eccles, J. S. (2001). Effects of mothers' parental efficacy beliefs and promotive parenting strategies on inner-city youth. *Journal of Family Issues, 22*, 944–972.

Arnett, J. J. (2007). Emerging adulthood: What is it, and what is it good for? *Child Development, 1*, 68–73.

Bair, A. N., & Steele, J. R. (2010). Examining the consequences of exposure to racism for the executive functioning of Black students. *Journal of Experimental Social Psychology, 46*(1), 127–132.

Bardone, A. M., Moffitt, T. E., Caspi, A., Dickson, N., Stanton, W. R., & Silva, P. A. (1998). Adult physical health outcomes of adolescent girls with conduct disorder, depression, and anxiety. *Journal of the American Academy of Child and Adolescent Psychiatry, 37*, 594–601.

Baron, R. M., & Kenny, D. A. (1986). The moderator mediator variable distinction in social psychological research – Conceptual, strategic, and statistical considerations. *Journal of Personality and Social Psychology, 51*, 1173–1182.

Belgrave, F. Z., Brome, D. R., & Hampton, C. (2000). The contribution of afrocentric values and racial identity to the prediction of drug knowledge, attitudes, and use among African American youth. *Journal of Black Psychology, 26*, 386–401.

Berglas, S., & Jones, E. E. (1978). Drug choice as a self-handicapping strategy in response to noncontingent success. *Journal of Personality and Social Psychology, 36*(4), 405.

Blanton, H., Gibbons, F. X., Gerrard, M., Conger, K. J., & Smith, G. E. (1997). Role of family and peers in the development of prototypes associated with substance use. *Journal of Family Psychology, 11*, 271–288.

Bodenhausen, G. V., Macrae, C. N., & Hugenberg, K. (2003). Social cognition. In T. Millon & M. J. Lerner (Eds.), *Handbook of psychology: Personality and social psychology* (pp. 257–282). Hoboken: Wiley.

Bolland, J. M., Bryant, C. M., Lian, B. E., McCallum, D. M., Vazsonyi, A. T., & Barth, J. M. (2007). Development and risk behavior among African American, Caucasian, and mixed-race adolescents living in high poverty inner-city neighborhoods. *American Journal of Community Psychology, 40*, 230–249.

Borrell, L. N., Kiefe, C. I., Williams, D. R., Diez-Roux, A. V., & Gordon-Larsen, P. (2006). Self-reported health, perceived racial discrimination, and skin color in African Americans in the CARDIA study. *Social Science & Medicine, 63*, 1415–1427.

Brody, G. H., Ge, X., Conger, R. D., Gibbons, F. X., Murry, V. M., Gerrard, M., & Simons, R. L. (2001). The influence of neighborhood disadvantage, collective socialization, and parenting on African American children's affiliation with deviant peers. *Child Development, 72*, 1231–1246.

Brody, G. H., Dorsey, S., Forehand, R., & Armistead, L. (2002). Unique and protective contributions of parenting and classroom processes to the adjustment of African American children living in single parent families. *Child Development, 73*, 274–286.

Brody, G. H., Kim, S., Murry, V. M., & Brown, A. C. (2004). Protective longitudinal paths linking child competence to behavioral problems among African American siblings. *Child Development, 75*, 455–467.

Brody, G. H., Chen, Y., Murry, V., Ge, X., Simons, R. L., Gibbons, F. X., Gerrard, M., & Cutrona, C. E. (2006a). Perceived discrimination and the adjustment of African American youths: A five-year longitudinal analysis with contextual moderation effects. *Child Development, 77*, 1170–1189.

Brody, G. H., Murry, V. M., Kogan, S. M., Gerrard, M., Gibbons, F. X., Molgaard, V., Brown, A. C., Anderson, T., Chen, Y., Luo, Z., & Wills, T. A. (2006b). The strong African American families program: A cluster-randomized prevention trial of long-term effects and a mediational model. *Journal of Consulting and Clinical Psychology, 74*, 356–366.

Brook, J. S., Balka, E. B., Brook, D. W., Win, P. E., & Gursen, M. D. (1998). Drug use among African Americans: Ethnic identity as a protective factor. *Psychological Reports, 83*, 1427–1446.

Brown, D. L. (2008). African American resiliency: Examining racial socialization and social support as protective factors. *Journal of Black Psychology, 34*, 32–48.

Buchanan, C. M., & Hughes, J. L. (2009). Construction of social reality during early adolescence: Can expecting storm and stress increase real or perceived storm and stress? *Journal of Research on Adolescence, 19*, 261–285.

Bynum, M. S., Burton, E. T., & Best, C. (2007). Racism experiences and psychological functioning in African American college freshmen: Is racial socialization a buffer? *Cultural Diversity and Ethnic Minority Psychology, 13*, 64–71.

Caldwell, C. H., Guthrie, B. J., & Jackson, J. S. (2006). Identity development, discrimination, and psychological well being among African American and Caribbean African American adolescents. In A. J. Schulz & L. Mullings (Eds.), *Gender, race, class, & health: Intersectional approaches* (pp. 163–191). San Francisco: Jossey-Bass.

Carver, C. S., Scheier, M. F., & Weintraub, J. K. (1989). Assessing coping strategies: A theoretically based approach. *Journal of Personality and Social Psychology, 56*, 267–283.

Chaiken, S., & Trope, Y. (Eds.). (1999). *Dual-process theories in social psychology*. New York: Guilford Press.

Chen, K. W., & Killeya-Jones, L. A. (2006). Understanding differences in marijuana use among urban African American and suburban white high school students from two U.S. Community samples. *Journal of Ethnicity in Substance Abuse, 5*, 51–73.

Cherry, V. R., Belgrave, F. Z., Jones, W., Kennon, D. K., Gray, F. S., & Phillips, F. (1998). NTU: An Africentric approach to substance abuse prevention among African American youth. *The Journal of Primary Prevention, 18*, 319–339.

Clark, R. (2004). Interethnic group and intraethnic group racism: Perceptions and coping in African American university students. *Journal of Black Psychology, 30*, 506–526.

Cleveland, M. J., Gibbons, F. X., Gerrard, M., Pomery, E. A., & Brody, G. H. (2005). The impact of parenting on risk cognitions and risk behavior: A study of mediation and moderation in a panel of African American adolescents. *Child Development, 76*, 900–916.

Coard, S. I., Wallace, S. A., Stevenson, H. C., & Brotman, L. M. (2004). Towards culturally relevant preventive interventions: The consideration of racial socialization in parent training with African American families. *Journal of Child and Family Studies, 13*, 277–293.

Conger, R. D., Conger, K. J., Elder, G. H., Lorenz, F. O., Simons, R. L., & Whitbeck, L. B. (1992). A family process model of economic hardship and adjustment of early adolescent boys. *Child Development, 63*, 526–541.

Corneille, M. A., & Belgrave, F. Z. (2007). Ethnic identity, neighborhood risk, and adolescent drug and sex attitudes and refusal efficacy: The urban African American girls' experience. *Journal of Drug Education, 37*(2), 177–190.

Curran, P. J., Stice, E., & Chassin, L. (1997). The relation between adolescent alcohol use and peer alcohol use: A longitudinal random coefficients model. *Journal of Consulting and Clinical Psychology, 65*, 130–140.

Curry, L. A., & Youngblade, L. M. (2006). Negative affect, risk perception, and adolescent risk behavior. *Journal of Applied Developmental Psychology, 27*, 468–485.

Cutrona, C. E., Russell, D. R., Hessling, R. M., Brown, P. A., & Murry, V. (2000). Direct and moderating effects of community context on the psychological well being of African American women. *Journal of Personality and Social Psychology, 79*, 1088–1101.

Dal Cin, S., Worth, K. A., Gerrard, M., Gibbons, F. X., Stoolmiller, M., Wills, T. A., & Sargent, J. D. (2009). Watching and drinking: Expectancies, prototypes, and friends' alcohol use mediate the effect of exposure to alcohol use in movies on adolescent drinking. *Health Psychology, 28*, 473–483.

DuBois, D. L., Burk-Braxton, C., Swenson, L. P., Tevendale, H. D., & Hardesty, J. L. (2002). Race and gender influences on adjustment in early adolescence: Investigation of an integrative model. *Child Development, 73*, 1573–1592.

Eisenberger, N. I., Lieberman, M. D., & Williams, K. D. (2003). Does rejection hurt? An fMRI study of social exclusion. *Science, 302*(5643), 290–292.

Ellis, B. J., & Boyce, W. T. (2011). Differential susceptibility to the environment: Toward an understanding of sensitivity to developmental experiences and context. *Development and Psychopathology, 23*(1), 1–5.

Fischer, A. R., & Shaw, C. (1999). African Americans' mental health and perceptions of racist discrimination: The moderating effects of racial socialization experiences and self-esteem. *Journal of Counseling Psychology, 46*, 395–407.

Fite, P. J., Colder, C. R., & O'Connor, R. M. (2006). Childhood behavior problems and peer selection and socialization: Risk for adolescent alcohol abuse. *Addictive Behaviors, 31*, 1454–1459.

Geronimus, A. T., Neidert, L. J., & Bound, J. (1993). Age patterns of smoking in US African American and white women of childbearing age. *American Journal of Public Health, 83*, 1258–1264.

Gerrard, M., Gibbons, F. X., Zhao, L., Russell, D. W., & Reis-Bergan, M. (1999). The effects of peers' alcohol consumption on parental influence: A cognitive mediational model. *Journal of Studies on Alcohol, 13*, 32–44.

Gerrard, M., Gibbons, F. X., Reis-Bergan, M., Trudeau, L., Vande Lune, L., & Buunk, B. P. (2002). Inhibitory effects of drinker and non-drinker prototypes on adolescent alcohol consumption. *Health Psychology, 21*, 601–609.

Gerrard, M., Gibbons, F. X., Stock, M. L., Vande Lune, L. S., & Cleveland, M. J. (2005). Images of smokers and willingness to smoke among African American pre-adolescents: An application of the prototype/willingness model of adolescent health risk behavior to smoking initiation. *Journal of Pediatric Psychology, 30*, 305–318.

Gerrard, M., Gibbons, F. X., Brody, G. H., Murry, V. M., Cleveland, M. J., & Wills, T. A. (2006). A theory- based dual focus alcohol intervention for pre-adolescents: The strong African American families program. *Psychology of Addictive Behaviors, 20*, 185–195.

Gerrard, M., Gibbons, F. X., Houlihan, A. E., Stock, M. L., & Pomery, E. A. (2008). A dual-process approach to health risk decision making: The prototype willingness model. *Developmental Review, 28*, 29–61.

Gerrard, M., Stock, M. L., Roberts, M. E., Gibbons, F. X., O'Hara, R., Weng, C.-Y., & Wills, T. A. (2012). Coping with racial discrimination: The role of substance use. *Psychology of Addictive Behaviors, 26*, 550–560.

Gibbons, F. X., & Gerrard, M. (1997). Health images and their effects on health behavior. In B. P. Buunk & F. X. Gibbons (Eds.), *Health, coping, and well-being: Perspectives from social comparison theory* (pp. 63–94). Hillsdale: Erlbaum.

Gibbons, F. X., Gerrard, M., & Lane, D. J. (2003). A social-reaction model of adolescent health risk. In J. M. Suls & K. A. Wallston (Eds.), *Social psychological foundations of health and illness* (pp. 107–136). Oxford: Blackwell.

Gibbons, F. X., Gerrard, M., Cleveland, M. J., Wills, T. A., & Brody, G. H. (2004a). Perceived discrimination and substance use in African American parents and their children: A panel study. *Journal of Personality and Social Psychology, 86*, 517–529.

Gibbons, F. X., Gerrard, M., Vande Lune, L. S., Wills, T. A., Brody, G., & Conger, R. D. (2004b). Context and cognitions: Environmental risk, social influence, and adolescent substance use. *Personality and Social Psychology Bulletin, 30*, 1048–1061.

Gibbons, F. X., Yeh, H., Gerrard, M., Cleveland, M. J., Cutrona, C., Simons, R. L., & Brody, G. H. (2007). Early experience with racial discrimination and conduct disorder as predictors of subsequent drug use: A critical period hypothesis. *Drug and Alcohol Dependence, 88*, S27–S37.

Gibbons, F. X., Houlihan, A. E., & Gerrard, M. (2009). Reason and reaction: The utility of a dual-focus, dual-processing perspective on promotion and prevention of adolescent health risk behaviour. *British Journal of Health Psychology, 14*, 231–248.

Gibbons, F. X., Etcheverry, P. E., Stock, M. L., Gerrard, M., Weng, C., Kiviniemi, M. T., & O'Hara, R. E. (2010a). Exploring the link between racial discrimination and substance use: What mediates, what buffers? *Journal of Personality and Social Psychology, 99*, 785–901.

Gibbons, F. X., Pomery, E. A., Gerrard, M., Sargent, J. D., Yeh, H., Wills, T. A., Dal Cin, S., Worth, K. A., & Stoolmiller, M. (2010b). Media as social influence: Racial differences in the effects of peers and media on adolescent alcohol cognitions and consumption. *Psychology of Addictive Behaviors, 24*, 649–659.

Gibbons, F. X., O'Hara, R. E., Stock, M. L., Gerrard, M., Weng, C.-Y., & Wills, T. A. (2012a). The erosive effects of racism: Reduced self-control mediates the relation between perceived racial discrimination and substance use in African American adolescents. *Journal of Personality and Social Psychology, 102*, 1089–1104.

Gibbons, F. X., Roberts, M. E., Gerrard, M., Li, Z., Beach, S. R. H., Simons, R. L., Weng, C.-Y., & Philibert, R. A. (2012b). The impact of stress on the life history strategies of African American adolescents: Cognitions, genetic moderation, and the role of discrimination. *Developmental Psychology, 48*, 722–739.

Gibbons, F. X., Gerrard, M., Weng, C.-Y., & Wills, T. A. (2013). *Early and cumulative effects of perceived racial dicrimination on African American adolescents' substance use: A latent growth curve analysis*. Unpublished data.

Gibbons, F. X., Kingsbury, J. H., Weng, C.-Y., Gerrard, M., Cutrona, C. E., Wills, T. A., & Stock, M. L. (2014). Effects of racial discrimination on health status and health behavior: A differential mediation hypothesis. *Health Psychology, 33*, 11–19.

Gibbons, F. X., Gerrard, M., Stock, M. L., & Finneran, S. (2015). The prototype/willingness model. In M. Connor, & P. Norman (Eds.), Predicting and changing health behaviour: Research and practice with social cognition models (3rd ed.). Cambridge: Cambridge University Press.

Guthrie, B. J., Young, A. M., Williams, D. R., Boyd, C. J., & Kintner, E. K. (2002). African American girls' smoking habits and day-to-day experiences with racial discrimination. *Nursing Research, 51*, 183–190.

Hockey, G. R. J., Maule, A. J., Clough, P. J., & Bdzola, L. (2000). Effects of negative mood states on risk in everyday decision making. *Cognition and Emotion, 14*, 823–856.

Holley, L. C., Kulis, S., Marsiglia, F. F., & Keith, V. M. (2006). Ethnicity versus ethnic identity: What predicts substance use norms and behaviors? *Journal of Social Work Practice in the Addictions, 6*, 53–79.

Hughes, D., & Johnson, D. (2001). Correlates in children's experiences of parents' racial socialization behaviors. *Journal of Marriage and Family, 63*, 981–995.

Hwang, W., & Goto, S. (2008). The impact of perceived racial discrimination on the mental health of Asian American and Latino college students. *Cultural Diversity and Ethnic Minority Psychology, 14*, 326–335.

Jessor, R., & Jessor, S. L. (1997). *Problem behavior and psychosocial development: A longitudinal study of youth*. New York: Academic.

Johnson, R. A., & Hoffman, J. P. (2000). Adolescent cigarette smoking in U.S. Racial/ethnic subgroups: Findings from the national education longitudinal study. *Journal of Health and Social Behavior, 41*, 392–407.

Johnston, L. D., O'Malley, P. M., & Bachman, J. G. (2003). *Monitoring the future national survey results on drug use: Overview of key findings, 2002* (NIH Publication No. 03-5374). Bethesda: National Institute on Drug Abuse.

Johnston, L. D., O'Malley, P. M., Bachman, J. G., & Schulenberg, J. E. (2008). *Various stimulant drugs show continuing gradual declines among teens in 2008, most illicit drugs hold steady.*

Ann Arbor: University of Michigan News Service. Retrieved 11/30/2009 from http://www.monitoringthefuture.org. Kaiser Family Foundation 2006.
Kandel, D. B. (1995). Ethnic differences in drug use: Patterns and paradoxes. In G. Botvin, S. Schinke, & M. Orlandi (Eds.), *Drug abuse prevention in multiethnic youth* (pp. 81–104). Thousand Oaks: Sage.
Kendall, P. C., & Wilcox, L. E. (1979). Self-control in children: Development of a rating scale. *Journal of Consulting and Clinical Psychology, 47*, 1020–1029.
Kessler, R. C. (1991). *UM-CIDI training guide for the national survey of health and stress, 1991–1992*. Ann Arbor: University of Michigan Institute for Social Research.
King, K. R. (2005). Why is discrimination stressful? The mediating role of cognitive appraisal. *Cultural Diversity and Ethnic Minority Psychology, 11*, 202–212.
Koopmans, G. T., & Lamers, L. M. (2005). Assessing the construct validity of three indicators of psychological distress in relation to perceived health and physical illness. *Social Psychiatry and Psychiatric Epidemiology, 40*, 1012–1018.
Landrine, H., & Klonoff, E. A. (1996). The schedule of racist events: A measure of racial discrimination and a study of its negative physical and mental health consequences. *Journal of Black Psychology, 22*, 144–168.
Laukkanen, E., Shemeikka, S., Notkola, I., Koivumaa-Honkanen, H., & Nissinen, A. (2002). Externalizing and internalizing problems at school as signs of health-damaging behaviour and incipient marginalization. *Health Promotion International, 17*, 139–146.
Lerner, J. S., & Keltner, D. (2001). Fear, anger, and risk. *Journal of Personality and Social Psychology, 81*, 146–159.
Levine, H. G., Gettman, J. B., & Siegel, L. (2010). *Targeting Blacks for marijuana: Possession arrests of African Americans in California, 2004–08*. California: Drug Policy Alliance.
Madon, S., Guyll, M., Spoth, D. L., Cross, S. E., & Hilbert, S. J. (2003). The self-fulfilling influence of mother expectations on children's underage drinking. *Journal of Personality and Social Psychology, 84*, 1188–1205.
Mays, V. M., Cochran, S. D., & Barnes, N. W. (2007). Race, race-based discrimination, and health outcomes among African Americans. *Annual Review of Psychology, 58*, 201.
Merikangas, K. R., He, J. P., Brody, D., Fisher, P. W., Bourdon, K., & Koretz, D. S. (2010). Prevalence and treatment of mental disorders among US children in the 2001–2004 NHANES. *Pediatrics, 125*(1), 75–81.
Michael, K., & Ben-Zur, H. (2007). Risk-taking among adolescents: Associations with social and affective factors. *Journal of Adolescence, 30*, 17–31.
Moffit, T. E. (1993). Adolescence-limited and life-course-persistent antisocial behavior – A developmental taxonomy. *Psychological Review, 100*, 674–701.
Moons, W. G., & Mackie, D. M. (2007). Thinking straight while seeing red: The influence of anger on information processing. *Personality and Social Psychology Bulletin, 33*, 706–720.
Murry, V. M., Berkel, C., Brody, G. H., Miller, S. J., & Chen, Y. F. (2009). Linking parental socialization to interpersonal protective processes, academic self-presentation, and expectations among rural African American youth. *Cultural Diversity & Ethnic Minority Psychology, 15*, 1–10.
Neblett, E. W., Philip, C. L., Cogburn, C. D., & Sellers, R. M. (2006). African American adolescents' discrimination experiences and academic achievement: Racial socialization as a cultural compensatory and protective factor. *Journal of Black Psychology, 32*, 199–218.
Pampel, F. C. (2008). Racial convergence in cigarette use from adolescence to the mid-thirties. *Journal of Health and Social Behavior, 49*, 484–498.
Paradies, Y. C. (2006). Defining, conceptualizing and characterizing racism in health research. *Critical Public Health, 16*, 143–157.
Pascoe, E. A., & Richman, L. S. (2009). Perceived discrimination and health: A meta-analytic review. *Psychological Bulletin, 135*, 531–554.

Pavalko, E. K., Mossakowski, K. N., & Hamilton, V. J. (2003). Does perceived discrimination affect health? Longitudinal relationships between work discrimination and women's physical and emotional health. *Journal of Health and Social Behavior, 44*, 18–33.

Pettit, B., & Western, B. (2004). Mass imprisonment and the life course: Race and class inequality in U.S. Incarceration. *American Sociological Review, 69*, 151–169.

Phinney, J. S. (1990). Ethnic identity in adolescents and adults: Review of research. *Psychological Bulletin, 108*(3), 499.

Phinney, J. S. (1992). Ethnic-differences in social identity formation in adolescence. *International Journal of Psychology, 27*, 206.

Pires, P., & Jenkins, J. M. (2007). A growth curve analysis of the joint influences of parenting affect, child characteristics and deviant peers on adolescent illicit drug use. *Journal of Youth and Adolescence, 36*, 169–183.

Pomery, E. A., Gibbons, F. X., Gerrard, M., Cleveland, M. J., Brody, G. H., & Wills, T. A. (2005). Families and risk: Prospective analyses of familial and social influences on adolescent substance use. *Journal of Family Psychology, 19*, 560–570.

Preacher, K. J., Rucker, D. D., & Hayes, A. F. (2007). Addressing moderated mediation hypotheses: Theory, methods, and prescriptions. *Multivariate Behavioral Research, 42*, 185–227.

Pugh, L. A., & Bry, B. H. (2007). The protective effects of ethnic identity for alcohol and marijuana use among African American young adults. *Cultural Diversity and Ethnic Minority Psychology, 13*(2), 187–193.

Reyna, V. F., & Farley, F. (2006). Risk and rationality in adolescent decision making: Implications for theory, practice, and public policy. *Psychological Science in the Public Interest, 7*, 1–44.

Richeson, J. A., & Shelton, J. N. (2007). Negotiating interracial interactions costs, consequences, and possibilities. *Current Directions in Psychological Science, 16*(6), 316–320.

Robles, T. F., Glaser, R., & Kiecolt-Glaser, J. K. (2005). Out of balance: A new look at chronic stress, depression, and immunity. *Current Directions in Psychological Science, 14*, 111–115.

Rydell, R. J., Mackie, D. M., Maitner, A. T., Claypool, H. M., Ryan, M. J., & Smith, E. R. (2008). Arousal, processing, and risk taking: Consequences of intergroup anger. *Personality and Social Psychology Bulletin, 34*, 1141–1152.

SAMHSA. (2008). *National household survey on drug abuse, 2007 and 2008*. Accessed online 11/03/09 from http://www.oas.samhsa.gov/nhsda.htm

Sargent, J. D., Wills, T. A., Stoolmiller, M., Gibson, J., & Gibbons, F. X. (2006). Alcohol use in motion pictures and its relation with early-onset teen drinking. *Journal of Studies on Alcohol, 67*, 54–65.

Schulenberg, J., Maggs, J. L., Dielman, T. E., Leech, S. L., Kloska, D. D., Shope, J. T., & Laetz, V. B. (1999). On peer influences to get drunk: A panel study of young adolescents. *Merrill-Palmer Quarterly, 45*, 108–142.

Schulenberg, J. E., Merline, A. C., Johnston, L. D., O'Malley, P. M., Bachman, J. G., & Laetz, V. B. (2005). Trajectories of marijuana use during the transition to adulthood: The big picture based on national panel data. *Journal of Drug Issues, 35*, 255–280.

Seaton, E. K., & Yip, T. (2009). School and neighborhood contexts, perceptions of racial discrimination, and psychological well-being among African American adolescents. *Journal of Youth and Adolescence, 38*, 153–163.

Sellers, R. M., Rowley, S. A., Chavous, T. M., Shelton, J. N., & Smith, M. A. (1997). Multidimensional Inventory of Black Identity: A preliminary investigation of reliability and constuct validity. *Journal of Personality and Social Psychology, 73*(4), 805.

Sellers, R. M., Smith, M. A., Shelton, J. N., Rowley, S. A., & Chavous, T. M. (1998). Multidimensional model of racial identity: A reconceptualization of African American racial identity. *Personality and Social Psychology Review, 2*(1), 18–39.

Sellers, R. M., Caldwell, C. H., Schmeelk-Cone, K. H., & Zimmerman, M. A. (2003). Racial identity, racial discrimination, perceived stress, and psychological distress among African American young adults. *Journal of Health and Social Behavior, 43*, 302–317.

Shaffer, D., Schwab-Stone, M., Fisher, P., Cohen, P., et al. (1993). The diagnostic interview schedule for children (DISC-R): IV. Preparation, field testing, inter-rater reliability and acceptability. *Journal of the American Academy of Child and Adolescent Psychiatry, 32*, 643–650.

Shelton, J. N., & Sellers, R. M. (2000). Situational stability and variability in African American racial identity. *Journal of Black Psychology, 26*, 27–50.

Sherman, J. W., Gawronski, B., & Trope, Y. (2014). *Dual-process theories of the social mind*. New York: The Guilford Press.

Simons, R. L., Chen, Y., Stewart, E., & Brody, G. (2003). Incidents of discrimination and risk for delinquency: A longitudinal test of strain theory with an African-American sample. *Justice Quarterly, 20*, 827–854.

Simons, R. L., Simons, L. G., Burt, C. H., Brody, G., & Cutrona, C. (2005). Collective efficacy, authoritative parenting and delinquency: A longitudinal test of a model integrating community- and family-level processes. *Criminology, 43*, 989–1030.

Simons, R. L., Simons, L. G., Burt, C. H., Drummund, H., Stewart, E., Brody, G. H., Gibbons, F. X., & Cutrona, C. (2006). Supportive parenting moderates the effect of discrimination upon anger; Hostile view of relationships, and violence among African American boys. *Journal of Health and Social Behavior, 47*, 373–389.

Smalls, C., White, R., Chavous, T., & Sellers, R. (2007). Racial ideological beliefs and racial discrimination experiences as predictors of academic engagement among African American adolescents. *Journal of Black Psychology, 33*, 299–330.

Stevenson, H. C., Cameron, R., Herrero-Taylor, T., & Davis, G. Y. (2002). Development of the teenager experience of racial socialization scale: Correlates of race-related socialization frequency from the perspective of African American youth. *The Journal of Black Psychology, 28*, 84–106.

Stock, M. L., Gibbons, F. X., Walsh, L. A., & Gerrard, M. (2011). Racial identification, racial discrimination, and substance use vulnerability among Black young adults. *Personality and Social Psychology Bulletin, 37*, 1349–1361.

Stock, M. L., Gibbons, F. X, Peterson, L. M., & Gerrard, M. (2013a). The effects of racial discrimination and general life stressors on the HIV-risk behaviors of African American adolescents and young adults. *Health Psychology, 32*, 543–550.

Stock, M. L., Gibbons, F. X., Gerrard, M., Houlihan, A. E., Simons, R. L., & Weng, C. (2013b). Racial identification, racial composition, and substance use vulnerability among African American adolescents and young adults. *Health Psychology, 32*, 237–247.

Substance Abuse and Mental Health Services Administration. (2008). *Results from the 2007 national survey on drug use and health: Detailed tables*. Rockville: Substance Abuse and Mental Health Services Administration.

Substance Abuse and Mental Health Services Administration. (2015). *2014 national survey on drug use and health: Detailed tables*. Rockville: Substance Abuse and Mental Health Services Administration.

Tanski, S., Demidenko, E., McClure, A., Sargent, J. D., & Gibbons, F. X. (2010). *Race, ethnicity and onset of teen alcohol use*. Unpublished data.

Tate, S. R., Patterson, K. A., Nagel, B. J., Anderson, K. G., & Brown, S. A. (2007). Addiction and stress in adolescents. In M. Al'absi (Ed.), *Stress and addiction: Biological and psychological mechanisms* (pp. 249–264). Burlington: Elsevier.

Terrell, F., Miller, A. R., Foster, K., & Watkins, C. E. (2006). Racial discrimination induced anger and alcohol use among African American adolescents. *Adolescence, 41*, 485–492.

Thornberry, T., Huzinga, D., & Loeber R. (1989). *Personal communication*. Adapted from measures developed for multisite study on causes and correlates of delinquency sponsored by the Office of Juvenile Justice and Delinquency Prevention.

Tourangeau, R., & Smith, T. W. (1996). Asking sensitive questions: The impact of data collection mode, question format, and question context. *Public Opinion Quarterly, 60*, 275–304.

Vega, W. A., & Gil, A. G. (1998). *Drug use and ethnicity in early adolescence*. New York: Plenum Press.

Wang, X. T. (2006). Emotions within reason: Resolving conflicts in risk preference. *Cognition & Emotion, 20*, 1132–1152.

Ware, J. E., Kosinski, M., & Keller, S. D. (1996). A 12-item short-form health survey: Construction of scales and preliminary tests of reliability and validity. *Medical Care, 34*, 220–233.

Watt, T. T. (2008). Race/ethnic differences in alcohol abuse among youth: An examination of risk-taking attitudes as a mediating factor. *Journal of Ethnicity in Substance Abuse, 3*, 33–47.

Watt, T. T., & Rogers, J. M. (2007). Factors contributing to differences in substance use among African American and White adolescents. *Youth & Society, 39*, 54–74.

Whaley, A. L., & McQueen, J. P. (2004). An Afrocentric program as primary prevention for African American youth: Qualitative and quantitative exploratory data. *The Journal of Primary Prevention, 25*, 253–269.

White, H. R., Nagin, D., Replogle, E., & Stouthamer-Loeber, M. (2004). Racial differences in trajectories of cigarette use. *Drug and Alcohol Dependence, 76*, 219–227.

Williams, K. D. (2007). Ostracism. *Annual Review of Psychology, 58*, 425–452.

Williams, K. D., & Jarvis, B. (2006). Cyberball: A program for use in research on interpersonal ostracism and acceptance. *Behavior Research Methods, 38*, 174–180.

Williams, D. R., & Mohammed, S. A. (2009). Discrimination and racial disparities in health: Evidence and needed research. *Journal of Behavioral Medicine, 32*, 20–47.

Williams, D. R., Spencer, M. S., & Jackson, J. S. (1999). Race, stress, and physical health. In R. J. Contrada & R. D. Ashmore (Eds.), *Self, social identity, and physical health: Interdisciplinary explorations* (pp. 71–100). Oxford: Oxford University Press.

Williams, D. R., Neighbors, H. W., & Jackson, J. S. (2003). Racial/ethnic discrimination and health: Findings from community studies. *American Journal of Public Health, 93*, 200–208.

Wills, T. A., & Cleary, S. D. (1996). How are social support effects mediated? A test with parental support and adolescent substance use. *Journal of Personality and Social Psychology, 71*(5), 937.

Wills, T. A., & Cleary, S. D. (1999). Peer and adolescent substance use among 6th–9th Graders: Latent growth analyses of influence versus selection mechanisms. *Health Psychology, 18*, 453–463.

Wills, T. A., Yaeger, A. M., & Sandy, J. M. (2003). Buffering effect of religiosity for adolescent substance use. *Psychology of Addictive Behaviors, 17*, 24–31.

Wills, T. A., Murry, V. M., Brody, G. H., Gibbons, F. X., Gerrard, M., Walker, C., & Ainette, M. G. (2007). Ethnic pride and self-control related to protective and risk factors: Test of the theoretical model for the strong African American families program. *Health Psychology, 26*, 50–59.

Windle, M., & Wiesner, M. (2004). Trajectories of marijuana use from adolescence to young adulthood: Predictors and outcomes. *Development and Psychopathology,* Special Issue: Transition from Adolescence into Adulthood, *16*, 1007–1027.

Wyatt, S. B., Williams, D. R., Calvin, R., Henderson, F. C., Walker, E. R., & Winters, K. (2003). Racism and cardiovascular disease in African Americans. *The American Journal of the Medical Sciences, 325*, 315–331.

Yoo, H. C., & Lee, R. M. (2008). Does ethnic identity buffer or exacerbate the effects of frequent racial discrimination on situational well-being of Asian Americans? *Journal of Counseling Psychology, 55*, 63–74.

Chapter 12
The Influence of Neighborhood Context on Exposure to and Use of Substances Among Urban African American Children

Michele Cooley-Strickland, Lindsay Bynum, Katherine Otte, Lingqi Tang, Robert S. Griffin, Tanya J. Quille, and Deborah Furr-Holden

Substance use is a widespread problem among adolescents. According to the 2008 *Monitoring the Future* survey, almost half (45 %) of American youth have smoked cigarettes by the end of high school, 21 % of whom had tried them before the beginning of eighth grade; 72 % have consumed alcohol by the end of high school, with 39 % having done so by eighth grade; and half (47 %) of American children have tried an illicit drug by the time they leave high school (Johnston et al. 2009). However, most of the research on adolescent substance use has been conducted on primarily Caucasian samples (Lambert et al. 2004), leaving a paucity of research on substance use among ethnic minority adolescents (De La Rosa et al. 1993; Wallace et al. 1999). It is important to investigate substance use among ethnic minorities

Michele R. Cooley-Strickland, Lindsay Bynum, Katherine Otte, Robert S. Griffin, Tanya J. Quille, and Deborah Furr-Holden, Department of Mental Health, Bloomberg School of Public Health, Johns Hopkins University.
Michele R. Cooley-Strickland and Lingqi Tang are at the Department of Psychiatry, NPI-Semel Institute for Neuroscience, University of California – Los Angeles.

M. Cooley-Strickland (✉)
Department of Mental Health, Bloomberg School of Public Health,
Johns Hopkins University, Baltimore, MD 21218, USA

Center for Culture and Health, Department of Psychiatry,
NPI-Semel Institute for Neuroscience, University of California – Los Angeles,
760 Westwood Plaza, Box 62, Los Angeles, CA 90024-1759, USA
e-mail: MCooley@mednet.ucla.edu

L. Bynum • K. Otte • R.S. Griffin • T.J. Quille • D. Furr-Holden
Department of Mental Health, Bloomberg School of Public Health,
Johns Hopkins University, Baltimore, MD 21218, USA

L. Tang
Center for Culture and Health, Department of Psychiatry,
NPI-Semel Institute for Neuroscience, University of California – Los Angeles,
760 Westwood Plaza, Box 62, Los Angeles, CA 90024-1759, USA

Y.F. Thomas, L.N. Price (eds.), *Drug Use Trajectories Among Minority Youth*,
DOI 10.1007/978-94-017-7491-8_12

because racial and ethnic minority groups exhibit disproportionately adverse social outcomes associated with drug use, including poverty, violence, crime and arrest (REF). Prior research has shown that African American adolescents are less likely to smoke cigarettes (Felton et al. 1999; Kann et al. 1996) and consume alcohol (Blum et al. 2000) than their Caucasian peers. However, prevalence rates for drug use initiation by race/ethnicity indicate that while African Americans are less likely than Caucasians to initiate smoking tobacco and drinking by age 13, they are at greater risk of initiating cocaine and marijuana use at earlier ages (i.e., 17.2 %, 11.1 %, and 1.3 % for smoking, marijuana, and cocaine initiation before 13 years of age respectively (Kann et al. 1996).

Prevalence data for early-onset substance abuse among African American children are rare. Nonetheless, there are limited descriptive statistics for substance use and progression. For example, a study of 454 elementary school children in a mixed urban-suburban community (mean age 11.8 years; 27 % African American) showed that 24 % reported trying one or two cigarettes, 7 % had smoked 4–5 times, and 1 % were smoking on at least a monthly basis. 30 % of the elementary school students reported drinking alcohol one or two times, 11 % reported drinking alcohol 4–5 times, and 2 % drank on a monthly basis. Regarding marijuana, 2 % had tried it once or twice, and less than 1 % reported using it regularly (Wills et al. 2001). Another study examined the prevalence of alcohol and drug use among children in Denver, Pittsburg, and Rochester (Huizinga et al. 1993). The authors found a high frequency of drug initiation prior to the teenage years. Denver's rates represented the highest-risk areas: 15.3 % of 7 year-old boys and 9.7 % of 7 year old girls reported drinking alcohol, while 1.2 % of 7 year-old boys, and 0.7 % of 7 year old girls reported having smoked marijuana (Huizinga et al. 1993). It has been documented that the transition from elementary to middle school, particularly between sixth and ninth grade, is a vulnerable time for the initiation of substance use (Horton 2007). The early onset of substance use increases the risk of substance use, and use related problems, later in life (Hoffman et al. 2006; Pitkanen et al. 2005; Wilson et al. 2005; Overstreet 2000). For example, early cigarette use increases the risk of later alcohol and marijuana use (Duncan et al. 1998; Griffin et al. 2002); while early alcohol or marijuana use increases the risk of later hard drug use (i.e. heroin and cocaine) (Office of Applied Studies 2004; Trimboli and Coumerlos 1998). In order to effectively prevent adolescent substance use, it is important to first understand the factors that influence early substance use (Fite et al. 2009), particularly from a culturally and contextually relevant perspective.

Developmental patterns of drug use may vary with age, gender, ethnicity, social class, and ecological, cultural and historical conditions (Kandel and Andrews 1987). Certain risk factors for drug initiation exert differential effects according to gender and ethnicity (Brunswick and Messeri 1984; Ellickson and Morton 1999). Factors from multiple domains (e.g., personal background, school achievement, family-peer orientation, health attitudes and behaviors) have been shown to influence smoking initiation in urban African American youth (Brunswick and Messeri 1984). Behavioral problems such as aggression and unsafe behaviors have also been associated with drug use initiation (Epstein et al. 1999) however, the only predictors of

African American youth's "hard" drug use were social influences promoting drug use and intentions to use them (Ellickson and Morton 1999). Although African Americans seem to have a later onset of marijuana use than other ethnicities, marijuana abuse and dependence rates are significantly higher for African Americans by age 20 (Reardon and Buka 2002). Several possible explanations for this pattern of use have been suggested, including: (1) access to substances at different ages due to differences in availability at home; (2) different age specific norms, expectations, and monitoring in homes and neighborhoods; and (3) the perception of different opportunity structures (e.g., fewer occupational and higher education opportunities for African American males leading to greater substance use with a later onset). Given the comparatively late onset of alcohol, tobacco, and other drug (ATOD) use among African American youth as compared with European Americans, there are limited data on risk and protective factors associated with African American children's ATOD use initiation.

Children's exposure to community violence is one factor that may contribute to drug use in African American youth both directly as a stress management mechanism and indirectly though mediators (e.g., aggression, impulsivity, inattentiveness, and disregulation of inhibition). Each of these factors may contribute to school failure, early drop-out rates, and association with negative peers, all risk factors for youth substance use (Loeber and Dishion 1983: Schwab-Stone et al. 1995). Other risk factors for substance abuse associated with exposure to community violence include external locus of control, hopelessness, and depression: all of which are particularly salient for girls (Sanders-Phillips 2003). To establish effective prevention interventions for drug use and its sequelae, it is important to understand drug use trajectories among younger African American youth as they are exposed to and initiate use of tobacco, alcohol, and other illicit substances. African American children who live in urban areas are particularly important to investigate given their disproportionate exposure to neighborhood and economic disadvantage.

There are several theories that potentially explain youth's initiation of substance use, some of which are presented here. The social development model (Catalano et al. 1996; Schaps and Soloman 2003), derived from social control theory (Hirschi 1969), posits that youth engage in substance use based on the beliefs, values, and actions of those with whom they have socially bonded (Abbey et al. 2006). The social ecology model (Kumpfer and Turner 1990) and problem behavior theory (Jessor 1992) both suggest multiple interrelated pathways to substance use, including individual, neighborhood, and contextual factors that influence the risk of adolescent substance use (Burlew et al. 2009). These factors include lack of [pro-]social bonding and poor social skills, familial communication problems, single-parent families, emotional and mental health problems, peer pressure, low educational commitment or failure at school, antisocial behavior, and delinquency (Vega et al. 1993; Spooner 1999; Biederman et al. 2000; Farrell et al. 2000; Challier et al. 2000). Exposure to community violence is common among these models and may be a significant risk factor for urban youth's substance use. For example, in the problem behavior model, community violence may serve as a proximal risk factor as a direct environmental risk, as well as a contributor to distraction and inattention at school.

It may be a distal, mediating factor in the social theories as younger youth imitate older youth in the community who exhibit power and authority though the display of violence or economic advantage in the illegal drug trade. Social disorganization theory also has relevance to adolescent substance use (Lambert et al. 2004) as some urban neighborhoods with high levels of social disorganization can increase youth's exposure to deviant peers and activities (Bowman et al. 2007; LaVeist and Wallace 2000; Wallace 1999). Urban neighborhood disadvantage (e.g., poverty, challenges to safety, illegal drug trade, gang activity) has been identified as a risk factor for "harder" drugs such as heroin and cocaine, and it seems to act as an important precursor to exposure to cocaine specifically (Crum et al. 1996). However, this drug specificity may be a function of cohort effects (i.e., the cocaine epidemic of the 1980s and 1990s). The more disadvantaged a neighborhood is, the more likely an adolescent living in that neighborhood will be exposed to hard drugs (Crum et al. 1996).

Illicit drug use has been shown to be more prevalent in African American neighborhoods as compared to those that are predominantly Caucasian (LaVeist and Wallace 2000; Wallace 1999). Relatedly, African American youth are more likely to report witnessing drug sales and drug activity within their neighborhoods (Wallace and Muroff 2002), as these activities are more visible in disadvantaged communities (Saxe et al. 2001). Increased visibility and exposure to substances over time may reduce negative perceptions of their use, and increase adolescents' likelihood of using them (Lambert et al. 2004). Additionally, an adolescent's *perception* of their neighborhood's characteristics may be more strongly associated with substance use than the actual characteristics of their neighborhood. This could be, in part, because perceived availability of drugs is a more salient issue than actual availability for adolescents who cannot purchase substances legally (Gibbons et al. 2007). Adolescents are less likely to use substances if that substance is perceived as hard to get (Knibbe et al. 2005). According to Gibbons et al. (2007), adolescents' substance use is more likely to be reactive than intentional. This supports the prototype-willingness model (Gibbons et al. 2003, 2006) which suggests that adolescent risk behavior is a reaction to "risk opportunity" (Gibbons et al. 2007). The more risk opportunity an adolescent encounters, the more likely they are to initiate substance use. Perceived neighborhood disorganization (i.e., violence, challenges to safety, drug activity, gang activity) is a demonstrated risk factor for adolescents' future substance use (Lambert et al. 2004). Conversely, attachment to one's neighborhood can also act as a protective factor in high poverty inner city neighborhoods (Bolland et al. 2007). In these neighborhoods, hopelessness was less of a contributor to risk behavior for African American adolescents as compared to Caucasians. Hopelessness may not be as great of a risk because there are often more African Americans in inner-city neighborhoods, providing a stronger sense of community and support than their Caucasian counterpart experience (Bolland et al. 2007).

Social factors at the individual, peer, and familial levels have been identified as either risk or protective factors for youth's substance use. Sobeck et al. (2000) found that sixth grade students with fewer positive peer relations, living in single parent households, and less knowledge about alcohol and drugs were at greater risk for initiating substance use. Individual-level factors such as lack of assertiveness, poor decision making skills, and peer pressure refusal were also found to place youths at an increased risk. African American adolescents' communication styles tend to be more assertive and power related than Caucasian youth, which may provide stronger resistance strategies to drugs (Moon et al. 1999).

A factor that may serve as both a protective and risk factor among urban African American families is parenting style. The authoritarian parenting style (Baumrind 1966) typically characterizes urban African American parents, and is most noted by attempts to "shape, control, and evaluate the behavior and attitudes of the child in accordance with a set standard of conduct, usually an absolute standard". This method of parenting is functional and protective in early childhood, but as the child moves through middle childhood into adolescence, the over-controlling style is decreasingly effective (Baumrind 1991). It has also been posited that parenting does not deteriorate over time to the extent previously thought. The literature suggests that the combination of the decreased ability of parents to monitor their urban children's whereabouts and activities, coupled with the stressors of poverty and violence, may create a vulnerability to substance use which at that time becomes amplified as compared with majority children (REF). Poverty and violence are especially stressful over time because of "weathering": the effect of cumulative stressors that are addictive and increase the likelihood of dysfunction over the lifespan (Geronimus 1992). As majority children develop the skills, abilities, and opportunities they gain may buffer them from continued and increased substance use. Conversely, minority children may begin to lose those skills and experience slowed development as a result of weathering from exposure to their environments and the perception of limited opportunities as the child ages and experiences more difficulties and blocks in educational achievement. Which factor – substance use or poor academic performance – precedes the other is uncertain (Bryant et al. 2000), however a decrease in school attendance and grades in a sample of middle and high school students was associated with community violence exposure (Bowen and Bowen 1999). A study by Abbey and colleagues (2006) found that after-school activities were protective for both urban and suburban girls, and increased after school involvement was linked with less substance use. However, for urban boys, after-school involvement had no effect and, for suburban boys, increased involvement led to increased substance use. These findings suggest that some types of after-school activities could create iatrogenic effects (e.g., increased exposure to substance using peers; Abbey et al. 2006; Gottfredson et al. 2004). For example, adolescent alcohol use has been shown to be increased by involvement in team sports (Eccles and Barber 1999). Conversely, strong bonds to school and family can act as a protective factor against nonmedical prescription drug use (Ford 2009). Parents serve a particularly important role in preventing adolescent substance use, as effective parenting and monitoring consistently shows an inverse relationship to substance use among adolescents (Wang et al. 2009). Adolescents who value a close relationship with their parents are less likely to engage in substance use because they believe it will harm these relationships (Ford 2009). Parental knowledge is a proven protective factor against youth's cigarette, marijuana, and alcohol use (Simons-Morton et al. 2001;

Barnes and Farrell 1992; Hoffman 2002). Just as positive family relationships can protect youth from using substances, negative family relationships are a risk factor. Bersamin et al. (2005) found that poor family management is positively associated with binge drinking. Peers could play an even more important role than parents, although parents influence peer choices (REF). Studies have consistently shown that peer substance use is among the strongest predictors of adolescent substance use (Wang et al. 2009; Hoffman et al. 2007). Adolescents who associate with substance using peers are more likely to use substances themselves (Simmons-Morton 2002; Simons-Morton et al. 2004; Bahr et al. 2005; Stanton et al. 2002). This provides further support for the problem behavior theory which suggests that associating with deviant peers increases one's likelihood of engaging in similar problem behaviors (Jessor and Jessor 1977).

Another important theory to consider is the stress reduction hypothesis, which suggests that substance use is a way of coping with stress (Lindenberg et al. 1994). Stress is a risk factor for substance use (Wallace et al. 1999) and the existing literature has shown that living in disadvantaged and poor neighborhoods increases one's risk of experiencing negative life events compared to those living elsewhere (Fang et al. 1998; Massey and Shibuya 1995). Scheier et al. (1999) found that perceived neighborhood stress predicted alcohol use among African American youth. Although drug use is a social behavior for many adolescents, it may also be considered a coping mechanism for stress relief. Youth who use alcohol, tobacco, and other substances as a coping mechanism are likely to become heavier drug users and are at risk for later drug dependence (Beauvais and Oetting 2002). In Beauvais and Oetting's 2002 preliminary study in which inner-city children's coping styles were assessed fifth grade African American students reported using substances when they were "faced with difficulties or felt tense." Using the Adolescent Coping for Problem Experiences (A-COPE; Patterson and McCubbin 1987) scale, 22.9 % of the 11- and 12-year olds reported smoking tobacco and 17.1 % reported drinking alcohol at least "sometimes" to help them "cope." Stress-coping can be categorized in two different ways, either problem focused (changing or removing a stressor) or emotion focused (managing affect related to or resulting from a stressor). Adolescents who used more emotion focused coping were more likely to use drugs (Wagner et al. 1999). In addition to stress, trauma, and resultant post-traumatic stress are associated with adolescent substance abuse (Kilpatrick et al. 2000). Several studies have shown that post-traumatic stress is a risk factor for the development and chronicity of depression and substance use (Bolton et al. 2000; Giaconia et al. 1995; Kilpatrick et al. 2000). Furthermore, substance use increases one's risk for experiencing subsequent traumatic events (Johnson et al. 2006). This sets the foundation for a cyclical effect of exposure to violence, trauma, and substance use among youth dwelling in disadvantaged urban neighborhoods.

The chronic exposure to neighborhood violence is one stressor that must be examined as requiring continual stress relief and therefore a contributing factor to long term drug use. It is important to examine not only the incidents, rates and types of exposure but also the effect of those exposures on behavioral functioning and academic performance as it mediates other problem behavior. Chronic exposure to crime ridden neighborhoods may also contribute to substance use as part of the cultural phenomena of weathering (Geronimus 1992). Research investigating the relationship between community violence exposure and stressors on youth's adjustment should target substance use as an outcome, using multiple measures. Children may not consider themselves "substance users" out of context (e.g., not associated with peers or distress). An advantage of studying children in adolescence over time permits the identification of moderating variables that influence onset of drug use. Prior research has established that individual-level characteristics are insufficient to account for ATOD use among youth; community-level contextual effects are critical to consider in understanding both group and individual-level behavior (Wilcox 2003). In comparison to family and peer context, neighborhood factors have been understudied in relation to youth's substance use (Lambert et al. 2004). Data from a prospective longitudinal study, the *Multiple Opportunities to Reach Excellence* (MORE) Project are used to investigate the individual, peer, familial, and neighborhood risk and protective factors associated with alcohol, tobacco, and other substance exposure – and use – among urban primarily African American children who have been exposed to varying levels (low, moderate, high) of neighborhood violence.

Method

Sample and Procedures

Sampling Design

Neighborhood crime is hypothesized to place youth at risk for exposure to community violence, as has been found in previous studies (e.g., Selner-O'Hagan et al. 1998). There are a total of 55 neighborhoods that link to Baltimore City Public Schools using data provided by the Baltimore City Data Collaborative (http://www.baltimorekidsdata.org). The Data Collaborative compiles agency databases and other informational rosters from sources such as the Baltimore City Health Department, Baltimore City Public Schools, Baltimore Police Department, Maryland Department of Health and Mental Hygiene, and Baltimore City Child Care Resource Center. Community boundaries were drawn considering the city's neighborhood and community organizations and existing census tract boundaries to create statistical profiles. All 55 communities were rank ordered from 1 to 55 based on their Baltimore City neighborhoods' homicide rates in 2002 (the most recent year for which data were available when the project was funded). Ten of those communities had zero homicides; the remaining 45 communities had between 1 and 164.3 homicides per 100,000 residents. The ten neighborhoods with no homicides in that year were placed in the "low" neighborhood violence stratum (i.e., 0 homicides per 100,000 residents). The "moderate" violence stratum consisted of the four communities in the middle of the distribution of homicide rates (i.e., 25.0–31.4

homicides per 100,000 residents), and the four communities with the highest homicide rates (i.e., 97.2–164.3 homicides per 100,000 residents) were placed into the "high" violence stratum. Within each stratum, the neighborhood with the two largest schools enrolling third through fifth graders became our target schools.

The principals at eight schools were contacted by the Project Investigator and/or Project Director to obtain their permission to partner together to conduct The MORE Project. Two principals declined, one in the high stratum reportedly because of his skepticism of research studies and feeling overburdened by current responsibilities, and the other from the moderate stratum who stated that it was her first year as principal and she did not have enough social capital among her school parents to ask them to participate in a research study. The elementary schools with the next largest student body in those neighborhood strata were contacted and their principals consented. A $1000 honorarium was presented to each participating school to thank them for partnering with The MORE Project and to defray associated expenses.

The two elementary schools in The MORE Project located in the high violence stratum are situated in one zip code, the two medium violence schools are located in another zip code, and the two low violence schools are in two other zip codes that have similar demographic characteristics within each stratum. Across strata, there are some differences. For example, the population density for the low violence community is less than the other strata, but we will be unable to "control for density" since it is fully collinear with strata and it is one of the distinguishing characteristics of urban versus inner-city life.

Participants

The MORE Project participants are comprised of 746 students, their parents/caregivers, and teachers. Recruitment spanned one and one-half academic years among 8–12 year old students who attended six urban public elementary schools located in three Baltimore, Maryland communities with low, moderate, and high levels of neighborhood crime. To avoid selection bias and human subjects concerns, race/ethnicity was not used as a selection criterion. The inclusion criteria for students at the time of recruitment were: (1) Enrolled as a full-time student in one of the six identified Baltimore City public elementary schools in the Fall of 2006 or 2007; (2) aged 8–12 years, inclusively; (3) speak English and live with an English-speaking parent/guardian. Exclusion criteria included: (1) Presence of serious medical or neurological illness (e.g., epilepsy, closed head trauma) or mental retardation that precluded completion of the interview; or (2) Does not live with at least one parent or legal guardian.

Initial recruitment began in January 2007 and yielded 490 eligible families who consented to participate and comprised Cohort 1; 427 (87.1 %) child interviews were conducted in the first semester of fielding the project. The teachers and caregivers of the Cohort 1 students were also interviewed, including 375 (88.2 %) teachers and 282 (66.4 %) parents/caregivers. In the following academic semester (the of Fall 2007), an additional 256 families consented and comprised Cohort 2. Cohort 2 consisted of third, fourth, and fifth grade students in the six participating schools who were not enrolled in The MORE Project during the previous year due to non-response or they were new student transfers. There were a total of 1119 eligible students across both cohorts representing a 67 % consent rate. School-level means and limited data on all students in grades 3, 4, and 5 were compared with those who consented. Comparing all students in the schools at the beginning of the academic year with those enrolled in the project, there was no difference in the proportion of males ($p > .05$). There was a slight difference in the proportion who were African American (86 % of participating students vs. 93 % of eligible, $p < .01$), but that 7 % difference should not affect the generalizability of the results to the population of all students in the selected schools. As such, it is believed that the families who consented to participate in The MORE Project are representative of those who were eligible. The current sample is comprised of the 364 African American students from cohort 1, along with their parents and teachers. At the time of consent, their mean age was 9.6 years old (SD = 1.06; range = 8 – 13 years); 53.3 % female; 2.7 % of the sample was in second grade, 35.2 % in the third grade, 33.0 % in the fourth grade, and 29.1 % in the fifth grade.

Participant Recruitment

The principals at each participating school designated a contact person, typically a member of the school's administrative staff, to assist The MORE Project with administrative requests. These designees provided the project with school-wide rosters for each second, third, fourth, and fifth grade classroom to identify the target participant pool. Once identified, rosters were used to personalize two consent packets for every eligible student (i.e., parental consent form, MORE Project lottery form, letter of support/commitment from their school's principal). One consent packet was mailed to the student's home address, the second was distributed in eligible classrooms to each student following a brief explanatory presentation by a MORE Project staff member. The mailings and classroom distributions were further supplemented with follow-up telephone calls to caregivers when there were working phone numbers. For parents/caregivers without telephones and/or for non-responsive families, specially trained consent gatherers attempted to make home visits to explain the study and obtain parental consent.

Data Collection

Annual student interviews were conducted at each school during school hours. First, each school's administrative contact was provided with a list of all students for whom parental consent had been obtained. Based on teacher convenience, attendance, and space availability (often a significant challenge), students were individually released from non-core classes to be interviewed by MORE Project staff. Interviewers briefly introduced The MORE Project and obtained child assent prior to the start of the interview. Rarely, a student might decline to leave a certain class or feel uncomfortable with an unfamiliar interviewer; a second attempt to interview the child was always successful and, of over one thousand child interviews, no child declined to participate in an interview.

Students were interviewed in private areas in the school (e.g., empty classrooms, break rooms). The child assessment measures were administered using a combined paper-pencil and computerized battery. Examples of paper-pencil administered instruments are the Wechsler scales (i.e., WIAT-II and WASI) which require individual administration following a standardized protocol. For other measures, interviewers read items from laptop screens; students could also read along simultaneously. Students' responses were entered by the interviewers. The computerized assessment battery was programmed using Sensus Multimedia version 2.0 software (Adaptive Technologies Group, Inc., 1994–1997). Sensus Multimedia is a Windows based program used to construct attractive, easy to follow interviews that facilitate accurate and efficient data collection. It comes with a fully integrated statistics and cross- tabulation package so data can be verified immediately. The average completion time for the child interview was 120 min (range=75–180 min), completed in one sitting including a light snack and brief break. Upon completion, each child is given a Wal-Mart gift card (Wave 1: $10; Wave 2: $15) as a token of appreciation and a letter to take to their caregiver notifying them that their child finished their interview and requesting them to schedule their parent interview.

The parents/caregivers of children whose interviews had been completed were called to schedule a telephone interview. For hard to reach caregivers, those without telephones, or those who preferred in-person interviews, parent interviews were conducted face-to-face either at The MORE Project offices at Johns Hopkins University or at their child's school. The entire parent interview was administered using a computerized battery and completed within an average of 60 min (range=40–180 min). Interviewers read each item and the possible answer choices to the caregiver; caregivers' responses were entered by the interviewers on the laptops. Methods for expressing appreciation to parents for their participation include: thank you notes, distribution of Baltimore City Resource Guides, and Wal-Mart gift card incentives (Wave 1: $40; Wave 2: $45). The Baltimore City Resource Guides were developed by MORE Project staff and contain information for families on a variety of social, educational, cultural, legal, employment, municipal, physical, and mental health resources. Retaining parents' commitment to the project has been through regular communication by newsletters, children's birthday cards, and other reminder post-cards distributed by mail or in-person via home visits.

Teacher and principal paper-pencil assessments were completed at the end of each school year. Teachers were given a folder containing an informational letter, a survey of general questions about their qualifications and the school/classroom environment, and individualized questionnaires for each consented child in their class (5–10 min per student). Wal-Mart gift cards or University checks were given as a token of appreciation ($5 per student; range=$20–100). Principals and vice-principals completed a brief survey about their teaching qualifications and their school's climate. Principals were given plaques; vice-principals and administrative staff were given certificates of appreciation. All received $25 Wal-Mart gift certificates as tokens of appreciation.

Measures

The MORE Project assessment instruments were selected for their age-appropriateness, psychometric properties, available norms, and when possible, appropriateness for use with ethnic minority youth. Organized by informant, the following describes each instrument included in this chapter.

Child Measures

Children's exposure to community violence is being assessed using child report of perceived events. Perceived exposure to violence to community violence may differ from objective accounts of events (e.g., police reports, indicators of neighborhood and social disorder). The Children's Report of Exposure to Violence (CREV; Cooley et al. 1995) is a widely used self-report questionnaire developed to assess children's lifetime exposure to community violence. Community violence is defined as deliberate acts intended to cause physical harm against persons in the community. The types of violent situations include being chased or threatened, beaten up, robbed or mugged, shot, stabbed, or killed. The original CREV has good two-week test–retest reliability ($r = .75$), internal consistency (overall $a = .78$), and construct validity (Cooley et al. 1995). An additional module was created in a previous project to assess youth's exposure to war and terrorism. Following the terrorist attacks in the United States in September 2001 and the initiation of the Iraqi war, this "world violence" module was designed to assess the frequency of children's perceived exposure to war and terrorism that may have occurred in their communities, their country, or elsewhere in the world (e.g., attacks on public transportation, chemical or biological attacks, bombs, war). As in the other CREV modules, frequency of exposure to world violence is through four modes (i.e., media, hearsay, direct witness, direct victimization). The CREV-Revised (CREV-R) is comprised of the original 29 items plus world violence items. Its Total score is derived by summing the responses (scored 0–4) on the 45 scored items for the Media, Reported/Hearsay, Witnessed, Victim, and World Violence subscales; higher scores indicate greater exposure. The potential range of scores is from 0 to 180 with higher scores indicating a higher level of reported violent events. The lifetime version of the CREV-R was used at Wave 1, but the past year version is being used in Waves 2 and 3 to determine chronicity/severity of violence exposure. The CREV-R has good reliability and validity as demonstrated in a preliminary study of school-based sample of third to fifth grade urban children using a paper–pencil version. Cronbach's a's for the computerized version of the lifetime CREV-R Total score with the World Violence module is 0.78, without it is 0.88. The past-year CREV-R Total score a is 0.89.

Children's views of their own executive functions are measured using the Behavior Inventory of Executive Function-Self Report Version (BRIEF-SR; Guy et al. 2004). Executive functions are a collection of processes that are responsible for guiding, directing, and managing cognitive, emotional, and behavioral functions. The BRIEF-SR comprises 80 items within eight nonoverlapping theoretically and empirically derived clinical scales that measure different aspects of executive functioning: Inhibit, Shift, Emotional Control, Monitor, Working Memory, Plan/Organize, Organization of Materials, and Task Completion. The clinical scales combine to form two indexes, Behavioral Regulation (BRI) and Metacognition (MI), and one summary composite, the Global Executive Composite (GEC). For the normative sample, internal consistency was moderate to high, with alpha coefficients ranging from .72 for those scales with fewer items to .96 for the full 80-item scale (i.e., GEC).

Substance exposure and use, the final hypothesized outcome, is assessed using the Baltimore Substance Use Scale (BSUS; Chilcoat et al. 1995; Chilcoat and Anthony 1996; Kellam and Anthony 1998). The BSUS was developed for use in longitudinal prospective community-epidemiological studies of students in third through eighth grades. It is an adaptation of Elliott and Huizinga's measure of substance use, which they developed for use in the National Survey of Delinquency and Drug Use (Elliot et al. 1985). Youth report on their knowledge, current and/or anticipated use of tobacco, alcohol, marijuana, crack cocaine, heroin, inhalants, and stimulants. Reliability coefficients were not calculated for the MORE Project because individual items are used to reflect intention and drug use patterns.

Youth's self-reported attitudes toward violence were assessed via five items derived from the Attitude toward Interpersonal Peer Violence Scale (Slaby and Guerra 1988). The scale indicates the perceived legitimization or appropriateness of aggressive responses to threat. Responses across the five items (e.g., "Its okay for me to hit someone if they hit me first") are averaged with higher scores indicating greater support for aggressive behavior. Prior research with this measure reported a's ranging from 0.75 (Dahlberg et al. 1998) to 0.85 (Bradshaw et al. 2012).

Students' perceptions of safety are assessed through the following three items: "I feel safe at my school," "I feel safe in my house," and "I feel safe in my neighborhood." Students indicate the extent to which they agree with each statement on a four-point Likert scale (higher scores indicating increased perception of safety). Prior research on this measure reported an a of 0.63 (Dahlberg et al. 1998) and in this project the reliability coefficient is 0.54.

Children's attitudes toward school are assessed via four items administered to the students from the Sense of School Membership Scale (Goodenow 1993). Youth indicate on a four-point Likert scale (higher scores indicating a more positive perception of school) the extent to which they agree with statements such as, "I feel like I belong at this school" and "The teachers here respect me." Prior research on this measure reported Cronbach's alphas ranging from 0.77 to 0.88 (Dahlberg et al. 1998; Goodenow 1993); the current a reliability coefficient is only 0.57.

The Youth Self-Report (YSR; Achenbach 1991), a widely used self-rating of competencies and problems over the past 6, indicates internalizing and externalizing syndromes. The YSR was normed on a large sample of youth of various ethnicities and socioeconomic levels. The test–retest reliabilities ranged from 0.47 to 0.79 and internal consistencies ranged from 0.71 to 0.95 (Achenbach). Although the published recommended minimum age for the YSR is 11 years, communication with a researcher at the YSR publication company (D. Jacobowitz, personal communication, June 24, 2003), Achenbach System of Empirically Based Assessment, clarified that the YSR may be used with younger children, but requires a fifth grade reading level. Standard practice, according to Mr. Jacobowitz, is to read the items aloud to elementary school students below the fifth grade. As such, MORE Project interviewers read the YSR aloud to all student participants. YSR standard scores (T-scores; Mean = 50; SD = 10) are used in the current project. The subscales used in the MORE Project include: Withdrawn, Somatic Complaints, Anxious/Depressed, Social Problems, Thought Problems, and Attention Problems. The Cronbach's a for the YSR Total score is 0.92.

Youth's perpetration of violence is assessed using DuRant's Youth Violence Perpetration Scale (DVPS). It is a brief self-report measure of different types of lifetime violent and aggressive acts across a range of severity (e.g., "Have you ever been in a gang fight?" "Have you ever carried a weapon?" "Have you ever hurt someone so badly they had to be treated by a doctor or nurse?"). Published psychometric properties are not available. Reliability estimates were calculated on the eight summed items (higher scores indicating more acts of violence perpetrated) and were low (Cronbach's a = 0.41).

Academic achievement in the MORE Project is assessed using the Wechsler Individual Achievement Test-Second Edition-Abbreviated (WIAT-II-A; Psychological Corporation 2001) consists of several subtests, two of which are used in this project: Word Reading and Numerical Operations. The WIAT-II-A efficiently assesses basic academic skills and intervention needs in young children through adults. The screener permits the calculation of age- and gradebased standard scores (higher score indicates increased level of reading and math ability) and was standardized using a large representative sample. The WIAT-II-A is widely used, and has demonstrated reliability and validity, with little evidence of practice effects (Psychological Corporation 2001).

Adverse Life Events in the child's life are assessed using the Multicultural Events Schedule for Adolescents (MESA; Gonzales et al. 1995). This scale was developed to assess major and minor life events that are specific to an inner city, multi-ethnic population (Gonzales et al. 1995). It was normed on African American and Caucasian youth, as well as English- and Spanish-speaking Mexican American adolescents. The MESA was derived from existing life events scales (e.g., Adolescent Perceived Events Scale, Compas et al. 1987; Adolescent Life Events Checklist, Johnson and McCutcheon 1982) and is comprised of 84 items that occur over the past 3 months. A Total life events score is based on the total number of events endorsed, with a higher score indicating more adverse life events and hassles. There are eight separate subscales: Family Trouble/Change; Family Conflict; Peer Hassles/

Conflict; School Hassles; Economic Stress; Perceived Discrimination; Language Conflicts; and Perceived Violence/Personal Victimization. The MESA has concurrent validity and adequate test–retest reliability (two weeks: $r=0.71$; Gonzales et al. 1995). In the current study, the Cronbach's a for the MESA total score is 0.90.

Children's Prosocial Coping is assessed by the child's self report on the ACOPE and SSRS, as well as teacher reports on the SSRS and TOCA. The Adolescent Coping for Problem Experiences (A-COPE; Patterson and McCubbin 1987) is a youth self-report measure that identifies major coping strategies and behaviors in dealing with general life stress (Schwarzer and Schwarzer 1996). The 54 items rated on a 5-point frequency scale comprise 12 subscales, although only seven of them are used in the MORE Project (i.e., Ventilating Feelings, Seeking Diversions, Solving Family Problems, Avoiding Problems, Seeking Spiritual Support, Investing in Close Friends, and Seeking Professional Support), with a higher score indicating increased coping skills. The A-COPE has been validated within a longitudinal study investigating health-risk behaviors and is appropriate for research on youth stress and health-risk behaviors (Schwarzer and Schwarzer 1996). In the current study, the internal consistency reliability for the A-COPE Total score is $a=0.72$.

General cognitive ability is assessed in the MORE Project using two of four subtests of the Wechsler Abbreviated Scale of Intelligence (WASI; Psychological Corporation 1999). The WASI was designed to provide a quick and accurate estimate of an individual's intellectual functioning for screening purposes (Psychological Corporation 1999). The Vocabulary subtest assesses expressive vocabulary, expressive knowledge, verbal knowledge, and fund of information; Matrix Reasoning involves gridded patterns to assess nonverbal reasoning ability (Psychological Corporation 1999). The raw scores for each subscale were converted into t scores which were then averaged to create a summary variable (higher scores indicating increased intellectual functioning). The published WASI test–retest reliability coefficients for the children's sample ranged from 0.86 to 0.93 for the Vocabulary subtest, 0.86–0.96 for Matrix Reasoning, and for the two subtests combined were from 0.85 to 0.88 (Psychological Corporation 1999).

Parent Measures

Demographic characteristics, including socioeconomic status, are reported by parents/caregivers using the Household Structure and Demographics questionnaire. It was created by researchers in the Baltimore Prevention Program at Johns Hopkins University for use in large school-based community-epidemiological studies is asked of parents/caregivers to provide family socio-demographic characteristics for each of the members of the household. It includes level of education, occupational status, ethnicity, employment status, age, and relationship to the target child. Additional information assessed includes self-reported total family income, the child's country of origin, the biological father's and mother's involvement in the child's caregiving, and the number of moves the family has made since the target child was born.

Parent reports of their child's internalizing symptoms and externalizing behaviors are assessed using a standardized questionnaire that parallels the Youth Self-Report form. The Child Behavior Checklist-4-18 (CBCL-4-18; Achenbach 1991) is a very widely used instrument (Crijnen et al. 1997) that yields parents' reports of children's competencies and problems in a standardized format. This parent-rated behavior problem checklist yields data on internalizing and externalizing behavior problems as well as social competence (i.e., activities, social, school functioning). The CBCL-4-18 is appropriate for parents of children aged 4–18 years. The behavior problem checklist items are grouped into behavioral syndromes that correspond to the Diagnostic and Statistical Manual, 4th Edition (DSM-IV; American Psychiatric Association 1994) diagnostic categories. The CBCL-4-18 was normed on nationally representative samples, with good to excellent internal consistency and inter-parental agreement (Doll 1998). In the MORE Project, the same subscales in the YSR are used in addition to Delinquent and Aggressive Behavior subscales. For the current study, the Cronbach's a for the CBCL-4-18 Total score is 0.94.

The Parenting Practices Scale (PPS) assesses parental involvement, monitoring, and discipline using a five-point Likert scale. Parents/caregivers are asked how frequently over the past month they engaged in positive, developmentally appropriate interactions and communications with their child for the Parental Involvement subscale (higher scores indicate an increased level of involvement) In the MORE Project, the Cronbach's as are 0.79.

Familial conflict and violence is assessed using the Conflict Tactics Scale—Form R (CTS1-Form R; Straus 1979, 1987, 1988, 1990), which is a 13-item parent report of intrafamilial violence used to resolve conflicts. Items are rated on a 6-point Likert scale; higher scores indicate more family conflict and higher levels of coerciveness. There are three subscales (Reasoning, Verbal aggression, and Violence), each of which the parent respondent rates: (a) their own behavior toward their partner (i.e., "participant"); and (b) their partner's behavior toward the participant (i.e., "partner"). No questions regarding parent/caregiver aggression toward their child were asked. The CTS-Form R has high internal consistency, face and concurrent validity, and acceptable construct validity (Straus 1979). In the current sample, the internal consistencies for the Reasoning, Verbal Aggression, and Violence subscales were Cronbach's alpa's = 0.69, 0.76, and 0.76, respectively, and for the Participant Parent Psychiatric Health is assessed using the Symptom Checklist-90-Revised (SCL-90-R; Derogatis 1977), a widely used self-report measure designed to assess a broad range of adult psychiatric symptom patterns. There are nine symptom scales and three global scales (i.e., Global Severity Index, Positive Symptom Distress Index; Positive Symptom Total). The nine symptom scales are: Somatization, Obsessive–Compulsive, Interpersonal Sensitivity, Depression, Anxiety, Hostility, Phobic Anxiety, Paranoid Ideation, and Psychoticism. The SCL-90-R has excellent reported reliability and validity (Derogatis and Savitz 2000). In the MORE Project sample, the SCL-90-R total score Cronbach's a is 0.98. Lower scores on the SCL-90-R are hypothesized as protective.

Teacher Measures

The Teacher Form of the SSRS (Gresham and Elliott 1990) individually assesses student's social skills and academic competence using this screening instrument that classifies social behavior in educational and family environments. The Teacher Form of the SSRS is comprised of subscales that assess social skills, problem behaviors and academic competence. The 57 items are rated using 3-point frequency and importance scales. The raw scores from the Social Skills Scale and the Problem Behaviors Scale are converted into age- and gender-normed standard scores (M = 100; SD = 15; Benes 1995) based on a large standardization sample that included regular and special education students, as well as a significant proportion of ethnic minority youth (Benes 1995). Higher academic competence scores indicate an increased level of academic ability. The teacher report is psychometrically sound and has good internal consistency, test–retest reliability, and validity (Gresham and Elliott 1990). The Cooperation, Assertion, Self-Control, Internalizing, Externalizing, and Hyperactivity subscales are used in the current study; the Cronbach's as for those scales range from 0.80 to 0.93 and the Cronbach's a for the Academic Competence scale as 0.96.

The Teacher Observation of Classroom Adaptation—Revised (TOCA-R; Werthamer-Larsson et al. 1991) is a brief measure of each child's adequacy of performance on the core tasks in the classroom as defined and assessed by the teacher. The teacher reports on the adequacy of each child's performance on a six-point scale on six basic tasks with higher scores indicating increased levels of each behavior: Accepting authority (aggressive behavior); social participation (shy or withdrawn behavior); self-regulation (impulsivity); motor control (hyperactivity); concentration (inattention); and peer likeability (rejection). In addition, the teacher reports on youth's academic performance, behavior, education, substance use, and mental health services s/he perceives each child needs or is receiving. The TOCA-R subscales included in the MORE Project include: Concentration, Aggression, Shy Behavior, Likeability, Hyperactivity, Impulsivity, Proactive Aggression, Oppositional Defiant, and Conduct Problems. Excluding the Shy Behavior scale, whose a was 0.51, the subscale Cronbach's as in the MORE Project range from 0.78 to 0.91.

Neighborhood Assessments

Characteristics of the neighborhood environment are assessed by the Neighborhood Inventory for Environmental Typology (NIfETy; Furr-Holden et al. 2008). The NIfETy method uses independent evaluators who go to the residential blocks of Baltimore neighborhoods to systematically assess physical and social disorder; indicators of violence, alcohol, and other drug exposure; and positive neighborhood characteristics. Built upon previous methods that assessed neighborhood context to inform child and family health (e.g., Caughy et al. 2001; McDonnell 2007; Raudenbush et al. 2003; Sampson and Raudenbush 1999, 2005), the NIfETy method involves an epidemiological approach to evaluate characteristics of residential neighborhoods that might indicate a change in crime, violence, victimization, and alcohol and other drug exposure in a manner that is quantifiable, replicable, and designed to be longitudinal (Furr-Holden et al. 2008). For the MORE Project, the city unit blocks in which the consented families resided were given to the NIfETy project investigators who sent trained field assessors to make evaluations using Palm OS Zire 31 Personal Digital Assistants (PDAs) programmed with Pendragon Form 5.0 software. For Cohort 1, 98.1 % of the families' neighborhood blocks were assessed. The assessments were conducted in the daytime.

There are 114 quantitative and 15 qualitative items that comprise seven domains assessed by the NIfETy that include positive/healthy and negative indicators: (1) Physical layout of the block face (e.g., length/width of block, alleys present [that run through to next street], dwelling count); (2) Types of structures (*Limitations.*, single family/detached homes, liquor stores, churches); (3) Adult activity (e.g., adults watching youth, adults in work uniforms, [male] adults sitting on steps); (4) Youth activity (e.g., youth playing, "corner kids/boys," dangerous youth activities; (5) Physical disorder and order (e.g., abandoned/vacant structures, new construction or renovation, police present); (6) Social disorder and order (e.g., outdoor community recreation outlets, homeless people, traffic); and (7) Violence and alcohol and other drug indicators (e.g., drug paraphernalia, memorials, obvious signs of drug selling). In an independent sample, internal consistency reliability for the Total NIfTEy scale was good (intra-class coefficient=0.84); a coefficients ranged from 0.27 to 0.90 for the subscales; and inter-rater reliability and validity were in the acceptable to good range (Furr-Holden et al. 2008).

Healthy Neighborhood Environment is a hypothesized neighborhood protective factor. One scale was created from the NIfETy data that solely included objective indicators of positive neighborhood conditions (e.g., parks, churches, children playing, detached homes). Secondly, parents reported positive aspects of their neighborhood environment using the Collective Efficacy Scale (Sampson et al. 1997). Collective efficacy is defined as "social cohesion among neighbors combined with their willingness to intervene on behalf of the common good" (Sampson et al. 1997). An extremely large survey of Chicago neighborhood residents showed that collective efficacy has a strong negative association with violence and high between-neighborhood reliability (Sampson et al. 1997). Twenty-eight items were summed to create a composite score with higher scores indicating higher levels of social cohesion. Lastly, the NIfETy scale was used to create two indicators of unhealthy neighborhood conditions: Neighborhood Violence (e.g., memorials, bullet casings, gang activity) and Unhealthy Neighborhood (e.g., obvious signs of drug selling, homeless people, unsupervised children).

Results

Based on a review of the literature and the conceptual foundation of the parent MORE Project (for details, see Cooley-Strickland et al. 2009), key variables were investigated as demographic, *Limitations.* protective, risk factors, or outcomes.

Specifically, demographic variables included child's age, gender, parent/guardian education and income, free/reduced school lunch status, school attendance, suspensions, expulsions, and history of residential moves. Hypothesized protective factors included prosocial coping skills, academic competence, cognitive ability estimates, standardized academic achievement scores, attitudes toward school (child likes school), feelings of safety (at home, school, neighborhood), parent-reported neighborhood collective efficacy, and parent involvement/monitoring. Hypothesized risk factors include the child's acceptance and use of violence, interparental conflict/aggression, child's experience of adverse life events, child's exposure to community violence, evidence of neighborhood violence and disorganization, symptoms of parental psychopathology, and child's externalizing behavior and internalizing symptoms. The outcomes examined included child's exposure to alcohol, tobacco, and other substances (via self-report or neighborhood-level independent assessment); child's self-reported use, and the child's reportedly being offered such substances.

Statistical analyses were conducted using SAS version 9. Between-group comparisons by neighborhood crime strata (Low versus Moderate versus High) were conducted (pairwise *t*-tests; Chi-square; Bonferroni post-hoc corrections: p's < .017; see Tables 12.1 and 12.2). To examine the association among the key variables, Pearson correlation coefficients were calculated for the entire sample (see Table 12.3). Separate logistic regression analyses were computed for the dichotomous outcome of children's substance use (any cigarette, alcohol, and/or other substance use versus no prior/current use; see Table 12.4). These analyses are computed using data from the first year of data collection ("wave 1). The remaining three tables contain repeated measures linear regression models to predict urban African American children's self-reported substance exposure, substance use, and neighborhood-level substance exposure over time (1-year period from Wave 1 to Wave 2: Tables 12.5, 12.6 and 12.7). Age and gender are covariates.

Table 12.1 presents the demographic characteristics of children residing in neighborhoods with varying levels of crime (i.e., crime strata: Low vs. Moderate vs. High). Statistically significant (p's < .017) results include that the children living in the high neighborhood crime strata are older than those in the low strata and have parents/guardians with lower levels of education (twice as many are high school drop-outs). Children residing in the high crime strata have parents/guardians with lower incomes than those in the moderate or low crime strata (there are almost two and three times, respectively, more families with annual incomes of less than $20,000 per year), and they receive significantly more free and/or reduced-priced school lunches. Lastly, children residing in the high crime strata reported having experienced more residential moves (at least four in their lifetimes) than children in the low neighborhood crime strata. There were no significant differences in gender, school attendance, or school suspensions/expulsion incidents across neighborhood strata (p's > .05).

Table 12.2 presents the hypothesized protective and risk factors, and outcomes, as compared across neighborhood crime strata. A statistical trend ($.10 < p > .05$) was found for WASI cognitive ability such that the children residing in the low crime

Table 12.1 Demographic characteristics among African American urban children by low, moderate, and high violence strata (wave 1)

			Violence strata			Test across strata[†]	
	Analytic N	Overall	Low (N = 101)	Moderate (N = 132)	High (N = 131)	Statistic	*p*
Child's age (years)	364	9.6 ± 1.1	9.4 ± 1.0	9.6 ± 1.0	9.7 ± 1.1[b]	$F_{[2363]}$ =3.3	0.038
Female	364	194 (53.3)	52 (51.5)	78 (59.1)	64 (48.9)	$Chis_{[2]}$ =2.95	0.229
Parent education, wave 1	218	4.3 ± 1.2	4.7 ± 1.4	4.3 ± 1.1	4.0 ± 1.0[b]	$F_{[2217]}$ =5.09	0.007
Parent education level, wave 1							
Less than high school	218	54 (24.8)	9 (14.8)	20 (25.0)	25 (32.5)[b]	$Chis_{[6]}$ =16.46	0.012
High school	218	68 (31.2)	20 (32.8)	24 (30.0)	24 (31.2)		
Some college	218	70 (32.1)	18 (29.5)	26 (32.5)	26 (33.8)		
College or higher	218	26 (11.9)	14 (23.0)	10 (12.5)	2 (2.6)		
Parent yearly income, wave-1							
Less than $4999	225	33 (14.7)	4 (6.3)	12 (14.3)	17 (22.1)[b c]	$Chis_{[12]}$ =38.51	<.001
$5000–$9999	225	21 (9.3)	2 (3.1)	6 (7.1)	13 (16.9)		
$10,000–$19,999	225	36 (16.0)	9 (14.1)	9 (10.7)	18 (23.4)		
$20,000–$29,999	225	43 (19.1)	11 (17.2)	22 (26.2)	10 (13.0)		
$30,000–$49,999	225	54 (24.0)	18 (28.1)	24 (28.6)	12 (15.6)		
$50,000–$99,999	225	31 (13.8)	15 (23.4)	9 (10.7)	7 (9.1)		
$100,000 or more	225	7 (3.1)	5 (7.8)	2 (2.4)	0 (0.0)		
Parent yearly income <$20,000	225	90 (40.0)	15 (23.4)	27 (32.1)	48 (62.3)[b c]	$Chis_{[2]}$ =25.48	<.001
Free/reduced school lunch	346	310 (89.6)	81 (82.7)	108 (87.8)	121 (96.8)[b c]	$Chis_{[2]}$ =12.45	0.002

(continued)

Table 12.1 (continued)

	Analytic N	Overall	Violence strata			Test across strata†	
			Low (N=101)	Moderate (N=132)	High (N=131)	Statistic	*p*
Free or reduced-price school lunch (proxy for poverty)							
None	344	36 (10.5)	17 (17.5)	15 (12.3)	4 (3.2)[b c]	$Chis_{[4]} = 18.48$	<.001
Free	344	264 (76.7)	63 (64.9)	91 (74.6)	110 (88.0)		
Reduced	344	44 (12.8)	17 (17.5)	16 (13.1)	11 (8.8)		
School attendance (percentage)	346	96.0±4.6	95.8±4.4	96.2±5.1	96.0±4.3	$F_{[2345]} = 0.23$	0.792
School suspension or expulsion incidents>=1	346	20 (5.8)	5 (5.1)	5 (4.1)	10 (8.0)	$Chis_{[2]} = 1.88$	0.391
4 or more household moves	364	127 (34.9)	25 (24.8)	49 (37.1)	53 (40.5)[b]	$Chis_{[2]} = 6.65$	0.036

Note. Analytic n's range from 218 to 364 due to missing data

Data are presented as n (%) for categorical variables and mean ± SD for continuous variables

†Chi-square or F-tests comparing differences across Low, Moderate, and High Violence Strata

[abc]Indicates significant pair-wise post-hoc comparisons using Bonferroni correction (i.e., $p < .017$: [a]Low vs. Moderate neighborhood crime strata; [b]Low vs. High neighborhood crime strata; [c]Moderate vs. High neighborhood crime strata)

Table 12.2 Pairwise comparisons among African American urban children by low, moderate, and high violence strata (Wave 1)

			Violence strata			Test across strata[†]	
	Analytic N	Overall	Low (N = 101)	Moderate (N = 132)	High (N = 131)	Statistic	*p*
Protective factors							
ACOPE prosocial coping, child	364	134.4 ± 15.1	133.7 ± 14.8	133.0 ± 14.9	136.4 ± 15.4	$F_{[2363]}$ = 1.91	0.15
SSRS academic competence, teacher	300	94.7 ± 14.3	94.9 ± 13.3	96.7 ± 15.0	92.7 ± 14.2	$F_{[2299]}$ = 2.11	0.123
WASI (intelligence), child	357	86.3 ± 18.4	90.1 ± 18.6	85.2 ± 17.6	84.6 ± 18.8	$F_{[2356]}$ = 2.94	0.054
WIAT-II-A (achievement) reading summary score	357	95.2 ± 15.6	99.8 ± 13.9	93.3 ± 14.8	93.4 ± 17.0[b c]	$F_{[2356]}$ = 6.41	0.002
WIAT-II-A (achievement) math summary score	357	92.5 ± 15.3	95.3 ± 14.7	92.8 ± 14.3	90.1 ± 16.2[b]	$F_{[2356]}$ = 3.35	0.036
Mean score reading/math WIAT-II-A (achievement)	357	93.8 ± 13.9	97.5 ± 12.3	93.0 ± 13.2	91.8 ± 15.1[b c]	$F_{[2356]}$ = 5.43	0.005
Attitudes toward school (like school), child	359	12.8 ± 2.0	12.8 ± 2.1	12.7 ± 1.9	12.8 ± 2.1	$F_{[2358]}$ = 0.13	0.878
Safety (feel safe), child	359	9.1 ± 1.7	9.3 ± 1.6	9.2 ± 1.7	8.8 ± 1.8	$F_{[2358]}$ = 2.22	0.11
Collective efficacy, parent	208	69.3 ± 17.0	64.7 ± 15.4	69.1 ± 16.1	73.3 ± 18.3[b]	$F_{[2207]}$ = 4.32	0.014
Parenting practices involvement scale, parent	208	42.6 ± 5.2	43.0 ± 5.0	42.9 ± 5.1	42.0 ± 5.5	$F_{[2207]}$ = 0.76	0.47
Risk factors							
Attitudes toward violence (is okay to fight), child	359	11.4 ± 2.4	10.9 ± 2.8	11.7 ± 2.3	11.4 ± 2.2	$F_{[2358]}$ = 2.97	0.053
CTS verbal aggression, parent	159	12.1 ± 9.5	10.7 ± 8.3	11.3 ± 8.5	14.2 ± 11.3	$F_{[2158]}$ = 1.98	0.142
MESA adverse life events, child report							
Total adverse life events	364	16.5 ± 10.3	16.1 ± 9.8	14.6 ± 9.7	18.7 ± 11.0[c]	$F_{[2363]}$ = 5.48	0.005
Family trouble/change	364	4.6 ± 2.8	4.2 ± 2.7	4.1 ± 2.6	5.4 ± 2.9[b c]	$F_{[2363]}$ = 9.61	<.001
Peer hassles scale	364	3.2 ± 2.6	3.1 ± 2.5	2.9 ± 2.7	3.5 ± 2.6	$F_{[2363]}$ = 1.47	0.232
Economic stress scale	364	1.5 ± 1.6	1.6 ± 1.6	1.2 ± 1.4	1.7 ± 1.6[c]	$F_{[2363]}$ = 3.65	0.027
Perceived discrimination	364	1.1 ± 1.2	1.0 ± 1.2	0.9 ± 1.2	1.3 ± 1.3[c]	$F_{[2363]}$ = 3.15	0.044
Violence/personal victim	364	2.3 ± 1.9	1.9 ± 1.8	2.0 ± 1.8	2.8 ± 2.0[b c]	$F_{[2363]}$ = 8.87	<.001

(continued)

Table 12.2 (continued)

	Analytic N	Overall	Violence strata			Test across strata[†]	
			Low (N = 101)	Moderate (N = 132)	High (N = 131)	Statistic	*p*
CREV community violence exposure, child report							
Total score (lifetime)	358	23.0 ± 12.6	23.0 ± 13.6	21.7 ± 11.6	24.4 ± 12.6	$F_{[2357]} = 1.48$	0.229
Media subscale (lifetime)	358	10.6 ± 4.1	10.5 ± 4.0	10.8 ± 4.4	10.3 ± 4.0	$F_{[2357]} = 0.38$	0.686
Reported Subscale (lifetime)	358	7.7 ± 6.2	8.0 ± 6.6	6.6 ± 5.6	8.6 ± 6.2[c]	$F_{[2357]} = 3.64$	0.027
Witnessed subscale(lifetime)	358	3.8 ± 4.4	3.5 ± 4.8	3.4 ± 3.7	4.4 ± 4.6	$F_{[2357]} = 2.03$	0.133
Victim subscale (lifetime)	358	1.0 ± 1.4	1.0 ± 1.5	1.0 ± 1.3	1.1 ± 1.4	$F_{[2357]} = 0.32$	0.729
World violence (lifetime)	358	6.8 ± 4.5	7.3 ± 5.2	6.1 ± 3.8	7.0 ± 4.5	$F_{[2357]} = 2.18$	0.115
DVPS vio. perpetration, child	358	1.1 ± 1.0	1.1 ± 1.2	1.0 ± 0.8	1.2 ± 0.9	$F_{[2357]} = 1.46$	0.234
Ever been in a physical fight?	358	202 (56.4)	47 (46.5)	74 (57.4)	81 (63.3)[b]	$Chis_{[2]} = 6.51$	0.039
NiFETy neighborhood violence (by resident block)	315	1.0 ± 1.4	0.3 ± 0.7	0.7 ± 1.1	1.9 ± 1.6[a b c]	$F_{[2314]} = 52.49$	<.001
NiFETy, unhealthy neighborhood (by block)	315	18.5 ± 8.3	11.4 ± 5.4	18.8 ± 7.6	24.2 ± 6.3[a b c]	$F_{[2314]} = 96.44$	<.001
SCL-90-R global severity index, parent	208	43.2 ± 11.7	42.4 ± 10.7	40.7 ± 9.1	46.8 ± 14.2[c]	$F_{[2207]} = 5.43$	0.005
SCL-90-R global severity index > =63, Parent	208	14 (6.7)	2 (3.4)	1 (1.3)	11 (15.7)[c]	$Chis_{[2]} = 13.81$	0.001
EXTERNALIZING behavior symptoms							
YSR externalizing T-Score, child report	350	46.5 ± 9.5	45.1 ± 9.3	47.8 ± 9.7	46.1 ± 9.2	$F_{[2349]} = 2.38$	0.094
YSR externalizing T-score > =60, child	350	30 (8.6)	5 (5.0)	15 (12.1)	10 (8.0)	$Chis_{[2]} = 3.71$	0.157
CBCL externalizing T-score, parent report	211	52.8 ± 10.0	51.3 ± 9.5	53.1 ± 10.2	53.6 ± 10.2	$F_{[2210]} = 0.93$	0.395
CBCL externalizing T-score > =60, parent	211	49 (23.2)	10 (16.4)	19 (24.1)	20 (28.2)	$Chis_{[2]} = 2.6$	0.272
TOCA-R externalizing, teacher report	292	39.8 ± 19.4	41.7 ± 17.2	34.1 ± 18.3	43.3 ± 21.1[a c]	$F_{[2291]} = 6.39$	0.002

INTERNALIZING symptoms							
YSR internalizing T-score, child report	350	54.5 ± 9.4	54.6 ± 9.2	54.3 ± 9.4	54.6 ± 9.6	$F_{[2349]}$ =0.04	0.962
YSR internalizing T-score>=60, child	350	117 (33.4)	38 (37.6)	38 (30.6)	41 (32.8)	$Chis_{[2]}$ =1.25	0.535
CBCL internalizing symptoms T- score, parent	211	47.8 ± 10.0	47.3 ± 9.9	47.4 ± 10.3	48.8 ± 9.8	$F_{[2210]}$ =0.46	0.634
CBCL internalizing symp-toms T-score>=60, parent	211	29 (13.7)	8 (13.1)	9 (11.4)	12 (16.9)	$Chis_{[2]}$ =0.99	0.611
TOCA-R internalizing scale, teacher report	293	19.8 ± 5.7	19.8 ± 4.9	18.1 ± 6.2	21.2 ± 5.5[c]	$F_{[2292]}$ =7.9	<.001
Outcomes							
NiFETy Alcohol & Drug Exposure (by Resident Block)	315	4.2 ± 2.8	2.4 ± 2.2	4.1 ± 2.9	5.6 ± 2.3[a b c]	$F_{[2314]}$ =41.21	<.001
Baltimore substance use scale (BSUS), child report							
BSUS substance exposure	356	2.6 ± 1.3	2.2 ± 1.2	2.6 ± 1.3	2.8 ± 1.4[b]	$F_{[2355]}$ =6.02	0.003
BSUS substance use	352	0.1 ± 0.3	0.1 ± 0.3	0.1 ± 0.2	0.1 ± 0.3	$F_{[2351]}$ =0.27	0.76
BSUS substance offer	356	0.2 ± 0.7	0.2 ± 0.5	0.2 ± 0.7	0.3 ± 0.7	$F_{[2355]}$ =1.02	0.362
Any substance use	352	22 (6.3)	5 (5.0)	8 (6.5)	9 (7.0)	$Chis_{[2]}$ =0.41	0.815
Any substance offer	356	50 (14.0)	11 (10.9)	16 (12.6)	23 (18.0)	$Chis_{[2]}$ =2.68	0.261

Note. Data are presented as n (%) for categorical variables and mean ± SD for continuous variables

[†]Chi-square or F-tests comparing differences across Low, Moderate, and High Violence Strata

[abc]Indicates significant pair-wise post-hoc comparisons using Bonferroni correction (i.e., $p < .017$: [a]Low vs. Moderate strata; [b]Low vs. High strata; [c]Moderate vs. High strata)

Table 12.3 Associations among select protective and risk factors for African American urban children

	Variable	c1	c2	c3	c4	c5	c6	c7	c8	c9	c10	c11	c12	c13	c14	c15
1	meanWIAT	–														
2	Attitudes to school	0.13*	–													
3	Perceived safe envi.	−0.02	0.51**	–												
4	BRIEF plan/organize	−0.06	−0.13*	−0.20**	–											
5	BRIEF behavior reg	−0.03	−0.15*	−0.18**	0.75**	–										
6	NIfETy healthy nei	0.00	0.01	−0.09	−0.02	−0.03	–									
7	DVPS violence Perp	0.05	−0.20**	−0.12*	0.17**	0.25**	−0.03	–								
8	MESA life events	−0.05	−0.27**	−0.23**	0.33**	0.34**	0.04	0.39**	–							
9	CTS respond total	0.01	−0.06	−0.02	0.12	0.07	0.09	−0.02	0.03	–						
10	SCL90 par sever. Ind	−0.08	0.01	0.08	0.18*	0.07	0.04	0.17*	0.19**	−0.01	–					
11	CREV commu vio exp total, past year	−0.04	−0.15**	−0.12*	0.17**	0.27**	0.03	0.35**	0.49**	0.05	0.05	–				
12	CREV commu vio exp total, lifetime	−0.01	−0.22**	−0.17**	0.19**	0.26**	0.01	0.38**	0.55**	0.05	0.01	0.92**	–			
13	CREV commu vio victimization, life	0.03	−0.20**	−0.17**	0.11	0.14*	0.02	0.31**	0.41**	0.03	−0.04	0.46**	0.53**	–		
14	BSUS substance exp	0.01	−0.11*	−0.07	0.25**	0.30**	0.01	0.23**	0.45**	0.07	0.07	0.39**	0.41**	0.19**	–	
15	BSUS substance use	−0.02	−0.04	−0.05	0.13*	0.14*	0.09	0.16**	0.26**	0.14	0.21**	0.13*	0.15**	0.13*	0.21**	–
16	NIfETy AlcDrg Exp	−0.07	0.03	−0.13*	0.09	0.03	0.29**	−0.03	0.13*	0.08	0.23**	0.06	0.06	−0.014	0.08	0.08

Note. Pearson Correlation Coefficients; analytic n's range from 159 to 364 due to missing data
[1]meanWIAT = WIAT-II-A (Achievement) Mean Reading and Math Summary Score; [2]Attitudes to School = Attitudes toward school; [3]Perceived Safe Envi. = Child's perception of Safety; [4]BRIEF Plan/Organize = BRIEF Planning/Organizational Subscale; [5]BRIEF Behavior Reg = BRIEF Behavioral Regulation Index; [6]NIfTEy Healthy Nei = NIfETy Healthy Neighborhood Environment Indicators; [7]DVPS Violence Perp = DVPS Total Score (Violence Perpetration); [8]MESA Life Events = MESA Total Adverse Life Events Score; [9]CTS Respond Total = CTS1 Conflict Tactics Respondent Total Score; [10]SCL-90 Par Sever. Ind = SCL-90R Parent Global Severity Index Score; [11]CREV Commu Vio Exp Total, Past Year = CREV Past Year Community Violence Exposure Total Score; [12]CREV Total, Life = CREV Lifetime Community Violence Exposure Total Score; [13]CREV Commu Vio Victimization, Life = CREV Lifetime Community Violence Victimization Score; [14]BSUS Substance Exp = BSUS Child Reported Exposure to Substances; [15]BSUS Substance Use = BSUS Child Reported Substance Use; [16]NIfETy AlcDrg Exp = NIfETy Neighborhood Alcohol and/or Drug Exposure

Table 12.4 Separate logistic regression models predicting urban African American children's self-reported substance use over time

Logit		Wave 1			Wave 2		
Fit	Predictor	N	Odds ratio 95 % CI	*p*-value	N	Odds ratio 95 % CI	*p*-value
1	Mean WIAT	351	1.30 (0.80, 2.13)	0.292	276	0.92 (0.54, 1.55)	0.748
2	Attitudes toward school	350	0.85 (0.55, 1.32)	0.477	277	0.80 (0.52, 1.22)	0.300
3	Perceived safe environment	350	0.81 (0.52, 1.26)	0.346	277	0.68 (0.44, 1.06)	0.090
4	BRIEF planning/organization	271	2.08 (1.14, 3.77)	0.016	275	1.75 (1.09, 2.83)	0.022
5	BRIEF behavior regulation index	271	2.26 (1.28, 3.97)	0.005	275	2.48 (1.51, 4.07)	0.000
6	NIfETy healthy neighborhood	309	1.23 (0.83, 1.84)	0.304	243	1.14 (0.68, 1.93)	0.613
7	DVPS violence perpetration	349	1.88 (1.21, 2.93)	0.005	277	1.94 (1.29, 2.93)	0.001
8	MESA total adverse life events	352	2.32 (1.55, 3.46)	0.000	277	2.28 (1.44, 3.60)	0.000
9	CTS parental conflict (respondent total)	156	1.77 (0.88, 3.58)	0.109	172	0.83 (0.49, 1.40)	0.476
10	SCL-90 parent global severity index	203	1.56 (1.08, 2.25)	0.019	226	0.83 (0.57, 1.22)	0.352
11	CREV total commu. violence exp, past year	349	1.46 (0.98, 2.15)	0.061	277	2.31 (1.50, 3.57)	0.000
12	CREV total commu. violence exp, lifetime	349	1.66 (1.12, 2.45)	0.011	274	1.22 (0.82, 1.83)	0.325
13	CREV commu. violence victimization, life	349	1.49 (1.04, 2.13)	0.029	274	1.05 (0.68, 1.62)	0.815

Note. Separate logistic regression models; one predictor per time. Adjusted covariates are female gender, age, and low crime strata. Odds ratio represents one standard deviation change in the predictor

[1]mean WIAT=WIAT-II-A (Achievement) Mean Reading and Math Summary Score; [2]Attitudes toward School=Attitudes toward school; [3]Perceived Safe Environment=Child's Perception of Safe Environment; [4]BRIEF Plan/Organize=BRIEF Planning/Organization Subscale; [5]BRIEF Behavior Regulation=BRIEF Behavioral Regulation Index; [6]NIfETy Healthy Neighborhood=NIfETy Healthy Neighborhood Environment Indicators; [7]DVPS Violence Perpetration=DVPS Total Score (Violence Perpetration); [8]MESA Total Adverse Life Events=MESA Total Adverse Life Events Score; [9]CTS Parental Conflict (Respondent Total)=CTS1 Conflict Tactics Respondent Total Score; [10]SCL-90 Parent Global Severity Index=SCL-90R Global Severity Index T-Score; [11]CREV Total Commu. Violence Exp, Past Year=CREV Past Year Community Violence Exposure Total Score; [12]CREV Total Commu. Violence Exp, Lifetime=CREV Lifetime Community Violence Exposure Total Score; [13]CREV Commu. Violence Victimization, Life=CREV Lifetime Community Violence Victimization Score

Table 12.5 Repeated measures linear regression models predicting urban African American children's self-reported substance exposure over time

	Predictor	Estimate	Standard Error	df	*t*-value	*p*-value
1	Age	0.006	0.072	202	0.08	0.938
2	Attitudes toward school	−0.111	0.038	259	−2.89	0.004
3	BRIEF Planning/Organization	0.020	0.009	168	2.19	0.030
4	CREV Total CommuVio Exp, PastYr	0.024	0.008	249	3.17	0.002
5	CREV CommuVio Victim, Lifetime	0.051	0.057	175	0.90	0.371
6	CTS parental conflict (Respondent)	−0.012	0.012	239	−0.99	0.324
7	SCL-90 parent global severity Index	−0.081	0.235	263	−0.34	0.731
8	DVPS violence perpetration	−0.031	0.082	263	−0.38	0.708
9	Female	−0.190	0.162	157	−1.17	0.243
10	Intercept	1.727	1.098	219	1.57	0.117
11	MESA total adverse life events	0.038	0.010	262	3.75	0.000
12	NIfETy healthy neighborhood	−0.040	0.031	263	−1.27	0.205
13	Perceived safe environment	0.106	0.048	253	2.19	0.029
14	Low crime strata	−0.367	0.155	155	−2.37	0.019
15	Wave 2	−0.328	0.130	178	−2.53	0.012

Note. Analysis of repeated measures; selected interaction × time (Wave 1, Wave 2) linear regression; n's range from 155 to 263

[1]Age=Age of student; [2]Attitudes toward School=Attitudes toward school; [3]BRIEF Plan/Organize=BRIEF Planning/Organization Subscale; [4]CREV Total Commu. Violence Exp, Past Year=CREV Past Year Community Violence Exposure Total Score; [5]CREV Total Commu. Violence Exp, Lifetime=CREV Lifetime Community Violence Exposure Total Score; [6]CTS Parental Conflict (Respondent Total)=CTS1 Conflict Tactics Respondent Total Score; [7]SCL-90 Parent Global Severity Index=SCL-90R Global Severity Index T-Score; [8]DVPS Violence Perpetration=DVPS Total Score (Violence Perpetration); [9]Female=Gender; [10]Intercept = ; [11]MESA Total Adverse Life Events=MESA Total Adverse Life Events Score; [12]NIfETy Healthy Neighborhood=NIfETy Healthy Neighborhood Environment Indicators; [13]Perce ived Safe Environment=Child's Perception of Safe Environment; [14]Low Crime Strata = ; [15]Wave 2=

strata earned intelligence quotient estimated scores that were marginally higher than the children residing in the high crime strata. Statistically significant results (p's<.017) include group differences between standardized achievement reading and mathematics WIAT-II-A scores such that the children living in the low neighborhood strata earned higher achievement scores than the residents of the high crime communities. The only other significant group difference was that parents in the high crime strata reported higher collective efficacy and reliance/dependability of their neighbors. There were no neighborhood strata differences in the child's prosocial coping skills, teacher reported academic competence, child's attitudes toward school or perceived safety (at home, school, neighborhood), or parent reported involvement and/or monitoring of their children.

Multiple hypothesized risk factors were compared across neighborhood crime strata; among the significant results (p's<0.017), children residing in the high crime strata reported experiencing more adverse life events than those in the other strata, including more family trouble/change, economic stress, perceived discrimination,

Table 12.6 Repeated measures linear regression models predicting urban African American children's neighborhood substance exposure over time

	Predictor	Estimate	Standard error	df	*t*-value	*p*-value
1	Age	0.180	0.132	196	1.36	0.175
2	Attitudes toward school	0.042	0.075	262	0.56	0.576
3	BRIEF planning/organization	0.016	0.017	171	0.97	0.335
4	CREV total commuVio exp, pastYr	0.020	0.015	263	1.36	0.174
5	CREV commuVio victim, lifetime	0.111	0.104	179	1.06	0.289
6	CTS parental conflict (respondent)	0.044	0.031	260	1.44	0.151
7	SCL-90 parent global severity index	2.589	0.807	259	3.21	0.002
8	DVPS violence perpetration	−0.232	0.158	248	−1.47	0.142
9	Female	−0.224	0.291	159	−0.77	0.443
10	Intercept	−0.477	2.055	212	−0.23	0.817
11	MESA total adverse life events	−0.004	0.020	245	−0.18	0.858
12	NIfETy healthy neighborhood	0.350	0.061	253	−5.77	0.000
13	Perceived safe environment	−0.099	0.095	263	−1.04	0.300
14	Low crime strata	−1.923	0.278	162	−6.91	0.000
15	Wave 2	0.689	0.580	177	−1.19	0.236
16	Wave 2*CTS parental conflict	−0.096	0.040	181	−2.38	0.018
17	Wave 2* SCL-90 parent severity	−2.524	0.911	207	−2.77	0.006

Note. Analysis of repeated measures; selected interaction × time (Wave 1, Wave 2) linear regression; n's range from 159 to 263

[1]Age = Age of student; [2]Attitudes toward School = Attitudes toward school; [3]BRIEF Plan/Organize = BRIEF Planning/Organization Subscale; [4]CREV Total Commu. Violence Exp, Past Year = CREV Past Year Community Violence Exposure Total Score; [5]CREV Total Commu. Violence Exp, Lifetime = CREV Lifetime Community Violence Exposure Total Score; [6]CTS Parental Conflict (Respondent Total) = CTS1 Conflict Tactics Respondent Total Score; [7]SCL-90 Parent Global Severity Index = SCL-90R Global Severity Index T-Score; [8]DVPS Violence Perpetration = DVPS Total Score (Violence Perpetration); [9]Female = Gender; [10]Intercept = ????? ; [11]MESA Total Adverse Life Events = MESA Total Adverse Life Events Score; [12]NIfETy Healthy Neighborhood = NIfETy Healthy Neighborhood Environment Indicators; [13]Perceived Safe Environment = Child's Perception of Safe Environment; [14]Low Crime Strata = ?; [15]Wave 2 = ?; [16] Wave 2*CTS Parental Conflict = ?; [17] Wave 2* SCL-90 Parent Severity = ?

violence/personal victimization, and total adverse life events. Children in the high crime strata reported having heard over their lifetimes about more community violence events occurring to familiar persons and strangers compared to those living in the moderate crime stratum. The high crime residents also reported having been in more physical fights than those in the low crime neighborhoods. Independent evaluations of neighborhood characteristics indicated that children residing in the low crime neighborhoods lived on blocks with the least violence; children in the high crime strata live in neighborhoods with the most evidence of violence and other indicators of disorganization and chaos. Parents/caregivers of these children in the high crime strata reported experiencing the most psychiatric symptomatology,

Table 12.7 Repeated measures linear regression models predicting urban African American children's substance use over time

	Predictor	Estimate	Standard error	df	*t*-value	*p*-value
1	Age	0.001	0.022	181	0.06	0.955
2	Attitudes toward school	−0.001	0.012	246	−0.09	0.929
3	BRIEF planning/organization	0.005	0.003	141	1.81	0.073
4	CREV total commuVio exp, pastYr	−0.001	0.003	173	−0.48	0.634
5	CREV commuVio victim, lifetime	0.011	0.024	261	0.48	0.633
6	CTS parental conflict (Respondent)	−0.003	0.004	238	−0.94	0.347
7	SCL-90 parent global severity index	0.294	0.122	200	2.41	0.017
8	DVPS violence perpetration	0.024	0.031	177	0.77	0.442
9	Female	0.034	0.051	130	0.67	0.502
10	Intercept	−0.375	0.344	206	−1.09	0.276
11	MESA total adverse life events	0.006	0.003	259	1.91	0.057
12	NIfETy healthy neighborhood	0.006	0.010	261	0.66	0.508
13	Perceived safe environment	0.007	0.015	229	0.46	0.646
14	Low crime strata	−0.068	0.049	125	−1.40	0.164
15	Wave 2	0.005	0.068	103	0.08	0.936
16	Wave 2*total commuVio exposure	0.011	0.004	134	3.09	0.002
17	Wave 2*CREV commuVio victim	−0.058	0.025	105	−2.38	0.019
18	Wave 2* SCL-90 parent severity	−0.307	0.131	122	−2.34	0.021
19	Wave 2*DVPS violence perpetration	−0.060	0.037	97.2	−1.61	0.112

Note. Note. Analysis of repeated measures; selected interaction x time (Wave 1, Wave 2) linear regression; n's range from 105 to 261

[1]Age=Age of student; [2]Attitudes toward School=Attitudes toward school; [3]BRIEF Plan/Organize=BRIEF Planning/Organization Subscale; [4]CREV Total Commu. Violence Exp, Past Year=CREV Past Year Community Violence Exposure Total Score; [5]CREV Total Commu. Violence Exp, Lifetime=CREV Lifetime Community Violence Exposure Total Score; [6]CTS Parental Conflict (Respondent Total)=CTS1 Conflict Tactics Respondent Total Score; [7]SCL-90 Parent Global Severity Index=SCL-90R Global Severity Index T-Score; [8]DVPS Violence Perpetration=DVPS Total Score (Violence Perpetration); [9]Female=Gender; [10]Intercept=?; [11]MESA Total Adverse Life Events=MESA Total Adverse Life Events Score; [12]NIfETy Healthy Neighborhood=NIfETy Healthy Neighborhood Environment Indicators; [13]Perceived Safe Environment=Child's Perception of Safe Environment; [14]Low Crime Strata=?; [15]Wave 2=?; [16]Wave 2*Total CommuVio Exposure=?; [17]Wave 2*CREV CommuVio Victim=?; [18]Wave 2* SCL-90 Parent Severity=?; [19]Wave 2*DVPS Violence Perpetration

including more parents with scores in the clinically significant range. Neither the parents nor students reported significant differences in child internalizing or externalizing symptoms by neighborhood crime strata, although their teachers did. Teachers reported that their students in the high crime neighborhoods exhibited the most externalizing behavior problems and internalizing symptomatology.

Although multiple outcomes related to children's exposure to and use of alcohol, tobacco, and other substances were explored by neighborhood crime strata, the only significant differences (p's < .05) were related to their exposure. Both independent assessors and self-reports of exposure to substances indicated that children living in the high crime neighborhood strata are exposed to the most alcohol, tobacco, and other drugs.

There were multiple significant associations (p's $< .05$) among select key variables in the correlation matrix (see Table 12.2), particularly those related to the hypothesized outcomes. Healthy neighborhood factors were not significantly related to any other variable except neighborhood alcohol and/or drug exposure. Independent assessors' evaluations of evidence of neighborhood alcohol and/or drug activity were negatively related to the children's reports of the safety of their homes, communities, and schools, and positively associated with their experience of adverse life events and parental psychopathology symptoms. Having a positive attitude toward school was inversely associated with self-reported exposure to substances and substance use. Problems with executive functioning, such as planning/organization and behavior regulation, were positively associated with urban African American children's substance exposure and use. Adverse life events, perpetration of violence, and community violence (past year, lifetime, and personal victimization) were each positively correlated with children's exposure to and use of alcohol, tobacco, or other substances.

The odds ratios for urban African American children's self-reported history of using any substances (alcohol, tobacco, or other illicit drugs) are presented separately by the first wave and one-year follow-up assessment (waves 1 and 2) in Table 12.3. Academic achievement scores, positive attitudes towards school, nor healthy neighborhood characteristics was significantly associated with substance use (each hypothesized protective factors). There was a statistical trend ($.10 < p > .05$) in wave 2 suggesting that urban African American children's perception of safety at home, school, and in their neighborhoods was potentially protective against their using substances (OR 0.68; 95 % CI: 0.44, 1.06; $p = 0.90$). Problems with executive functioning significantly predicted substance use in both waves such that urban African American children with poor planning and organizational skills and behavior regulation problems were at about twice the risk of drinking alcohol, smoking cigarettes, or using illicit drugs. Specifically, significant predictors of substance use were planning/organization problems (Wave 1: OR = 2.08; 95 % CI: 1.14, 3.77; Wave 2: OR = 1.75; 95 % CI: 1.09, 2.83; p's $= .02$) and behavior regulation problems (Wave 1: OR = 2.26; 95 % CI: 1.28, 3.97; $p < .01$; Wave 2: OR = 2.48; 95 % CI: 1.51, 4.07; $p = .00$). Other hypothesized risk factors that were predictive of African American children's substance use included their perpetration of violence (Wave 1: OR = 1.88; 95 % CI: 1.21, 2.93; $p < .01$; Wave 2: OR = 1.94; 95 % CI: 1.29, 2.93; p's $< .00$), their experience of adverse life events (Wave 1: OR = 2.32; 95 % CI: 1.55, 3.46; $p < .00$; Wave 2: OR = 2.28; 95 % CI: 1.44, 3.60; p's $< .00$), parental psychopathology symptoms (Wave 1: OR = 1.56; 95 % CI: 1.08, 2.25; $p < .02$), and exposure to community violence (i.e., in the past year: Wave 1: OR = 1.46; 95 % CI: 0.98, 2.15; $p = .06$; Wave 2: OR = 2.31; 95 % CI: 1.50, 3.57; p's $< .00$); over the lifetime: Wave 1: OR = 1.66; 95 % CI: 1.12, 2.45; $p < .01$); victimization: Wave 1: OR = 1.49; 95 % CI: 1.04, 2.13; $p < .03$).

Further examination predictors of urban African American children's exposure to and use of substances over 1-year (waves 1–2) was conducted using repeated measures linear regression models. Tables 12.5, 12.6 and 12.7 present the results of different outcomes: self-reported substance exposure, neighborhood level indicators of exposure to substance use activity, and child's self-reported substance use, respectively. Self-reported exposure to substances was significantly (p's $< .05$) predicted by poor attitudes toward school; executive functioning problems (i.e., planning/organization); past year exposure to total community violence; adverse life events; perceived sense of environmental safety; and living in middle or high neighborhood crime strata. However, the interactions by time were not significant (p's $> .05$).

There were fewer and largely different predictors for neighborhood level indicators of children's exposure to substances than those associated with children's self-reports. Specifically, parental psychopathology symptoms, healthy neighborhood characteristics, and living in middle or high neighborhood crime strata (p's $< .05$). There were interactions at wave 2, such that parental conflict and parental psychopathology were significantly negatively associated with exposure to neighborhood indicators of substance use over time. Regarding predictors of urban African American children's substance use, parental psychopathology was a significant predictor; marginal associations were problems with executive functioning (i.e., planning, organization, $p = .07$) and adverse life events ($p = .057$). The interactions with substance use at wave 2 were total community violence exposure (positive association), community violence victimization and parental psychopathology symptoms (negative associations).

Discussion

Residential mobility was explored as a demographic factor, but also could be considered a potential risk factor for gateway substance use. Previous research on ethnic minority groups has compared frequent movers (i.e., those moved more than four times in the past 5 years) to those who have never moved. Frequent movers were more likely to smoke cigarettes and marijuana (Lee 2007). Among Hispanic adolescents, frequent relocations, being female and older (ages 14–17) placed them at enhanced risk for gateway substances (Lee 2007). In the current study, urban African American children who experienced frequent residential moves were also more likely to reside in the highest crime neighborhoods, thus placing them at enhanced risk for exposure to substances, crime, and violence. Caregivers of these inner-city African American youth should be encouraged to provide stable residential environments for their children, to help minimize the stressors and exposures to.

Limitations

There are multiple limitations to the current study, including lack of peer informants and lack of toxicology screens to indicate substance use. However, the children in this sample are still young and the inclusion of their peers as informants will become increasingly important as the youth become adolescents. A recent study reported that peer substance use is an important indicator of potential to use substances, as youth more readily report their friends' substance use than their own (REF). Given the young ages of this sample and their very low reported rates of substance use, toxicology screens may be premature. Again, their importance will increase over time.

Acknowledgements The authors would like to acknowledge those who have assisted with The MORE Project research. We thank our collaborators: Drs. Megan Bair-Merrit, Catherine Bradshaw, Nicholas Ialongo, Phillip Leaf, and Elizabeth Stuart, and Mr. Scott Hubbard. This endeavor is only successful with the ongoing support and cooperation of the Baltimore City Public School System and our six partner schools. The administrators and staff at these schools have provided access and guidance, allowing us to learn from them. We sincerely thank the Baltimore City students and their families who share their lives with us for the betterment of others. We thank the dedicated MORE Team (Alisa Failor, Amber Williams, Andrew Castro, Corina Morga, Dan Kim, Debbie Skillman, Katherine Otte, Maria Mangione, Max Yusem, MerryJessica Fuerst, Michael Nemeti, Stephanie Dobrowolski, Tyshanda Scott, and Winn Bolton). Support and funding for The MORE Project comes from a grant from the National Institute on Drug Abuse to M. Cooley (1 R01 DA018318).

References

Abbey, A., Jacques, A. J., & Hayman, L. W. (2006). Predictors of early substance use among African American and Caucasian youth from urban and suburban communities. *Merrill-Palmer Quarterly, 52*, 305–326.

Achenbach, T. M. (1991). *Manual for the youth self-report and 1991 profile*. Burlington: University of Vermont, Department of Psychiatry.

Bahr, S. J., Hoffman, K. P., & Yang, X. (2005). Parental and peer influences on the risk of adolescent drug use. *Journal of Primary Prevention, 26*, 529–551.

Barnes, G. M., & Farrell, M. P. (1992). Parental support and control as predictors of adolescent drinking, delinquency, and related problem behaviors. *Journal of Marriage and the Family, 54*, 763–776.

Baumrind, D. (1966). Effects of authoritative parental control on child behavior. *Child Development, 37*, 887–907.

Baumrind, D. (1991). The influence of parenting style on adolescent competence and substance use. *The Journal of Early Adolescence, 11*, 56–95.

Beauvais, F., & Oetting, E. (2002). Variances in the etiology of drug use among ethnic groups of adolescents. *Public Health Reports, 11*, 8–14.

Benes, K. M. (1995). Review of the social skills rating system. In J. C. Conoley & J. C. Impara (Eds.), *The twelfth mental measurements yearbook* (pp. 965–967). Lincoln: Buros Institute of Mental Measurements.

Bersamin, M., Paschall, M., & Flewelling, R. (2005). Ethnic differences in relationships between risk factors and adolescent binge drinking: A national study. *Prevention Science, 6*, 127–137.

Biederman, J., Farone, S. V., Monteaux, M. C., & Feighner, J. A. (2000). Patterns of alcohol and drug use in adolescents can be predicted by parental substance use disorders. *Pediatrics, 106*, 792–797.

Blum, R. W., Beuhring, T., Shew, M. L., Bearinger, L. H., Siering, R. E., & Resnick, M. D. (2000). The effects of race/ethnicity, income, and family structure on adolescent risk behaviors. *American Journal of Public Health, 90*, 1879–1884.

Bolland, J. M., Bryant, C. M., Lian, B. E., McCallum, D. M., Vazsonyi, A. T., & Barth, J. M. (2007). Development and risk behavior among African American, Caucasian, and mixed-race adolescent living in high poverty inner-city neighborhoods. *American Journal of Community Psychology, 40*, 230–249.

Bolton, D., O'Ryan, D., Udwin, O., Boyle, S., & Yule, W. (2000). The long-term psychological effects of a disaster experienced in adolescence: II: General psychopathology. *Journal of Child Psychology and Psychiatry, 41*, 513–523.

Bowen, N., & Bowen, G. (1999). Effects of crime and violence in neighborhoods and schools on the school behavior and performance of adolescents. *Journal of Adolescent Research, 14*, 319–342.

Bowman, M. A., Prelow, H. M., & Weaver, S. R. (2007). Parenting behaviors, association with deviant peers and delinquency in African American adolescents: A mediated-moderation model. *Journal of Youth and Adolescence, 36*, 517–527.

Bradshaw, C. P., Waasdorp, T. E., & Leaf, P. J. (2012). Effects of school-wide positive behavioral interventions and supports on child behavior problems. *Pediatrics, 130*(5), e1136–e1145.

Brunswick, A., & Messeri, P. (1984). Causal factors in onset of adolescents' cigarette smoking: A prospective study of urban African American youth. In H. Shaffer, & B. Stimmel (Eds.), *The addictive behaviors* (pp. 35–52). New York: Hawthorne Press.

Bryant, A., Schulenberg, J., Bachman, J., O'Malley, P., & Johnston, L. (2000). Understanding the links among school misbehavior, academic achievement, and cigarette use: A national panel study of adolescents. *Prevention Science, 1*, 71–87.

Burlew, A. K., Johnson, C. S., Flowers, A. M., Peteet, B. J., Griffith-Henry, K. D., & Buchanan, N. D. (2009). Neighborhood risk, parental supervision and the onset of substance use among African American adolescents. *Journal of Child and Family Studies, 18*, 680–689.

Catalano, R. F., Kosterman, R., Hawkins, J. D., Newcomb, M. D., & Abbott, R. D. (1996). Modeling the etiology of adolescent substance use: A test of the social development model. *Journal of Drug Issues, 26*, 429–455.

Caughy, M. O., O'Campo, P. J., & Patterson, J. (2001). A brief observational measure for urban neighborhoods. *Health & Place, 7*, 225–236.

Challier, B., Chau, N., Predine, R., Choquet, M., & Legras, B. (2000). Associations of family environment and individual factors with tobacco, alcohol, and illicit drug use in adolescents. *European Journal of Epidemiology, 16*, 33–42.

Chilcoat, H. D., Dishion, T. J., & Anthony, J. C. (1995). Parent monitoring and the incidence of drug sampling in urban elementary school children. *American Journal of Epidemiology, 141*(1), 25–31.

Chilcoat, H. D., & Anthony, J. C. (1996). Impact of parent monitoring on initiation of drug use through late childhood. *Journal of the American Academy of Child & Adolescent Psychiatry, 35*(1), 91–100.

Compas, B. E., Davis, G. E., Forsythe, C. J., & Wagner, B. M. (1987). Assessment of major and daily stressful events during adolescence: The adolescent perceived events scale. *Journal of Consulting and Clinical Psychology, 55*, 534–541.

Cooley-Quille, M. R., Turner, S. M., & Beidel, D. C. (1995). The emotional impact of children's exposure to community violence: A preliminary study. *Journal of the American Academy of Child and Adolescent Psychiatry, 34*, 1362–1368.

Cooley-Strickland, M. R., Quille, T. J., Griffin, R., Stuart, E., Bradshaw, C., & Furr-Holden, D. (2009). Community violence and youth: Affect, behavior, substance use, and academics. *Clinical Child and Family Psychology Review, 12*, 127–156.

Corporation, P. (1999). *Wechsler abbreviated scale of intelligence manual*. San Antonio: Author.

Crum, R. M., Lille-Blanton, M., & Anthony, J. C. (1996). Neighborhood environment and opportunity to use cocaine and other drugs in late childhood and early adolescence. *Drug and Alcohol Dependence, 43*, 155–161.

Crijnen, A. A., Achenbach, T. M., & Verhulst, F. C. (1997). Comparisons of problems reported by parents of children in 12 cultures: Total problems, externalizing, and internalizing. *Journal of the American Academy of Child & Adolescent Psychiatry, 36*(9), 1269–1277.

Dahlberg, L. L., Toal, S. B., & Behrens, C. B. (Eds.). (1998). *Measuring violence-related attitudes, beliefs, and behaviors among youth: A compendium of assessment tools*. Atlanta: National Center for Injury Prevention and Control of the Centers for Disease Control and Prevention.

De La Rosa, R. M., Adrados, J. R., & Milburn, N. (1993). Introduction and overview. National Institute on Drug Abuse Research Monograph Series #130: *Drug abuse among minority youth: Methodological issues and research advances* (NIH Publication Number 93–3479, pp. 1–7). Rockville: National Institute on Drug Abuse.

Derogatis, L. R. (1977). *SCL-90-R: Administration. Scoring and procedures manual*. Baltimore: Clinical Psychometrics Research.

Derogatis, L. R., & Savitz, K. L. (2000). The SCL-90-R and brief symptom inventory in primary care. In M. Maruish (Ed.), *Handbook of psychological assessment in primary care settings* (pp. 297–334). Mahwah: Lawrence Erlbaum Associates, Inc.

Duncan, S. C., Duncan, T. E., & Hops, H. (1998). Progressions of alcohol, cigarette and marijuana use in adolescence. *Journal of Behavioral Medicine, 21*, 375–388.

Doll, B. (1998). Review of the child behavior checklist. In *The thirteenth mental measurements yearbook*. Lincoln: Buros Institute of Mental Measurements.

Eccles, J. S., & Barber, B. L. (1999). Student council, volunteering, basketball, or marching band: What kind of extracurricular involvement matters? *Journal of Adolescent Research, 14*, 10–43.

Ellickson, P., & Morton, S. (1999). Identifying adolescents at risk for hard drug use: Racial/ethnic variations. *Journal of Adolescent Health, 25*, 382–395.

Elliott, D. S., Huizinga, D., & Menard, S. (2012). *Multiple problem youth: Delinquency, substance use, and mental health problems*. Springer Science & Business Media.

Epstein, J. A., Botvin, G. J., Griffin, K. W., & Diaz, T. (1999). Role of ethnicity and gender in polydrug use among a longitudinal sample of inner-city adolescents. *Journal of Alcohol and Drug Education, 45*, 1–12.

Fang, J., Madhaven, S., Bossworth, W., & Alderman, M. H. (1998). Residential segregation and mortality in New York city. *Social Science and Medicine, 47*, 469–476.

Farrell, A. D., Kung, E. M., White, K. S., & Valois, R. F. (2000). The structure of self-reported aggression, drug use, and delinquent behaviors during early adolescence. *Journal of Clinical Child Psychology, 29*, 282–292.

Felton, B., Parsons, M. A., Ward, D. S., Pate, E. R., Saunders, R. P., Dowda, M., & Trost, S. (1999). Tracking of avoidance of alcohol use and smoking behavior in a fifth grade cohort over three years. *Public Health Nursing, 16*, 32–40.

Fite, P. J., Wynn, P., Lochman, J. E., & Wells, K. C. (2009). The influence of neighborhood disadvantage and perceived disapproval on early substance use initiation. *Addictive Behaviors, 34*, 769–771.

Ford, J. A. (2009). Nonmedical prescription drug use among adolescents : The influence of bonds to family and school. *Youth & Society, 40*, 336–352.

Furr-Holden, C. D. M., Smart, M. J., Pokorni, J. P., Ialongo, N. S., Holder, H., & Anthony, J. C. (2008). The NIfETy method for environmental assessment of neighborhood-level indicators of alcohol and other drug exposure. *Prevention Science, 9*, 245–255.

Geronimus, A. T. (1992). The weathering hypothesis and the health of African American women and children. *Ethnicity and Disease, 2*, 207–221.

Giaconia, R. M., Reinherz, H. Z., Silverman, A. B., Pakiz, B., Frost, A. K., & Cohen, E. (1995). Traumas and posttraumatic stress disorder in a community population of older adolescents. *Journal of the American Academy of Child and Adolescent Psychiatry, 34*, 1369–1380.

Gibbons, F. X., Gerrard, M., & Lane, D. J. (2003). A social-reaction model of adolescent health risk. In J. M. Suls, & K. A. Wallston (Eds.), *Social psychological foundations of health and illness* (pp. 107–136). Oxford: African Americanwell.

Gibbons, F. X., Gerrard, M., Reimer, R. A., & Pomery, E. A. (2006) Online health decision-making: a subrational approach. In D de Ridder, & J. de Wit (Eds.), *New perspectives on health behavior: The role of self-regulation* (pp. 45–70). Hoboken: Wiley.

Gibbons, F. X., Reimer, R. A., Gerrard, M., Yeh, H., Houlihan, A. E., Cutrona, C., Simons, R., & Brody, G. (2007). Rural-urban differences in substance use among African-American adolescents. *The Journal of Rural Health, 23*, 22–28.

Goodenow, C. (1993). The psychological sense of school membership among adolescents: Scale development and educational correlates. *Psychology in the Schools, 30*, 79–90.

Gottfredson, D. C., Gerstenblith, S. A., Soule, D. A., Womer, S. C., & Lu, S. (2004). Do after school programs reduce delinquency? *Prevention Science, 5*, 253–266.

Gonzales, N. I., Gunnoe, M. L., Samaniego, R., & Jackson, K. M. (1995, June). Validation of the Multicultural Events Schedule for Urban Adolescents (MESA). In *Fifth biennial conference of the society for community research in action,* Chicago, IL.

Gresham, F. M., & Elliott, S. N. (1990). *Social skills rating system manual*. Circle Pines: American Guidance Service.

Griffin, K., Botvin, G., Scheier, L., & Nichols, T. (2002). Factors associated with regular marijuana use among high school student: A long-term follow-up study. *Substance Use and Misuse, 37*, 225–238.

Hirschi, T. (1969). *Causes of delinquency*. Berkeley: University of California Press.

Hoffman, J. P. (2002). The community context of family structure and adolescent drug use. *Journal of Marriage and the Family, 64*, 314–330.

Hoffman, B. R., Sussman, S., Unger, J. B., & Valente, T. W. (2006). Peer influences in adolescent cigarette smoking: A theoretical review of the literature. *Substance Use and Misuse, 41*, 103–155.

Hoffman, B. R., Monge, P. R., Chou, C. P., & Valente, T. W. (2007). Perceived peer influence and peer selection on adolescent smoking. *Addictive Behaviors, 32*, 1546–1554.

Horton, G. (2007). Racial differences in the effects of age of onset on alcohol consumption and development of alcohol-related problems among males from mid-adolescence to young adulthood. *Journal of Ethnicity in Substance Abuse, 6*, 1–13.

Huizinga, D., Loeber, R., & Thornberry, T. (1993). Longitudinal study of delinquency, drug use, sexual activity, and pregnancy among children and youth in three cities. *Public Health Reports, 108*, 90–96.

Jessor, R. (1992). Risk Behavior in adolescence: A psychosocial framework for understanding and action. In D. E. Rogers & E. Ginzburg (Eds.), *Adolescents at risk: Medical-social perspectives* (pp. 19–34). Boulder: Westview Press.

Jessor, R., & Jessor, S. L. (1977). *Problem behavior and psychosocial development: A longitudinal study of youth*. New York: Academic Press.

Johnson, J. H., & McCutcheon, S. (1982). Assessing life events in older children and adolescents: Preliminary findings with the life events checklist. In I. Sarason & C. Spielberger (Eds.), *Stress and anxiety*. New York: Norton.

Johnson, S. D., Striley, C., & Cottler, L. B. (2006). The association of substance use disorders with trauma exposure and PTSD among African American drug users. *Addictive Behaviors, 31*, 2063–2073.

Johnston, L. D., O'Malley, P. M., Bachman, J. G., & Schulenberg, J. E. (2009). *Monitoring the Future national results on adolescent drug use: Overview of key findings, 2008* (NIH Publication No. 09–7401). Bethesda: National Institute on Drug Abuse.

Kandel, D. B., & Andrews, K. (1987). Processes of adolescent socialization by parents and peers. *Substance Use & Misuse, 22*, 319–342.

Kann, L., Warren, C. W., Harris, W. A., Collins, J. L., Williams, B. I., Ross, J. G., & Kolbe, L. J. (1996). Youth risk behavior surveillance B United States, 1995. *Morbidity and Mortality Weekly Report, 45*, 1–26.

Kilpatrick, D. G., Acierno, R., Saunders, B., Resnick, H. S., Best, C. L., & Schnurr, P. P. (2000). Risk factors for adolescent substance abuse and dependence: Data from a national sample. *Journal of Consulting and Clinical Psychology, 68*, 19–30.

Knibbe, R. A., Joosten, J., Derrickx, M., Choquet M., Morin D., Monshouwer K., & Vollebergh W. (2005). Perceived availability of substances, substance use and substance-related problems: a cross-national study among French and Dutch adolescents. *Journal of Substance Use*, Special issue: A nationwide Dutch addiction program, *10*, 151–136.

Kumpfer, K. L., & Turner, C. W. (1990). The social ecology model of adolescent substance abuse: Implications for prevention. *The International Journal of the Addictions, 25*, 435–463.

Lambert, S. F., Brown, T. L., Phillips, C. M., & Ialongo, N. S. (2004). The relationship between perceptions of neighborhood characteristics and substance use among urban African American adolescents. *American Journal of Community Psychology, 34*, 205–218.

LaVeist, T. A., & Wallace, J. (2000). Health risk and inequitable distribution of liquor stores in African American neighborhoods. *Social Science and Medicine, 51*, 613–617.

Lindenberg, C. S., Reiskin, H. K., & Gendrop, S. C. (1994). The social stress model of substance abuse among childbearing-age women: A review of the literature. *Journal of Drug Education, 24*, 253–268.

Loeber, R., & Dishion, T. (1983). Early predictors of male delinquency: A review. *Psychological Bulletin, 94*, 68–99.

Massey, D., & Shibuya, K. (1995). Unraveling the tangle of pathology: The effect of spatially concentrated joblessness on the well-being of African Americans. *Social Science Research, 24*, 352–366.

McDonnell, J. R. (2007). Neighborhood characteristics, parenting, and children's safety. *Social Indicators Research, 83*, 177–199.

Moon, D. G., Hecht, M. L., Jackson, K. M., & Spellers, R. E. (1999). Ethnic and gender differences and similarities in adolescent drug use and refusal of drug use. *Substance Use & Misuse, 34*, 1059–1083.

Office of Applied Studies. (2004). *Results from the 2003 National Survey on Drug Use and Health: National Findings* (Office of Applied Studies, NSDUH Series H-25, DHHS Publication No. SMA 04–3964). Rockville: Substance Abuse and Mental Health Services Administration.

Overstreet, S. (2000). Exposure to community violence: Defining the problem and understanding the consequences. *Journal of Child and Family Studies, 9*, 7–25.

Patterson, J. M., & McCubbin, H. I. (1987). Adolescent coping style and behaviors: Conceptualization and measurement. *Journal of Adolescence, 10*, 163–186.

Pitkanen, T., Lyyra, A., & Pulkkinen, L. (2005). Age of onset of drinking and the use of alcohol in adulthood: A follow-up study from age 8–42 for females and males. *Addiction, 100*, 652–661.

Psychological Corporation. (2001). *Wechsler individual achievement test screener*. San Antonio: Author.

Raudenbush, S., Johnson, C., & Sampson, R. J. (2003). A multivariate, multilevel rasch model with application to self-reported criminal behavior. *Sociological Methodology, 29*, 1–41.

Reardon, S., & Buka, S. (2002). Differences in onset and persistence of substance abuse and dependence among whites, African Americans, and Hispanics. *Public Health Reports, 117*, 51–59.

Sampson, R. J., & Raudenbush, S. W. (1999). Systematic social observation of public spaces: A new look at disorder in urban neighborhoods. *American Journal of Sociology, 105*, 603–651.
Sampson, R. J., & Raudenbush, S. W. (2005). Neighborhood stigma and the perception of disorder. *Focus, 24*, 7–11.
Sampson, R. J., Raudenbush, S. W., & Earls, F. (1997). Neighborhoods and violent crime: A multilevel study of collective efficacy. *Science, 277*, 918–924.
Sanders-Phillips, K. (2003). *Social and psychological factors influencing drug use in adolescents. Preliminary findings*. Unpublished raw data & preliminary findings.
Saxe, L., Kadushin, C., Beveridge, A., Livert, D., Tighe, E., Rindskopf, D., Ford, J., & Brodsky, A. (2001). The visibility of illicit drugs: Implications for community-based drug control strategies. *American Journal of Public Health, 91*, 1987–1994.
Schaps, E., & Soloman, D. (2003). The role of the school's social environment in preventing student drug use. *The Journal of Primary Prevention, 23*, 299–328.
Scheier, L. M., Botvin, G. J., & Miller, N. L. (1999). Life events, neighborhood stress, psychosocial functioning, and alcohol use among urban minority youth. *Journal of Child and Adolescent Substance Abuse, 9*, 19–50.
Schwab-Stone, M. E., Ayers, T. S., Kasprow, W., Voyce, C., Barone, C., Shriver, T., & Wiessberg, R. P. (1995). No safe haven: A study of violence exposure in an urban community. *Journal of the American Academy of Child and Adolescent Psychiatry, 34*, 1343–1352.
Schwarzer, R., & Schwarzer, C. (1996). A critical survey of coping instruments. In M. Zeidner & N. Endler (Eds.), *Handbook of coping: Theory, research, applications* (pp. 107–132). New York: Wiley.
Selner-O'Hagan, M. B., Kindlon, D. J., Buka, S. L., Raudenbush, S. W., & Earls, F. J. (1998). Assessing exposure to violence in urban youth. *Journal of Child Psychology and Psychiatry, and Allied Disciplines, 39*, 215–224.
Simmons-Morton, B. G. (2002). Prospective analysis of peer and parent influences on smoking initiation among early adolescents. *Prevention Science, 3*, 275–283.
Simons-Morton, B., Haynie, D. L., Crump, A. D., & Saylor, K. E. (2001). Peer and parent influences on smoking and drinking among early adolescents. *Health education & behavior : the official publication of the Society for Public Health Education, 28*, 95–107.
Simons-Morton, B. G., Chen, R., Abroms, R., & Haynie, D. L. (2004). Latent growth curve analyses of peer and parent influences on smoking progression among early adolescents. *Health Psychology, 23*, 612–621.
Slaby, R. G., & Guerra, N. G. (1988). Cognitive mediators of aggression in adolescent offenders: I. Assessment. *Developmental Psychology, 24*, 580–588.
Sobeck, J., Abbey, A., Agius, E., Clinton, M., & Harrison, K. (2000). Predicting early adolescent substance use: Do risk factors differ depending on age of onset? *Journal of Substance Abuse, 11*, 89–102.
Spooner, C. (1999). Causes and correlates of adolescent drug abuse and implications for treatment. *Drug and Alcohol Review, 18*, 453–475.
Stanton, B., Li, X., Pack, R., Cottrell, L., Harris, C., & Burns, J. M. (2002). Longitudinal influence of perceptions of peer and parental factors on African American adolescent risk involvement. *Journal of Urban Health, 79*, 536–548.
Straus, M. A. (1979). Measuring intrafamily conflict and violence: The conflict tactics (CT) scales. *Journal of Marriage and the Family, 41*, 75–88.
Straus, M. A. (1987). *Is child abuse increasing? Evidence from the National Family Violence Resurvey*. Paper presented at the conference on child abuse, American Association for Protecting Children, Austin, Texas.
Straus, M. A. (1988). *Measuring psychological and physical abuse of children with the conflict tactics scale*. Durham: Family Research Laboratory, University of NH.
Straus, M. A. (1990). The conflict tactics scale and its critics: An evaluation and new data on validity and reliability. In M. A. Straus & R. J. Gelles (Eds.), *Physical violence in American fami-*

lies: Risk factors and adaptations to violence in 8, 145 families (pp. 75–91). New Brunswick: Transaction Publishing.

Trimboli, L., & Coumerlos, C. (1998). Cannabis and crime: Treatment programs for adolescent cannabis use. *Crime and Justice Bulletin, 41*, 1–16.

Vega, W. A., Zimmerman, R. S., Warheit, G. J., Apospori, E., & Gil, A. G. (1993). Risk factors for early adolescent drug use in four ethnic and racial groups. *American Journal of Public Health, 83*, 185–189.

Wagner, E. F., Myers, M. G., & McIninch, J. L. (1999). Stress-coping and temptation-coping as predictors of adolescent substance use. *Addictive Behaviors, 24*, 769–779.

Wallace, J. M. (1999). Explaining race differences in adolescent and young adult drug use: The role of racialized social systems. *Drugs & Society, 14*, 21–36.

Wallace, J., & Muroff, J. (2002). Preventing substance abuse among African American children and youth: Race differences in risk for exposure and vulnerability. *The Journal of Primary Prevention, 22*, 235–261.

Wallace, J. M., Forman, T. A., Guthrie, B. J., Bachman, J. G., O'Malley, P. M., & Johnston, L. D. (1999). The epidemiology of alcohol, tobacco, and other drug use among Youth. *Journal of Studies on Alcohol, 60*, 800–809.

Wang, J., Simmons-Morton, B. G., Farhart, T., & Luk, J. W. (2009). Socio-demographic variability in adolescent substance use: Mediation by parents and peers. *Prevention Science, 10*, 387–396.

Werthamer-Larsson, L., Kellam, S., & Wheeler, L. (1991). Effect of first-grade classroom environment on shy behavior, aggressive behavior, and concentration problems. *American Journal of Community Psychology, 19*, 585–602.

Wilcox, P. (2003). An ecological approach to understanding youth smoking trajectories: Problems and prospects. *Addiction, 98*(s1), 57–77.

Wills, T. A., Sandy, J. M., & Yaeger, A. M. (2001). Time perspective and early-onset substance use: A model based on stress–coping theory. *Psychology of Addictive Behaviors, 15*, 118–125.

Wilson, N., Syme, S. L., Boyce, W. T., & Battistich, V. A. (2005). Adolescent alcohol, tobacco and marijuana use: The influence of neighborhood disorder and hope. *American Journal of Health Promotion, 20*, 11–19.

Chapter 13
Trajectories of Substance Use: Academic Performance and Graduation Differentials Among African American, Hispanic, and White Secondary School Students

Marie-Claude Jipguep-Akhtar, Roderick J. Harrison, and Jennifer Goode

Introduction

In an increasingly competitive, global economy the consequences of dropping out of high school are devastating to individuals, communities and national economies. At an absolute minimum, adults need a high school diploma if they are to have any reasonable opportunities to earn a living wage (Orfield et al. 2004). Every year in the United States, about one million people who should graduate from high school don't, condemning them to a lifetime of low incomes and limited opportunities. Students who fail to graduate from high school are also less likely to have access to the country's political and social opportunities (Small 2006). Existing research suggests that there are marked racial/ethnic differences in academic performance and graduation differentials among secondary school students in the US, due in part to the inverse relationships between educational attainment and achievement, as measured by years and level of education and by standardized proficiency tests on one hand, and drug use, on the other hand (Fothergill et al. 2008; Agrawal et al. 2005; Fothergill and Ensminger 2006; Brier 1995; Jeynes 2002).

Researchers have proposed several explanations for these relationships. Some have argued that academic problems precede substance initiation; others have claimed that students initiate or accelerate substance use behaviors as a way to cope with anxiety over academic failure. Substance use varies with involvement in other problematic behaviors, such as violence, and conventional behaviors, including,

M.-C. Jipguep-Akhtar (✉)
Department of Sociology and Anthropology, Howard University, Washington, DC, USA
e-mail: Mjipguep-akhtar@howard.edu

R.J. Harrison
2M Research Services, Washington, DC, USA

J. Goode
U.S. Census Bureau, Washington, DC, USA

Y.F. Thomas, L.N. Price (eds.), *Drug Use Trajectories Among Minority Youth*,
DOI 10.1007/978-94-017-7491-8_13

school achievement. Other researchers have claimed that delinquency and drug use precede dropping out, and should be modeled as predictors of dropping out of school (Cox et al. 2007; Beman 1995; Brunswick and Messeri 1984; Donovan and Jessor 1978; Battin-Pearson et al. 2000; Mandell et al. 2002). Although existing recent research suggests that substance use plays a role in racial/ethnic differentials in academic performance and graduation rates, the evidence is mixed. According to Wallace (1999), as African American and white young people (e.g., secondary school students) make the transition into young adulthood, there is evidence that drug use declines significantly among white young adults while it continues to increase among African American young adults.

The purpose of this chapter is to conduct an initial investigation into whether lower academic performance is associated with higher rates of substance use and with lower gradation prospects for adolescents, and to identify which of these relationships might be stronger for African Americans and Hispanics than for Whites. The analysis explores the possible associations between early academic performance and the onset of substance use, and then the relationship between substance use and academic outcomes for the respondents. We also seek to identify the possible relationships that individual risk factors, neighborhood and peer influences, parenting style, and urban versus rural residence might have with substance use and educational outcomes. We begin to examine these relationships using the publically available 1997 National Longitudinal Survey of Youth (NLSY) and data from annual re-interviews of NLSY 2007.

Methods

Data

Data for this chapter come from the National Longitudinal Survey of Youth (NLSY), which in 1997, interviewed a nationally representative sample of 8984 respondents, who were 12–16 years old as of December 31, 1996 (NLS Web Investigator 2009). Interviewers screened 75,291 households in 147 primary sampling units (comprised of metropolitan areas, counties, or groups of counties) to identify 6819 households with eligible respondents; 1862 households included more than one respondent. The respondents were drawn from two samples: a nationally representative sample of 6748 respondents, and a supplemental sample comprised of 1354 African Americans and 980 Hispanics. The survey, sponsored by the Bureau of Labor Statistics, conducted 1 h of Computer Assisted Personal Interviews (CAPI) with each of the sampled youths and one of the youth's parents in 1997. The youths have been re-interviewed annually since then, with attrition reducing the sample to 7418 in 2007, for an 82.6 % retention rate (Whitney et al. 2000). The data from the 1997 through 2007 interviews is publically available online (NLS Web Investigator 2009). Due to both the supplemental sample and attrition of the sample over time, custom weights must be downloaded from National Web Investigator website (NLS Web Investigator 2009) and merged into data extracted from the site.

The NLSY was designed by the Bureau of Labor Statistics to follow the transition of the youths from adolescence and school to adulthood and labor market experiences. The NLSY therefore contains very detailed data on the educational and employment histories of the respondents, and also extensive measures of family, neighborhood, and school characteristics that are thought to affect the transitions of interest. Given this, the 1997 NLSY provides a rich source of data for initial explorations of relationships that might hold between substance use and academic performance amongst white, African American, and Hispanic students. The initial interviews conducted in 1997 asked respondents about their grades in 8th grade, and whether they had ever used alcohol or marijuana. In each subsequent year, respondents were asked if they had used alcohol, marijuana, or illegal substances since their last interview, and in the past 30 days, and if so, how frequently. The 1997 NLSY also included several questions on whether there were gangs in the respondent's neighborhood or school, and on the percentage of his/her peers who smoked cigarettes, or used alcohol or marijuana. Each year's interview also had an extensive set of questions on family and parental relationships, including, in the 1997 interview a question on the parenting style of the mother and father with whom each respondent resided. In addition to very detailed information on the courses, grades, and proficiency test scores of the each respondent in each year, educational outcomes were also measured for each respondent, including whether or not they completed high school with a diploma or GED, and for those who attended or completed college extensive detail on their coursework and grades.

This exploratory study of the NLSY examines only a few of the hundreds of variables that could eventually be used to trace, in detail, how academic performance and substance use might affect one another and the longer range academic and employment trajectories of the respondents. The small set of variables that we examine in this paper nevertheless succeeds in sketching sufficiently strong relationships between academic performance and substance use among the respondents, and sufficiently distinct differences between White, African American and Hispanic youth to suggest the value of further research on the relationships and the considerably different trajectories on which they seem to launch student from these populations.

Concepts and Variables

We wished to explore possible associations between *early academic performance* and the onset of substance use, and then the relationship between substance use and academic outcomes for the respondents. We also sought to identify the possible relationships that individual risk factors, neighborhood and peer influences, parenting style, and urban versus rural residence might have with substance use and academic outcomes. The 1997 NLSY offered several straightforward measures for building simple, initial models for examining these relationships. The 1997 NLSY interview asked respondents if they had ever used marijuana. About 60 % of respondents (61.3 %) indicated that they had, including 63.8 % of Whites, 55.7 % of African Americans, and 58.2 % of Hispanics. This variable seemed well suited for an initial examination of correlates of substance use.

Three variables were selected to represent academic and school-behavior risk factors: (1) grades before the 8th grade, with values ranging from C's and D's, to mostly A's or skipped a grade, and a category for those not yet in the 8th grade; (2) Whether the respondent ever repeated a grade prior to 1997, with yes and no response options; and (3) whether the respondent has been suspended from school (yes or no). Whether the respondent had smoked or drank alcohol were used as measures of substance-use related personal risk factors, and whether the respondent had a gang in the neighborhood or school was selected to represent a neighborhood risk factor. Several peer group indicators were selected for examination, including the percent of peers who: belong to a gang, use illegal drugs, cut class or school, smoke, or have one or more drinks per month. Parental influence was represented by the residential mother's parenting style, with response categories of authoritative, uninvolved, permissive and authoritarian.

The 1997 NLYS also included a variable identifying the respondent's area of residence in 1997 as urban or rural residence, but residence was not known in a small number (380) of the cases. A race-ethnicity variable was constructed from the responses to the race and Hispanic origin questions. Respondents who did not report as Hispanic on the Hispanic origin question were classified in the race category they reported. The sample included too few American Indians and Alaskan Natives, Asians and Pacific Islanders, and those who reported *Something Else* to support separate analyses. They were collapsed into a category for *Other*, and results for this group are not discussed in this paper.

Several other indicators of personal and neighborhood risk factors and of parental influence were also examined, but had to be dropped from the logistic regressions either because they created too many missing cases in the regression (e.g., % of peers who have had sexual intercourse, residential father's parenting style), or because the equations failed to converge when they were included (e.g., indexes for delinquent behavior and school behavioral problems).

First, we observed that marijuana use within the past 30 days (current use) does exhibit a cross-over effect in the NLSY97 sample (see Fig. 13.1). In 1997, about 44 % of white and of Hispanic youths reported that they had used marijuana in the past month, compared to only 38 % of African Americans. By 1999, however, the percentage of African Americans' using marijuana had grown comparable to the percentage of Whites at about 69 %. In each subsequent year through 2006, the percentage of African American current users grew more rapidly than the percentage for Whites, creating a growing gap in current usage.

In 2006, about 84 % of African Americans and 77 % of Whites had used marijuana in the past 30 days. These percentages declined to 80 % of African Americans and 73 % of Whites in 2007, as most in each group had transitioned from adolescence to young adulthood. The percentage of current users among Hispanics

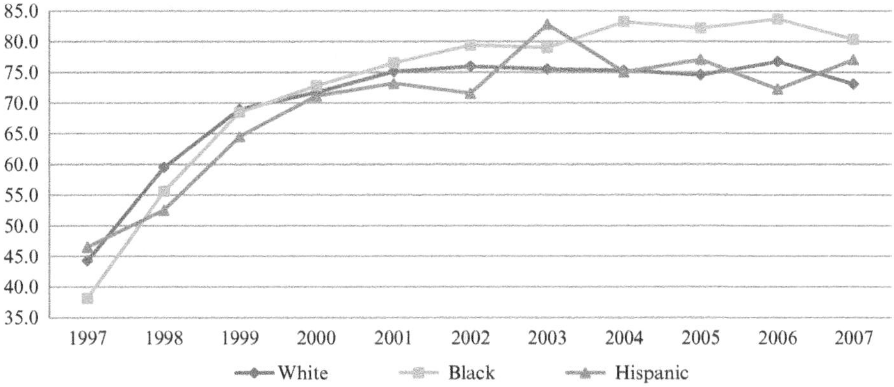

Fig. 13.1 Percent of White, African American and Hispanic students sampled in 1997 who used Marijuana in past 30 days of interview

fluctuates in ways that suggest sampling variability, but a 3 year moving average would suggest a trajectory closer to that of Whites than to that of African Americans throughout the period.

Results

This chapter explored the relationships between *academic performance*; *personal, peer and neighborhood resources; urban-rural residence*; *parenting style*; and *race-ethnicity* and *substance use* in the initial NLSY1997 interview, and then the associations between these variables, subsequent use of marijuana or illegal substances between 1998 and 2007, and failure to complete high school with a diploma. Having used marijuana before 1997 and not completing high school with a diploma are the dependent variables in the logistic regressions presented. We consider it critical to emphasize that in this initial exploration of these relationships, we have not attempted to identify the temporal order of the independent and dependent variables in ways that would support causal interpretations.

Therefore, we can determine whether, for example, lower grades prior to the 8th grade is correlated or associated with use of marijuana prior to the 1997 interview, or whether use of illegal substances in prior years is associated with dropping out. However, since we do not know when a respondent used marijuana prior to the 1997 interview, we cannot determine whether poorer academic performance increased the likelihood of trying or using marijuana, or if use of marijuana or other substances, or if unmeasured factors associated with both led to lower grades in the years before the eighth grade. Similarly, although we know that the risk factors measured in 1997 are temporally prior to completing high school or dropping out, we do not know when the respondent left or completed school, and hence cannot know whether illegal substance use in a given year preceded or followed that event as an "outcome."

Use of Marijuana Prior to 1997 Interview

To explore how having ever used marijuana prior to the initial interview in 1997 might be related to academic performance, school behavior, and selected peer, parental and neighborhood factors, we estimated logistic regression models of these characteristics upon having ever used marijuana as the dependent variable. We estimated a model for the total population of respondents, which included a race-ethnicity variable among the independent variables, and then separate models for Whites, African Americans, and Hispanics to examine differences in the relationships amongst the variables for these groups.

The pseudo-R^2's for the regressions indicate that the models explain substantial portions of the variance in use of marijuana prior to the 1997 interview (see Table 13.1). The Cox and Snell R^2 for all students was .378, and the Nagelkerke R^2 .593. These measures were higher for Hispanics (a Cox and Snell R^2 of .420 and a Nagelkerke R^2 of .638) and for Whites (.395 and .614, respectively), and were lower for African Americans (.288 and .495). This suggests that together, the associations of academic performance and school behavior measures, individual, neighborhood and peer risk factors, parenting style, and urban and rural residence with use of marijuana prior to 1997 were somewhat stronger for Hispanics and Whites than for African Americans. The results for each of these factors are presented below.

Academic Performance and School Behavior

The results of the logistic regression for all students suggest that academic performance and use of marijuana prior to the 1997 interview were related. Respondents who had been suspended from school before 1997 were more than twice as likely (Exp(B)=2.16) as students with no suspensions to also have used marijuana. The relationship between suspension from school and marijuana use prior to 1997 was substantially stronger for Whites (2.24), and especially for Hispanics (3.07), than for African Americans (1.51). However, respondents who repeated grade prior to

Table 13.1 Logistic regression fit statistics for Marijuana use prior to the 1997 NLSY interview

	−2 log likelihood	Cox & Snell R square	Nagelkerke R square
All respondents	7323618.57	.378	.593
Whites	4965580.95	.395	.614
African Americans	1199916.26	.299	.497
Hispanics	845798.616	.420	.638

1997 were about 26 % more likely (Exp(B) = 1.26) to have also used marijuana before 1997, but this association was far stronger for African Americans (1.67) than for Whites (1.19) or Hispanics (1.04). In comparison with suspension from school and repeating a grade prior to the 1997 interview, one's grades before the 8th grade were weakly related to the likelihood of having used marijuana. Respondents with grades of C's and D's were only about 6 % more likely than those who skipped a grade or had mostly A's to have used marijuana prior to 1997. Those with grades between half B's and C's and half A's and B's were between 17 and 27 % less likely to have used marijuana.

There were very important differences across the race-ethnic groups, however, in the relationship between grades and use of marijuana before 1997. Grades before the 8th grade had no relationship at all to use of marijuana before 1997 among Whites. In notable contrast, African Americans and Hispanics with C's and D's before 8th grade were, respectively 36 % (Exp(B) = 1.36) and 49 % (Exp(B) = 1.49) more likely to have also used marijuana before the 1997 interview. Both African Americans and Hispanics with half A's and B's were far less likely (Exp(B) = 0.49 and 0.60) than those who had skipped a grade or earned A's to have used marijuana. Hispanics with half B's and C's (Exp(B) = 1.19) and mostly B's (1.07) were respectively 19 % and 7 % more likely to have used marijuana than those with A's, but African Americans in these grade ranges were about 15 % less likely (Exp(B) = 0.85 and 0.84) to have done so. The results thus suggest that lower academic performance, as measured by having repeated a grade before 1997 and having received C's and D's (and for Hispanics, grades below mostly B's) is more strongly related to use of marijuana prior to the 1997 interview among African American and Hispanic than among white respondents. In contrast, school behavior that led to a suspension before 1997 is more highly associated with marijuana use among Whites than among African Americans or Hispanics (Table 13.2).

Substance-Related Personal Risk Factors

Students who had ever smoked or drank alcohol were several times more likely to also have smoked marijuana, and these represented by far the strongest relationships in the model. Among all respondents, students who had smoked tobacco before the 1997 interview were over 12 times more likely to have also smoked marijuana (Exp(B) = 12.56), and those who had drank alcohol were nearly seven times more likely (Exp(B) = 6.91) to have also used marijuana before the 1997 interview.

The relationship between having smoked cigarettes and having used marijuana prior to the 1997 interview was even higher for Whites and Hispanics (Exp(B) = 18.06 and 15.61, respectively), but was considerably weaker for African Americans (Exp(B) = 5.61). The associations between tobacco and marijuana use, and also for alcohol and marijuana use prior to the 1997 interview (Exp(B) = 5.70) were nevertheless still by far the largest for African Americans. The association between alcohol and marijuana use was comparable for Hispanics (Exp(B) = 5.95), and that for Whites somewhat higher (Exp(B) = 7.63).

Table 13.2 Logistic regression on ever used Marijuana before 1997

	All students				Whites				African Americans				Hispanics			
	B	S.E.	Sig.	Exp(B)	B	S.E.	Sig.	Exp(B)	B	S.E.	Sig.	Exp(B)	B	S.E.	Sig.	Exp(B)
Academic performance and school behavior																
Grades before 8th grade																
Mostly A's or skipped a grade			.000		−.010	.001	.000	.990			.000				.000	
Not Yet 8th grade	−.236	.005	.000	.790					.010	.010	.356	1.010	−.797	.015	.000	.451
C's and D's	.057	.003	.000	1.058					.306	.008	.000	1.358	.402	.009	.000	1.494
Half B's and C's	−.186	.003	.000	.831					−.163	.008	.000	.850	.173	.009	.000	1.188
Mostly B's	−.212	.003	.000	.809					−.175	.009	.000	.840	.070	.011	.000	1.073
Half A's and B's	−.321	.004	.000	.725					−.724	.014	.000	.485	−.519	.013	.000	.595
Ever repeat a grade prior to 1997	.233	.002	.000	1.262	.177	.003	.000	1.194	.515	.004	.000	1.674	.040	.006	.000	1.041
R has been Suspended from school	.770	.002	.000	2.160	.806	.003	.000	2.240	.411	.005	.000	1.508	1.121	.006	.000	3.069
Substance use related personal risk																
R has smoked	2.531	.003	.000	12.562	2.894	.004	.000	18.063	1.725	.005	.000	5.612	2.748	.007	.000	15.614
R has drank alcohol	1.933	.003	.000	6.910	2.032	.003	.000	7.633	1.740	.005	.000	5.696	1.784	.007	.000	5.954
Neighborhood, school, and peer risk factors																
% of peers who belong to a gang	−.147	.001	.000	.863	−.205	.002	.000	.815	−.078	.002	.000	.925	−.133	.003	.000	.876
% of peers who use illegal drugs	.703	.001	.000	2.019	.724	.001	.000	2.062	.567	.002	.000	1.763	.849	.003	.000	2.337
% of peers who cut class/school	−.089	.001	.000	.915	−.085	.001	.000	.919	−.068	.002	.000	.935	−.202	.003	.000	.817

% of peers who smoke	–.196	.001	.000	.822	–.231	.001	.000	.794	–.036	.002	.000	.965	–.184	.003	.000	.832
% of peers who drink 1+/month	.187	.001	.000	1.205	.257	.001	.000	1.293	.073	.002	.000	1.076	–.040	.003	.000	.960
R has gang in neighborhood or school	.272	.002	.000	1.312	.248	.002	.000	1.281	.472	.005	.000	1.603	.538	.007	.000	1.712
Parental influence: residential mother's parenting style																
Authoritative			.000				.000				.000				.000	
Uninvolved	.343	.003	.000	1.410	.289	.004	.000	1.335	.358	.008	.000	1.430	.788	.009	.000	2.198
Permissive	.103	.002	.000	1.108	.010	.003	.000	1.010	.442	.006	.000	1.556	.001	.007	.857	1.001
Authoritarian	.226	.003	.000	1.254	.280	.004	.000	1.323	.311	.007	.000	1.365	.284	.008	.000	1.328
Residence																
Rural			.000				.000				.000				.000	
Urban	.252	.002	.000	1.286	.240	.002	.000	1.272	.503	.007	.000	1.653	–.468	.010	.000	.626
Not known	–.212	.005	.000	.809	–.547	.006	.000	.579	.239	.013	.000	1.269	.558	.017	.000	1.747
Race/ethnicity																
White			.000													
African American	.026	.003	.000	1.027												
Hispanic	.082	.003	.000	1.085												
Other	–.404	.006	.000	.667												

Neighborhood and Peer Risk Factors

The Cox and Snell R^2 (.395) and the Nagelkerke R^2 (.611) suggest moderately strong relationships between grades, school, peer and neighborhood risk factors, and ever having used marijuana. Over 25 % of the cases were list-wise deleted from the regression, however, and models with such additional variables of interest as parental education and income, and other neighborhood characteristics could not be estimated.

Having gangs in the neighborhood (Exp(B) = 1.31), and having higher percentages of one's peers who used illegal drugs (Exp(B) = 2.02) or who had gotten drunk 1 or more times in the past month (Exp(B) = 1.21) were all associated with higher likelihoods of having used marijuana before the 1997 interview. On the other hand, having higher percentages of one's peers who belonged to a gang, who cut classes or school, or who smoked cigarettes were each modestly associated with lower likelihoods of having used marijuana prior to 1997 (Exp(B) = 0.86, 0.92, and 0.82, respectively).

For Whites, the associations with marijuana use prior to 1997 of having a gang in the neighborhood (Exp(B) = 1.28), and having higher percentages of peers who used illegal substances (Exp(B) = 2.06) or got drunk one or more times in the past month (Exp(B) = 1.29), and so were the associations with the percentages of peers who belonged to gangs, cut classes or school, or smoked cigarettes Exp(B) = 0.82, 0.92, and 0.79, respectively. For African Americans, however, the relationship between using marijuana prior to 1997 and having a gang in the neighborhood was stronger than for Whites and all respondents (Exp(B) = 1.60), while those involving the percentages of peers in gangs (Exp(B) = 1.76) and peers who had gotten drunk once or more in the past month (Exp(B) = 1.08), were weaker. The associations of the percentages of peers who were in gangs, who cut classes or school, or who smoked cigarettes with lower likelihoods of having used marijuana were also weaker.

Having a gang in the neighborhood (Exp(B) = 1.71) and higher percentages of peers who used illegal substances (Exp(B) = 2.34) were more closely associated with marijuana use prior to 1997 for Hispanics than for the other groups, but having higher percentages of peers who had gotten drunk once or more in the past month was associated with a slightly lower likelihood of having used marijuana (Exp(B) = 0.96). As with the other groups, Hispanic respondents with higher percentages of peers who belonged to a gang, who cut classes or school, or who smoked cigarettes were less likely to have used marijuana prior to 1997 (Exp(B) > 0.81 and < 0.88).

Parenting Style

Among all respondents, compared to those with mother's with authoritative parenting styles, those with mothers who were uninvolved were 41 % (Exp(B) = 1.41) more likely to have used marijuana prior to the 1997 interview. Children of authoritarian mothers were about 25 % more likely to have used marijuana (Exp(B) = 1.25) than children of authoritative mothers, while those with permissive mothers were about ten percent more likely (Exp(B) = 1.11). Having uninvolved mothers had similar relationships with prior marijuana use by white (Exp(B) = 1.34) and African American respondents (Exp(B) = 1.43), but was much more strongly associated with prior marijuana use among Hispanics (Exp(B) = 2.20). On the other hand, African Americans with permissive mothers (Exp(B) = 1.56) were considerably more likely to have used marijuana before the 1997 interview than comparable Whites and Hispanics (Exp(B) = 1.01 and 1.00, respectively). For all three groups, respondents with authoritarian mothers were about 32–37 % more likely to have used marijuana before 1997 than those with authoritative mothers.

Urban-Rural Residence

After controlling for all the factors discussed above, urban respondents were nearly 30 % more likely (Exp(B) = 1.29) than rural respondents to have used marijuana prior to the 1997 interview. The association was similar for Whites (Exp(B) = 1.27), but stronger for African Americans (Exp(B) = 1.65), and urban Hispanics were about 37 % less likely (Exp(B) = 0.63) than rural Hispanics to have used marijuana before the 1997 interview.

Race-Ethnicity

Race-ethnicity had very small relationships with marijuana use prior to the 1997 interview after controlling for all the factors examined above. African Americans were only 3 % more likely and Hispanics 9 % more likely (Exp (B) = 1.03 and 1.09, respectively) than Whites to have used marijuana prior to the 1997 interview. Since a higher percentage of white respondents than of African American and Hispanic respondents had used marijuana prior to the 1997, the small positive association of being African American or Hispanic with prior use suggests that the factors examined in the regression account for most of the racial-ethnic differences, and African Americans and Hispanics who are comparable to Whites on these factors are actually slightly more likely to have used marijuana prior to the 1997 interview.

Leaving High School Without a Diploma

The logistic regression on respondents' use of marijuana before the 1997 interview indicated that substance-related personal risk factors – having smoked cigarettes or used alcohol – were by far the strongest correlates for all groups, but that the relationships, especially for smoking cigarettes, were far weaker for African Americans than for Whites or Hispanics. Academic performance, including having repeated a grade before 1997 and getting C's and D's had solid associations with use of marijuana prior to 1997 for African Americans and Hispanics, but grades were not related to prior use by Whites. On the other hand, having been suspended from school before 1997 was more strongly associated with prior use of marijuana by Whites than by African Americans or Hispanics. Having gangs in the neighborhood and having higher percentages of peers who used illegal drugs were strongly associated with greater likelihoods of prior marijuana use in all groups, but the relationships were strongest for Hispanics. Uninvolved and authoritarian parenting styles were also associated with use of marijuana prior to the 1997 interview, with the largest relationship holding among Hispanic respondents with uninvolved mothers. Marijuana use prior to the 1997 interview was higher among urban than rural Whites and African Americans, but lower amongst Hispanics.

The results strongly support the conjecture that in addition to substance-related personal risk factors, academic performance, neighborhood and peer risk factors, parenting styles, and urban-rural residence are all associated with marijuana use before the 1997 interview, and that many of these factors are more strongly associated with marijuana use among African Americans and Hispanics than among Whites. We next sought to examine whether these same factors, and substance use measured in the interview years following 1997 might also have relationships to such educational outcomes as leaving high school without a diploma.

To explore this, we finally estimated a logistic regression of the effects of the variables in the first model and then of subsequent substance use upon the likelihood of dropping out. The dependent variable distinguished respondents who earned a high school diploma with those who left school without one or earned a GED. In addition to the variables included in the first regression, we added a variable that identified the year in which respondents who had never before used marijuana or illegal substances first did so. This variable assigned the value of 1997 to those who reported using marijuana prior to the 1997 interview. Those who had not used marijuana in before 1997 but who reported in the 1998 interview that they had used marijuana or an illegal substance since the date of the last interview were assigned a value of 1998. Those who had still never used a substance in 1998, but who reported in the 1999 interview that they had used marijuana or an illegal substance since the 1998 interview were assigned a score of 1999, and so on (Table 13.3).

The pseudo-R^2's for the logistic regressions for leaving high school without a diploma were lower than those for having used marijuana before the 1997 interview, but were still reasonably strong. For Cox and Snell R^2 was .247 for all respondents

Table 13.3 Logistic regression fit statistics for Leaving High School without a diploma

	−2 log likelihood	Cox & Snell R square	Nagelkerke R square
All respondents	9.089E+06	.247	.403
Whites	5.532E+06	.239	.414
African Americans	1.866E+06	.253	.366
Hispanics	1.337E+06	.247	.367

and for Hispanics, .239 for Whites and .253 for African Americans. The Nagelkerke R^2 for all respondents was .403 and rose to .414 for Whites. It was about .366 for African American and Hispanics.

Academic Performance and School Behavior

While substance-related personal risk factors were by far the most highly correlated with marijuana use prior to the 1997 interview, the most important factors in dropping out are academic. The second strongest set of relationships, however, are for having used marijuana before 1997, or having initiated use of marijuana or illegal substances in years thereafter.

Respondents who repeated a grade after 1997 (Exp(B)=4.07), or who repeated more than one grade (Exp(B)=4.02), were about four times more likely to drop out of high school than those who did not repeat a grade. The relationships were similar for Whites (Exp(B)=3.98 and 3.70), but somewhat stronger for African Americans (Exp(B)=4.18 and 4.42), and weaker for Hispanics (Exp(B)=3.46 and 3.76). The association between dropping out and having repeated a grade before 1997 is substantial – about 35 % for all respondents – but much lower, as one might expect, than the association with repeating a more recent grade. The relationship of repeating a grade before 1997 with dropping out was similar for Whites, African Americans and Hispanics (Exp(B)>1.31 and <1.41) .

Grades before 8th grade had very strong relationships to dropping out. Those who were not yet in the 8th grade (Exp(B)=3.68), or who received C's and D's (Exp(B)=3.37), were more than three times more likely to drop out than those who received A's or skipped a grade. Respondents with half B's and C's (Exp(B)=2.14), and mostly B's (Exp(B)=1.26) were also substantially more likely to drop out than students with A's before the 8th grade. However, those with half A's and B's were less likely to drop out. (Exp(B)=0.36). For white respondents, the associations between each grade category and the likelihood of dropping out were similar to those of all respondents, but African Americans (Exp(B)=3.86) and Hispanics (Exp(B)=4.88) with C's and D's were at greater risk, and Hispanics with mostly B's had lower risks of dropping out. Having been suspended from school before 1997 was also strongly related to dropping out, increasing the odds more than 2.5 times for all respondents (Exp(B)=2.50) and for African Americans (Exp(B)=2.47), even more for Whites (Exp(B)=2.81), but only twofold for Hispanics (Exp(B)=1.97).

Substance-Related Personal Risk Factors

Having smoked tobacco before 1997 remains a risk factor that increases the likelihood of dropping out by 64 % (Exp(B)=1.64) for all respondents, and by 81 % (Exp(B)=1.82) for Whites. The association is weaker for African Americans (Exp(B)=1.43) and for Hispanics (Exp(B)=1.15). Having used alcohol before

1997 increases the likelihood of high school completion, however (Exp(B) = 0.724), though less for African Americans (Exp(B) = 0.94), and more for Hispanics (Exp(B) = 0.64).

Neighborhood and Peer Risk Factors

Respondents who have a gang in their neighborhood or school are about 25 % more likely to drop out than those who don't (Exp(B) = 1.25). The relationship is a bit weaker for African Americans (Exp(B) = 1.19) and stronger for Hispanics (Exp(B) = 1.44) than for Whites (Exp(B) = 1.28). Having higher percentages of peers who reported having smoked tobacco before 1997 was also associated with higher risks of dropping out (Exp(B) = 1.21), and the relationship did not vary greatly across race-ethnic groups (Exp(B) > 1.11 and < 1.31).

Parenting Style

The children of uninvolved mothers were nearly 60 % more likely to drop out of school (Exp(B) = 1.58) than children of authoritative parents. The relationship was slightly lower for African Americans (Exp(B) = 1.43), however. Having permissive or authoritarian mothers was associated with slightly higher risks of dropping out for white respondents (Exp(B) = 1.08), but with substantially higher risks for Hispanics (Exp(B) = 1.53 and 1.11). African Americans with permissive and authoritarian mothers were actually more likely to complete high school than those with authoritative mothers.

Urban-Rural Residence

After controlling for all the factors in the regression, urban respondents were about 40 % more likely to drop out than rural respondents (Exp(B) = 1.41), and the differential was higher for African Americans (Exp(B) = 1.80) than for Whites (Exp(B) = 1.34) or Hispanics (Exp(B) = 1.29).

Use of Marijuana or Illegal Substances Prior to and After 1997

Among the most important results of the regression are the very strong associations between dropping out and having used marijuana prior to the 1997 interview, or having initiated use of marijuana or illegal substances in the period since the last interview. Respondents who used marijuana before the 1997 interview or who initiated the use of marijuana or illegal drugs before the 2000 interview were 2.3 times as likely to drop out as those who had never used these substances. These associations were even higher for Whites (Exp(B) > 2.5 and < 2.8) up to 2000 and for

African Americans using marijuana or illegal substances by 1997 or 1998 (Exp(B)=2.69 and 2.81), or by 2000 and 2001 (Exp(B)=2.31 and 2.39) The associations were considerably lower for Hispanics (Exp(B)>1.1 and <1.9) through 2000, however. Initiating the use of marijuana or illegal drugs in 2001 or 2002 is associated with a 1.7–1.8 times higher likelihood of dropping out, and then falls to 1.13 in 2003 and 0.79 in 2004 before rising again to 1.33 in 2005 and the 1.8–1.9 range in 2006 and 2007.

The relationships also decline and rise for each race-ethnic group after 2000, but the patterns are far less consistent. The results suggest that are strong and consistent relationships between not completing a high school diploma and the use of marijuana or illegal substances prior to 1997 and up through the interview in 2000, and important, but less consistent relationships in the interview years following (Table 13.4).

Discussion and Conclusion

This chapter undertook a preliminary exploration of the possibility that low academic performance, substance-related risk factors, neighborhood and peer risk factors, uninvolved and authoritarian parenting styles and urban residence might be associated with higher likelihoods that respondents would have used marijuana prior to the first NLSY interview in 1997, and that use of marijuana in the and ensuing interview years might in turn be associated with poorer educational and employment outcomes, here represented by leaving high school without a diploma.

These two sets of possible relationships were examined using logistic regressions on use of marijuana prior to the 1997 interview, and on leaving school without a high school diploma. The results, though only exploratory, provide fairly consistent evidence that the factors examined have reasonably strong relationships with marijuana use prior to the 1997 interview and with the risks of dropping out or earning a GED instead of a diploma.

Substance-related personal risk factors have by far the strongest relationships with marijuana use prior to 1997, but the conjectured relationship of marijuana use to having repeated a grade, and low grades prior to the 8th grade were supported, especially for African American and Hispanic students. On the other hand, among Whites grades had no relationships to marijuana use prior to 1997, and having been suspended from school was more closely associated with marijuana use than it was for African Americans and Hispanics. Such neighborhood and peer effects as having a gang in the neighborhood, and the percent of peers who smoked cigarettes, used illegal substances or got drunk one or time times in the past month, also had important relationships to marijuana use prior to 1997, as conjectured. Relationships were also found with uninvolved, permissive, and authoritarian parenting styles by mothers, and except for Hispanics, after controlling for all other variables in the model, urban residents were more likely than rural to have used marijuana prior to 1997.

Table 13.4 Logistic regression on earning a HS diploma or more

	All students				Whites				African Americans				Hispanics			
	B	S.E.	Sig.	Exp(B)	B	S.E.	Sig.	Exp(B)	B	S.E.	Sig.	Exp(B)	B	S.E.	Sig.	Exp(B)
Academic performance and school behavior																
Grades before 8th grade																
Mostly A's or skipped a grade			.000				.000				.000				.000	
Not yet 8th grade	1.302	.003	.000	3.677	1.199	.004	.000	3.316	1.350	.008	.000	3.857	1.586	.009	.000	4.884
C's and D's	1.216	.003	.000	3.374	1.209	.004	.000	3.350	1.067	.007	.000	2.907	1.082	.008	.000	2.950
Half B's and C's	.763	.003	.000	2.144	.693	.004	.000	1.999	.893	.007	.000	2.442	.714	.008	.000	2.043
Mostly B's	.228	.004	.000	1.257	.296	.005	.000	1.344	.213	.009	.000	1.237	−.133	.010	.000	.875
Half A's and B's	−1.011	.006	.000	.364	−1.118	.007	.000	.327	.139	.013	.000	1.150	−.979	.017	.000	.376
Ever repeat a grade prior to 1997	.299	.002	.000	1.348	.340	.003	.000	1.405	.303	.004	.000	1.353	.268	.005	.000	1.307
R has been suspended from school	.916	.002	.000	2.500	1.032	.003	.000	2.808	.902	.004	.000	2.465	.679	.005	.000	1.971
Every repeated a grade							.000								.000	
Repeated one grade since 1997	1.405	.003	.000	4.074	1.381	.004	.000	3.979	1.429	.006	.000	4.176	1.242	.007	.000	3.462
Repeated 2/ more grades since 1997	1.390	.002	.000	4.015	1.308	.003	.000	3.698	1.486	.005	.000	4.421	1.324	.005	.000	3.758

Substance use related personal risk																
R has smoked	.492	.002	.000	1.635	.596	.003	.000	1.815	.355	.004	.000	1.426	.137	.006	.000	1.147
R has drank alcohol	−.323	.002	.000	.724	−.351	.003	.000	.704	−.063	.004	.000	.939	−.441	.006	.000	.644
Neighborhood, school, and peer risk factors																
% of peers who belong to a gang	.219	.001	.000	1.245	.249	.001	.000	1.283	.171	.002	.000	1.187	.361	.002	.000	1.435
% of peers who use illegal drugs	−.049	.001	.000	.952	−.023	.001	.000	.978	−.089	.002	.000	.915	−.144	.003	.000	.866
% of peers who cut class/school	−.004	.001	.000	.996	−.009	.001	.000	.991	−.066	.002	.000	.936	.044	.002	.000	1.045
% of peers who smoke	.191	.001	.000	1.211	.270	.001	.000	1.310	.111	.002	.000	1.118	.118	.002	.000	1.125
% of peers who drink 1+/month	−.120	.001	.000	.887	−.198	.001	.000	.820	.022	.002	.000	1.022	−.086	.003	.000	.918
R has gang in neighborhood or school	.074	.002	.000	1.077	.118	.002	.000	1.125	−.200	.004	.000	.819	.105	.005	.000	1.110
Parental influence: residential mother's parenting style																
Authoritative			.000				.000				.000				.000	
Uninvolved	.459	.003	.000	1.583	.519	.003	.000	1.680	.354	.006	.000	1.425	.445	.007	.000	1.560
Permissive	.006	.002	.003	1.006	.084	.003	.000	1.088	−.551	.004	.000	.576	.423	.005	.000	1.526
Authoritarian	−.049	.003	.000	.952	.076	.004	.000	1.079	−.279	.006	.000	.757	.105	.007	.000	1.111

(continued)

Table 13.4 (continued)

	All students				Whites				African Americans				Hispanics			
	B	S.E.	Sig.	Exp(B)	B	S.E.	Sig.	Exp(B)	B	S.E.	Sig.	Exp(B)	B	S.E.	Sig.	Exp(B)
Used marijuana or an illegal substance																
Never			.000				.000				.000				.000	
By 1997	.845	.003	.000	2.329	.925	.004	.000	2.521	.990	.006	.000	2.692	.657	.008	.000	1.929
By 1998	.842	.003	.000	2.321	.987	.004	.000	2.683	1.036	.007	.000	2.817	.320	.008	.000	1.377
By 1999	.822	.003	.000	2.275	1.029	.004	.000	2.797	.695	.007	.000	2.004	.442	.009	.000	1.556
By 2000	.838	.004	.000	2.312	1.028	.005	.000	2.795	.838	.007	.000	2.312	.128	.010	.000	1.137
By 2001	.526	.004	.000	1.693	.549	.006	.000	1.731	.951	.007	.000	2.589	−.037	.010	.000	.964
By 2002	.611	.005	.000	1.842	.962	.006	.000	2.617	.373	.010	.000	1.452	.259	.012	.000	1.296
By 2003	.115	.006	.000	1.122	.258	.008	.000	1.295	−.013	.012	.272	.987	−.801	.018	.000	.449
By 2004	−.233	.009	.000	.792	−.109	.012	.000	.897	.035	.019	.067	1.035	−.644	.021	.000	.525
By 2005	.286	.008	.000	1.331	.817	.010	.000	2.263	.112	.017	.000	1.119	−19.7	333.54	.953	.000
By 2006	.598	.010	.000	1.818	−.390	.018	.000	.677	.727	.018	.000	2.069	1.426	.021	.000	4.160
By 2007	.644	.008	.000	1.904	1.027	.011	.000	2.792	−.187	.018	.000	.829	.587	.017	.000	1.798
Residence			.000				.000				.000				.000	
Rural	−.018	.002	.000	.983	−.068	.002	.000	.934	.054	.005	.000	1.055	.332	.008	.000	1.393
Urban	.343	.004	.000	1.409	.292	.005	.000	1.339	.585	.009	.000	1.796	.258	.016	.000	1.294
Not known	−4.444	.004	.000	.012	−4.78	.005	.000	.008	−3.91	.010	.000	.020	−4.08	.012	.000	.017
Race/Ethnicity																
White			.000													
African American	.104	.002	.000	1.109												
Hispanic	.209	.003	.000	1.233												
Other	−.019	.005	.000	.981												

Measures of academic performance and school behavior were the most strongly associated with dropping out, but neighborhood, peer, and substance-related personal risk factors, parenting styles and urban residence were also associated with this outcome. The most important finding, however, was that use of marijuana prior to 1997, and initiating the use of marijuana or illegal substances in the years up to the 2000 interview had strong associations – second only to the academic and school behavior measures – with failing to earn a high school diploma.

These findings are only exploratory and preliminary. Far more detailed and complex analyses must be conducted to build and test more theoretically coherent models of substance use among NLSY respondents. An important priority of future analyses should be reconstructing marijuana and substance use by the respondent's age or grade instead of by interview date. This would permit, and would be essential to establishing, temporal or developmental and therefore causal order in the relationships between substance use and failure to complete high school, as well as a wide range of other educational and labor market outcomes measured in the NLSY.

The grades respondents received in each year of their schooling and their performance on proficiency tests could also be examined, allowing one to trace the effects of academic performance in a given year upon the use of substances in the next, and vice-versa. The NLSY also contains extensive batteries of measures on family relationships, problematic school behaviors, delinquency and involvement in the criminal justice system, and family and peer support systems, that can be used to identify and develop models incorporating the most influential protective and risk factors into the analysis. The important, if limited, initial contribution of the exploratory research presented here has been to confirm the merits and potential value of undertaking such future research.

References

Agrawal, A., Gardner, C. O., Prescott, C. A., & Kendler, K. S. (2005). The differential impact of risk factors on illicit drug involvement in females. *Social Psychiatry and Psychiatric Epidemiology, 40*, 454–466.

Battin-Pearson, S., Newcomb, M. D., Abbott, R. T., Hill, F. G., Catalano, R. F., & Hawkins, D. J. (2000). Predictors of early high school dropout: A test of five theories. *Journal of Educational Psychology, 92*(3), 568–582.

Beman, D. S. (1995). Risk factors leading to adolescent substance abuse. *Adolescence, 30*(117), 201–208.

Brier, N. (1995). Predicting antisocial behavior in youngsters displaying poor academic achievement: A review of risk factors. *Journal of Development and Behavioral Pediatrics, 16*, 271–276.

Brunswick, A. F., & Messeri, P. A. (1984). Origins of cigarette smoking in academic achievement, stress and social expectations: Does gender make a difference? *The Journal of Early Adolescence, 4*, 353–370.

Cox, R. G., Zhang, L., Johnson, W. D., & Bender, D. R. (2007). Academic performance and substance use: Findings from a state survey of public high school students. *Journal of School Health, 77*, 109–115.

Donovan, J. E., & Jessor, R. (1978). Adolescent problem drinking: Psychosocial correlate in a national sample study. *Journal Studies on Alcohol, 39*(9), 1506–1524.

Fothergill, K. E., & Ensminger, M. E. (2006). Childhood and adolescent antecedents of drug and alcohol problems: A longitudinal study. *Drug and Alcohol Dependence, 82*, 61–76.

Fothergill, K. E., Ensminger, M. E., Green, K. M., Crum, R. M., Robertson, J., & Juon, H.-S. (2008). The impact of early school behavior and educational achievement on adult drug use disorders: A prospective study. *Drug and Alcohol Dependence, 92*, 191–199.

Jeynes, K. H. (2002). The relationship between the consumption of various drugs by adolescents and their academic achievement. *American Journal of Drug and Alcohol Abuse, 28*(1), 15–35.

Mandell, D. J., Hill, S. L., Carter, L., & Brandon, R. N. (2002). *The impact of substance use and violence/delinquency on academic achievement for groups of middle and high school students in Washington.* Washington Kids Count Human Services Policy Center, Evans School of Public Affairs, University of Washington, Seattle, WA. [On-line].

NLS Web-Investigator. (2009). [On-line].

Orfield, G., Losen, D., Wald, J., & Swanson, C. (2004). *Losing our future: How minority youth are being left behind by the graduation rate crisis.* Cambridge, MA: The Civil Rights Project at Harvard University. Contributors: Advocates for Children of New York, The Civil Society Institute.

Small, C. (2006). *The children derailed from the train tracks: An analysis of African-American male academic achievement in the American K-12 school system.* M.P.A. thesis, Florida State University, Tallahassee, FL.

Wallace, J. M., Jr. (1999). Explaining race differences in adolescent and young drug use: The role of racialized social systems. *Drugs and Society, 14*(1/2), 21–36.

Whitney, M., Pedlow, S., Krishnamurty, P., & Wolter, K. (2000). *National longitudinal survey of youth 1997 (NLSY97) technical sampling report.* Chicago: NORC, 2000. [On-line].

Chapter 14
The Influence of Informal Social Control Processes on Drug Trajectories and Delinquent Behavior Among Mexican American Gang Members

Avelardo Valdez, Jarron M. Saint Onge, Alice Cepeda, and Charles Kaplan

Introduction

Mexican American youth represent one of the fastest growing subpopulations with persons of Mexican origin comprising approximately 65 % of the Hispanic population in the United States (U.S. Census Bureau 2008). Mexican Americans also represent a disproportionate percentage of both substance using and gang populations (Valdez and Sifaneck 2004). In part, this may be due to both a young Mexican American age composition that reflects immigration patterns and an age-graded proclivity towards adolescent substance use and gang membership. In addition, due to socioeconomic disadvantages, Mexican American youth may be at an elevated risk for compromised social bonds and subsequent delinquent behaviors. This chapter employs a life course perspective to examine the extent that adolescent informal social control processes mediate the relationship between gang memberships and adolescent delinquent behavior in a population of young Mexican Americans.

Substance use and delinquent behavior are not typically isolated or acute behaviors but rather take place within defined social contexts, geographic locations, and social relationships. Substance use coincides with and/or parallels dynamic trajectories of criminal or delinquent behavior. Moreover, substance use often follows systematic patterns along age-specific developmental trajectories characterized by persistent affliction, relapse, and health consequences. Recent life course research has framed substance use and addiction as a form of chronic illness as opposed to previous classifications as acute problems and this approach recognizes and incorporates the relevance of developmental trajectories, transitions, and pathways over

A. Valdez (✉) • A. Cepeda • C. Kaplan
School of Social Work, University of Southern California, Los Angeles, CA, USA
e-mail: avelardv@usc.edu

J.M. Saint Onge
Department Sociology, University of Kansas, Lawrence, KS, USA

Y.F. Thomas, L.N. Price (eds.), *Drug Use Trajectories Among Minority Youth*,
DOI 10.1007/978-94-017-7491-8_14

time (Elder 1985; Caspi et al. 1992). Much like other chronic illnesses, substance use and delinquency must be conceptualized within a broader social context with special attention given to key developmental stages. While this has particular implications for treatment and intervention strategies, it also emphasizes the salience of substance use, delinquency, and social controls through the life course (McLellan et al. 2000; Hser et al. 2007).

Life Course and Social Control

Recent developmental and life course theories of criminal and deviant behavior focused on social control have gained much attention. Social control broadly refers to the regulation of behavior and is a factor that varies across people's lives which may account for stability or change in antisocial behavior. Institutions of both formal and informal social control vary across the lifespan. Whereas formal social control focuses on legal sanctions, informal social controls emerge from role-relationships across key social institutions such as families, schools, or neighborhood associations (Kornhauser 1978). Informal social control emphasizes the structure of interpersonal bonds linking individuals to one another and to social institutions. One of the key assumptions for this project is that delinquent behaviors are more likely to occur when an individual has a low level of attachment and bonds to society. Parental, neighborhood, and school-level controls are important deterrents for both direct delinquency and potentially serious peer-level contexts for delinquency.

One of the most prominent theories using the life course perspective is that of Sampson and Laub's (2008) age-graded theory of informal social control. They stress that while the persistence in crime is explained by a lack of social controls, structured activities, and purposive human agency, these characteristics vary by age over the life course. These age gradients contain life events or transitions which occur over a larger life trajectory (Elder 1985).

While social controls and behaviors are age-graded, the consequences tend to follow defined trajectories. For instance, continuities in antisocial behavior beginning in childhood are often continued throughout adulthood across such domains as alcohol abuse, domestic violence, and crime. Delinquent or antisocial behavior follows long-term trajectories that can be interpreted as pathways of development over the lifespan. And salient life events and social bonds in adulthood (i.e. attachment to the labor force and cohesive marriage) explain variations in criminal behavior independent of prior differences in criminal propensity.

Criminal behavior typically follows an age-graded transition, positioned within the larger structure of informal and formal social controls. A life course perspective compels researchers to identify critical factors that contribute to behaviors at specific times during the life course. Accordingly, we emphasize the role of informal social control processes that influence adolescent behaviors through parental and school contexts. We suggest that the level of informal social control during adolescence is likely to have lasting impacts on young adulthood and adulthood

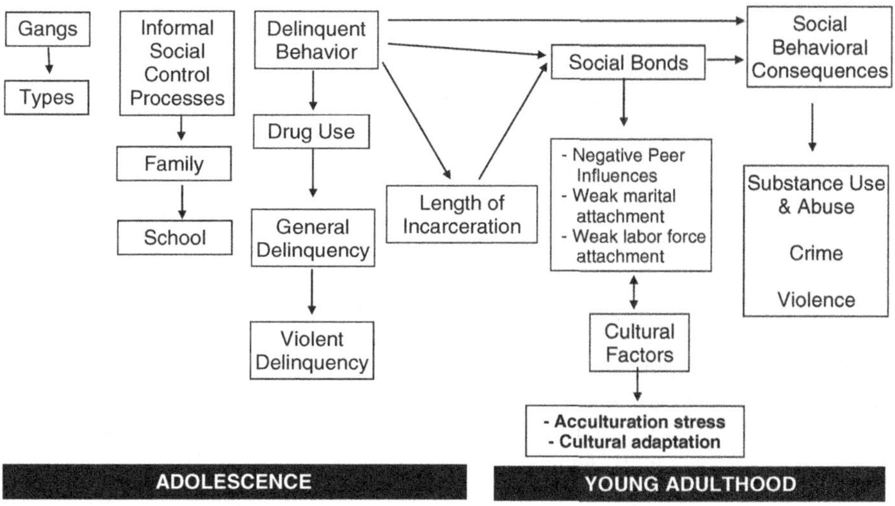

Fig. 14.1 Consequences of gang membership: a life course approach

behaviors. We aim to identify factors across adolescent lives that account for stability or change in antisocial behavior among gang involved youth. Figure 14.1 represents the conceptual model that guided this research.

The model indicates that family and school-level connections are associated with delinquent behaviors and subsequent social and behavioral consequences. We use data from 150 Mexican American adolescent gang members to understand how informal social controls are associated with substance use behaviors and gang membership.

AIMS

Focusing on key components of a life course approach we identify two primary aims. First, we seek to understand the association between informal social control processes with delinquency, violence, and substance use. Second, we aim to understand the association of individual and family characteristics, family and school social control processes, general and violent delinquency, other risk behaviors, and lifetime prevalence and frequency of drug use with length of gang membership.

Methods

The sample is collected from a research project aimed to identify and distinguish the relationship between gang violence and drug use among 26 male adolescent gangs in San Antonio, Texas. Contrary to school and institutional based samples used in previous life course research, our sample is drawn from a community-based sample in a distinct disadvantaged urban context. Accordingly, we expect our sample to be more likely to be overrepresented among those who experience adverse health, behavioral, and social consequences related to drug use such as HIV/AIDS, injuries, violence, crime, and/ or incarceration.

Using a multi-methods approach including questionnaires, ethnographic methods, focus groups, and life history intensive interviews, we draw on a sample of 150 Mexican American male gang members between the ages of 16–20 years old. The respondents have a mean age of 18.56 with an average education level of 9.28 years. Thirty-one percent have children, 33 % were employed at the time of the interview, and 15 % are currently or have ever been married. Additionally, nearly 70 % of the sample has lived in public housing at some point in their life. The great majority of the sample has been arrested at least once in their lifetime (87.5 %).

We find a high level of gang membership among our respondents, with the average number of years in the gang at approximately 5 years, with 16.4 % of our sample indicating 4 years of gang membership. This sample shows a much higher percentage of experienced gang members than previous studies. Table 14.1 shows the results of comparing the length of time in gang and lifetime prevalence of delinquency and drug use for our San Antonio sample with a sample of gang members in Rochester from a study conducted by Terrance Thornberry. Thornberry et al. (2003) find that approximately 50 % of gang members indicated 1 year of membership, compared to 6.3 % in our sample. In contrast to the Rochester sample where 100 % were in the gang for 4 years or less, in the San Antonio only 44 % were members for 4 years or less. However, we find very similar levels of delinquency among the two samples, with 88.1 % and 91.9 % of the San Antonio sample indicating engagement in general and violent delinquency respectively, compared to 98.1 and 90.6 % in the Rochester sample. Differences between the two samples appear in drug use,

Table 14.1 Comparisons of Rochester and San Antonio male gang samples on length of gang membership (%) and lifetime prevalence of delinquency, drug use and drug sales (%)

	Rochester (N=152)	San Antonio (N=160)
# of years		
One year	50.4	6.3
Two years	28.0	13.2
Three years	14.3	8.2
Four years	7.3	16.4
Prevalence		
General delinquency	98.1	88.8
Violent delinquency	90.6	91.9
Drug use	65.1	99.4
Drug sales	39.5	51.0
Arrests	54.6	76.3

drug sales and arrest. The San Antonio gang members had higher levels of drug use (99.4 %), drug sales (51.0 %), and arrests (76.3 %).

Variables

Gang Membership

Gang membership is categorized according to a gang's organizational structure and involvement in drug dealing (Valdez and Sifaneck 2004). We find that 77 % of the San Antonio sample is connected to street gangs, with the remaining 23 % attached to drug-related gangs. Street gangs are traditional territorial-based gangs that are not typically identified by organized crimes. Drug-related gangs are hierarchically organized, criminal, and tied to the drug market. Delinquency in drug-related gangs is more systematic and is focused less on territorial issues and more focused on maintaining and not calling attention to the drug trade.

Individual and Family Characteristics

Individual and family characteristics were measured in this study. Individual characteristics include current employment, number of children, marital status (single) and ever arrested. The individual characteristics appear above in the description of the sample. Family characteristics include having relatives ever involved in illegal activities, having relatives using drugs, attend church with family, and family ever lived in public housing. Almost three-fourths of the sample had relatives engaged in illegal activities when they were growing up (73.1 %). More than half had relatives who used drugs (51.1 %). Nearly half attended church with their family (44.4 %). More two-thirds have lived in public housing in their lifetime (67.5 %).

Family Social Control Processes

Family Social Control Processes include parental supervision, attachment to parent, a family attachment activities index, and harsh discipline. Respondents were asked to identify how often they had seen or talked with the person who raised them during the last year. Responses were measured using a 5-point scale ranging from "0 = never" to "4 = everyday". Attachment to parent assessed what kind of relationship the respondent had with the person that raised them. The 4-point scale ranged from 0 = no relationship to 3 = good relationship. Respondents were also asked if they had taken part in each of 10 family activities during the last year (0 = No, 1 = Yes). The items were summed to create a composite score for participation in family activities ranging from 0 to 10. The final construct in the family social control process of harsh discipline was measured by asking respondents whether compared to other people their age, if they thought they were physically abused, beaten, or hit more by their family at home. A 3-point measure ranging from 0 = less to 2 = more was used.

We find a high level of parental supervision, with 69 % of the sample indicating that they experience everyday parental supervision. Similarly, we find that only 26 % of our sample indicates that their attachment to their parents was bad, strained, or lack of relationship. The family attachment activities index report a mean value of 3.89, with a standard deviation of 2.28. Harsh discipline was relatively low in the sample (mean 0.71, sd 0.75). Over 80 % reported that compared to other people their age they had the same or less amount of harsh disciple from their parents.

School Social Control Processes

School social control processes includes skipping school, suspended from school, expelled from school, and school performance. Skipping school is coded on a 5-point scale ranging from frequent attendance (3 or more times a week) to never attend. When asked if they had ever been suspended from school, over 60 % of respondents indicated suspension from school at least once and 53 % indicated that they have been expelled from school. School performance used a 4-point scale from 0 = poor to = excellent. We find a mean value of 0.70, with only 15 % of the sample indicating good or excellent performance.

Lifetime Prevalence, Current Prevalence, and Frequency of Substance Use

Lifetime prevalence, current prevalence, and frequency of substance use are measured for seven substances including alcohol, marijuana, cocaine, heroin, speedball (heroin and cocaine mix), psychedelics, and benzodiazepines. Lifetime prevalence was assessed using a standard drug measure that identifies whether an individual has ever used any of an array of these substances. Current prevalence employed the same measure for the past month. Respondents were also asked to report the number of days per month they used each substance to measure frequency of use. Overall, there are high levels of lifetime prevalence of substance use. Exceptionally high lifetime use was reported for alcohol (98 %), marijuana (98 %) and cocaine (90 %). More than half of the sample reported lifetime use for heroin (57 %), psychedelics (58 %) and benzodiazepines (74 %). Additionally, a large percentage reported lifetime use for speedball (44 %) (injected heroin and cocaine). And Current use (past 30 days) was also high for alcohol (83 %), marijuana (75 %), and cocaine (53 %) with one-quarter reporting heroin use. Marijuana showed the highest frequency of use (mean days per month 20.3, sd 11.35). The lowest was psychedelics which showed 1.8 days per month (sd 0.13). Frequency of heroin use in the past month was notably high with a mean of 12.19 days (sd 11.16). This frequency was higher than alcohol (mean 11.73, sd 9.70), cocaine (mean 7.62, sd 7.73) and benzodiazepines (mean 3.76, sd 4.50). The mean frequency of speedball use was also relatively high at 6.09 days (sd 8.42).

General Delinquency, Violent Delinquency and Violence Risk

General delinquency, violent delinquency and violence risk was assessed using self-reported measures. The general delinquency measure was adapted from the National Youth Survey (NYS). Items ask respondents how often they have been involved in each of 19 activities during the last year. Responses range from 0 = never to 5 = very often. Items include actions such as running away, selling drugs, shoplifting, and vandalism. A reliability alpha of .77 for this index was observed for our sample. Similar to the delinquency index, the violent delinquency index includes six violent acts they engaged in during the last year ranging from never to very often. Items include: fighting, bringing a weapon to school, arson, violent acts (drive-bys, assault), armed robbery, and carjacking. The alpha coefficient for this index was .63. For our violence risk measure, we used the Plutchik Feelings and Acts of Violence Scale (PFAV). The PFAV measures the degree to which respondents engaged in violent acts when they were growing up. The scale consists of a 12-item questionnaire with a proven high reliability and validity. Eleven questions are measured on a 4-point scale. The final item requires a "yes" or "no" response. The violence risk score is calculated as the total numeric weight on the twelve items. Scores range cumulatively from a low of 0 to a high of 34. Respondents scoring at least five affirmative answers ("affirmative" is defined as any response other than "never") are considered potentially violent according to the scale. Alpha coefficient levels for the PFAV was .85 in the sample. About 90 % of the sample engaged in general delinquency (88.8 %) and violent delinquency (91.9 %). Violence risk was also elevated with a mean of 17.40 on the PFAV scale (sd 3.79).

Other Risk Factors

Other risk factors involved specific behaviors that are associated with gang involvement. Almost all of the respondents had fired a gun in gang fight (88.1 %), and used weapons to harm someone (83.1 %) in their lifetime. Half of the sample sold drugs in the last 3 months (50.6 %) and almost one-fifth had been shot while selling drugs (18.6 %).

Results

The first step in the analyses is an examination of the bivariate correlations between both general and violent delinquency and the informal social control processes. Bivariate correlation analysis reveals significant relationships between some of the family and school informal social control processes and adolescent outcome variables. For instance, a stronger attachment to a parent was associated with a decrease in the violent delinquency index score (r=−.169, $p<0.05$). Also, we find that ever having been expelled from school was associated with a higher score on the violent delinquency index (r=.203, $p<0.05$) and the violence risk scale (PFAV) (r=.245. $p<0.05$) Table 14.2.

Next, we examine the correlations between substance use prevalence and each of the informal social control processes (table not shown). First, we find that lifetime use of inhalants was negatively correlated with parent supervision, with less parental supervision associated with a higher likelihood of ever using inhalants. (r=−.160). School attachment variables are found to be correlated with prevalence of drug use. Lifetime cocaine use has a modest positive relationship with suspensions in school (r=.182). Similarly, lifetime benzodiazepines (r=.200) and alcohol (r=.205) use were correlated with the school attachment variable of skipping school. In addition, use of inhalants was positively correlated with ever being expelled from high school. Lifetime use of acid and speedball was negatively correlated with school performance (r=−.187).

Substance use frequency is also associated with informal social control processes. Harsher discipline is positively correlated with alcohol use in last 30 days (r=.189) and parental supervision was inversely correlated with frequency of speedball (r=−.519). School attachment as evidenced by not skipping school is negatively associated with frequency of alcohol (r=−.178) and marijuana (r=−.268) use. Never being suspended was negatively correlated with frequency of alcohol (r=−.264) and school performance was inversely related to the frequency of adolescent marijuana (r=−.209).

Bivariate Odds Ratios between Risk Factors and Length of Time in Gang

Building on the associations between the variables, we aim to understand the relationship of both risk factors and informal social controls and gang membership. In Tables 14.3 and 14.4 we present an examination of the bivariate relationships between informal social control and substance use risk factors and individual characteristics with subsequent length of time in a gang. Length of time in a gang was coded into less than 5 years and 5 years or more. We present the odds ratios from bivariate logistic regressions to provide simple estimates of the strength of the associations.

An examination of individual-level characteristics shows that being arrested has a strong relationship with gang membership. We find those gang members who have ever been arrested are 3 times as likely to be in a gang for a longer period of time.

Risk factors include the delinquency indexes, individual delinquent behaviors, and drug use. While the general and violent delinquency do not appear to significantly affect the likelihood of a respondent being in a gang for a longer period of

Table 14.2 Correlation matrix of general and violent delinquency and informal social control processes (n = 160)

	1	2	3	4	5	6	7	8	9	10	11
General and violent delinquency											
1. General illegal delinquent activity	1										
2. Violent activities	.653**	1									
3. Violence risk (PFAV)	.296**	.368**	1								
Family social control processes											
4. Parental supervision	−0.084	−0.03	−0.055	1							
5. Attachment to parent	−0.162	−.169*	−0.079	0.117	1						
6. Attachment activities	−0.028	−0.148	0.094	−0.042	0.024	1					
7. Harsh discipline	−0.045	−0.092	−0.072	−.165*	−0.084	.178*	1				
School social control processes											
8. Skipping school	−0.058	−0.08	−0.123	−0.093	0.127	−.082	0.074	1			
9. School suspension	−0.054	−0.067	−0.035	0.142	−.182*	.149	0.63	.220*	1		
10. Expelled from school	0.083	.203*	.245*	−0.041	1.149	−.153	−0.127	−0.087	−.217*	1	
11. School performance	−0.099	−0.125	−0.156	−0.051	0.084	.137	0.51	.289**	2.48*	−0.059	1

*P<.05, **P<0.1

Table 14.3 Bivariate odds ratios between individual and family characteristics, informal social control processes risk factors, delinquency/violence, other risk behaviors and length of time in gang

Individual characteristics	Odds ratio
Currently employed	1.71
Have children	1.55
Single	.19**
Ever arrested	3.04
Family characteristics	
Growing up relative involved in illegal activities	1
Growing up relative used drugs	0.86
Attend church with family	0.91
Family ever lived public housing	0.6
Family social control processes	
Parental supervision	0.55
Attachment to parent	2.26
Harsh discipline	0.97
School social control processes	
Attachment to school	0.57
School performance	1.15
General and violent delinquency	
General delinquency index	0.62
Violent delinquency index	0.79
Violence risk	1.03
Other risk behaviors	
Ever fired gun in gang fight	3.82*
Ever arrested for armed robbery/assault	1.24
Sold drugs in last 3 months	1.7
Shot at while selling drugs	1.47

*P<.05, **P<.01

time, the opposite was found for more serious risk behaviors. For instance, firing a gun in a gang fight indicates the likelihood of a longer duration in a gang by almost four times as much. A similar trend is observed for the remaining three indicators. This data represents the severity of risk behaviors as respondents immerse themselves in gang life for a longer period of time.

The lifetime prevalence of cocaine and heroin are significantly associated with increased time in gang membership. Importantly, we see that lifetime cocaine use appears to be the most important risk factor for predicting length of time in gang among this sample. Similarly, the frequency of cocaine use also has a strong positive relationship with increased length of gang membership. We also find that frequency of marijuana use has a negative association with length in gang.

Finally, we provide results from the informal social control variables. Surprisingly, neither family nor school social control process had a statistically significant rela-

Table 14.4 Bivariate odds ratios between lifetime and substance use risk factors and length of time in gang

Lifetime use	Odds ratio
Alcohol	0.28
Marijuana	0.16
Crack	0.81
Cocaine	6.26**
Heroin	2.47+
Speedball	0.62
Acid	1.35
Inhalants	1.17
Benzodiazepine	0.66
Other Opiates	2.97+
Amphetamine	0.6
Current use (past month)	
Alcohol	0.56
Marijuana	.32**
Crack	0.83
Cocaine	2.53*
Heroin	1.54
Speedball	0.49
Acid	2.16
Inhalants	0.46
Benzodiazepine	1.59
Amphetamine	.17*

*P<.05. **P<.01

tionship. Though not significant, the most remarkable was counterintuitive from the point of the view of life course theory. Attachment to parent increased the odds more than two times of being a member of the gang for 5 or more years.

Discussion

This chapter used a life course approach to explore the risk and protective factors associated with gang trajectory among a marginalized group of Mexican American gang members. We found that length of gang membership in this San Antonio population is considerably longer than reported in other seminal gang studies, such as Terrence Thornberry's longitudinal investigation in Rochester. This longer exposure to gang life is associated with more serious risk behaviors, such as firing guns, selling drugs, shot at while selling drugs, and arrests. As found in similar studies, we contend that engaging in these serious risk behaviors facilitate the development of specific norms and values which encourage the continuation of the gang trajectory (Decker and Van Winkle 1996; Thornberry et al. 2003). This longer gang trajectory may have more serious negative consequences as the adolescent transitions to young adulthood as reflected in our theoretical model.

The life course approach emphasizes the role of both family and school informal control processes in shaping behaviors over long periods of time. An unanticipated finding from our study was that these processes proved not be significant. This may be explained by the unconventional characteristics of these disadvantaged Mexican American families. As our data show, almost three quarters of the respondents were raised in households where relatives grew participated in illegal activities and more than half ha relatives that used drugs. Moore (1994) and others have identified "cholo" families within the Mexican American community that are characterized by generations of drug use, criminality, incarceration and street connections. Instead of being supporters of societal norms and conventional morality, members in these types of families may not discourage an adolescent from remaining in the gang (Covey et al. 1992). This is revealed by the trend that stronger attachment to parents is positively associated with a longer trajectory in the gang. These counterintuitive findings are further reinforced by the absence of a statistically significant association between the length of time in the gang and school informal control processes, such as attachment to educational institutions. Furthermore, these findings reflect cholo familes' negative perceptions and hostility toward schools and other secondary institutions such as local governments, criminal justice and organizational religion (Valdez and Kaplan 2007).

In contrast to what we would expect from life course theory, the drug trajectory was highly associated with length of time in the gang. The strongest finding in our analysis was that lifetime prevalence of cocaine was more than six times more likely to be associated with long-term gang members. Current cocaine use (past month) was also significantly associated with longer time in the gang. The cocaine trajectory from initial use to current use seems to parallel closely the gang trajectory. Interestingly, our findings show an inverse relationship between both the marijuana and amphetamine trajectories and length of time in the gang. Current use of marijuana and amphetamines were associated with shorter time in the gang. Although not statistically significant, lifetime use of marijuana was also highly associated with shorter time in the gang. Finally, there was a strong trend for the heroin trajectory and other opium trajectories to be associated with longer time in the gang. Overall, this reveals that a harder drug use trajectory (cocaine, heroin) is tightly interwoven with the life course trajectory of gang involvement.

What may explain some of the differences between our findings and those reported in the gang literature are sampling strategies. For example, in Thornberry's Rochester study there was no long term gang members (5 year or more) compared to our sample. This difference might be attributed to relying on school based samples which are likely to underestimate the extent of the gang trajectory and bias studies that use a life course perspective. For example, school based gang studies might be less likely to capture the intersection of the gang and hard drug trajectories found in this study. However, despite the methodological strengths of the study, certain limitations should be mentioned. Our analysis has largely been descriptive and the research design was cross-sectional. The bivariate results that we reported need be tested in more extensive multivariate analyses controlling for confounding variables. Furthermore, future research should be longitudinal in design in order to fully understand the consequences of gang membership in the life course.

This study illustrates the need to develop interventions that are more relevant for this population of disadvantaged Mexican American youth. Family based interventions are needed to help families reconnect with social institutions (i.e. schools, health system, conventional employment etc.). The families' role in the lives of these adolescents is distinct from those in other Hispanic groups. It has generally been argued that Mexican American families form a supportive tight-knit group characterized by protective factors. In the Westside neighborhood of San Antonio, parents are supportive of these delinquent youth through frequent parental supervision. However, these youth live in a family context where crime and drug use are normative behaviors. The adolescents in this study have family members who engage in illegal activities and illicit drug use. Given this context, drug use and subsequent gang membership may not be actively discouraged. Without prosocial adults in their lives, these adolescents are not likely to alter their peer networks and disengage from these delinquent activities. Any intervention must consider this cultural context of the "cholo" family. Furthermore, drug use and participation in the drug trade cannot be separated from gang membership. Every respondent in this study reported some type of current drug use with long-term gang members reporting more "hard" drug use (i.e. cocaine and heroin). Presented with low household incomes and the economic opportunity afforded by gang membership it is not surprising that adolescents join gangs. A neighborhood targeted intervention focused on community employment opportunities for adolescents and young adults may help discourage young people from joining gangs.

Acknowledgements This Research was supported by the National Institutes of Health, National Institute on Drug Abuse, Grants R01DA08604 and R01DA023857.

References

Caspi, A., Elder, G. H, Jr. et al. (1992). Childhood personality and the prediction of life-course patterns. In L. Robins & M. Rutter (Eds.), *Straight and devious pathways from childhood to adulthood* (pp. 13–35). New York: Cambridge Press.

Covey, H. C., Menard, S. et al. (1992). Juvenile gang violence. In H. C. Covey (Ed.), *Juvenile gangs* (pp. 27–47). Springfield: Charles C. Thompson Publications.

Decker, S. H., & Van Winkle, B. (1996). *Life in the gang: family, friends, and violence*. New York: Cambridge University Press.

Elder, G. H, Jr. (1985). Perspectives in the Life Course. In G. H. Elder Jr. (Ed.), *Life course dynamics* (pp. 23–49). Ithaca: Cornell University Press.

Hser, Y. I., Longshore, D., et al. (2007). The life course perspective on drug use: A conceptual framework for understanding drug use trajectories. *Evaluation Review, 31*(6), 515.

Kornhauser, R. R. (1978). *Social sources of delinquency*. Chicago: University of Chicago Press.

McLellan, A. T., Lewis, D. C., et al. (2000). Drug dependence, a chronic medical illness: Implications for treatment, insurance, and outcomes evaluation. *JAMA, 284*(13), 1689.

Moore, J. W. (1994). The chola life course: Chicana heroin users and the barrio gang. *International Journal of the Addictions, 29*(9), 1115–1126.

Sampson, R. J., & Laub, J. H. (2008). A general age-graded theory of crime: Lessons learned and the future of life-course criminology. In D. P. Farrington (Ed.), *Integrated developmental and life-course theories of offending: Advances in criminological theory* (Vol. 14). New Brunswick: Transaction.

Thornberry, T. P., Krohn, M. D., et al. (2003). *Gangs and delinquency in developmental perspective*. Cambridge: University of Cambridge Press.

U.S. Census Bureau. (2008). *American Community Survey demographic and housing estimates: 2006–2008*. Washington, DC: U. S. Department of Commerce.

Valdez, A., & Kaplan, C. (2007). Conditions that increase drug market involvement: The invitational edge and the case of Mexicans in South Texas. *Journal of Drug Issues, 37*(4), 893–917.

Valdez, A., & Sifaneck, S. J. (2004). "Getting high and getting by": Dimensions of drug selling behaviors among American Mexican gang members in south Texas. *Journal of Research in Crime and Delinquency, 41*(1), 82.

Chapter 15
Latino Youth Substance Use in States with Emerging Immigrant Communities

Charles R. Martinez, Heather H. McClure Jr., and J. Mark Eddy

While immigrants typically arrive in the U.S. in the hope of building a better life for themselves and their families, many encounter circumstances upon their arrival that make reaching such a goal difficult. The process of adapting to a culture that is different from that in the country of origin, or "acculturating," can be extremely stressful in general, with acculturative stressors varying by the contexts within which immigrants settle. Social structure characteristics of the geographic areas in which Latino families reside impact the nature, complexity, and magnitude of potential contextual stressors, including those related to cultural adaptation (Vega and Gil 1999), which in turn can have unique and profound effects for behavioral outcomes. In states that historically have been centers of Latino settlement, longer residence in the U.S. and greater acculturation have been linked to heightened risks for adolescent substance use initiation and abuse, and other deleterious health outcomes, such as depression (Gil et al. 2000). However, for the past several decades, a substantial number of Latino immigrants have settled outside of these states, and little has been written about the impact of acculturative and other contextual stressors on these new arrivals.

Portes and Zhou (1993) refer to the process by which social structure differences shape the acculturation experience as segmented assimilation. In geographic regions where significant Latino social, cultural, political, and business enclaves already exist

C.R. Martinez (✉) H.H. McClure Jr.
Department of Educational Methodology, Policy, and Leadership,
University of Oregon, Eugene, OR, USA

Center for Equity Promotion, University of Oregon,
Eugene, OR, USA
e-mail: charlesm@uoregon.edu

J.M. Eddy
Partners for our Children, School of Social Work, University of Washington,
Seattle, WA, USA

Y.F. Thomas, L.N. Price (eds.), *Drug Use Trajectories Among Minority Youth*,
DOI 10.1007/978-94-017-7491-8_15

(e.g., South Florida, California, New York), maintenance of one's language and culture of origin is possible and functional. However, Latino communities in most "emerging" states have little infrastructure to support a dual cultural society. Preliminary studies in emerging states indicate that many immigrant families in such locales operate within social contexts that may favor intense psychosocial stress (McClure et al. 2008, 2010a, b), and may contribute to a variety of poor outcomes for Latino youth, including substance use (Martinez et al. 2009; Martinez 2006). The limited evidence suggests that epidemiological studies focusing on Latinos in states with established immigrant communities—and the interventions that have been developed based on the data from these studies—may fail to reflect unique vulnerabilities for immigrants in emerging states (Martinez 2006; Prado et al. 2008).

In this chapter, we examine differences in the experiences of Latino immigrant families residing in one of two state groupings—states with emerging immigrant communities versus states with established immigrant communities—and explore how these communities may moderate or mediate the relationship between contextual stressors and Latino youth substance use. As the ethnic label "Latino" reflects an extremely broad grouping of individuals (e.g., Martinez and Eddy 2005; Szapocznik et al. 2007), we explore within-group heterogeneity through a comparison by state grouping of risk and protective factors and substance use prevalence for Latino youth and families. We close with a consideration of the implications of these findings for adolescent substance use interventions for Latino youth living within emerging immigration states.

New Trends in Latino Immigration

For most of the twentieth century, Latino populations have been concentrated primarily in California, Texas, Florida, Illinois, New Jersey, and New York (Capps and Fortuny 2006). Together, these states accounted for approximately 75 % of the U.S. Latino population (U.S. Census 1990). Beginning in about 1990, immigrants of Latin American and Spanish-speaking Caribbean ancestry in particular increasingly settled outside of these "Big 6" states, opting instead to establish new lives in 22 states with relatively little recent experience with large scale immigrant populations. According to U.S. Census data, immigrant populations in these states experienced a 92 % or greater increase (Capps and Fortuny 2006), marking them as distinct from states with substantially less immigrant population change. Over 50 % of these immigrants were Latino (Capps 2007; Fortuny et al. 2009).

This population movement was motivated by a number of factors. Changes in the labor market in a variety of countries resulted in record-high immigration into the U.S. (Capps et al. 2002). As the job markets became full in the Big 6 states, new markets were needed. In states with a large agricultural base, many new immigrants were drawn by pre-existing community ties, for instance, between towns in Mexico and farming communities in the U.S. that have long relied on migrant Mexican labor (e.g., Stephen 2007). In several other states, Latino cultural ties were even longer lasting, and again provided the foundation for establishing new communities. For example, many new growth states in the west (e.g., Oregon, Colorado, Nevada, Utah) were part of, or closely bordered by, Mexican territory during the nineteenth century, and had small but long-established Mexican-origin communities. Over time, many of these communities have become increasingly diverse, while also serving as critical sites for the orientation of Latino newcomers. Besides these reasons, Latino immigrants have settled in new states for the typical reasons that bring other new arrivals: to be closer to family who have already come to the U.S., to pursue better employment opportunities, to gain affordable housing, and to live in more peaceful neighborhoods and schools. As a result of factors such as these, Latino populations—typically including large numbers of foreign-born—grew dramatically in emerging states, whose majority populations tend to be U.S. born and monolingual English-speaking (Prado et al. 2008), and whose educational, social service and health systems may have been poorly prepared to support Latino newcomers, particularly if foreign-born, non-English speaking, and poor (Capps et al. 2002).

Segmented Assimilation: Social Contexts in Emerging versus Established Immigrant States

A number of elements, including social structure characteristics, family environments, and youth behaviors, tend to increase the risk that youth will use substances or protect youth from using substances. These risk and protective factors do not operate in isolation, but interact with each other to influence whether or not a youth will use and abuse substances (discussed in greater detail below). Community characteristics may increase the number of stressful life events (Latkin et al. 2007), and level of psychological distress experienced by community members (Galea et al. 2007; Inman and Yeh 2007; Yeh et al. 2005), and diminish individuals' psychological coping resources (Wilson 1996), making substance use more likely as a way to cope with stress (Boardman et al. 2001).

Researchers have identified differences between social contexts within states with established versus emerging immigrant communities. In addition to new growth states having less developed service infrastructures (e.g., bilingual education, medical interpreters in health settings, immigrant organizations), they may limit legal immigrants' access to the social safety net in ways that "Big 6" states historically have not (Capps et al. 2002). The sociodemographic characteristics of new immigrants themselves also may interact in important ways with broader social and economic contexts to increase risk or protective factors for families and youth. When new immigrants have lower educational levels, limited English language proficiency, and lower incomes, they may establish lives in neighborhoods, schools, and jobs that may contain more psychosocial stressors. These settings may be marked by unequal access to institutions, services and resources, and may offer fewer safeguards against economic, social, educational, and even physical vulnerability (Blau and Stearns 2002; LaVeist et al. 2007; National Research Council 2004; Turner and Hench 2003; Wallace 1999).

The unique issues immigrants can encounter in emerging states are best illustrated by an example of a specific state. Oregon appears to be relatively typical of the states with emerging immigrant communities in both the sociodemographic characteristics of new Latino immigrants, and in the social contexts that can shape immigrants' efforts to establish new lives. Approximately 80 % of Latino adults are recent immigrants (i.e., 10 years or less U.S. residency; Martinez et al. 2009; Mendoza and McClure 2009). However, generational histories vary widely in different areas of the state, with some areas having much greater concentrations of U.S. born residents.

At least 85 % of Oregon Latino children under age 18 are U.S. born (U.S. Census 2009a), with the result that most Latino families are of mixed legal status. The available, though limited, evidence suggests that this pattern is found across new growth states with the great majority of Latino immigrant families of mixed-status (Fortuny et al. 2009). An estimated 80 % of undocumented immigrants are of Latin American ancestry, and the largest share of immigrants who are undocumented reside in new growth states (Passel et al. 2004), with an unknown number of undocumented Latino immigrants in Oregon. About 95 % of Latinos in Oregon trace their family roots to the country of Mexico, with most of the remainder having origins in Central and South American countries (Martinez and Eddy 2005; Cheriel et al. 2009). Though immigrants' nativity varies among the 22 states of focus, the great majority of immigrants in these states also are of Mexican ancestry (Capps 2007).

Many Latino immigrants in Oregon trace their roots to rural areas in their home countries with limited infrastructure where opportunities for education were limited. As a result, approximately 70 % of Oregon Latino immigrants report attending school up to the ninth grade or less (Cheriel et al. 2009; Farquhar et al. 2008a,b; McClure et al. 2010a). A similar pattern can be noted for new growth states as a whole, with approximately 36 % of Latino immigrant parents reporting less than a high school degree (Urban Institute 2009). A comparison of Census data on Latino parents' educational attainment levels (Urban Institute 2009) reveals that more parents in new growth states have not completed high school (35.6 % versus 28.6 %), and fewer parents have a degree from a 4-year college (12.6 % versus 16.5 %). Though similar patterns were found among Whites in new growth compared with "Big 6" states, there appears to be little difference in the prevalence of those who did not finish high school by state grouping (5.1 % versus 3.9 %). Further, there appear to be substantially more White college graduates (+9 %) in "Big 6" versus new growth states, with a smaller difference (+4 %) by state grouping in Latino college graduation rates. In sum, these data reflect widespread findings that Latinos, on the whole, have lower levels of educational attainment compared with Whites in the U.S. They also suggest that Latinos living in new growth states have substantially less formal education on average than Latinos in "Big 6" states.

Similar discrepancies between new growth and "Big 6" states can be observed in regards to language use. While U.S. Census data (U.S. Census 2009a) indicate that 59 % of the Spanish-speaking population in Oregon speaks English less than "very well," data collected by researchers working in conjunction with community-based organizations suggest that a much higher proportion of the recently immigrated adult population is monolingual Spanish-speaking (Farquhar et al. 2008a; Mendoza and McClure 2009; Oregon Social Learning Center, 2007). When Census data on linguistic isolation[1] in Latino immigrant households are compared, a greater share of Latino immigrant households are linguistically isolated in new growth than in "Big 6" states (29.5 % versus 24.7 %; The Urban Institute 2009).

Educational attainment and language proficiency are both highly related to employment opportunities and income, and Oregon appears to be typical of states with emerging immigrant communities in having a high proportion (60 %) of working poor[2] (Fortuny et al. 2009). Indeed, the proportion of Latino immigrant families who are working poor is noticeably higher in new growth states than in "Big 6" states (54.1 % versus 44.8 %), a pattern also observed for White working poor households (23.9 % versus 16.4 %; The Urban Institute 2009). Importantly, the share of working poor households among Whites is approximately half that for Latinos.

Latino immigrant families residing in new immigrant growth states appear to share key social structure characteristics and may confront similar contextual stressors. These stressors can combine to produce a social milieu that can promote intense psychosocial and physiological stress for both parents and youth. In Oregon and similar states, there may be few formal and informal buffers (e.g., the support of established immigrant enclaves, and linguistically and culturally competent institutions and services) against such stressors. One consequence of a lack of stress buffers is that a greater proportion of Latino youth and parents in emerging immigrant states may experience poor emotional, behavioral and educational outcomes. For example, one indicator of the disproportionate toll of these stressors on Latino school-age children in new growth states is secondary school graduation rates. Though there is a small difference (4 %) in graduation rates among Whites in these two state groupings, Latino students in emerging immigrant states appear to have substantially lower graduation rates than Latino students in "Big 6" states (47 % versus 55 %; Greene & Winters, 2002). Social structural characteristics that include a high proportion of monolingual Spanish speakers, lower levels of educational

[1] Linguistically isolated households are those in which no person age 14 and older either speaks only English at home or speaks a language other than English at home and also speaks English very well. All members of such a household are considered linguistically isolated, even though children under 14 who speak only English may live there.

[2] Low-income working families—families with total family income below 200 % of the federal poverty level that have high or moderate work effort. Family work effort is classified as high if any adult reports at least 1800 h of work in the prior year (approximately equal to 35 h of work a week for 52 weeks a year); as medium if adults average at least 1000 h or the total hours worked is at least 1800 h, but no adult reports 1800 h of work in the prior year; and as low if neither criteria is met (Acs and Nichols 2005).

attainment, and greater poverty, when combined with social contextual and institutional barriers, have important implications for immigrant families' expectations and acculturation experiences upon arrival in the U.S., for the disruption of powerful community and family-based cultural protective factors (discussed below), and for related youth problem behaviors, such as youth substance use.

Youth Substance Use

The use and the abuse of substances per se have been frequently linked to poor behavioral and physical health for Latino youth, as reflected in outcomes such as disproportionately high school dropout rates (e.g., Arias et al. 2003; Wallace et al. 1995; Oregon Department of Education 2010; Oregon Youth Authority 2009). Earlier use of substances is related to a higher likelihood of abuse and dependence during adulthood (Anthony and Petronis 1995; Warner et al. 2006).

A variety of national surveys have been conducted that examine differences in use between Latino and other youth, most notably the Monitoring the Future (MTF) survey (Johnston et al. 2008), the National Survey on Drug Use and Health (NSDUH; SAMHSA), and the Youth Risk Behavior Survey (YRBS; CDC). In regards to prevalence, these various national epidemiological surveys, which rely on somewhat different methods and designs, show some similarities as well as notable differences. For example, the MTF and YRBS data show that Latino youth have the highest lifetime rates of alcohol, cigarette, and marijuana use when compared with White and African American youth (Table 15.1), while NSDUH data show White youth having higher lifetime rates than either Latino or African American youth. MTF is the only study that documented higher past-month use of these three substances for Latinos compared to African American and White youth, with the YRBS only identifying Latino alcohol use as higher than other groups, and the NSDUH indicating that the past-month use of these substances among Whites exceeded that for either Latino or African American youth.

Prior MTF drug use rankings showing earlier initiation by Latinos, with Whites initiating later and eventually surpassing Latinos, were attributed by some researchers to methodological issues related to the school-based sampling frame and disproportionate Latino school dropout by 12th grade (Johnston et al. 2009). However, current MTF and YRBS findings showing that drug use among older Latino secondary students exceeds that for Whites raises doubt about the potential role of school dropout on prevalence rates (Prado et al. 2008). It is notable that the NSDUH, a household-based survey that may be substantially less influenced by school dropout rates, is unique among these three studies in showing *decreasing* rates of Latino youth substance use over time, with simultaneously increasing rates among White youth (Table 15.1).

Beyond these general findings, the question of importance here is whether Latino youth in states with emerging immigrant communities are at higher risk for substance use and related poor behavioral and academic outcomes than youth in

Table 15.1 Proportion of youngsters indicating lifetime and past-month substance use in national epidemiological studies: comparisons among racial/ethnic groups[a]

National survey on drug use and health (2007)				Monitoring the future (2007)			Youth risk behavior survey (2007)		
	Hispanic	White	Black	Hispanic	White	Black	Hispanic	White	Black
Age/grade:	12–17 years			8th, 10th, 12th grade			9th–12th grades		
Sample size	4667	15,053	3848	12,900	57,300	7440	2373	8467	2120
Lifetime use									
Alcohol	39.1	41.9	34.0	62.7	54.0	57.7	77.9	76.1	69.1
Cigarettes	20.8	26.7	17.8	36.8	31.8	33.7	53.3	50.0	50.3
Marijuana	14.4	17.3	15.4	33.9	26.8	30.4	38.9	38.0	39.6
Past-month use									
Alcohol	15.2	18.2	10.1	33.5	27.5	29.4	47.6	47.3	34.5
Cigarettes	6.7	12.2	6.1	13.3	11.4	11.8	16.7	23.2	11.6
Marijuana	5.7	7.3	5.8	14.1	11.3	12.8	18.5	19.9	21.5

[a]Only three racial/ethnic subgroups are presented here, due to limitations in the MTF and YRBS datasets. The NHSDA does provide additional details on other ethnic subgroups for interested readers

"Big 6" states. To examine this question, we examined data from the YRBS. General substance use prevalence trends identified by the YRBS were similar to those of the MTF (as opposed to the NSDUH which had unique trends in comparison to the other two datasets), and YRBS state-level data disaggregated by race/ethnicity were accessible through state departments of health. Thus, YRBS data were examined to detect possible differences among Latino, White and African American youth by different state groupings organized by emerging versus established immigrant communities. Particular attention was given to comparisons of Latino and White youth, as we hypothesize that White youth who are mostly U.S. born, English-speaking, and part of a demographic majority in most states, on average, will be exposed to lower levels of social contextual stressors than will Latino youth, *particularly in new growth states*, with potential implications for lower drug use prevalence among White youth when compared with Latino youth. Finally, a caveat related to limitations in our methodological approach: our results are descriptive as this investigation is intended as a first step toward the identification of potential trends in state-reported prevalence rates; a more thorough examination of differences between state groupings would require the analysis of complete state-level data sets.

In nearly all new growth and "Big 6" states, a greater proportion of Latino than White or African American youth initiated use of alcohol, cigarettes, and marijuana before age 13 (see Tables 15.2, 15.3 and 15.4). In relation to current use of alcohol (within the last 30 days; Table 15.2), of 18 new growth states reporting data for Latino youth, Latinos in nearly three-quarters of these states (72 %), reported current alcohol use that exceeded that for White students.

Table 15.2 Alcohol use among Latino, African American, and White youth in states with emerging compared with established immigrant communities

	Alcohol											
	Before age 13				Current				Binge			
	State	White	Black	Hispanic	State	White	Black	Hispanic	State	White	Black	Hispanic
Youth Risk Behavior Survey National Statistics	N/A	21.5	26.7	29.0	N/A	43.7	34.5	47.6	N/A	29.8	12.5	26.8
States with emerging immigrant communities												
Alabama YRBS (2007)		26.5	40.9	37.2								
Arkansas YRBS (2007)	26.0	26.0	26.0	26.0	42.0	43.0	35.0	44.0	25.0	26.0	12.0	29.0
Arizona YRBS (2007)	23.5	19.3	21.9	27.4	45.6	44.0	31.3	50.5	30.4	31.0	9.0	31.7
Colorado YRBS (2007)	28.1	25.2	N/A	32.6	48.7	47.5	N/A	60.3	31.8	32.7	N/A	37.9
Delaware YRBS (2007)	25.0	22.7	24.8	28.5	45.2	54.5	29.2	45.5	25.4	34.8	9.4	24.9
Georgia YRBS (2007)	23.9	21.6	24.9	29.7	37.7	44.6	29.2	37.1	19.0	27.5	9.2	19.1
Idaho YRBS (2007)	23.0	20.5	N/A	34.9	42.5	40.8	N/A	53.4	30.4	29.6	N/A	37.2
Indiana YRBS (2007)	21.9	20.8	21.9	31.4	43.9	44.9	29.3	49.4	28.2	30.0	10.7	34.9
Iowa YRBS (2007) **Latino sample size too small**												
Kansas YRBS (2007)	23.3	20.8	29.9	35.5	42.2	42.3	34.5	50.7	27.1	28.3	12.7	30.3
Kentucky YRBS (2007)	25.1	24.8	21.1	35.9	40.6	41.0	35.2	46.3	27.1	28.7	13.7	27.4
Minnesota Student Survey (2007)	21.0	15.0	20.0	24.5	35.8	35.8	30.0	38.0	21.5	21.5	17.5	25.0
Mississippi YRBS (2007) **Latino sample size = 23**												
Nebraska YRBS (2005)	23.9	21.4	N/A	32.4	42.9	42.1	N/A	43.4	29.8	29.1	N/A	30.7
Nevada YRBS (2007)	24.6	22.9	23.8	26.8	37.0	40.5	26.9	34.6	21.1	25.1	14.5	18.8
North Carolina YRBS (2007)	19.7	18.0	20.2	31.7	37.7	43.0	27.2	38.7	21.1	26.4	10.1	23.9
Oklahoma YRBS (2007)	23.3	20.7	22.9	32.2	43.1	43.1	40.5	45.0	27.9	29.4	13.3	27.2
Oregon Healthy Teens (2005)	29.4*	27.7	30.8	35.8	47.4*	38.7	44.3	38.7	28.1	20.1	20.9	16.8
South Carolina (2007) **Hispanic sample size = 43												
Tennessee YRBS (2007)	22.3	22.3	20.5	79.0	36.7	39.1	25.7	47.0	21.7	25.6	10.2	30.1
Utah YRBS (2007)	13.0	11.8	N/A	11.2	17.0	15.7	N/A	26.5	11.7	10.4	N/A	16.7
Washington Healthy Youth Survey (2008)	N/A	N/A	N/A	N/A	36.3	36.6	33.9	38.4	22.2	21.6	22.9	25.4
Averages:		21.6		32.9		41.0		43.8		26.5		27.1
States with established immigrant communities												
California (2007)	22.9	19.7	23.8	27.7	26.2	27.4	22.7	30.4	14.7	16.6	12.0	16.7
Florida YRBS (2007)	24.2	22.0	24.5	28.8	42.3	49.2	29.1	42.8	22.8	29.7	9.9	23.2
Illinois YRBS (2007)	23.8	21.5	26.7	29.0	43.7	47.8	28.5	50.1	28.0	33.6	10.8	29.8
New Jersey (2007)	18.6	16.2	23.3	26.3	51.0	57.0	39.2	56.7	31.1	38.6	16.5	30.0
New York YRBS (2007)	22.9	21.4	24.3	28.3	43.7	48.1	30.1	46.5	24.9	29.5	10.6	28.6
Texas YRBS (2007)	27.8	25.5	30.1	30.1	48.3	51.3	42.2	48.7	29.0	34.0	14.8	29.5
Averages:		21.1		28.4		46.8		45.9		30.3		26.3
Hispanic/Latino > White												

In 61 % of states with emerging immigrant communities, Latino youth reported higher prevalence rates for binge drinking (within the last 30 days, 5 or more drinks within a couple of hours) than Whites. When examining data for "Big 6" states, Latino youth had a higher prevalence of current alcohol use than Whites in two states (California and Illinois), and in California, Latino youth reported rates of binge drinking that were comparable to those for White youth (at 16.7 % versus 16.6 %). A comparison of average current alcohol and binge drinking rates by state groupings suggests, however, that White youth in "Big 6" states reported more current alcohol use and binge drinking than White youth in new growth states, while only slight differences are apparent by state grouping for Latino youth. Though this finding raises questions about which factors might have contributed to an increase in White youths' alcohol use in "Big 6" states, equally compelling is the question about which factors might have prevented a parallel *decrease* in Latino youths' current alcohol use and binge drinking rates in new growth states.

Table 15.3 Cigarette use among Latino, African American, and White youth in states with emerging compared with established immigrant communities

	Cigarettes							
	Before age 13				Current			
	State	White	Black	Hispanic	State	White	Black	Hispanic
Youth Risk Behavior Survey National Statistics	N/A	14.4	12.5	14.3	N/A	23.2	11.6	16.7
States with emerging immigrant communities								
Alabama YRBS (2007)	16.7	13.7	19.0	21.5				
Arkansas YRBS (2007)	18.0	16.0	16.0	16.0	20.0	22.0	14.0	16.0
Arizona YRBS (2007)	14.7	11.9	12.4	14.5	22.2	23.3	11.4	18.0
Colorado YRBS (2007)	13.4	11.7	N/A	17.6	18.1	17.7	N/A	21.6
Delaware YRBS (2007)	14.8	14.6	11.2	21.5	20.2	26.6	8.5	20.4
Georgia YRBS (2007)	14.5	17.2	9.5	17.1	18.6	25.0	9.5	20.3
Idaho YRBS (2007)	13.9	12.2	N/A	23.3	20.0	19.2	N/A	23.6
Indiana YRBS (2007)	16.0	15.1	14.2	24.7	22.5	23.1	15.6	24.0
Iowa YRBS (2007) **Latino sample size too small**								
Kansas YRBS (2007)	14.0	13.1	10.0	21.2	20.6	21.0	15.9	20.6
Kentucky YRBS (2007)	23.8	24.0	19.5	27.7	26.0	27.0	14.4	32.1
Minnesota Student Survey (2007)	13.5	12.8	21.3	22.4	18.8	18.3	16.3	22.0
Mississippi YRBS (2007) **Latino sample size = 23**								
Nebraska YRBS (2005)	16.5	14.0	N/A	25.6	21.8	20.9	N/A	25.0
Nevada YRBS (2007)	12.6	13.5	13.5	11.6	13.6	16.0	7.4	11.2
North Carolina YRBS (2007)	17.3	17.3	15.7	23.0	22.5	26.7	14.8	20.2
Oklahoma YRBS (2007)								
Oregon Healthy Teens (2005)	12.7	10.4	13.4	15.7	18.8	12.6	11.0	8.5
South Carolina (2007) **Hispanic sample size = 43								
Tennessee YRBS (2007)	17.9	19.2	12.1	26.4	25.5	30.6	9.0	30.6
Utah YRBS (2007)	8.3	7.0	10.7	16.6	7.9	6.9	5.3	9.9
Washington Healthy Youth Survey (2008)	9.8	9.0	14.3	10.4	17.2	17.7	18.3	13.4
Averages:		14.0		19.8		20.9		19.8
States with established immigrant communities								
California (2007)	12.6	11.0	15.2	14.8	9.9	11.5	9.3	10.1
Florida YRBS (2007)	13.1	13.8	9.8	15.2	15.9	21.4	5.0	15.2
Illinois YRBS (2007)	13.4	12.6	11.6	19.4	19.9	24.1	8.0	19.2
New Jersey (2007)	8.8	6.9	14.8	10.9	18.7	20.5	14.5	21.7
New York YRBS (2007)	11.1	10.7	10.0	14.1	13.8	16.1	5.8	15.7
Texas YRBS (2007)	14.3	13.9	11.4	15.9	21.1	26.5	10.8	19.3
Averages:		11.5		15.1		20.0		16.9
Hispanic/Latino > White								

Latino youth reported higher current cigarette use than White youth in seven out of 17 new growth states reporting data (41 %; Table 15.3).

In states with established immigrant communities, in only one state (New Jersey) did Latino youth report higher current cigarette use (within the last 30 days) than Whites (21.7 % vs. 20.5 %). There were no noticeable differences in White youths' prevalence of current cigarette use by state grouping, though Latino youths' average prevalence was higher in new growth compared with "Big 6" states (19.8 % vs. 16.9 %). A comparison of current marijuana use (within the last 30 days) among Latinos and Whites in new growth states shows that of 18 states reporting data, Latinos in 13 states (72 %) reported higher use than Whites (Table 15.4).

Table 15.4 Marijuana use among Latino, African American, and White youth in states with emerging compared with established immigrant communities

	Marijuana							
	Before age 13				Current Use			
	State	White	Black	Hispanic	State	White	Black	Hispanic
Youth Risk Behavior Survey National Statistics	N/A	N/A	N/A	N/A	N/A	19.9	21.5	18.5
States with emerging immigrant communities								
Alabama YRBS (2007)	9.3	6.3	11.7	12.1				
Arkansas YRBS (2007)	10.0	8.0	12.0	13.0	16.0	14.0	20.0	18.0
Arizona YRBS (2007)	13.0	9.2	13.1	14.3	22.0	20.1	17.5	21.1
Colorado YRBS (2007)	12.4	8.9	N/A	18.7	22.8	22.3	N/A	27.1
Delaware YRBS (2007)	10.1	8.6	10.1	13.9	25.1	27.2	23.4	25.7
Georgia YRBS (2007)	8.1	7.3	9.1	6.5	19.6	20.6	19.6	14.6
Idaho YRBS (2007)	8.2	6.8	N/A	13.7	17.9	17.0	N/A	19.3
Indiana YRBS (2007)	9.1	7.6	14.6	15.6	18.9	17.0	31.2	21.9
Iowa YRBS (2007) **Latino sample size too small**								
Kansas YRBS (2007)	8.3	6.0	16.5	15.0	15.3	13.6	21.8	20.8
Kentucky YRBS (2007)	10.2	9.3	13.7	16.8	16.4	15.3	23.9	23.9
Minnesota Student Survey (2007)	5.3	3.5	12.3	10.0	14.8	13.5	20.5	19.8
Mississippi YRBS (2007) **Latino sample size = 23**								
Nebraska YRBS (2005)	7.0	5.2	N/A	12.1	17.5	15.0	N/A	19.2
Nevada YRBS (2007)	8.4	7.0	11.8	10.4	15.5	15.6	19.1	14.0
North Carolina YRBS (2007)	8.3	7.2	8.3	12.5	19.1	19.1	19.7	14.5
Oklahoma YRBS (2007)	8.1	6.3	13.7	12.7	15.9	13.7	25.2	14.1
Oregon Healthy Teens (2005)	*9th-	7.9	15.7	18.3	**9th	14.7	21.3	40.6
South Carolina (2007) **Hispanic sample size = 43								
Tennessee YRBS (2007)	10.7	9.7	11.5	25.4	19.4	17.8	23.6	25.8
Utah YRBS (2007)	7.6	6.5	N/A	9.9	8.7	7.5	N/A	14.2
Washington Healthy Youth Survey (2008)	N/A	N/A	N/A	N/A	21.3	20.8	29.3	18.8
Averages:		7.3		13.9		16.9		20.7
States with established immigrant communities								
California (2007)	7.5	6.3	11.5	9.0	12.7	13.8	15.4	13.9
Florida YRBS (2007)	8.8	8.9	7.9	9.6	18.9	22.6	13.0	17.7
Illinois YRBS (2007)	8.7	7.2	10.7	12.4	20.3	21.3	21.1	19.3
New Jersey (2007)	5.9	4.5	10.2	8.5	19.4	20.6	18.0	21.0
New York YRBS (2007)	7.3	6.5	8.2	8.7	18.6	19.9	16.9	19.1
Texas YRBS (2007)	9.4	7.8	10.7	11.0	19.3	19.5	18.5	19.5
Averages:		6.9		9.9		19.6		18.4
Hispanic/Latino > White								

In "Big 6" states, Latinos reported rates of current marijuana use (within the last 30 days) that were comparable to those for Whites in California (13.9 % vs. 13.8 %) and New Jersey (21.0 % vs. 20.6 %). Latino youth in new growth states report higher average marijuana use than Latino youth in "Big 6" states (20.7 % vs. 18.4 %), while this pattern appears to be reversed among Whites (16.9 % vs. 19.6 %).

Although Whites may surpass Latinos by 12th grade in terms of drug use prevalence nationwide (Johnston et al. 2009), a comparison of states with emerging versus established immigrant communities reveals a more complex picture. In many new growth states, aggregate data drawn from 9th through 12th grade students show that Latino substance use exceeds that of Whites, a pattern that is less evident in "Big 6" states. In addition, these data provide evidence of Latino early initiation of drug use, a worrisome finding given that the consequences of drug initiation by age 13 may be more severe than the consequences of drug initiation at older ages (Anthony and Petronis 1995; Warner et al. 2006). Taken together, these findings suggest that, in new growth states, prevalence rates for Latino substance use throughout adolescence are higher compared to the majority population in these states, with Latino youngsters initiating use earlier, activating an earlier trajectory toward more severe outcomes, including risk for school dropout and incarceration (Hawkins et al. 1992; Martinez et al. 2003; Wallace et al. 1995).

Linking Context and Substance Use: A Developmental Model

Psychosocial stressors related to social structure characteristics impact key elements of the early family environment such as parent-child relationships, parental affection, and family conflict (Reid et al. 2002). These, in turn, influence youngsters' development of emotional and cognitive responses to environmental challenges (e.g., psychosocial stressors), which are predictive of concurrent and subsequent emotional adjustment (Bruce et al. 2009; Cicchetti and Rogosch 2007; DeCaro and Worthman 2008a, b; Dozier et al. 2006; Gunnar and Donzella 2002; Pears and Fisher 2005). Poor adjustment in childhood or early adolescence, including aggressive and antisocial behavior, is predictive of late adolescent and adult substance use (Block et al. 1988; Kellam et al. 1983). While these are the common "malleable" issues often discussed by prevention researchers interested in psychosocial interventions, a variety of other issues, including socioeconomic and cultural protective factors, are of great importance to consider for immigrants in emerging states.

Socioeconomic Status and Parent Education

Extensive evidence exists for an inverse relationship between socioeconomic status (SES, typically defined as parent education, occupation, and household income) and adolescent substance use among White adolescents (Goodman and Huang 2002; Lowry et al. 1996; Winkleby et al. 1999). The relationship between SES and Latino youth substance use, however, is less clear, particularly within emerging immigrant contexts. Analyses of national datasets (e.g., Third National Health and Nutrition Examination Survey and the National Health Interview Survey) showed that increased education was associated with decreased cigarette use among both Mexican American and African American youth, and that increased income was associated with decreased alcohol use (Lowry et al. 1996; Winkleby et al. 1999). Analyses of data from the 1995 National Longitudinal Study of Adolescent Health, however, showed exactly the opposite relationships for non-White youth (Goodman and Huang 2002). Finally, parental education was more consistently associated with substance use than household income (Goodman and Huang 2002), indicating that SES is a multidimensional construct within which income, education, occupation and perceived social status, among others, may exert important and independent effects upon youth drug use (Goodman et al. 2001).

Socioeconomic characteristics of Latino immigrant families may play a crucial role in shaping acculturation experiences and related psychosocial stress exposure of parents and youth in neighborhoods, workplaces, schools, and other domains of life. In new growth states, this stress exposure may be further exacerbated by a lack of institutional and community-based buffers. In this way, acculturation and attendant stressors may exert more proximal effects on substance use outcomes among youth in Latino immigrant families, particularly in new growth states.

Acculturation, Discrimination and Latino Immigrant Families

Acculturation is multifaceted, and includes such factors as linguistic proficiency, language use, nativity and culture-related behavioral preferences, among others (Berry 1998). Importantly, acculturation serves as a marker for other psychosocial processes (e.g., ethnic identity processes, experiences of structural barriers such as poverty, and difficulty obtaining work authorization) that link acculturative stressors to adolescent outcomes.

As mentioned, adaptation to U.S. norms and institutions, a critical part of the acculturation process, involves a complex and potentially stressful process for youth and their families. The process of acculturation begins immediately upon arrival for immigrant families, but operates differently for adolescents and adults. Adolescents tend to acquire language proficiency in English and acculturate to cultural norms much more quickly than do their parents. The evidence is mixed, however, as to what elements of Latino youths' acculturation experience increase their risk for poor outcomes, such as substance use (Valencia and Johnson 2008). Recent studies indicate that adolescents who lose their orientation toward Latino social norms, language, and cultural values, regardless of their acceptance of U.S. dominant cultural norms, are at greater risk of substance use than those youngsters who remain oriented towards Latino cultural values and practices (Coatsworth et al. 2005; Martinez 2006; Sullivan et al. 2007). Increasing evidence also indicates that interactions between acculturation and discrimination can boost risk of youth substance use.

Contrada and colleagues (2001) define ethnic discrimination as unfair treatment received because of one's "ethnicity," referring to various groupings of individuals based on culture of origin (Brondolo et al. 2009b). Numerous studies have illustrated the ways in which discrimination and stigma, social processes linked to the reproduction of inequality and exclusion (Paradies 2006), are an intrinsic part of life in the U.S. and affect minority groups on a daily basis (Szalacha et al. 2003). To date, no studies have been conducted comparing reported exposure to discrimination among Latinos in states with established versus emerging immigrant communities, though research in Oregon suggest that the prevalence of discrimination reported by Latinos is comparable or higher than those in states with established immigrant communities (e.g., DeGarmo and Martinez 2006; Martinez et al. 2008; McClure et al. 2010a, b).

Research indicates that the higher the level of reported discrimination, the greater the risk of adolescent drug use (Bennett et al. 2005; Okamoto et al. 2009; Vermeiren et al. 2003), with ethnic/racial minority youth more likely to be affected by discrimination than Whites (Johnston et al. 2008). In Oregon and elsewhere, Latino youth who are immigrants or children of immigrants may be exposed to "double discrimination" stemming from injurious stereotypes regarding ethnic/racial minority groups in general, and immigrants—especially those of Mexican ancestry—in particular (Stephen 2007). Further, the greater the English proficiency of Latino youth, the higher their potential exposure to certain types of discrimination stress; for instance, they may increasingly understand when someone discriminates against them in English, while their non-English speaking parents may be protected from this stressor through linguistic isolation (Pérez et al. 2008).

Increasingly, studies are documenting the ways in which perceived discrimination varies by nativity, age upon arrival in the U.S., and generational status (Dominguez et al. 2009; Finch et al. 2000), with Latino youth who are immigrants or children of immigrants reporting more discrimination and greater related stress than more acculturated adolescents (Edwards and Romero 2008). In Oregon-based studies of Latino middle-school aged youth who are immigrants or children of immigrants, approximately half of all youth participants consistently reported experiencing discrimination (DeGarmo and Martinez 2006; Martinez and Eddy 2004), a finding echoed in qualitative studies with Latino youth elsewhere in the state (Gonzales-Berry et al. 2006). Numerous researchers have pointed to alcohol and other drug use as a coping mechanism for psychosocial stress, including that related to acculturation and perceived discrimination (Borrell et al. 2007; Broman et al. 2000; Brondolo et al. 2009a; Guthrie et al. 2002; Rospenda et al. 2008; Rutledge and Sher 2001), and it remains to be seen whether Latino youth in Oregon and similar new immigrant settlement states are attempting to cope with potentially higher levels of discrimination and acculturative stress (when compared with Latino youth in "Big 6" states) through drinking and other drug use (Gil et al. 2000).

Acculturation Gaps Within Latino Families

Differences between parents and youngsters in their levels and rates of acculturation can create "acculturation gaps" (also referred to as "differential acculturation") that can contribute to additional stress for families. These gaps are particularly acute within emerging immigrant communities where the assimilation forces are great and there is diminished access to culturally familiar enclaves. The propensity of youth in immigrant families to assimilate more quickly to U.S. norms, can create strain in families as parents become increasingly concerned that their children are shedding traditional values and becoming too "Americanized" and youth are concerned that their parents fail to understand the assimilation demands that they encounter to be more accepted within their peer groups. The resulting stress can disrupt effective parenting practices, which, in turn, can create vulnerability for substance use and related problems among adolescents (Kurtines and Szapocznik 1996; Martinez 2006; Martinez et al. 2009; Murray et al. 1987; Pantin et al. 2003; Murry et al. 2008). Research with Latino families also show that a variety of culturally-specific protective family and parenting factors related to *familismo* (e.g., a value system emphasizing family pride, respect, cohesion, consistent discipline, and parental support; Sabogal et al. 1987; Stanton-Salazar 2001; Vega 1990; Harwood et al. 2002) can buffer the negative effects of acculturation and related stress on Latino youngsters' disposition to deviance, drug use, and depression and conduct problems (Martinez 2006; Martinez and Eddy 2005; Prado et al. 2008). Such findings underscore that parenting factors are important proximal predictors of behavioral outcomes for Latino youngsters, and further can mediate the effects of acculturative stressors on those outcomes.

An Integrative Theoretical Model

Social learning theory posits that adolescent adjustment is predicted to be influenced most proximally by parenting practices and most distally by contextual factors (Reid and Eddy 1997). Within this framework, the contextual factors noted above (e.g., SES, education, acculturation) are thought to exert their effects on youngster adjustment indirectly, through their effects on parent adjustment and parenting practices. When negative contexts impinge on a family, many aspects of parenting practices can suffer and adolescent adjustment can be negatively affected. Thus, the effects of contextual factors on youngsters' adjustment are hypothesized to be mediated directly by parent adjustment and parenting practices. However, when parents are immigrants who are simultaneously parenting while adapting to challenging new sociocultural contexts, stress can increase leading to related negative outcomes for parents and adolescents. Our model also draws upon ecodevelopmental theory, adapted by (Szapocznik and Coatsworth 1999) from (Bronfenbrenner 1979), a theory that has been supported by extensive empirical research among immigrant and ethnic minority populations (Kulis et al. 2005; Martinez and Eddy 2005; Pantin et al. 2004; Prado et al. 2008; Sullivan et al. 2007). This model considers both the mesosystem and exosystem influences (i.e., social context stressors) that may set the stage for disruptions in functioning for adolescents (Fig. 15.1).

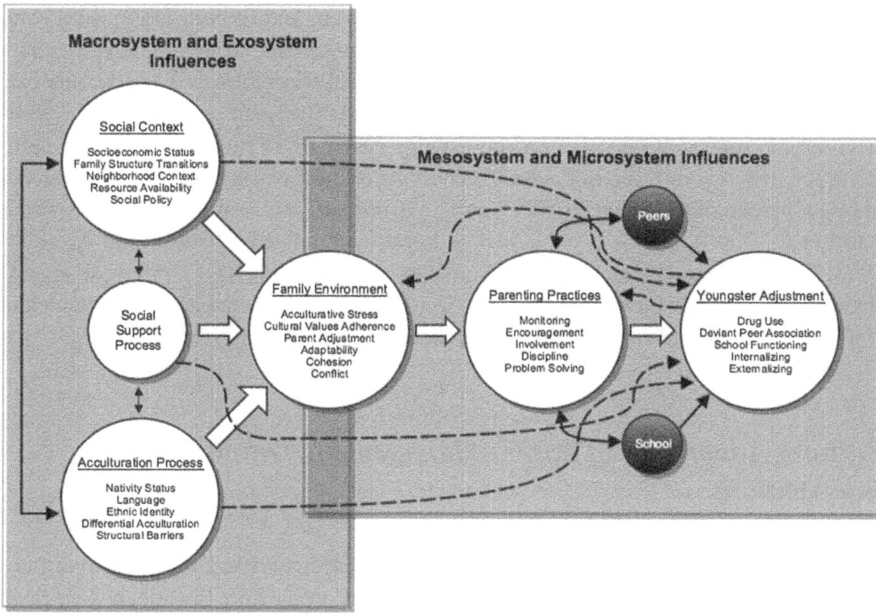

Fig. 15.1 Developmental model of Latino youth adjustment

Implications of the Developmental Model for Prevention in Emerging States

One consistent theme of the popular development models that describe the etiology of substance use and abuse is the central influence that family has on developmental pathways, particularly for Latinos (see Szapocnik et al. 2007; Prado et al. 2008; Martinez et al. 2003). Yet, surprising few interventions have been developed that target substance use within culturally specific contexts for Latino families. In addition, little research has examined the impacts of culturally specific interventions in comparison with more universal (or mainstream) approaches, and a debate in the field persists about whether cultural adaptation is even necessary (Castro et al. 2004; Lau 2006; Miranda et al. 2005).

A large of body of research supports the general efficacy of family-based interventions for drug abuse prevention and treatment. Reviewers of the literature consistently conclude that parent interventions, in particular, impact adjustment outcomes for children across a broad spectrum of behavioral, social, and emotional problems, including substance use (Kazdin 1987; Lipsey and Wilson 1993; Weisz et al. 1995). Parent Management Training (PMT) represents a group of interventions with a long track record of empirical support (Brestan and Eyberg 1998; Reid et al. 2002). For example, in a review of 82 studies on psychosocial treatment of conduct disordered children and adolescents, the PMT model was at the core of the only two interventions that were found to meet the most stringent criteria of a "well-established" intervention (Brestan and Eyberg 1998). PMT interventions have demonstrated efficacy in reducing a wide variety of related youth problems such as out of home placements, police contacts, and days institutionalized (Chamberlain 1990; Eddy et al. 2000), problem behaviors at home and school (Dishion and Andrews 1995; Forgatch and DeGarmo 1999; Martinez and Forgatch 2001), physical aggression among students on the playground (Reid et al. 1999), depression (Forgatch and DeGarmo 1999), and substance use (Dishion and Andrews 1995).

Cultural Adaptation of Parent Management Training Interventions

The theoretical underpinnings, structure, and content of PMT interventions have been developed and evaluated primarily with children from European American backgrounds (Forehand and Kotchick 1996). While some emerging research demonstrates the promise of PMT interventions to improve outcomes among culturally diverse populations (e.g., Webster-Stratton et al. 2001), numerous researchers continue to espouse the need to develop and evaluate the efficacy of culturally specified PMT interventions programs (Castro et al. 2004; Forehand and Kotchick 1996; Kumpfer et al. 2002; Lau 2006; Martinez and Eddy 2005). Unfortunately, there is limited literature on the effects of PMT intervention programs for specific ethnic subgroups, though numerous researchers have suggested that the principles of PMT may be particularly suitable across ethnic groups because of its inherent flexibility in how interventions are implemented with individual families and because of the strong family orientation within many culturally diverse communities (Kumpfer et al. 2002).

Informal and formal reviews of studies that have been conducted to evaluate the efficacy of culturally specific PMT interventions are mixed. Some reviews have indicated equivalent or somewhat better outcomes for culturally adapted versus non-adapted programs, while others suggest that there is little evidence for the greater efficacy of culturally adapted interventions overall (Castro et al. 2004; Griner and Smith 2006; Kazdin 1993; Lau 2006). While critical questions remain about the incremental efficacy of culturally adapted PMT interventions in comparison to non-adapted approaches, no pattern of evidence suggests that culturally adapted programs fare *worse* in terms of efficacy, and, more importantly, there is evidence that such programs increase the participation, completion, and community uptake of the intervention (Harachi et al. 1997; Castro et al. 2004).

Culturally Specific Preventive Interventions in Emerging Immigrant Community Contexts

Within a culturally specific preventive intervention context, cultural variables are frequently viewed as essential to bolster protective factors within families that buffer them against risk processes (Szapocznik et al. 2007). Recent reviews of culturally specific interventions for Latino families with children at risk of substance use have identified a broad range of prevention programs that have demonstrated efficacy by addressing such culturally specific factors (Castro et al. 2006; Prado et al. 2008; Szapocznik et al. 2007). For example, in a recent review, Szapocnik et al. (2007) identified four Latino adolescent substance use prevention programs that involved randomized controlled trials and were found to be efficacious: *Family Effectiveness Training* (Szapocznik et al. 1989), *Familias Unidas* (Pantin et al. 2003), *Keepin' it REAL* (Kulis et al. 2005), and *Nuestras Familias* (Martinez and Eddy 2005). Among the programs that are family-based, two of those were developed and tested within the context of emerging Latino immigrant communities.

Family Effectiveness Training (FET)

The early roots of FET were developed in South Florida in the 1970s, at a time when that region was experiencing rapid and steep growth of the Latino immigrant population, in the context of a larger social environment ill-prepared to accommodate rapid growth—much like the new immigrant growth states of 2009. FET is grounded in a structural systems family therapy model and was designed to address intergenerational conflict, acculturative stress, and family disorganization in Cuban families with youngsters at risk of substance use (Szapocznik et al. 1986, 1989). Importantly, as part of FET, parents learn effective parenting skills including family communication, positive encouragement, problem solving, and conflict resolution. During the course of the program, interventionists also use Bicultural Effectiveness Training (BET) to help families address intergenerational and intercultural conflict arising from acculturation gaps between parents and youngsters (Szapocznik et al. 1989). Randomized studies have shown that FET produced benefits to youngster problem behavior as reported by parents, and youngster report of self-concept relative to a minimum contact control condition (Szapocznik et al. 1989). In another study, the bicultural effectiveness training component of FET was evaluated independently in a comparison with structural family therapy (Szapocznik et al. 1986). BET was as effective as structural family therapy in improving questionnaire-assessed family interaction patterns, adolescent behavior problems, and psychopathology, and was more effective in reducing acculturation and bicultural gaps for Cuban families.

Nuestras Familias: Andando Entre Culturas

The *Nuestras Familias* intervention (Martinez and Eddy 2005) is theoretically grounded in social interaction learning theory (Reid et al. 2002) and ecodevelopmental theory (Szapocznik and Coatsworth 1999; see above). The 12-week, group-delivered intervention program is a culturally specific adaptation of PMT (Forgatch and Martinez 1999), and is designed to promote healthy adjustment in families and prevent substance use and other behavioral adjustment problems. The intervention is built upon a parent empowerment framework that actively works to strengthen self-efficacy among Latino immigrant parents who often have experienced disempowerment through their course of adaptation to life in the U.S. in the emerging immigrant community context. The program is designed to support general parenting skills in encouragement, parental monitoring, discipline, and problem solving while utilizing a framework that recognizes and supports the culturally specific application of these skills within Latino families. Moreover, the intervention focuses on a host of culturally specific contexts that are critical for healthy adjustment among immigrant Latino families (e.g., addressing differential acculturation, dealing with racism and structural barriers within the educational and other systems). In a recent randomized controlled trial, *Nuestras Familias* was shown to be efficacious in increasing parent reports of general parenting, skill encouragement, and effect parenting practices. The intervention was also shown to improve parent report of youth academic success, to reduce parent report of youth aggression and other externalizing behavior problems, and to reduce youth report of future likelihood of tobacco use. Importantly, the intervention focused entirely on parents, so outcomes for youth are particularly noteworthy as the theoretical model positions parents as the most influential and proximal source of change for improving youth outcomes.

While the literature suggests the need for more culturally specific interventions that are contextualized for the unique challenges faced by Latino families within emerging immigrant community contexts, too few programs exist that integrate concerns for the effects of segmented assimilation or of unique social and cultural contexts—and moderators of risk within these contexts—into the structure, theoretical basis, and content of the intervention.

Looking to the Future

Anticipated demographic change in the U.S. as a whole adds urgency to the need for this research. By 2050, the Census predicts that one in three people in the United States will be Latino (U.S. Census 2009b). As part of this demographic shift, Latinos, and especially the children of Latino immigrants, will increasingly play key roles in economic, political and social sectors, for instance, supplying the labor market and thus supporting (through social security taxes) a very large elderly and primarily non-Latino population (Pew Hispanic Center 2005). If the Latino population continues to include a high proportion of foreign-born individuals and their children who settle in states that are unprepared to support their successful integration, existing disparities in outcomes, such as Latino youth drug use, may persist and possibly worsen.

The coming years will be a critical time for the study of contributors to Latino youth drug use in social contexts that are poorly represented in the current literature, and prevention interventions designed in response to within-group differences may be of critical importance. Further, these interventions can be informed by community-based promising practices that have emerged over generations as Latino families have navigated adaptation challenges that come with life in states with emerging immigrant communities. While the shift toward evidence-based practices is important in ensuring dissemination of high-quality and effective services, our communities will benefit from enhancing support for research efforts that allow for the validation of these essential community-based practices, and for their ultimate incorporation into evidence-based prevention and intervention programs that are responsive to differential effects of segmented assimilation in the U.S. on Latinos, and particularly on those who are foreign-born. Programs that support the health and well-being of immigrant families in new growth states, such as those highlighted in this chapter, will ultimately further the life chances of children of immigrants, a group of young people that all of us in the U.S. are relying upon to succeed.

Acknowledgement Support for this project was provided by Grants Nos. R01 DA017937 and R01 DA01965 (Charles R. Martinez, Jr., Principal Investigator) from the National Institute on Drug Abuse, U.S. PHS, and by Grant No. R01 MH054248 (J. Mark Eddy, Principal Investigator) from the National Institute of Mental Health, US PHS.

References

Acs, G., & Nichols, A. (2005). *Working to make ends meet: Understanding the income and expenses of America's low-income families* (Low-Income Working Families Paper 2). Washington, DC: The Urban Institute.

Anthony, J. C., & Petronis, K. R. (1995). Early-onset drug use and risk of later drug problems. *Drug and Alcohol Dependence, 40*(1), 9–15.

Arias, E., Anderson, R. N., Kung, H. C., Murphy, S. L., & Kochanek, K. D. (2003). Deaths: Final data for 2001. *National Vital Statistics Reports, 52*(3), 1–116.

Bennett, G. G., Wolin, K. Y., Robinson, E. L., Fowler, S., & Edwards, C. L. (2005). Perceived racial/ethnic harassment and tobacco use among African American young adults. *American Journal of Public Health, 95*(2), 238–240.

Berry, J. W. (1998). Acculturation and health: Theory and research. In S. S. Kazarian & D. R. Evans (Eds.), *Cultural clinical psychology: Theory, research, and practice* (pp. 39–57). New York: Oxford University Press.

Blau, J. R., & Stearns, E. (2002). Adolescent integrity: Race and ethnic differences. *Critical Sociology, 28*(1–2), 145–167.

Block, J., Block, J. H., & Keyes, S. (1988). Longitudinally foretelling drug use in adolescence: Early childhood personality and environmental precursors. *Child Development, 59*, 336–355.

Boardman, J. D., Finch, B. K., Ellison, C. G., Williams, D. R., & Jackson, J. S. (2001). Neighborhood disadvantage, stress, and drug use among adults. *Journal of Health and Social Behavior, 42*, 151–165.

Borrell, L. N., Jacobs, D. R. J., Williams, D. R., Pletcher, M. J., Houston, T. K., & Kiefe, C. I. (2007). Self-reported racial discrimination and substance use in the coronary artery risk development in adults study. *American Journal of Epidemiology, 166*(9), 1068–1079.

Brestan, E., & Eyberg, S. (1998). Effective psychosocial treatments of conduct-disordered children and adolescents: 29 years, 82 studies, and 5,272 kids. *Journal of Clinical Child Psychology, 27*, 180–189.

Broman, C. L., Mavaddat, R., & Hsu, S. (2000). The experience and consequences of racial discrimination. *Journal of African American Psychology, 26*, 165–180.

Brondolo, E., Brady, N., Pencille, M., Betty, D., & Contrada, R. J. (2009a). Coping with racism: A selective review of the literature and a theoretical and methodological critique. *Journal of Behavioral Medicine, 32*(1), 64–88.

Brondolo, E., Gallo, L. C., & Myers, H. F. (2009b). Race, racism, and health: Disparities, mechanisms, and interventions. *Journal of Behavioral Medicine, 32*, 1–8.

Bronfenbrenner, U. (1979). *The ecology of human development*. Cambridge, MA: Harvard University Press.

Bruce, J., Fisher, P. A., Pears, K. C., & Levine, S. (2009). Morning cortisol levels in preschool-aged foster children: Differential effects of maltreatment type. *Developmental Psychobiology, 51*(1), 14–23.

Capps, R. (2007, April 20). *Demographics of children in immigrant families. Immigrants and health policy: Implications for SCHIP reauthorization. Congressional briefing*. Washington, DC: The Urban Institute.

Capps, R., & Fortuny, K. (2006). Immigration and child and family policy. *A report retrieved* from http://www.urban.org/publications/311362.html

Capps, R., Fix, M. E., & Passel, J. S. (2002). *The dispersal of immigrants in the 1990s* (Policybrief #2). Retrieved from http://www.urban.org/publications/410589.html

Castro, F. G., Barrera, M., Jr., & Martinez, C. R., Jr. (2004). The cultural adaptation of prevention interventions: Resolving tensions between fidelity and fit. *Prevention Science, 5*(1), 41–45.

Castro, F. G., Barrera, M., Jr., Pantin, H., Martinez, C. R., Jr., Felix-Ortiz, M., Rios, R., et al. (2006). Substance abuse prevention intervention research with Hispanic populations. *Drug and Alcohol Dependence, 84S*, S29–S42.

Chamberlain, P. (1990). Comparative evaluation of specialized foster care for seriously delinquent youths: A first step. *Community Alternatives: International Journal of Family Care, 2*(2), 21–36.

Cheriel, C., Huguet, N., Gupta, S., McClure, H., Leman, R. F., & Ngo, D. L. (2009). Arthritic pain among Latinos: Results from a community-based survey. *Arthritis and Rheumatism, 61*(11), 1491–1496.

Cicchetti, D., & Rogosch, F. A. (2007). Personality, adrenal steroid hormones, and resilience in maltreated children: A multilevel perspective. *Development and Psychopathology, 19*, 787–809.

Coatsworth, J. D., Maldonado-Molina, M., Pantin, H., & Szapocznik, J. (2005). A person-centered and ecological investigation of acculturation strategies in Hispanic immigrant youth. *Journal of Community Psychology, 33*(2), 157–174.

Contrada, R. J., Ashmore, R. D., Gary, M. L., Coups, E., Egeth, J. D., Sewell, A., et al. (2001). Measures of ethnicity-related stress: Psychometric properties, ethnic group differences, and associations with well-being. *Journal of Applied Social Psychology, 31*, 1775–1820.

DeCaro, J. A., & Worthman, C. M. (2008a). Culture and the socialization of child cardiovascular regulation at school entry in the U.S. *American Journal of Human Biology, 20*, 572–583.

DeCaro, J. A., & Worthman, C. M. (2008b). Return to school accompanied by changing associations between family ecology and cortisol. *Developmental Psychobiology, 50*, 183–195.

DeGarmo, D. S., & Martinez, C. R., Jr. (2006). A culturally informed model of academic well-being for Latino youth: The importance of discriminatory experiences and social support. *Family Relations, 55*, 267–278.

Dishion, T. J., & Andrews, D. W. (1995). Preventing escalation in problem behaviors with high-risk young adolescents: Immediate and 1-year outcomes. *Journal of Consulting and Clinical Psychology, 63*(4), 538–548.

Dominguez, T. P., Strong, E. F., Krieger, N., Gillman, M. W., & Rich-Edwards, J. W. (2009). Differences in the self-reported racism experiences of U.S.-born and foreign-born African American pregnant women. *Social Science & Medicine, 69*(2), 258–265.

Dozier, M., Manni, M., Gordon, M. K., Peloso, E., Gunnar, M. R., Stovall-McClough, K. C., et al. (2006). Foster children's diurnal production of cortisol: An exploratory study. *Child Maltreatment, 11*, 189–197.

Eddy, J. M., Reid, J. B., & Fetrow, R. A. (2000). An elementary-school based prevention program targeting modifiable antecedents of youth delinquency and violence: Linking the interests of families and teachers (LIFT). *Journal of Emotional and Behavioral Disorders, 8*(3), 165–176.

Edwards, L. M., & Romero, A. J. (2008). Coping with discrimination among Mexican descent adolescents. *Hispanic Journal of Behavioral Sciences, 30*(1), 24–39.

Farquhar, S., Samples, J., Ventura, S., Davis, S., Abernathy, M., McCauley, L., et al. (2008a). Promoting the occupation health of indigenous farmworkers. *Journal of Immigrant and Minority Health, 10*(3), 269–280.

Farquhar, S., Shadbeh, N., Samples, J., Ventura, S., & Goff, N. (2008b). Occupational conditions and well-being of indigenous farmworkers. *American Journal of Public Health, 98*(11), 1–4.

Finch, B. K., Kolody, B., & Vega, W. A. (2000). Perceived discrimination and depression among Mexican-origin adults in California. *Journal of Health and Social Behavior, 41*(3), 295–313.

Forehand, R., & Kotchick, B. A. (1996). Cultural diversity: A wake-up call for parent training. *Behavior Therapy, 27*(2), 187–206.

Forgatch, M. S., & DeGarmo, D. S. (1999). Parenting through change: An effective prevention program for single mothers. *Journal of Consulting and Clinical Psychology, 67*(5), 711–724.

Forgatch, M. S., & Martinez, C. R., Jr. (1999). Parent management training: A program linking basic research and practical application. *Journal of the Norwegian Psychological Society, 36*, 923–937.

Fortuny, K., Capps, R., Simms, M., & Chaudry, A. (2009). *Children of immigrants: National and State characteristics* (Brief no. 9). Washington, DC: The Urban Institute.

Galea, S., Ahern, J., Tracy, M., & Vlahov, D. (2007). Neighborhood income and income distribution and the use of cigarettes, alcohol and marijuana. *American Journal of Preventive Medicine, 32*(6S), S195–S202.

Gil, A. G., Wagner, E. F., & Vega, W. A. (2000). Acculturation, familism and alcohol use among Latino adolescent males: Longitudinal relations. *Journal of Community Psychology, 28*(4), 443–458.

Gonzales-Berry, E., Mendoza, M., & Plaza, D. (2006). Segmented assimilation of one-and-a-half generation Mexican youth in Oregon. *Latino(a) Research Review, 6*(1–2), 94–118.

Goodman, E., & Huang, B. (2002). Socioeconomic status, depressive symptoms and adolescent substance use. *Archives of Pediatrics & Adolescent Medicine, 156*, 448–453.

Goodman, E., Adler, N. E., Kawachi, I., Frazier, A. L., Huang, B., & Colditz, G. A. (2001). Adolescents' perceptions of social status: Development and evaluation of a new indicator. *Pediatrics, 108*(2), E31.

Greene, J. P., & Winters, M. A. (2002). *Public school graduation rates in the United States* (Civic Report, No. 31). New York: The Manhattan Institute for Policy Research.

Griner, D., & Smith, T. (2006). Culturally adapted mental health intervention: A meta-analytic review. *Psychotherapy: Theory, Research, Practice, Training, 43*(4), 531–548.

Gunnar, M., & Donzella, B. (2002). Social regulation of the cortisol levels in early human development. *Psychoneuroendocrinology, 27*, 199–220.

Guthrie, B. J., Young, A. M., Williams, D. R., Boyd, C. J., & Kinter, E. K. (2002). African American girls' smoking habits and day-to-day experiences with racial discrimination. *Nursing Research, 51*(3), 183–190.

Harachi, T. W., Catalano, R. F., & Hawkins, J. D. (1997). Effective recruitment for parenting programs within ethnic minority communities. *Child and Adolescent Social Work Journal, 14*(1), 23–39.

Harwood, R., Leyendecker, B., Carlson, V., Asencio, M., & Miller, A. (2002). Parenting among Latino families in the U.S. In M. H. Bornstein (Ed.), *Handbook of parenting: Social conditions and applied parenting* (Vol. 4, pp. 21–46). Mahwah: Lawrence Erlbaum Associates.

Hawkins, J. D., Catalano, R. F., & Miller, J. Y. (1992). Risk and protective factors for alcohol and other drug problems in adolescence and early adulthood: Implications for substance abuse prevention. *Psychological Bulletin, 112*(1), 64–105.

Inman, A. G., & Yeh, C. J. (2007). Asian American stress and coping. In F. T. L. Leong, A. G. Inman, A. Ebreo, L. H. Yang, L. Kinoshita, & M. Fu (Eds.), *Handbook of Asian American psychology* (pp. 323–339). Los Angeles: Sage.

Johnston, L. D., O'Malley, P. M., Bachman, J. G., & Schulenberg, J. E. (2008). *Future national results on adolescent drug use: Overview of key findings* (No. 09-7401). Bethesda: National Institute on Drug Abuse.

Johnston, L. D., O'Malley, P. M., Bachman, J. G., & Schulenberg, J. E. (2009). Demographic subgroup trends for various licit and illicit drugs, 1975–2008 (Monitoring the future occasional paper no. 71). From http://www.monitoringthefuture.org/

Kazdin, A. E. (1987). Treatment of antisocial behavior in children: Current status and future directions. *Psychological Bulletin, 102*(2), 187–203.

Kazdin, A. E. (1993). Adolescent mental health: Prevention and treatment programs. *American Psychologist, 48*(2), 127–141.

Kellam, S. G., Brown, C. H., Rubin, B. R., & Ensminger, M. E. (1983). Paths leading to teenage psychiatric symptoms and substance use: Developmental epidemiological studies in Woodlawn. In S. B. Guze, F. J. Earls, & J. E. Barrett (Eds.), *Childhood psychopathology and development* (pp. 17–47). New York: Raven.

Kulis, S., Marsiglia, F. F., Elek, E., Dustman, P., Wagstaff, D. A., & Hecht, M. (2005). Mexican/Mexican American adolescents and keepin' it REAL: An evidence-based substance use prevention program. *Children and Schools, 27*(3), 133–145.

Kumpfer, K. L., Alvarado, R., Smith, P., & Bellamy, N. (2002). Cultural sensitivity and adaptation in family-based prevention interventions. *Prevention Science, 3*(3), 241–246.

Kurtines, W. M., & Szapocznik, J. (1996). Structural family therapy in contexts of cultural diversity. In E. Hibbs & R. Jensen (Eds.), *Psychosocial treatment research with children and adolescents* (pp. 671–697). Washington, DC: American Psychological Association.

Latkin, C., Curry, A. D., Hua, W., & Davey, M. A. (2007). Direct and indirect associations of neighborhood disorder with drug use and high-risk sexual partners. *American Journal of Preventive Medicine, 32*(6S), S234–S241.

Lau, A. S. (2006). Making the case for selective and directed cultural adaptations of evidence-based treatments: Examples from parent training. *Clinical Psychology: Science and Practice, 13*(4), 295–310.

LaVeist, T. A., Thorpe, R. J., Mance, G. A., & Jackson, J. (2007). Overcoming confounding of race with socio-economic status and segregation to explore race disparities in smoking. *Addiction, 102*(Suppl. 2), 65–70.

Lipsey, M. W., & Wilson, D. B. (1993). The efficacy of psychological, educational and behavioral treatment: Confirmation from meta-analysis. *American Psychologist, 48*(12), 1181–1209.

Lowry, R., Kann, L., Collins, J. L., & Kolbe, L. J. (1996). The effect of socioeconomic status on chronic disease behaviors among U.S. adolescents. *JAMA, 276*, 792–797.

Martinez, C. R., Jr. (2006). Effects of differential family acculturation on Latino adolescent substance use. *Family Relations, 55*(3), 306–317.

Martinez, C. R. Jr., & Eddy, J. M. (2004). *From best practices to best processes: Issues in cultural adaptation of prevention and treatment programs*. Manuscript in preparation.

Martinez, C. R., Jr., & Eddy, J. M. (2005). Effects of culturally adapted Parent Management Training on Latino youth behavioral health outcomes. *Journal of Consulting and Clinical Psychology, 73*(5), 841–851.

Martinez, C. R., Jr., & Forgatch, M. S. (2001). Preventing problems with boys' noncompliance: Effects of a parent training intervention for divorcing mothers. *Journal of Consulting and Clinical Psychology, 69*(3), 416–428.

Martinez, C. R., Jr., Eddy, J. M., & DeGarmo, D. S. (2003). Preventing substance use among Latino youth. In W. J. Bukoski & Z. Sloboda (Eds.), *Handbook of drug abuse prevention: Theory, science, and practice* (pp. 365–380). New York: Kluwer Academic/Plenum Publishers.

Martinez, C. R., Jr., McClure, H. H., & Eddy, J. M. (2008). Latino immigrant children and families: Demographics, challenges, and promise. In B. Bussell (Ed.), *Understanding the immigrant experience in Oregon: Research, analysis, and recommendations from University of Oregon Scholars*. Eugene: University of Oregon Press.

Martinez, C. R., Jr., McClure, H. H., & Eddy, J. M. (2009). Language brokering contexts and behavioral and emotional adjustment among Latino parents and adolescent. *Journal of Early Adolescence, 29*(1), 71–98.

McClure, H. H., Martinez, C. R., Jr., Eddy, J. M., Jimenez, R., Isiordia, L. E., McDade, T. W., & Snodgrass, J. J. (2008). Stress and lifestyle change among recent Latino immigrants in Oregon. *American Journal of Human Biology, 20*(2), 201–211.

McClure, H. H., Martinez, C. R., Jr., Snodgrass, J. J., Eddy, J. M., Jiménez, R., Isiordia, L. E., & McDade, T. W. (2010a). Discrimination-related stress, blood pressure, and Epstein Barr Virus antibodies among Latin American immigrants in Oregon. *Journal of Biosocial Science, 42*(4), 433–461.

McClure, H. H., Snodgrass, J. J., Martinez, C. R., Jr., Eddy, J. M., Jiménez, R. A., & Isiordia, L. E. (2010b). Discrimination, psychosocial stress, and health among Latin American immigrants in Oregon. *American Journal of Human Biology, 22*(3), 421–423.

Mendoza, M., & McClure, H. H. (2009). *Centro LatinoAmericano: Contributing to the integration of Latino families in Lane County*. Eugene/Lane County: Department of Health and Human Services/Centro LatinoAmericano.

Miranda, J., Bernal, G., Lau, A., Kohn, L., Hwang, W.-C., & LaFromboise, T. (2005). State of the science on psychosocial interventions for ethnic minorities. *Annual Review of Clinical Psychology, 1*, 113–142.

Murray, D. M., O'Connell, C. M., Schmid, L. A., & Perry, C. L. (1987). The validity of smoking self-reports by adolescents: A reexamination of the bogus pipeline procedure. *Addictive Behaviors, 12*, 7–15.

Murry, V. M., Harrell, A. W., Brody, G. H., Chen, Y.-F., Simons, R. L., Black, A. R., et al. (2008). Long-term effects of stressors on relationship well-being and parenting among rural African American women. *Family Relations, 57*, 117.

National Research Council. (2004). *Measuring racial discrimination*. Washington, DC: The National Academies Press.

Okamoto, J., Ritt-Olson, A., Soto, D., Baezconde-Garbanati, L., & Unger, J. B. (2009). Perceived discrimination and substance use among Latino adolescents. *American Journal of Health Behavior, 336*(6), 718–727.

Oregon Department of Education. (2010). *Dropout rates in Oregon high schools*. Retrieved from http://www.ode.state.or.us/search/page/?=1

Oregon Social Learning Center Latino Research Team. (2007). *Adolescent Latino acculturation study* (Unpublished data). Eugene: Oregon Social Learning Center Latino Research Team.

Oregon Youth Authority. (2009). *Issue brief*. Salem: Oregon Youth Authority.

Pantin, H., Coatsworth, J. D., Feaster, D. J., Newman, F. L., Briones, E., Prado, G., et al. (2003). Familias Unidas: The efficacy of an intervention to promote parental investment in Hispanic immigrant families. *Prevention Science, 4*(3), 189–201.

Pantin, H., Schwartz, S. J., Sullivan, S., Prado, G., & Szapocznik, J. (2004). Ecodevelopmental HIV prevention programs for Hispanic adolescents. *American Journal of Orthopsychiatry, 74*(4), 545–558.

Paradies, Y. C. (2006). Defining, conceptualizing and characterizing racism in health research. *Critical Public Health, 16*(2), 143–157.

Passel, J. S., Capps, R., & Fix, M. (2004). *Undocumented immigrants: Facts and figures (a fact sheet)*. Washington, DC: The Urban Institute.

Pears, K., & Fisher, P. A. (2005). Developmental, cognitive, and neuropsychological functioning in preschool-aged foster children: Associations with prior maltreatment and placement history. *Journal of Developmental and Behavioral Pediatrics, 26*, 112–122.

Pérez, D. J., Fortuna, L., & Alegría, M. (2008). Prevalence and correlates of everyday discrimination among U.S. Latinos. *Journal of Community Psychology, 36*(4), 421–433.

Pew Hispanic Center. (2005). *Hispanics and the social security debate*. Washington, DC: Pew Hispanic Center.

Portes, A., & Zhou, M. (1993). The new second generation: Segmented assimilation and its variants. *The Annals of the American Academy, 530*, 74–96.

Prado, G., Szapocznik, J., Maldonado-Molina, M. M., Schwartz, S. J., & Pantin, H. (2008). Drug use/abuse prevalence, etiology, prevention, and treatment in Hispanic adolescents: A cultural perspective. *Journal of Drug Issues, 28*(1), 5–36.

Reid, J. B., & Eddy, J. M. (1997). The prevention of antisocial behavior: Some considerations in the search for effective interventions. In D. M. Stoff, J. Breiling, & J. D. Master (Eds.), *The handbook of antisocial behavior* (pp. 343–356). New York: Wiley.

Reid, J. B., Eddy, J. M., Fetrow, R. A., & Stoolmiller, M. (1999). Description and immediate impacts of a preventive intervention for conduct problems. *American Journal of Community Psychology, 27*(4), 483–517.

Reid, J. B., Patterson, G. R., & Snyder, J. (Eds.). (2002). *The Oregon model: Understanding and altering the delinquency trajectory*. Washington, DC: American Psychological Association.

Rospenda, K. M., Richman, J. A., & Shannon, C. A. (2008). Prevalence and mental health correlates of harassment and discrimination in the workplace: Results from a National study. *Journal of Interpersonal Violence, 24*(5), 819–843.

Rutledge, P. C., & Sher, K. J. (2001). Heavy drinking from the freshman year into early young adulthood: The roles of stress, tension-reduction drinking motives, gender and personality. *Journal of Studies on Alcohol, 62*, 457–466.

Sabogal, F., Marin, G., Otero-Sabogal, R., Marin, B. V., & Perez-Stable, E. J. (1987). Hispanic familism and acculturation: What changes and what doesn't? *Hispanic Journal of Behavioral Sciences, 9*, 397–412.

Stanton-Salazar, R. D. (2001). *Manufacturing hope and despair: The school and kin support networks of U.S.-Mexican youth*. New York: Teachers College Press.

Stephen, L. (2007). *Transborder lives: Indigenous Oaxacans in Mexico, California, and Oregon*. Durham: Duke University Press.

Sullivan, S., Schwartz, S. J., Prado, G., Pantin, H., Huang, S., & Szapocznik, J. (2007). A bidimensional model of acculturation for examining differences in family functioning and behavior problems in Hispanic immigrant adolescents. *Journal of Early Adolescence, 27*, 405–430.

Szalacha, L. A., Erkut, S., Garcia Coll, C., Fields, J. P., Alarcon, L., & Ceder, I. (2003). Perceived discrimination and resilience. In S. S. Luthar (Ed.), *Resilience and vulnerability: Adaptation in the context of childhood adversities* (pp. 414–435). Cambridge: Cambridge University Press.

Szapocznik, J., & Coatsworth, J. D. (1999). An ecodevelopmental framework for organizing the influences on drug abuse: A developmental model of risk and protection. In M. D. Glantz & C. R. Hartel (Eds.), *Drug abuse: Origins & interventions* (pp. 331–366). Washington, DC: American Psychological Association.

Szapocznik, J., Rio, A. T., Perez-Vidal, A., Kurtines, W., Hervis, O., & Santisteban, D. (1986). Bicultural effectiveness training (BET): An experimental test of an intervention modality for families experiencing intergenerational/intercultural conflict. *Hispanic Journal of Behavioral Sciences, 8*(4), 303–330.

Szapocznik, J., Santisteban, D., Rio, A. T., Perez-Vidal, A., Santisteban, D., & Kurtines, W. M. (1989). Family effectiveness training: An intervention to prevent drug abuse and problem behaviors in Hispanic adolescents. *Hispanic Journal of Behavioral Sciences, 11*(1), 4–27.

Szapocznik, J., Prado, G., Burlew, A. K., Williams, R. A., & Santisteban, D. A. (2007). Drug abuse in African American and Hispanic adolescents: Culture, development and behavior. *Annual Review of Clinical Behavior, 3*, 77–105.

Turner, W. L., & Hench, M. J. (2003). African-American substance use epidemiology and prevention issues. In Z. Sloboda & W. J. Bukoski (Eds.), *Handbook of drug abuse prevention: Theory, science and practice*. New York: Kluwer Academic/Plenum Publishers.

U.S. Census. (1990). *Summary Tape File 1 (STF 1) – 100-percent data*. Retrieved January 2, 2010, from: www.census.gov

U.S. Census. (2009a). *2005–2007 American Community Survey, 3-year estimates*. Retrieved September 25, 2009, from http://factfinder.census.gov/

U.S. Census. (2009b). *2005–2007 American Community Survey, 3-year estimates*. Retrieved September 25, 2009, from http://factfinder.census.gov/servlet/DTTable?_bm=y&-context=dt&-ds_name=ACS_2007_3YR_G00_&-CONTEXT=dt&-mt_name=ACS_2007_3YR_G2000_C05003I&-tree_id=3307&-geo_id=04000US41&-search_results=01000US&-format=&-_lang=en

Urban Institute. (2009). *Children of immigrants data tool*. Urban Institute tabulations from the integrated public use microdata series datasets drawn from the 1990 U.S. Census of Population and Housing, 5 percent sample, and the 2005 and 2006 U.S. Census Bureau American Community Surveys. Retrieved October 15–18, 2009, from http://datatool.urban.org/charts/datatool/pages.cfm

Valencia, E. Y., & Johnson, V. (2008). Acculturation among Latino youth and the risk for substance use: Issues of definition and measurement. *Journal of Drug Issues, 38*(1), 37–68.

Vega, W. A. (1990). Hispanic families in the 1980s: A decade of research. *Journal of Marriage and the Family, 52*, 1015–1024.

Vega, W. A., & Gil, A. G. (1999). A model for explaining drug use behavior among Hispanic adolescents. *Drugs and Society, 14*(1–2), 57–74.

Vermeiren, R., Schwab-Stone, M., Deboutte, D., Leckman, P. E., & Ruchkin, V. (2003). Violence exposure and substance use in adolescents: Findings from three countries. *Pediatrics, 111*(3), 535–540.

Wallace, J. M. (1999). Explaining race differences in adolescent and young adult drug use: The role of racialized social systems. *Drugs and Society, 14*(1/2), 21–36.

Wallace, J. M., Jr., Bachman, J. G., O'Malley, P. M., & Johnston, L. D. (1995). Racial/ethnic differences in adolescent drug use: Exploring possible explanations. In G. J. Botvin, S. Schinke, & M. A. Orlandi (Eds.), *Drug use prevention with multiethnic youth* (pp. 59–80). Thousand Oaks: Sage.

Warner, L. A., Valdez, A., Vega, W. A., De La Rosa, M. R., Turner, R. J., & Canino, G. (2006). Hispanic drug abuse in an evolving cultural context: An agenda for research. *Drug and Alcohol Dependence, 84S*, S8–S16.

Webster-Stratton, C., Reid, J., & Hammond, M. (2001). Social skills and problem-solving training for children with early-onset conduct problems: Who benefits? *Journal of Child Psychology & Psychiatry, 42*(7), 943–952.

Weisz, J. R., Weiss, B., Han, S. S., Granger, D. A., & Morton, T. (1995). Effects of psychotherapy with children and adolescents revisited: A meta-analysis of treatment outcome studies. *Psychological Bulletin, 117*, 450–468.

Wilson, J. Q. (1996). Foreword. In G. L. Kelling & C. M. Coles (Eds.), *Fixing broken windows: Restoring order and reducing crime in our communities*. New York: Simon & Schuster.

Winkleby, M. A., Robinson, T. N., Sundquist, J., & Kraemer, H. C. (1999). Ethnic variation in cardiovascular disease risk factors among children and young adults: Findings from the Third National Health and Nutrition Examination Survey, 1988–1994. *JAMA, 281*, 1006–1013.

Yeh, C. J., Arora, A. K., & Wu, K. A. (2005). A new theoretical model of collectivistic coping. In P. T. P. Wong & L. C. J. Wong (Eds.), *Handbook of multicultural perspectives on stress and coping* (pp. 53–70). New York: Springer.

Chapter 16
Racial and Ethnic Influences on Mental Health and Substance Use Disorders: The Case of Caribbean Blacks

Krim K. Lacey, James S. Jackson, and Niki Matusko

Introduction: Drugs Use Among Blacks

Consistently, studies show that substance use rates are higher among Whites (Merline et al. 2004; Breslau et al. 2007; Kandel et al. 1997; Xueqin and Shive 2000; Kessler et al. 2005). Even given these findings, there remains a perception that Blacks use (and misuse) drugs at higher rates than other racial groups. These perceptions may be the result of the long-term health and behavioral consequences drugs have had on Blacks and their communities (Turner and Wallace 2003; Broman et al. 2008). For example, mortality rates for chronic disease associated with substance use were twice as high for Blacks as compared to whites (DHHS 2003). New HIV/AIDS cases among Blacks were also attributable to drug needle use or sex with a drug needle user (DHHS 2005).

Concerns over the impact of drug use and abuse in Black communities came at a time of noticeable growth of the U.S. Black population due in large part to migration from the Caribbean. Caribbean Blacks currently comprise one of the largest and fastest growing subgroups within the Unites States (Keane et al. 2008; Takeuchi et al. 2007). Little, however, is known about how context, environmental factors, immigration and nativity influence substance use and misuse among this population. A growing body of research suggests that these are important factors in health and mental health outcomes (Broman et al. 2008; Breslau et al. 2009; Breslau and Chang 2006; Williams et al. 2007).

Substance Abuse Among Blacks Across the Diaspora: Krim K. Lacey, Dawne M. Mouzon, Ishtar Govia, Niki Matusko, Ivy Forsythe-Brown, Jamie Miller Abelson & James S. Jackson. Substance Use and Misuse, accepted for publication by Taylor & Francis LLC (http://www.tandfonline.com).

K.K. Lacey (✉) • J.S. Jackson • N. Matusko
Institute for Social Research, Program for Research on Black Americans,
University of Michigan, Ann Arbor, MI, USA
e-mail: krimlacey@gmail.com

Y.F. Thomas, L.N. Price (eds.), *Drug Use Trajectories Among Minority Youth*,
DOI 10.1007/978-94-017-7491-8_16

Our understanding of substance use and misuse among the U.S. Black population is further complicated by the fact very few studies have examined within group factors. The tendency of research to group ethnic minorities into one large category can obscure significant differences (Nazroo 2001). Disaggregation will assist in furthering our understanding of substance use and misuse patterns among groups within the Black population. Additionally, comparative studies are needed between migrating U.S. Blacks and those residing within their homelands to broaden our understanding of the role of geographic residence and migration on substance use and misuse.

More studies on substance use among the adult Black population are also needed. Although drugs have adversely affected the lives of all members of minority communities, adolescent youth remain the primary emphasis of most drug studies (Boardman et al. 2001). In this chapter we explore the effects of nativity, environmental conditions, length of stay and mental health disorders on substance abuse among adult U.S. Blacks and individuals of Caribbean descent.

Structural and Social Stressors and Health

Research demonstrates that social and structural conditions are linked to physical and mental health outcomes (Baum et al. 2006; Darity 2003; Williams et al. 1997, 2003; Williams 1999; Acevedo-Garcia et al. 2003). Structural inequalities, such as poverty and segregation, limit access to quality health care and subsequently affect health outcomes (Acevedo-Garcia 2000). The social consequences of inequality produce psychological and emotional stress and can have a lasting effect on the well-being and life course of individuals (Jackson et al. 2004).

Racial and ethnic minorities, particularly Blacks, are more likely to experience these structural and social stressors given their disadvantaged social standing in society (Turner and Wallace 2003; Peterson and Krivo 1999; Krivo et al. 1998). Blacks remain materially disadvantaged and geographically segregated especially in poor, core urban and rural areas (Massey et al. 1991; Massey and Eggers 1990). Low education attainment and high unemployment rates have been known to plague minority communities. Moreover, racial and ethnic minorities historically have been confronted with discrimination and racism in their social interactions.

Migrant populations face similar and other stressors as they relocate to new environments. In pursuit of a better life and improved socioeconomic opportunities (Bhugra 2004), they may underestimate some of the challenges that native minority groups contend with on a daily basis. Migrating to a new environment in and of itself is a major life change, which can be challenging to any individual. Along with this, some migrants may confront issues of cultural marginality, family separation, poverty, language barriers, loss of status and discrimination, which can encourage substance use and misuse (Johnson et al. 2002).

For other migrants, their stress may be compounded while searching for housing and employment in their new homes. Obtaining high-income jobs may especially pose a greater challenge if they lack experience or the educational background necessary to compete with others on the job market. And while some migrants might have the experience and credentials required for their profession, they may not be given the same value in the United States as in their homeland. The lack of available opportunities can leave some feeling disappointed and discouraged. There is evidence that suggests immigrants who are frustrated with opportunities in American society are more likely to feel depressed (cf. Vega and Rumbaut 1991; Latkin and Curry 2003), which is consistently associated with substance use and misuse (Williams and Adams-Campbell 2000; Weiss et al. 1992; McDowell and Clodfelter 2001; Chen et al. 2002). The social isolation resulting from their relocation can also have lasting effects on their mental health and consequentially lead to substance abuse (Bhurga 2004).

Coping and Drug Use

When living under stressful conditions, individuals seek ways of coping with the resulting experienced stress; and this may come in the form of drug use (VanGeest and Johnson 1997). The stress associated with poor economic conditions, exacerbated by lower educational attainment and racism are believed to make individuals more vulnerable to addictive behaviors, such as drug use (Galea et al. 2004; Kessler et al. 1999; Boardman et al. 2001; Williams and Williams-Morris 2000; Brady 2004).

A recent study found that unemployment, poor education and low-income among African Americans were linked to drug abuse (Broman et al. 2008). Other findings indicate a moderate relationship between disadvantaged neighborhoods and drug behavior, indirectly through increased social stressors and higher levels of psychological distress (Boardman et al. 2001; Galea et al. 2004). Research conducted among college students further illuminates the relationship between stress and substance misuse. Broman (2005) found that life and traumatic stress were associated with increased substance use. More studies are needed to understand the effects of social conditions on substance use and misuse among ethnic minorities and immigrant minority populations.

The extant literature suggests that migration is a risk factor for mental health disorders (Bhugra 2004; Breslau and Chang 2006; Nazroo 2001; Williams et al. 2007). A growing body of literature reveals that in many instances foreign-born migrants have better health than their native-born counterparts (Breslau and Chang 2006). The results of these studies are consistent with those that specifically focus on substance use and misuse. Immigrants to the United States were found to have lower rates of substance use and misuse. Broman and colleagues (2008) found that foreign-born Blacks were less likely to misuse drugs. Studies on older African Americans and individuals of Caribbean descent within the U.S. indicate lower levels of tobacco use among individuals of Caribbean decent (Keane et al. 2008). Johnson and colleagues (2002) found immigrants to the U.S. reported lower substance use prevalence than native-born citizens. However, other studies indicate that migrants are at greater risk for mental health and health related consequences. A study of first generation migrants to England and Wales found that rates of alcohol related consequences were marginally higher for Caribbeans than native British (Harrison et al. 1997). Evidence of increased alcohol use was also found for Indian and Black Caribbean communities among ethnic minorities in England (Rassool 2006). Although not the topic of this study, data from the Health Survey for England have identified large differences among ethnic groups in terms of psychotic and non-psychotic mental disorders, with Caribbean blacks consistently faring worse than other groups (Nazroo 2001; Nazroo et al. 2007).

Few studies have examined how length of time in the U.S. influences substance use and misuse by individuals of Caribbean descent residing within the United States. Research suggests that lengthier exposure to the U.S. context is associated with substance abuse and misuse (Johnson et al. 2002; Gfroere and Tan 2003; Blake 2001).

Chapter Objectives

This chapter addresses three questions: (1) Do rates of substance disorder differ between Black Americans and those residing in the Caribbean? (2) What are the factors that influence substance disorders across cohorts? (3) What roles do race/ethnicity, nativity, length of time in the United States, and geographic location play in lifetime substance disorders?

Approach

Data

Our study draws from multiple data sources that used probability sampling frames collected within the United States (2003), Guyana (2005) and Jamaica (2005).

United States The National Survey of American Life (NSAL) is the most comprehensive study to date conducted on ethnic minorities residing in the same context and the first national probability sample ever collected on Caribbean Blacks (Jackson et al. 2004). Collection of the data in the U.S. began in February 2001 and ended in March 2003. Face-to-face interviewing was the primary method of collection with an additional 10–20 % of the sample collected by telephone interviews. The average length of interviews on African Americans was 2 h and 20 min, slightly shorter than interviews conducted with Caribbean Blacks that typically lasted for 2 h and 43 min. In total, 6082 interviews were completed: 3570 of which were African Americans,

1621 Caribbean Blacks, and 891 non-Hispanic Whites who resided in Black populated regions of 10 % or more.

Guyana Questionnaires were the data collection method for this sample. The questionnaires were administered by indigenous interviewers to a probability sample of study participants in both the rural, suburban and urban parts of the island. Altogether 2068 questionnaires were completed between July and December 2005. The sample included Blacks 55.2 %, East Indian 34.7 %, and Mixed/Other 10.1 %.

Jamaica Face-to-face interviewing was used to collect the sample in Jamaica. Data collection for this study began in August and was completed in December 2005. Unlike U.S. and Guyanese samples, interviews were limited to a probability sample of those who resided in the urban Kingston metropolitan area. There were 1216 participants included in the sample. The sample consisted of 97.4 % Blacks, 1.3 % Asians, 1.4 % who were classified as "Other."

Measures

Socio-Demographic Variables The predictors included age, gender, marital status, household income, employment status, and education. The measures used in this study were for the most part consistent across samples except for household income and educational attainment. These variables differed slightly because of differences in systems between the United States and the Caribbean. For example, household income in the NSAL was obtained by the question, "Which comes closest to the total income you and (your family) had in 2000?" Participants were provided options in increments ranging from no income to more than one million dollars. Jamaica and Guyana study participants were asked, "Now, thinking about you and your family's total income from all sources, how much did you (and all the members of your family living there) receive in the year 2004 before taxes?" Household income for this group was collected on a continuous scale. In this study income was reported in quintiles; the lowest quintile representing the lowest income category and vice versa. Educational level in the NSAL was measured by asking participants, "What is the highest level of education you have completed, if they had obtained a high school diploma, college degree or certificate, and if they had any other form of schooling (e.g. high school, college) (Nazroo et al. 2007). Within the Guyana and Jamaica sample participants were asked, "What is the highest level of education you have completed (e.g. primary, high school, technical school, college/university, beyond 4 years) and if they had achieved their 'O' and 'A' levels.

Nativity Nativity status is a binary variable that measured whether participants had been born within or outside the United States. Only NSAL participants were administered this question.

Length of Time in the United States Length of time in the United States is a categorical variable that measured whether participants were born in the United States and the number of years they lived within the United States. Again, only NSAL participants were administered this question. This variable was not included in multivariate analysis because of potential colinearity concerns.

Mental Health Disorders Two mental health disorders, major depressive episode (MDE), and the major dependent variable, lifetime substance abuse, were included in this study. Substance use included the use of all drugs with the exception of tobacco. Criteria for these disorders were based on a slightly modified version of the World Health Organization Composite International Diagnostic Interview (WHO CIDI) as defined by the DSM IV.

Analytic Approach

We performed bivariate and multivariate analyses. Chi-square tests of significance were used to assess relationships and prevalence rates of substance abuse. Hierarchical logistic regression was employed to examine the contributions of explanatory factors (e.g. age, income) at each stage on substance abuse (Nazroo et al. 2007). These procedures were conducted across samples. The final step of the multivariate analysis was presented in the table of this study and discussed. The NSAL corrected for sample design. Post-stratification weights were applied to all samples. Significance level was set at the .05 level.

Study Outcomes

Sample Characteristics

Table 16.1 shows that participants' average age varied slightly across groups. African-Americans on average were older (M=42.3, SD=0.5) as compared to other cohorts. Females were over-represented, with the exception of Caribbean Americans where males made up a slightly larger (50.9 %) portion of the sample. U.S. participants (e.g. both African Americans and Caribbean Americans) had larger proportions in the highest household income quintile category (20.4 % vs. 31.6 %). Caribbean Americans, however, fared better among all groups. The educational attainment level was also higher for U.S. Blacks compared to the Caribbean samples. Among the groups, U.S. Caribbean Blacks attained the highest levels of education. A relatively large percentage of these participants had some form of training or education beyond the high school level (49.1 %). In contrast, a majority of Guyanese

Table 16.1 Characteristics of sample

Percentage (except for age)	NSAL		Caribbean samples	
Characteristics	African American (2003)	Caribbean American (2003)	Guyana (2005)	Jamaica (2005)
Mean age	42.3	40.3	40.5	37.2
Gender				
Male	44.0	50.9	48.2	29.8
Female	56.0	49.1	51.8	70.2
Equivalised income				
Bottom quintile	21.7	14.0	14.0	19.7
Second quintile	18.2	14.6	30.0	24.1
Middle quintile	21.1	20.8	23.4	1.5
Fourth quintile	18.7	19.1	22.4	43.6
Highest quintile	20.4	31.6	10.2	11.1
Education level				
Primary/some high school	24.2	21.2	54.0	26.1
High school graduate	37.9	29.7	29.7	51.8
College-vocation-technical	38.0	49.1	16.3	22.1
Employment status				
Employed	66.8	75.2	53.7	45.2
Unemployed	10.1	8.8	10.8	29.6
Not in the labor force	23.1	16.0	35.5	25.2
Marital status				
Married	32.9	37.6	34.2	19.5
Partner	8.7	12.6	16.0	13.6
Sep-div-widow	26.8	18.9	18.6	8.5
Never married	31.6	30.9	31.2	58.5
Length of time in the U.S.				
US born	–	35.8	–	–
<6 years	–	7.9	–	–
6–10 years	–	19.9	–	–
11–20 years	–	28.2	–	–
>20 years	–	8.3	–	–
[N]	3570	1621	2068	1216

Note. Statistics are unweighted

(54.0 %) had a primary or some high school level of education, while the Jamaican sample (51.8 %) consisted mostly of high school graduates. Across all cohorts, most individuals were employed. African Americans (32.9 %), Caribbean Americans (37.6 %), and Guyanese (34.2 %) reported higher rates of marriage. This was opposite of what was found in Jamaica where a higher proportion of respondents was not married (58.5 %). Finally, the highest proportion of U.S. Caribbean participants were born within the United States (35.8 %), followed by a relatively large percent that migrated 11–20 years previously (28.2 %), between 6 and 10 years (19.9 %), more than 20 years (8.3 %), and for less than 6 years (7.9 %).

Drug Use Analyses by Race, Ethnicity, Gender, Nativity and Length of Stay

Table 16.2 presents prevalence rates of substance abuse by socio-demographic factors, nativity and length of stay. Overall, rates of substance misuse were highest within the United States among African Americans and Caribbean Blacks (11.5 % vs. 9.6 %, ns). Differences were found in substance misuse between age groups. African Americans between the age of 45–59 had higher prevalence of substance

Table 16.2 Prevalence of substance abuse by socio-demographic factors, nativity and immigration

	African American (2003)	Caribbean American (2003)	Guyana (2005)	Jamaica (2005)
Lifetime substance abuse	11.5	9.6	4.7	2.7
Age				
18–29	8.4	15.2	3.5	1.9
30–44	12.6	8.0	6.5	2.0
45–59	15.0	3.6	4.2	2.8
>59	8.5	9.8	3.2	4.6
χ^2	25.68**	34.86	9.05*	3.20
P value	.008	.180	.029	.362
Gender				
Male	18.1	16.4	8.7	5.2
Female	6.3	2.8	1.0	1.2
χ^2	120.63***	90.46***	68.00***	18.16***
P value	.000	.000	.000	.000
Equivalised income				
Bottom quintile	15.3	13.2	4.8	4.2
Second quintile	11.5	4.6	3.1	2.0
Middle quintile	14.7	9.1	5.0	0.0
Fourth quintile	8.6	11.8	7.8	2.1
Highest quintile	6.9	9.5	2.4	1.5
χ^2	40.24***	12.71	15.94**	4.58
P value	.000	.662	.003	.333

(continued)

Table 16.2 (continued)

	African American (2003)	Caribbean American (2003)	Guyana (2005)	Jamaica (2005)
Education level				
Primary/some high school	18.0	15.2	4.9	3.8
High school graduate	9.6	4.5	4.7	2.2
College-vocation-technical	9.2	10.3	4.5	1.1
χ^2	48.03***	28.71	.134	4.60
P value	.000	.337	.935	.100
Employment status				
Employed	11.1	9.1	6.7	2.8
Unemployed	14.8	9.2	5.8	2.8
Not in the labor force	11.2	12.4	1.5	1.3
χ^2	4.43	2.66	26.74***	2.05
P value	.149	.687	.000	.359
Marital status				
Married	9.6	8.2	3.4	0.8
Partner	13.4	17.0	5.1	4.2
Sep-div-widow	14.1	5.6	5.5	4.9
Never married	10.8	10.8	5.4	2.1
χ^2	12.12*	22.20	4.10	7.74
P value	.021	.577	.251	.052
Nativity				
US born	11.6	19.5	–	–
Foreign	1.9	4.2	–	–
χ^2	7.84**	105.31**		
P value	.004	.002		
Length of stay				
US born	–	19.5	–	–
<6 years	–	0.5	–	–
6–10 years	–	11.1	–	–
11–20 years	–	1.0	–	–
20 years	–	5.9	–	–
χ^2	–	107.99***	–	–
P value	–	.002	–	–

Note. *$p<.05$, **$p<.01$, *** $p<.001$

misuse (15 %, $p<.01$). In the Caribbean, and among Guyanese, rates of substance misuse were highest among individuals within the 30–44 age group (6.5 %, $p<.05$).

Across all samples, males had higher rates of substance misuse ($p<.001$). Differences were found between household income and substance use for some groups, but not others. African Americans within the lowest household quintile income category were among those who had higher rates of substance misuse (15.3 %, $p < .001$). The opposite was found for Guyanese, where prevalence rates were higher among those in the upper quintile (fourth) category (7.8 %, $p < .01$).

Substance misuse differences were also found by educational attainment, employment, and marital status among the groups. A higher proportion of substance abuse was found among low educational achieving African Americans (18 %, $p < .001$). Employed Guyanese were found to have higher prevalence rates of substance misuse (6.7 %, $p < .001$). Also, higher rates of substance misuse were found among separated, divorced, or widowed African American (14.1 %, $p < .05$) and among Jamaicans (4.9 %, $p < .05$).

Both U.S. born African Americans and Caribbean Blacks had elevated levels of substance use ($p < .01$). The rates were noticeably higher for U.S. born Caribbean Blacks than it was for U.S. born African Americans (19.5 % vs. 11.6 %). An association was also observed between length of time in the United States and substance misuse. Specifically, individuals of Caribbean descent born in the U.S. had elevated levels of substance misuse (19.5 %, $p < .01$).

Logistic Regressions Predicting Substance Use

Table 16.3 presents the odd ratios of substance abuse across cohorts. Our findings indicate that while gender was a consistent predictor ($p < .001$) when other relevant factors are controlled, different factors predicted substance misuse for different groups. For instance, the odds of misusing drugs significantly increased among African Americans between the ages of 30–44 (AOR = 1.69, 95 % CI = 1.02, 2.80; $p = .042$) and 45–59 (AOR = 2.05, 95 % CI = 1.17, 3.60; $p = .014$). By contrast, there were reduced odds of misusing drugs among U.S. born Caribbeans within the 45–59 age range (AOR = 0.06, 95 % CI = 0.04, 0.79; $p = .035$).

Although race and ethnicity were not included in other models, it proved to influence substance misuse in Guyana. Indo-Guyanese (AOR = 2.71, 95 % CI = 1.46, 5.04; $p = .002$) and Mixed Guyanese (AOR = 3.17, 95 % CI = 1.09, 9.25; $p = .034$) were more likely to misuse drugs when compared to Black Guyanese.

The analysis further revealed lower drug abuse among higher household income African Americans. There were reduced odds of misusing drugs among the fourth (AOR = 0.43, 95 % CI = 0.24, 0.76; $p = .005$) and highest quintile (AOR = 0.30, 95 % CI = 0.19, 0.48; $p = .001$) participants. Higher income (e.g., fourth quintile) U.S. born Caribbeans were also at reduced odds (AOR = 0.16, 95 % CI = 0.04, 0.73; $p = .020$) of abusing drugs. Among U.S. foreign-born Caribbeans, effects were detected among two household income categories. Specifically, the odds of misusing drugs were significantly reduced among participants within the second (AOR = 0.18, 95 % CI = 0.03, 0.91; $p = .039$) and the highest (AOR = 0.23, 95 %, 0.06, 0.95; $p = .042$) quintile household income groups. No effects were detected in models where participants resided in the Caribbean.

Table 16.3 Socio-demographic factors and Nativity Predicting Substance Abuse

	African American	US born Caribbeans	US foreign born Caribbeans	Guyana	Jamaica
Age					
18–29	1	1	1	1	1
30–44	1.69*	0.42	3.08	2.46	0.94
45–59	2.05*	0.06*	1.08	1.29	1.62
>59	0.97	0.30	3.63	1.42	3.11
Gender					
Male	1	1	1	1	1
Female	0.21***	0.05***	0.09***	0.10***	0.15***
Race/ethnicity					
Black	–	–	–	1	–
Indo	–	–	–	2.71**	–
Mixed	–	–	–	3.17*	–
Equivalised income					
Bottom quintile	1	1	1	1	1
Second quintile	0.78	0.73	0.18*	0.55	0.60
Middle quintile	0.82	0.45	0.19	0.86	n.a
Fourth quintile	0.43**	0.16*	2.27	1.54	2.00
Highest quintile	0.30***	0.23	0.23*	0.49	0.81
Education level					
Primary/some high school	1	1	1	1	1
High school graduate	0.48***	0.15**	1.46	0.95	0.40
College-vocation-technical	0.59**	0.78	0.67	0.70	0.14*
Employment status					
Employed	1	1	1	1	1
Unemployed	1.08	1.73	0.37	1.94	1.97
Not in labor force	0.83	2.47	3.09*	0.46	4.13
Marital status					
Married	1	1	1	1	1
Partner	1.34	0.13	26.17***	1.63	5.07
Sep-div-widow	1.37*	0.33	4.53*	2.58	2.92
Never married	1.07	0.09*	3.18	1.98	1.28
MDE (Yes)	3.84***	11.85***	1.65	3.32*	2.71

Note. *$p<.05$, **$p<.01$, *** $p<.001$

Participants' education and employment was also predictive of substance misuse among African Americans, U.S. born Caribbeans, and Jamaicans. The odds of misusing drugs were significantly reduced for high school graduating African Americans (AOR = 0.48, 95 % CI = 0.34, 0.67; p = .000) and U.S. born Caribbeans (AOR = 0.15, 95 % CI = 0.43, 0.52; p = .004). College educated African Americans (AOR = 0.59, 95 % CI = 0.41, 0.83; p = .003) and Jamaicans (AOR = 0.14, 95 % CI = 0.03, 0.72; p = .019) were also at lower odds of misusing drugs. Furthermore, the odds of misusing drugs increased for U.S. foreign-born Caribbeans who were not in the labor force (AOR = 3.09, 95 % CI = 0.98, 9.79; p = .054).

Relationship status predicted drug misuse among African Americans, U.S. born and foreign-born Caribbeans. The odds of misusing drugs increased for separated, divorced or widowed African Americans (AOR = 1.37, 95 % CI = 1.06, 1.78; p = .017) and U.S. foreign Born Caribbeans (AOR = 4.53, 95 % CI = 1.07, 19.25; p = .041). There was also a significant increase in drug misuse among partnered U.S. foreign-born Caribbean respondents (AOR = 26.17, 95 % CI = 5.41, 126.79; p = .000). By contrast, the odds of misusing drugs were reduced among never married U.S. born Caribbeans (AOR = 0.09, 95 % CI = 0.11, 0.64; p = .019).

With the exception of Jamaicans and U.S. foreign-born Caribbeans, the analyses reveal associations among some group between major depressive episode (MDE) and substance misuse. Specifically, African Americans (AOR = 3.84, 95 % CI = 2.82, 5.21; p = .000), U.S. born Caribbeans (AOR = 11.85, 95 % CI = 4.45, 31.60; p = .000) and Guyanese (AOR = 3.32, 95 % CI = 1.13, 9.73; p = .029) were all at increased odds of abusing drugs when they met criteria for major depressive episode.

Summary and Future Direction

This study explored the effects of contextual influences, nativity, immigration and geographic location on substance abuse using representative samples from the United States, Guyana and Jamaica. Our findings revealed that rates of substance abuse were in general higher in the United States than the Caribbean, but were greater among African Americans. Nativity and length of time in the United States were also associated with substance misuse. Rates of substance abuse were particularly high for those who were born in the United States. Foreign-born participant were less likely to misuse drugs. These results suggest that exposure to the U.S. context may play a role in substance misuse. At the same time, it supports the fact that nativity may offer some protection against mental health disorders.

Consistent with other studies we found that gender predicted substance misuse across samples (Broman et al. 2008). In fact, women across cohorts were less likely to misuse drugs. This finding may signify cultural prohibitions that may develop about women who engage in these activities; overall women may be less likely than men to participate in risky behaviors.

Other contributing factors to substance misuse were age, household income, educational attainment, relationship status and major depressive episode; and these all affected the groups differently. Specifically, African Americans within the 30–44 and 45–59 age range were more likely to abuse drugs. U.S. born Caribbeans between the ages of 45–59 were also more likely to misuse drugs. This finding is inconsistent with other studies that suggest that drug use is more likely to occur at the earlier stages of individuals' lives.

Moreover, African Americans and U.S. Caribbean born Blacks with relatively higher household incomes were less likely to misuse drugs. This differed slightly for U.S. foreign-born Caribbean, where both higher and lower household income participants were less likely to misuse drugs. U.S. sampled participants and Jamaicans with high school or college education were also less likely to abuse drugs. Additionally, separated, divorced and widowed African American and U.S. born foreign born Caribbeans were more likely to misuse drugs as well as partnered U.S. foreign-born Caribbeans. There was however, a decrease in drug use among never married U.S. born Caribbeans.

Finally, comorbid major depressive episodes (MDE) and drug misuse was observed among African American, U.S. born Caribbean and Guyanese; individuals that met criteria for major depressive episode were more likely to misuse a drug, which has been found in other studies (e.g. McDowell and Clodfelter 2001). This finding suggests that individuals may engage in drug use or self medicate to deal with their depression; a possibility that needs to be explored in future research.

There are several limitations to these analyses. To begin, measures differ slightly across samples or were not available in all samples. The study did not include all possible drugs and other substances, and we believe that religious and cultural choices may have affected their use across countries. Also, while the U.S. and Guyanese samples included participants from both urban and rural regions, the Jamaican sample was limited to the urban area of Kingston. In addition, other countries in the Caribbean may differ greatly from the two countries included in this study. Jamaica and Guyana were selected because of their relatively high rates of migration to the U.S. Had we collected data from another country (e.g. Trinidad) which has relatively low rates of migration, the results may have differed. This limitation actually begs for the need for larger and more comprehensive studies across different Caribbean countries.

Nonetheless, these limitations should not overshadow the strengths of this research. This chapter makes a contribution to the literature as one of the first empirical analyses to test models on substance use and misuse across national samples of ethnically different Blacks within the U.S. It is also one of the first known studies to make comparisons between Caribbean Blacks living in their homeland, even though our analyses were limited to Jamaica and Guyana, and those residing in the United States; making it possible to examine cross cultural, contextual, migratory, and geographical differences in substance use and misuse patterns. The study further highlights the fact that substance use is a complex set of behaviors influenced by race and ethnic background, nativity, and timing of migration; thus, highlighting the need for more comparative studies on the migration destinations of Black peoples from the Caribbean. Finally, more studies are also needed to examine the affect of discrimination and acculturative stress on the health and mental health of Black migrants in different geographical and cultural national contexts.

References

Acevedo-Garcia, D. (2000). Residential segregation and the epidemiology of infectious diseases. *Social Sciences & Medicine, 51*, 1143–1161.

Acevedo-Garcia, D., Lochner, K. A., Osypuk, T. L., & Subramanian, S. V. (2003). Future directions in residential segregation and health research: A multilevel approach. *American Journal of Public Health, 93*(2), 215–221.

Baum, A., Garofalo, J. P., & Yali, A. M. (2006). Socioeconomic status and chronic stress: Does stress account for SES effects on health? *Annals of the New York Academy Sciences, 896*, 131–144.

Bhugra, D. (2004). Migration and mental health. *Acta Psychiatrica Scandinavica, 109*, 243–258.

Blake, S. M. (2001). Recency of immigration, substance use, and sexual behavior among Massachusetts Adolescents. *American Journal of Public Health, 91*(5), 794–798.

Boardman, J. D., Finch, B. K., Ellison, C. G., Williams, D. R., & Jackson, J. S. (2001). Neighborhood disadvantage, stress, and drug use among adults. *Journal of Health and Social Behavior, 42*(2), 151–165.

Brady, K. T., Back, S. E., & Coffey, S. F. (2004). Substance abuse and postraumatic stress disorder. *Current Directions in Psychological Science, 13*(5), 206–209.

Breslau, J., & Chang, D. (2006). Psychiatric disorders among foreign-born and US-born Asian-Americans in a US national survey. *Social Psychiatry and Psychiatric Epidemiology, 41*(12), 943–950. doi:10.1007/s00127-006-0119-2.

Breslau, J., Borges, G., Hagar, Y., Tancredi, D., & Gilman, S. (2009). Immigration to the USA and risk for mood and anxiety disorders: Variation by origin and age at immigration. *Psychological Medicine, 39*, 1117–1127.

Breslau, J., Aguilar-Gaxiola, S., Borges, G., Kendler, K. S., Su, M., & Kessler, R. C. (2007). Risk for psychiatric disorder among immigrants and their descendants: Evidence from the National Comorbidity Survey-Replication. *The Journal of Nervous and Mental Disease, 194*(12), 1–7. doi:10.1097/01.nmd.0000243779.35541.c6.

Broman, C. L. (2005). Stress, race and substance use in college. *College Student Journal, 39*(2), 340–352.

Broman, C., Neighbors, H., Delva, J., Torres, M., & Jackson, J. S. (2008). Prevalence of substance use disorders among African Americans and Caribbean Blacks in the national survey of American life. *American Journal of Public Health, 98*(6), 1107–1114.

Chen, C., Wagner, F. A., & Anthony, J. C. (2002). Marijuana use and the risk of major depressive episode: Epidemiological evidence from the United States National Comorbidity Survey. *Social Psychiatry and Psychiatric Epidemiology, 37*, 199–206.

Darity, W. A. (2003). Employment discrimination, segregation, and health. *American Journal of Public Health, 93*(2), 226–231.

Galea, S., Nandi, A., & Vlahov, D. (2004). The social epidemiology of substance use. *Epidemiologic Reviews, 26*, 26–36. doi:10.1093/epirev/mxh007.

Gfroerer, J., & Tan, L. L. (2003). Substance use among foreign youths in the United States: Does the length of residence matter. *American Journal of Public Health, 98*(11), 1892–1895.

Harrison, L., Sutton, M., & Gardiner, E. (1997). Ethnic differences in substance use and alcohol-use-related mortality among first generation migrants to England and Wales. *Substance Use and Misuse, 32*(7&8), 849–876.

Jackson, J. S., Torres, M., Caldwell, C. H., Neighbors, H. W., Nesse, R. M., Taylor, R. J., Trierweiler, S. J., & Williams, D. R. (2004). The national survey of American life: A study of racial, ethnic, and cultural influences on mental disorders and mental health. *International Journal of Methods in Psychiatric Research, 13*, 196–207.

Johnson, T. P., VanGeest, J. B., & Cho, Y. I. (2002). Migration and substance use: Evidence from the U.S. national health survey interview. *Substance Use & Misuse, 37*(8–10), 941–972. doi:10.1081/JA-120004160.

Kandel, D., Chen, K., Warner, L. A., Kessler, R. C., & Grant, B. (1997). Prevalence and demographic correlates of symptoms of last year dependence of alcohol, nicotine, marijuana and cocaine in the U.S. population. *Drug and Alcohol Dependence, 44*, 11–29.

Keane, F., Tappen, R. M., Williams, C. L., & Rosselli, M. (2008). Comparison of African-American and Afro-Caribbean older adults' self-reported health status, function, and substance use. *Journal Black Psychology, 35*(1), 44–62.

Kessler, R. C., Mickelson, K. L., & Williams, D. R. (1999). The prevalence, distribution, and mental health correlates of perceived discrimination in the United States. *Journal of Health and Social Behavior, 40*(3), 208–230.

Kessler, R. C., Berglund, P., Demler, O., Jin, R., & Merikangas, K. R. (2005). Lifetime prevalence and age-of-onset distribution of DSM-IV disorders in the National Comorbidity Survey Replication. *Archives of General Psychiatry, 62*(6), 593–602.

Krivo, L. J., Peterson, R. D., Rizzo, H., & Reynolds, J. R. (1998). Race, segregation, and the concentration of disadvantage: 1980–1990. *Social Problems, 45*(1), 61–80.

Latkin, C. A., & Curry, A. D. (2003). Stressful neighborhoods and depression: A prospective study of the impact of neighborhood disorder. *Journal of Health and Social Behavior, 44*(1), 34–44.

Massey, D. S., & Eggers, M. L. (1990). Ecology of inequality: Minorities and the concentration of poverty, 1970–1980. *The American Journal of Sociology, 95*(5), 1153–1188.

Massey, D. S., Gross, A. B., & Eggers, M. L. (1991). Segregation, the concentration of poverty, and the life chances of individuals. *Social Sciences Research, 20*, 397–420.

McDowell, D. M., & Clodfelter, R. C. (2001). Depression and substance: Considerations of etiology, comorbidity, evaluation, and treatment. *Psychiatric Annals, 31*(4), 244–251.

Merline, A. C., O'Malley, P. M., Schulenberg, J. E., Bachman, J. G., & Johnston, L. D. (2004). Substance use among adults 35 years of age: Prevalence, adulthood predictors, and impact of adolescent substance use. *American Journal of Public Health, 94*(1), 96–102.

Nazroo, J. Y. (2001). *Ethnicity, class and health*. London: Policy Studies Institute, Westminster University.

Nazroo, J. Y., Jackson, J., Karlsen, S., & Torres, M. (2007). The Black diaspora and health inequalities in the US and England: Does where you go and how you get there make a difference? *Sociology of Health & Illness, 29*, 811–830.

Peterson, R. D., & Krivo, L. J. (1999). Racial segregation, the concentration of disadvantage, and Black and White homicide victimization. *Sociological Forum, 14*(3), 465–493.

Rassool, G. H. (2006). Substance abuse in black and minority ethnic communities in the United Kingdom: A neglected problem. *Journal of Addictions Nursing, 17*(2), 127–132.

Takeuchi, D. T., Alegria, M., Jackson, J. S., & Williams, D. R. (2007). Immigration and mental health: Diverse findings in Asian, Black, and Latino populations. *American Journal of Public Health, 97*(1), 11–12.

Turner, W. L., & Wallace, B. (2003). African American substance use: Treatment, prevention, and treatment. *Violence Against Women, 9*(5), 576–589. doi:10.1177/1077801202250452.

US Department of Health and Human Services, National Institutes of Health, National Institute on Drug Abuse (2003) Drug use among racial/ethnic minorities, revised. Bethesda: Author.

US Department of Health and Human Services, National Institutes of Health, National Institute on Drug Abuse. (2005). *Drug abuse and HIV/AIDS*. Bethesda: Author.

VanGeest, J. B., & Johnson, T. P. (1997). Substance use patterns among homeless migrants and non-migrants in Chicago. *Substance Use & Misuse, 32*(7&8), 877–907. doi:10.3109/10826089709055863.

Vega, W. A., & Rumbaut, R. G. (1991). Ethnic minorities and mental health. *Annual Reviews of Sociology, 17*, 351–383.

Weiss, R. D., Griffin, M. L., & Mirin, S. M. (1992). Drug abuse as self-medication for depression: An empirical study. *The American Journal of Drug and Alcohol Abuse, 18*(2), 121–129.

Williams, D. R. (1999). Race, socioeconomic status, and health: The added effects of racism and discrimination. *Annals of New York Academy of Sciences, 896*, 173–188.

Williams, C. D., & Adams-Campbell, L. L. (2000). Addiction behaviors and depression among African Americans residing in a public housing community. *Addictive Behaviors, 25*(1), 45–56.

Williams, D. R., & Williams-Morris, R. (2000). Racism and mental health: The African American experience. *Ethnicity and Health, 5*(3/4), 243–268.

Williams, D. R., Yu, Y., Jackson, J. S., & Anderson, N. B. (1997). Racial differences in physical and mental health: Socio-economic status, stress and discrimination. *Journal of Health Psychology, 2*, 335–350.

Williams, D. R., Neighbors, H. W., & Jackson, J. S. (2003). Racial/ethnic discrimination and health: Findings from community studies. *American Journal of Public Health, 93*(2), 200–208.

Williams, D. R., Haile, R., Gonzalez, H. M., Neighbors, H., Baser, R., & Jackson, J. S. (2007). The mental health of Black Caribbean Immigrants: Results from the National Survey of American Life. *American Journal of Public Health, 97*(1), 52–59.

Xueqin, G., & Shive, S. (2000). A comparative analysis of perceived risks and substance abuse among ethnic groups. *Addictive Behaviors, 25*(3), 361–371.

Chapter 17
Drug Abuse Preventive Interventions for Hispanic Youth: State of the Science and Implications for Future Research

Guillermo Prado, David Cordova, Nicole Cano, Margaret Arzon, Hilda Pantin, and C. Hendricks Brown

Drug use represents a major public health problem facing America's youth (Johnston et al. 2011a). The effects of drug use on physical health in adolescents are severe. For example, approximately 75 % of all deaths in the U.S. can be attributed to accidents, assaults/homicides, intentional harm/suicides, HIV/AIDS, and chronic lower respiratory disease, all of which have been associated with drug use (Arias et al. 2003). In addition to these physical consequences, drug use among adolescents has a myriad of social consequences. For example, drug use in adolescents has been attributed to both criminal activity and elevated school dropout rates. Adolescent drug use also has long-term consequences such as the development of psychiatric disorders, including drug abuse and dependence (Gil et al. 2004). Hispanic adolescents are disproportionately affected by drug use and some of its associated consequences (e.g., school dropout rates) when compared to mainstream society. In fact, drug use and its consequences largely contribute to the health disparities that exist between Hispanics and other segments of the population. Thus, preventing/reducing drug use among Hispanic youth is essential to eliminating the health disparities that exist between Hispanics and other segments of the population. In spite of the fact that Hispanic adolescents experience significant drug use health disparities, a dearth of evidence-based interventions exist for preventing drug use behaviors among this population. This chapter will review the epidemiology (both prevalence and etiology) of drug use in Hispanic youth, the state of intervention science among this population, and provide recommendations for future research to work towards

---

G. Prado (✉) • N. Cano • M. Arzon • H. Pantin
Miller School of Medicine, University of Miami, Miami, FL, USA
e-mail: gprado@med.miami.edu

D. Cordova
School of Social Work, University of Michigan, Ann Arbor, MI, USA

C.H. Brown
Departments of Psychiatry and Behavioral Sciences and Medical Social Sciences, Northwestern University, Chicago, IL, US

Y.F. Thomas, L.N. Price (eds.), *Drug Use Trajectories Among Minority Youth*,
DOI 10.1007/978-94-017-7491-8_17

narrowing and ultimately eliminating the health disparities in drug use that exist between Hispanic youth and other segments of the population.

Prevalence of Drug Use Among Hispanic Youth

Hispanic adolescents comprise an important at-risk population for drug use. Hispanic youth report disproportionate use of illicit drug use. Results from the latest school-based Monitoring the Future survey (Johnston et al. 2011a) indicate that Hispanic 8th grade adolescents report higher annual and current use across all drug categories (with the exception of amphetamines) than both non-Hispanic Whites and African Americans. For example, 18.2 % of Hispanic 8th graders report annual illicit drug use compared to 15.7 % of African Americans and 14.1 % of non-Hispanic whites (Johnston et al. 2011b). Although Hispanic 10th graders also report higher rates of illicit drug use than both non-Hispanic whites and African Americans, by the 12th grade Hispanic youth report lower rates of use than non-Hispanics whites (although still higher than African Americans).

Several explanations have been postulated for this change in drug use trends between the 10th and 12th grades. One possible explanation is that Hispanic adolescents drop out of school at considerably higher rates relative to both African Americans and non-Hispanic whites. Thus, compared to both non-Hispanic whites and African Americans, a larger percentage of the drug-prone segment of Hispanics may drop out before 12th grade (Greene and Forster 2003). However, evidence from non-school-based surveys may contradict the high school dropout hypothesis. Specifically, the population-based National Household Survey on Drug Use and Health (Substance Abuse and Mental Health Services Administration 2005) has found that during late adolescence and emerging adulthood, non-Hispanic whites have the highest rates of drug use, followed by Hispanics. Notewothy, however, is that this same population-based household survey has found that Hispanics have higher rates of substance abuse/dependence (9.5 %) than both non-Hispanic whites (9 %) and African Americans (8 %). Alternatively, the change in rankings between the 10th and 12th grades could also be attributed to the fact that Hispanics tend to initiate drug use earlier, and that non-Hispanic whites initate use later and surpass Hispanic use rates by 12th grade.

Etiology of Drug Use: Risk and Protective Processes

An understanding of adolescent development and of the etiology of drug use is needed to prevent/reduce drug use among adolescents. Adolescent development is embedded in a complex set of interdependent and interrelated contexts that predispose and/or protect adolescents from drug use. These contexts influence and are influenced by developmental trajectories as they interact and develop over time

(Szapocznik and Coatsworth 1999). A number of theorists affirm that culture is an important aspect of context that affects development (e.g., Bronfenbrenner 1979, 1986). An interest in the "cultural patterns" that may be found in Hispanic youth and their families stems from the belief that beyond race and ethnicity, an understanding of cultural processes that are proximal to daily life is needed. Thus, to understand the etiology of Hispanic drug use and consequently to prevent/reduce drug use in this population, it is important to understand and note that cultural, familial, school, peer, and neighborhood trajectories influence each other over time, and that these mutual influences serve to direct and redirect adolescent development (Cicchetti and Rogosch 2002). Contemporary views of risk and protection, such as ecodevelopmental theory (Pantin et al. 2003a, b; Prado et al. 2010) take into consideration the multiple social contexts that influence development, the interrelations among those contexts, the changing nature of each context, and how these elements affect risk for the development of drug use and other problem behaviors (Cicchetti and Richters 1993). Within this and other frameworks, culture is viewed as a larger system that encompasses and shapes many of the more proximal influences on adolescent drug use.

Ecodevelopmental Theory

Ecodevelopmental theory is the next generation of risk and protective factors models that borrows from and extends Bronfenbrenner's social-ecological framework (see Schwartz et al. 2006a, for a more extended discussion). Thus, risk and protective factors are organized by the multiple influences on adolescent development according to their proximity to the adolescent, presented here from furthest to closest: *macrosystems*, are the broad social and philosophical ideologies that define a particular culture, such as cultural values and ideals; *exosystems*, refer to contexts in which the adolescent does not participate directly but that impact the functioning of important members of the adolescent's life, such as parents' occupational and financial stress and systems of support; *mesosystems*, are the systems comprised of the interactions between important members of the different contexts in which the adolescent participates directly, such as parental monitoring of peers; and *microsystems*, refer to contexts in which the adolescent participates directly, such as the family and peers.

From this perspective ecodevelopmental theory posits that cultural phenomena in the macrosystem, such as the mismatch between a family's culture of origin and that of the host country, produce a "trickle-down" effect by contributing to exosystemic problems such as parental isolation, which in turn may cut parents off from systems of parental support (i.e., exosystemic problems) as well as their adolescents' peer networks (i.e., mesosystemic problems) (Pantin et al. 2003a, b, 2004). Moreover, clinical literature suggests that when immigrant parents are unfamiliar with the culture of their new homeland, they tend to remain isolated and not to reach out for social support (Leon and Dziegielewski 2000), which, in turn, may inhibit supportive and involved parenting (Simons et al. 1992). Lack of supportive and involved parenting in the adolescent's life (microsystemic problem) and monitoring of peer networks (mesosystemic problem) then increases the likelihood of behavior problems (Coatsworth et al. 2002), and association with deviant peers (Ary et al. 1999). Consequently, behavior problems and association with substance abusing peers, coupled with low parental monitoring of adolescents' social activities, also increase the likelihood of substance use (Pettit et al. 1999).

Although these same contextual processes are important predictors of problematic outcomes across all ethnic groups in general (e.g., Barrera et al. 2001), in Hispanic families, these processes may be set in motion and exacerbated through cultural mechanisms such as immigration and acculturation related stressors. These cultural and acculturation-related processes remain salient for years after immigration, such that they affect children born after the family's arrival as well as children who immigrate with the family. As a result, understanding the role of cultural change in Hispanic families – and the sequelae of these changes – is an important step towards preventing drug use in Hispanic adolescents.

Cultural Risk and Protective Processes

The most consistent culture-specific finding in the risk and protection literature on Hispanic adolescents is the relationship between drug use and acculturation. Acculturation is a cultural process that is proximal to the daily life of Hispanic adolescents and their families. Acculturation has been conceptualized as the process of adaptation that occurs when two cultures come into contact (Redfield et al. 1936). As applied to immigrants, acculturation is change in cultural practices, values, and identifications that occur as a result of contact with (and expectations from) the receiving society (Cabassa 2003; Schwartz et al. 2006b). Acculturation is multidimensional in at least two different ways (Escobar and Vega 2000). First, it refers both to receiving-culture acquisition and to heritage-culture retention; and second, it encompasses a number of domains, including behaviors (e.g., language use), values (e.g., individualism and collectivism), and identifications (e.g., attachment to the United States and to one's country of origin).

This multidimensionality is not acknowledged in many studies on acculturation and health risk behavior. In many studies "acculturation" is used to index a one-dimensional process ranging from endorsement of heritage-culture practices to endorsement of American cultural practices. "Acculturated" individuals are viewed as having discarded their culture of origin. Moreover, the approach adopted in many of these studies considers only cultural behaviors such as language use, media preferences, and choice of friends. Studies using this approach to acculturation have found that, among Hispanic adolescents and emerging adults, more "highly acculturated" individuals are more likely to use drugs and alcohol (Gil et al. 2000; Ramirez et al. 2004), and to engage in unsafe sex (Ford and Norris 1993). Hispanic adolescents who speak mostly Spanish, associate primarily with Spanish speakers, and engage in Hispanic cultural practices are less likely to use drugs and alcohol (Allen et al. 2008) than their "more acculturated" peers. The message, more or less, is that acculturation may be hazardous to one's health. This phenomenon has been termed the "immigrant paradox" (Alegria et al. 2008).

A major limitation in most studies on acculturation and drug use (as well as on other health outcomes), however, is their reliance on unidimensional conceptions of acculturation. Thus, it is not clear whether the "immigrant paradox" is due to acquiring receiving cultural practices, loss of heritage cultural practices, or both. More contemporary approaches in cultural studies rely on bidimensional models where American-culture acquisition is treated separately from heritage-culture retention (e.g., Phinney 2003; Ryder et al. 2000). Studies relating bidimensional models of acculturation to health outcomes have suggested that heritage-culture retention is protective against many health risk behaviors, regardless of whether or not the person acquires American cultural practices, values, and identifications (e.g., Schwartz et al. 2010). In addition to cultural risk and protective processes, family, school, and peer risk and protective processes are salient in the prevention of Hispanic adolescent drug use.

Family Risk and Protective Processes

Clearly, many of the effects of culture on drug use in Hispanic adolescents operate through family risk and protective processes. Although the factors that influence these processes may differ across ethnic groups, the effects of these processes on adolescent outcomes are strongly consistent across groups (e.g., Barrera et al. 2001; Brook et al. 1997). An extant amount of research has established the important role that families play in adolescent development (Steinberg and Morris 2001), including drug use (e.g., Broman et al. 2006; Bahr et al. 2005). Positive family processes such as parental investment for the adolescent (Crosby et al. 2001; Rodgers 1999) and parent-adolescent communication (Brody and Ge 2001; O'Sullivan et al. 1999) are powerful protective processes against adolescent drug use. These processes can also be compromised during the acculturation process (Martinez 2006; Szapocznik and Kurtines 1993).

Parental disinvestment may represent an important concern in Hispanic families, partly due to the influence of acculturation on the Hispanic family and partly because of the prominence of family in most Hispanic cultures. For instance, when Hispanic immigrant parents with limited English proficiency first arrive in the United States, they are faced with the daunting task of raising children in an unfamiliar and foreign culture. Consequently, these parents are likely to confront numerous obstacles, including cultural incompatibilities between the receiving culture and the immigrant's culture of origin, social isolation, and marginalization from important systems of support and thereby potentially place their children at risk for drug use and other problematic behaviors (Pantin et al. 2003a, b, 2004). These challenges may pose the greatest difficulties for low-income parents, who often do not have access to supportive resources that can assist them in the transition to a new homeland. In Latin American countries, the family is generally prioritized above the individual (e.g., Marin et al. 1987). Values such as respect for adults, conformity, and a sense of duty to parents – which are regarded as important aspects of parent-child relationships in Latin American countries (Santisteban et al. 2002) – may conflict with those commonly endorsed in American society, where the individual is generally prioritized over the family. Moreover, there is a robust social-psychological literature demonstrating that individuals perceived as "foreigners" because of linguistic, cultural, or ethnic differences may be ostracized and marginalized from the mainstream cultural group (e.g., Mummendey et al. 2001). Therefore, as a result of linguistic and cultural incompatibilities between Hispanic and American culture, Hispanic immigrants often find themselves isolated from sources of support in the United States, even in predominantly Hispanic neighborhoods (Leon and Dziegielewski 2000; Pantin et al. 2003a, b, 2004). Social isolation, coupled with the stresses of daily living, long work hours resulting from downward economic mobility, and lack of support from family and community, may increase the likelihood that parents will become frustrated and overwhelmed and consequently disinvest from their adolescents' lives.

Peer Risk and Protective Processes

There is evidence which shows that risk-taking adolescents often affiliate with peers who take similar risks (Brooks-Gunn and Furstenberg 1989; Jessor and Jessor 1977). In fact, peer influence is a significant predictor of adolescent drug use (Epstein et al. 1999; Marshal and Chassin 2000). Although peers are an important source of influence for all adolescents, peers may play an especially important role in the lives of Hispanic immigrant adolescents – whose parents are not equipped to socialize them about the receiving culture. As a result, peers become the primary vehicle through which Hispanic immigrant adolescents learn to navigate through and adjust to their new receiving culture. Consequently, given the especially high rates of drug use among Hispanic adolescents, whether or not they associate with deviant peers is a powerful predictor of drug use in this population.

School Risk and Protective Processes

Schools play a vital role in the lives of children and adolescents. Adolescents who are not bonded to (or lack interest in) school, are at higher risk for drug use, when compared to their peers who are bonded or have an interest in school (Henry et al. 2005; Simons-Morton et al. 1999). Moreover, adolescents who drop out of school are at increased risk of using drugs (Ellickson et al. 1998; Guagliardo et al. 1998). School is an especially important domain for Hispanic adolescents, given that almost 40 % of Hispanics 25 years of age and older report not having graduated high school (Greene and Forster 2003). Having reviewed the contextual mechanisms that predict drug use in Hispanics, we now turn to a review of prevention programs that have targeted some of these mechanisms. All of these interventions have integrated cultural components at least to some extent (Szapocznik et al. 2007).

Drug Use Preventive Interventions for Hispanic Adolescents

Although there is no clear consensus on whether and the extent to which culture plays a role in the prevention of Hispanic adolescent drug use (Castro et al. 2004, 2006), a review of the literature on Hispanic adolescent drug abuse suggests that culture is an integral component of adolescent drug abuse prevention intervention programs (Szapocznik et al. 2007). The role of culture in prevention is to bolster culturally-rooted protective processes and reduce culturally-rooted risk processes. An example of promoting a culturally-rooted protective process might be capitalizing on the collectivist nature of the Hispanic culture by reinvesting parents in the lives of their adolescents. An example of ameliorating a culturally rooted risk process might be to reintroduce Hispanic culture to adolescents as a way to reduce the negative effects of acculturation on family processes, on the perceived importance of family, and on exposure to negative peer influences.

Although the prevalence data show that Hispanics are more likely to report drug use, a dearth of preventive intervention programs have been developed and evaluated that have been found to be efficacious in preventing/reducing drug use among Hispanic youth. In fact, in a review of the literature Szapocznik et al. (2007) identified only four prevention interventions that have been found to be efficacious in preventing/reducing Hispanic adolescent drug use or risk processes associated with drug use. The four efficacious intervention programs identified by Szapocznik et al. (2007) were: Family Effectiveness Training (Szapocznik et al. 1978a, b), Keepin' it REAL (Kulis et al. 2005), Nuestras Familias (Martinez and Eddy 2005), and Familias Unidas (Prado et al. 2007; Pantin et al. 2009). Of those four, two targeted risk and protective processes for drug use (including family conflict, general parenting, and intentions to use substances), but did not have drug use as an actual outcome. Only the two interventions (keepin it REAL, Familias Unidas) that had drug use as an outcome are reviewed in this chapter.

Familias Unidas

Familias Unidas (Pantin et al. 2003a, b, 2009; Prado et al. 2006, 2007) is a Hispanic-specific, family-based preventive intervention aimed to reduce risks and increase protection for HIV risk behaviors, including substance use and unprotected sexual behavior in Hispanic adolescents. Familias Unidas is guided by ecodevelopmental theory (Szapocznik and Coatsworth 1999) and informed by culturally specific models developed for Hispanic populations in the United States (Szapocznik and Kurtines 1993; Szapocznik and Williams 2000). Consistent with ecodevelopmental theory, Familias Unidas aims to prevent drug use, cigarette use, alcohol use, and unprotected sexual behavior by increasing positive parenting, family support of the adolescent, parental involvement, as well as by improving general parent-adolescent communication and parent-adolescent communication specific to substance use, unsafe sexual behavior, and HIV (Pantin et al. 2004).

Familias Unidas is delivered through both multi-parent groups and family visits that place parents in the change agent role. Parent groups are the main activity of the intervention, which bring parents together for the purposes of establishing parental investment, increasing parental support, and providing a context for parent participation in a conjoint skills learning process. Family visits create an opportunity for parents to transfer the competencies learned in the group sessions to their adolescent, foster more nurturing and supportive relationships, and increase parent-child communication, all in the context of family.

Familias Unidas has been evaluated and found to be efficacious in preventing and reducing substance use (including alcohol use, cigarette use, and illicit drug use) and unprotected sexual behavior in three randomized clinical trials (Pantin et al. 2003a, b, 2009; Prado et al. 2007, 2010). Because of the young age of the sample at baseline (12 years old) and the short follow-up (3-months post-intervention) of the first Familias Unidas efficacy study (Pantin et al. 2003a, b), no significant effects on drug use were observed (although effects were observed on family functioning and conduct problems). Hence, the results of Pantin et al. (2003a, b) are not reviewed here.

Prado and colleagues (2007) evaluated the efficacy of Familias Unidas in preventing adolescent alcohol, cigarette, and illicit drug use as well as unsafe sexual behavior, relative to an HIV preventive intervention and a cardiovascular preventive intervention. Participants in the study were 128 boys and 138 girls. This second study used a 3 (Condition) × 5 (Time) randomized controlled design. Participants were assessed at baseline, randomized, and reassessed at 6, 12, 24, and 36 months post baseline. This study evaluated the efficacy of Familias Unidas plus PATH, an HIV preventive intervention relative to (a) PATH plus English for Speakers of Other Languages (ESOL) and (b) ESOL plus HEART, a cardiovascular health promotion intervention. The ESOL + PATH condition focused specifically on increasing parent-adolescent communication about sex and HIV risks (Krauss et al. 2000). However, unlike Familias Unidas + PATH, it did not target positive parenting or general communication skills, which may be a necessary prerequisite to initiating parent-adolescent discussions about sexuality and HIV (Rodgers 1999). Similarly, the ESOL + HEART condition focused specifically on increasing parent-adolescent communication regarding risk factors for cardiovascular disease (including cigarette use).

The results of the study showed that (a) Familias Unidas + PATH was efficacious in preventing and reducing cigarette use relative to both control conditions; (b) Familias Unidas + PATH was efficacious, relative to ESOL + HEART, in reducing illicit drug use; and (c) Familias Unidas + PATH was efficacious, relative to ESOL + PATH, in reducing unsafe sexual behavior. As reported in the article reporting the study's primary findings, the proportion of adolescents reporting current cigarette use in Familias Unidas + PATH decreased from 3.3 to 1.4 %, whereas it increased in both ESOL + PATH (from 1.2 to 10 %) and ESOL + HEART (from 3.3 to 14.3 %) Fig. 17.1.

Pantin and colleagues (2009) evaluated the efficacy of Familias Unidas in preventing adolescent substance use as well as unsafe sexual behavior relative to a prevention as usual control condition. This study consisted of an indicated sample of Hispanic youth who were selected because they had clinical levels of behavior problems.

Participants in the present study were 136 boys and 77 girls and their primary caregivers. At baseline, the majority (74.2 %) of the participating adolescents self-reported having at least one major psychiatric disorder using the Diagnostic Interview Schedule for Children predictive scales (DISC-DPS: Lucas et al. 2001).

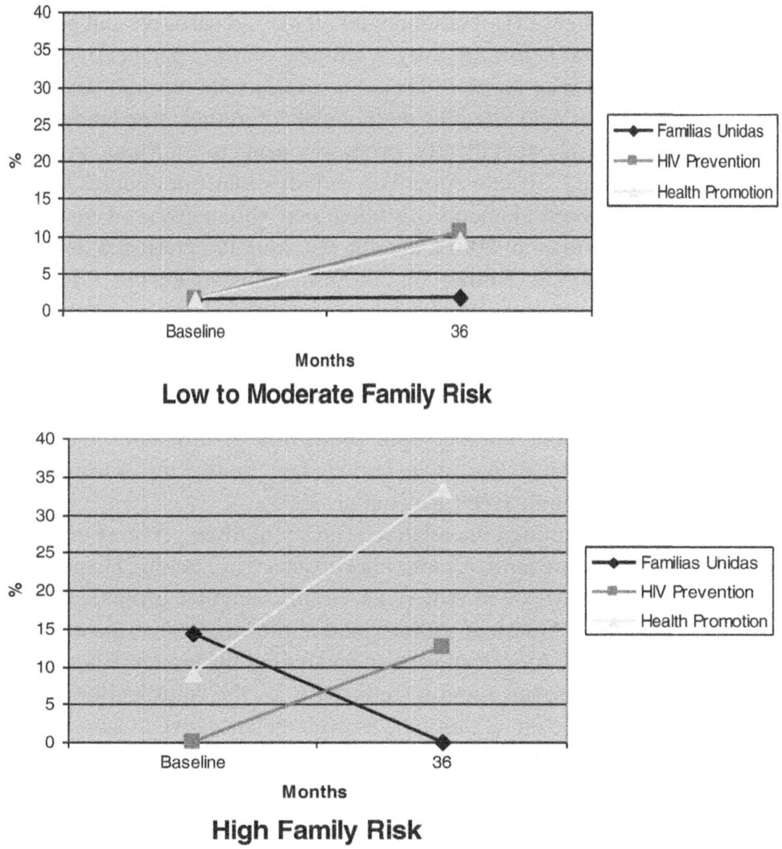

Fig. 17.1 Percent reporting past 90 days cigarette use by condition and family risk

This study used a 2 (Condition) × 4 (Time) randomized controlled design. Participants were assessed at baseline, randomized, and reassessed at 6, 18, and 30 months post baseline. As with the prior studies, Familias Unidas was found to improve family functioning compared to the control condition. The results also showed that Familias Unidas was efficacious, relative to the prevention as usual in reducing substance use, which included drug use. The results also showed that Familias Unidas was efficacious, relative to the prevention as usual control condition in reducing the frequency of unprotected sexual behavior and in preventing the incidence of adolescent externalizing problem behaviors.

Keepin' it REAL

keepin' it REAL is a 4-month school-based intervention, which works with the youth at school, and is designed to disseminate culturally grounded prevention messages to prevent substance use (Kulis et al. 2005). keepin' it REAL is guided by the ecological risk and resilience (Bogenschneider 1996) and narrative and performance frameworks (Fisher 1987). Informed by a cultural context perspective, keepin' it REAL aims to prevent drug abuse in Hispanic adolescents through the integration of cultural norms, values, and effective methods of communication among Hispanic culture. keepin' it REAL specifically promotes aspects of Mexican American, European American, and African American cultures that run counter to drug and alcohol use. These cultural elements are integrated with aspects of resistance and life skills models (Botvin et al. 2001), and the resulting integration is delivered as a classroom-based curriculum (Kulis et al. 2005). In doing so, keepin' it REAL seeks to foster cultural identification, emphasize an antidrug norms and behaviors perspective, as well as further develop the skill set necessary to promote positive decision-making and antidrug resistant skills. Hence, keepin' it REAL was developed to incorporate traditional ethnic values and practices that diminish intrapersonal risk for and promote intrapersonal protection against substance use (i.e., communication, competence, narrative knowledge, motivating norms, and drug resistance skills; Castro et al. 1999; Kulis et al. 2005).

Kulis et al. (2005) examined the relative efficacy of three versions of the keepin' it REAL intervention in preventing/reducing substance use among Hispanic adolescents: one grounded in Mexican American culture, another grounded in African American and European American culture, and a third that consisted of half of the lessons from the Mexican-American version and half of the lessons from the African American/European American version (referred to as the "multicultural" version). The efficacy of these three interventions was evaluated, relative to a no intervention control group, in (a) preventing and reducing substance use by enhancing cultural identification, (b) promoting personal anti-drug norms, and (c) increasing decision-making and resistance skills.

This study used a 4 (Condition) × 2 (Time) randomized controlled design. A total of 35 schools in Arizona were randomized into one of the four conditions, and students were assessed at pre and post intervention (which corresponded to 14 months post-intervention). The sample for this study consisted of 3402 7th grade students who reported their race or ethnicity as Mexican American, Mexican, or Chicano. The mean age at baseline of participants was 12.52 years ($SD = .64$ years) and the majority of the sample was male (51.5 %). Approximately nine out of ten families were identified as low income families. Participants predominately reported Spanish as their primary language (Kulis et al. 2005).

Results indicated that, compared with students in the control condition, those students who participated in the Mexican American or "multicultural" versions of keepin' it REAL reported significantly less substance use and marijuana use in the past 30 days. Additionally, students in the Mexican American version of keepin' it REAL reported greater refusal confidence, fewer intentions to accept offers of substance use, and less peer use. No significant differences in any of the outcomes emerged between the African American/European version of the intervention and the control group. Thus, these findings provide preliminary evidence that culturally specific interventions are more efficacious than generic interventions.

Recommendations for Advancing the Field

Although the drug abuse prevention field has received a fair amount of attention by scientists, community practitioners, and politicians alike, it is estimated that less than 1 % of efficacious interventions have been disseminated and are currently being implemented in communities. Given that the number of drug abuse preventive interventions for Hispanic adolescents is limited, the state of intervention science for Hispanic youth is considerably behind that of non-Hispanic Whites. Therefore, there remains the need to move interventions along the stages of intervention development (Rounsaville et al. 2001). For example, now that the *efficacy* of keepin' it REAL (Kulis et al. 2005) and Familias Unidas (Prado et al. 2007; Pantin et al. 2009) have been established, it is important to evaluate the effectiveness of these interventions. Our research team is currently evaluating the effectiveness of Familias Unidas in a population based, representative sample of Hispanic youth in Miami-Dade County. In this study, we take the view that an effectiveness trial should test a defined intervention (in this case Familias Unidas) that is delivered by intervention agents within community settings in a manner that could ultimately be used for large scale implementation. We also take the view that effectiveness trials should be conducted with rigorous randomized trial designs, trained clinicians, and intervention manuals (Brown et al. 2008; Flay 1986).

This effectiveness study uses a 2 (Condition) × 4 (Time) randomized controlled design, where participants will be assessed at baseline, randomized, and reassessed at 6, 18, and 30 months post baseline. A total of 31 youth and their primary caregivers from 24 schools will comprise the study sample. Adolescents and their families will be randomized to either Familias Unidas or the Community Practice control condition. Unlike our efficacy study, where facilitators were highly skilled clinicians working for the research team, in this effectiveness study facilitators will consist of school mental health counselors (i.e., school Trust counselors and school social workers) at each of the 24 participating middle schools. We have partnered with Miami-Dade Public Schools under a subcontract to this effectiveness study. Using school counselors to deliver the intervention will test whether the intervention will be effective when delivered by school mental health counselors within the public school system as would occur during large scale implementation and will maximize the chances that this intervention, if effective, will be implemented and sustained in community practice. Equally important in this study, will be to examine those factors and processes associated with the successful adoption and implementation of the intervention. In summary, moving efficacious drug abuse preventive interventions, such as keepin' it REAL (Kulis et al. 2005) and Familias Unidas (Prado et al. 2007; Pantin et al. 2009) from efficacy to effectiveness to implementation is essential to reducing the drug abuse health disparities that exist between Hispanics and other segments of the population.

A second recommendation for advancing the field and reducing Hispanic drug use is to examine the mechanisms of action (i.e., mediators) of intervention efficacy/effectiveness across development. For example, research with Hispanic early adolescents has demonstrated improvements in family functioning account for reductions in substance use (Prado et al. 2007; Pantin et al. 2009), but it is possible that the efficacy of interventions for early elementary school children may be explained by reductions in delinquency and academic failure (Barrera et al. 2002). In this regard though, it is important to conduct long-term follow up of interventions that are delivered at a stage of development where drug use is too low or not reported at all because of the age of the sample (e.g., early elementary school).

Such long-term follow up will yield important information as to whether targeting risk and protective factors associated with Hispanic drug use has an impact on later drug use. Equally important is to examine whether interventions that have an effect on adolescent drug use sustain their effect into emerging adulthood or if they have an effect on the consequences of drug use (e.g., HIV risk behaviors) at a later stage of development. While such research has been conducted with non-Hispanic Whites and other minorities (Furr-Holden et al. 2004; Kellam et al. 2008), research of this nature for the Hispanic population is lacking.

A third recommendation for advancing the prevention field and preventing/reducing Hispanic drug use is to understand for whom are preventive interventions most (and least) efficacious/effective for. Some research has examined the extent to which the efficacy of drug abuse preventive interventions for Hispanic youth varies as a function of demographic and environmental factors (Martinez and Eddy 2005; Prado et al. 2009a, b). For example, Martinez and Eddy (2005) found differential effects by nativity status (i.e., U.S. born vs. foreign born) on risk and protective processes associated with drug use in an evaluation study of their parenting intervention for Hispanic youth. Research on whether gender moderates the effects of interventions for Hispanic youth is much more limited. For example, no interventions have reported differential effects by gender for drug abuse preventive interventions with Hispanic youth. It is unclear whether effects of this nature have not been reported because studies have shown no such effects or because no such analyses have been conducted.

Research on whether genetic factors or on whether gene by environment interactions moderates the efficacy of preventive interventions is also lacking. In fact, we could not find a single study among Hispanics that examined whether and to what extent genetics moderates the impact of an intervention on drug use for this population. Although still limited, some research has been conducted in this area with other populations. For example, Brody et al. (2009) found that a parenting intervention decreased the likelihood of substance use among a sample of rural African American adolescents who had a genetic predisposition to substance use. It would be important to examine whether such findings generalize to other populations, such as Hispanics, where family is critical to positive development.

Drug use preventive interventions are efficacious and effective only to the extent that participants are effectively recruited, engaged, retained and actively participate in the intervention. Unfortunately, many Hispanics encounter multiple contextual challenges (Parra-Cardon et al. 2009) that prevent them from fully participating in drug use prevention services. Consequently, many preventive interventions which target populations that are most affected by substance use, including Hispanics, are often characterized by low participation rates (Perrino et al. 2001). Therefore, a fourth recommendation for advancing the prevention field and reducing/preventing Hispanic drug use is to develop more easily transportable versions of efficacious preventive interventions that target drug use. For example, the field of prevention science could capitalize on current technological advances, such as the internet, to combat drug use. In fact, internet-based prevention interventions for adolescents have shown promising results for drug abuse (Newton et al. 2009, 2010; Schwinn et al. 2010). However, there currently are no studies evaluating the efficacy of internet-based drug use preventive interventions for Hispanic adolescents (Ybarra and Bull 2007).

The field of drug abuse prevention for Hispanic populations could be advanced by examining the effects of drug use preventive intervention on brain activity through the use of novel methods such as magnetic resonance imaging (MRI). For example, MRI has been used to examine the effects of cell phone radiofrequency signal exposure on brain glucose metabolism (Volkow et al. 2011), dopaminergic pathways in obese adults (Volkow et al. 2011), and cocaine on vascular and cellular changes of the brain (Yuan et al. 2011). Thus, MRI could potentially be used to determine whether and the extent to which having participated in a drug use preventive intervention affects brain glucose metabolism, a marker of brain activity (Volkow et al. 2011). The use of such novel techniques could aid in forging basic and behavioral sciences, and thereby facilitate transdisciplinary approaches aimed at better understanding the complexity of drug abuse prevention in Hispanic populations.

Finally, in line with both using novel methods and transdisciplinary approaches to prevent drug abuse in Hispanic adolescents, the field of prevention science can also be advanced by examining systemic mechanisms that may have an effect on drug abuse among this population. For example, scientists recently have taken a keen interest in examining the effects of physical activity on drug abuse (Volkow 2011). Research on the effects of exercise on the brain in animal models have suggested that exercise may facilitate the development of blood vessels in the brain, build connections between cells, help damaged neural tissue, and increase resiliency to stress, all of which may influence and be influenced by drug abuse (Volkow 2011). That is, physical exercise may promote changes in the brain which in turn may prevent/reduce drug abuse. Thus, drug abuse prevention scientists and, for example, basic scientists and physical health experts, can forge collaborative relationships to examine whether and the extent to which these findings extend to human models, in that way advancing the field of drug abuse prevention in Hispanic adolescents.

Conclusions

Preventing/reducing drug use is essential to eliminating the health disparities that exist between Hispanics and other segments of the population. Although Hispanics report drug use at disproportionate levels, the state of intervention science is lacking for this population. In addition to a need for demonstrating the efficacy of drug use preventive interventions for Hispanic youth, there is a need to (a) establish the long-term efficacy of these interventions, (b) translate these interventions from laboratory based interventions to real-world interventions, (c) examine those factors and processes that will increase the likelihood that such interventions are successfully adopted and implemented in community settings, (d) understand those mechanisms that explain intervention efficacy/effectiveness, and (e) understand whether and to what extent environmental, genetic, and biological factors independently or jointly interact to predict intervention efficacy/effectiveness.

Acknowledgement This article was supported by NIDA grants # 1R01DA025192-01A1 and 1R01DA025894-01as well as NIDA and NIAAA grant #3R01DA025192-01A1S1 awarded to Guillermo Prado.

References

Alegria, M., Canino, G., Shrout, P. E., Woo, M., Duan, N., Vila, D., et al. (2008). Prevalence of mental illness in immigrant and non-immigrant U.S. Latino groups. *American Journal of Psychiatry, 165*, 359–369.

Allen, M. L., Elliott, M. N., Fuligni, A. J., Morales, L. S., Hambarsoomian, K., et al. (2008). The relationship between Spanish language use and substance use behaviors among Latino youth: A social network approach. *Journal of Adolescent Health, 43*, 372–379.

Arias, E., Anderson, R. N., Kung, H. C., Murphy, S. L., & Kochanek, K. D. (2003). Deaths: Final data for 2001. *National Vital Statistics Reports, 52*(3), 1–116.

Ary, D. V., Duncan, T. E., Duncan, S. C., & Hops, H. (1999). Adolescent problem behavior: The influence of parents and peers. *Behaviour Research and Therapy, 37*(3), 217–230.

Bahr, S. J., Hoffmann, J. P., & Yang, X. (2005). Parental and peer influences on the risk of adolescent drug use. *Journal of Primary Prevention, 26*, 529–551.

Barrera, M., Jr., Biglan, A., Ary, D., & Li, F. (2001). Replication of a problem behavior model with American Indian, Hispanic, and Caucasian youth. *Journal of Early Adolescence, 21*, 133–157.

Barrera, M., Jr., Biglan, A., Taylor, T. K., Gunn, B. K., Smolkowski, K., et al. (2002). Early elementary school intervention to reduce conduct problems: A randomized trial with Hispanic and non-Hispanic children. *Prevention Science, 3*(2), 83–94.

Bogenschneider, K. (1996). An ecological risk/protective theory for building prevention programs, policies, and community capacity to support youth. *Family Relations, 45*, 127–138.

Botvin, G. J., Griffen, K. W., Diaz, T., & Ifill-Williams. (2001). Drug abuse prevention among minority adolesecents: Posttest and one-year follow-up of a school-based preventive intervention. *Prevention Science, 2*, 1–13.

Brody, G. H., & Ge, X. (2001). Linking parenting processes and self regulation to psychological functioning and alcohol use during early adolescence. *Journal of Family Psychology, 15*, 82–94.

Brody, G. H., Beach, S. R. H., Philibert, R. A., Chen, Y.-f., Lei, M.-K., Murry, V. M., et al. (2009). Parenting moderates a genetic vulnerability factor on longitudinal increases in youths' substance use. *Journal of Consulting and Clinical Psychology, 77*, 1–11.

Broman, C. L., Reckase, M. D., & Freedman-Doan, C. R. (2006). The role of parenting in drug use among African American, Latino, and white adolescents. *Journal of Ethnicity in Substance Abuse, 5*(1), 39–50.

Bronfenbrenner, U. (1979). *The ecology of human development: Experiments by nature and design*. Cambridge, MA: Cambridge University Press.

Bronfenbrenner, U. (1986). Ecology of the family as a context for human development. *The American Psychologist, 32*, 513–531.

Brook, J. S., Whiteman, M., Balka, E. B., Win, P. T., & Gursen, M. D. (1997). African-American and Puerto Rican drug use: A longitudinal study. *Journal of the American Academy of Child and Adolescent Psychiatry, 36*, 1260–1268.

Brooks-Gunn, J., & Furstenberg, F. F. (1989). Adolescent sexual behavior. *American Psychologist, 44*, 249–257.

Brown, C. H., Wang, W., Kellam, S. G., Muthen, B. O., Petras, H., Toyinbo, P., et al. (2008). Methods for testing theory and evaluating impact in randomized field trials: Intent-to-treat analysis for integrating the perspectives of person, place, and time [Supplement 1]. *Journal of Drug and Alcohol Dependence, 95*, S74–S104.

Cabassa, L. J. (2003). Measuring acculturation: Where we are and where we need to go. *Hispanic Journal of Behavioral Sciences, 25*, 127–146.

Castro, F. G., Proescholdbell, R. J., Abeita, L., & Rodriguez, D. (1999). Ethnic and cultural minority groups. In B. S. McCrady & E. E. Epstein (Eds.), *Addictions: A comprehensive guidebook* (pp. 499–526). New York: Oxford Press.

Castro, F. G., Barrera, M., Jr., & Martinez, C. R., Jr. (2004). The cultural adaptation of prevention interventions: Resolving tensions between fidelity and fit. *Prevention Science, 5*, 41–45.

Castro, F. G., Barrera, M., Jr., Pantin, H., Martinez, C. R., Jr., Felix-Ortiz, M., Rios, R., et al. (2006). Substance abuse prevention intervention research with Hispanic populations. *Drug and Alcohol Dependence, 84*, S29–S42.

Cicchetti, D., & Richters, J. E. (1993). Developmental considerations in the investigation of conduct disorder. *Development and Psychopathology, 5*(1–2), 331–344.

Cicchetti, D., & Rogosch, F. A. (2002). A developmental psychopathology perspective on adolescence. *Journal of Consulting and Clinical Psychology, 70*(1), 6–20.

Coatsworth, J. D., Pantin, H., & Szapocznik, J. (2002). Familias Unidas: A family-centered ecodevelopmental intervention to reduce risk for problem behavior among Hispanic adolescents. *Clinical Child and Family Psychology Review, 5*, 113–132.

Crosby, R., DiClemente, R., Wingood, G., Cobb, B., Harrington, K., Davies, S., et al. (2001). HIV/STD-protective benefits of living with mothers in perceived supportive families: A study of high-risk African American female teens. *Preventive Medicine, 33*(3), 175–178.

Ellickson, P., Bui, K., Bell, R., & McGuigan, K. A. (1998). Does early drug use increase the risk of dropping out of high school? *Journal of Drug Issues, 28*, 357–380.

Epstein, J. A., Botvin, G. J., Baker, E., & Diaz, T. (1999). Impact of social influences and problem behavior on alcohol use among inner-city Hispanic and African American adolescents. *Journal of Studies on Alcohol, 60*, 595–604.

Escobar, J., & Vega, W. (2000). Mental health and immigration's AAAs: Where are we and where do we go from here? *The Journal of Nervous and Mental Disease, 188*, 736–740.

Fisher, W. R. (1987). *Human communication as narration: Toward a philosophy of reason, value, and action*. Columbia: University of South Carolina Press.

Flay, B. R. (1986). Efficacy and effectiveness trials (and other phases of research) in the development of health promotion programs. *Preventive Medicine, 15*, 451–474.

Ford, K., & Norris, A. E. (1993). Knowledge of AIDS transmission, risk behavior, and perceptions of risk among urban, low-income, African-American and Hispanic youth. *American Journal of Preventive Medicine, 9*, 297–306.

Furr-Holden, C. D. M., Ialongo, N. S., Anthony, J. C., Petras, H., & Kellam, S. G. (2004). Developmentally inspired drug prevention: Middle school outcomes in a school-based randomized prevention trial. *Drug and Alcohol Dependence, 73*, 149–158.

Gil, A. G., Wagner, E. F., & Vega, W. A. (2000). Acculturation, familism, and alcohol use among Latino adolescent males: Longitudinal relations. *Journal of Community Psychology, 28*, 443–458.

Gil, A. G., Wagner, E. F., & Tubman, J. G. (2004). Culturally sensitive substance abuse intervention for Hispanic and African American adolescents: Empirical examples from the alcohol treatment targeting adolescents in need (ATTAIN) project. *Addiction, 99*, 140–150.

Greene, J. P., & Forster, G. (2003). *Public high school graduation and college readiness rates in the United States*. Manhattan Institute, September 2003.

Guagliardo, M. F., Huang, Z., Hicks, J., & D'Angelo, L. (1998). Increased drug use among old-for-grade and dropout urban adolescents. *American Journal of Preventive Medicine, 15*, 42–48.

Henry, K. L., Swaim, R. L., & Slater, M. D. (2005). Intraindividual variability of school bonding and adolescents' beliefs about the effect of substance use on future aspirations. *Prevention Science, 6*, 101–112.

Jessor, R., & Jessor, S. L. (1977). *Problem behavior and psychosocial development-A longitudinal study of youth*. New York: Academic.

Johnston, L. D., O'Malley, P. M., Bachman, J. G., & Schulenberg, J. E. (2011a). *Monitoring the future national results on adolescent drug use: Overview of key findings, 2010*. Ann Arbor: Institute for Social Research, The University of Michigan.

Johnston, L. D., O'Malley, P. M., Bachman, J. G., & Schulenberg, J. E. (2011b). *Demographic subgroup trends for various licit and illicit drugs, 1975–2010* (Monitoring the future occasional paper no. 74). Ann Arbor: Institute for Social Research. Available: http://www.monitoringthefuture.org/

Kellam, S. G., Brown, C. H., Poduska, J., et al. (2008). Effects of a universal classroom behavior management program in first and second grades on young adult behavioral, psychiatric, and social outcomes. *Drug and Alcohol Dependence, 95*, S5–S28.

Krauss, B. J., Godfrey, C., Yee, D., Goldsamt, L., Tiffany, J., Almeyda, L., et al. (2000). Saving our children from a silent epidemic: The PATH program for parents and preadolescents. In W. Pequegnat & J. Szapocznik (Eds.), *Working with families in the era of HIV/AIDS* (pp. 89–112). Thousand Oaks: Sage.

Kulis, S., Marsiglia, F. F., Elek, E., Dustman, P., Wagstaff, D. A., & Hecht, M. L. (2005). Mexican/Mexican American adolescents and Keepin' it REAL: An evidence-based substance abuse prevention program. *Children & Schools, 27*, 133–145.

Leon, A. M., & Dziegielewski, S. F. (2000). Engaging Hispanic immigrant mothers: Revisiting the time-limited psycho-educational group model. *Crisis Intervention and Time-Limited Treatment, 6*, 13–27.

Lucas, C. P., Zhang, H., Fisher, P. W., Shaffer, D., Regier, D. A., et al. (2001). The DISC Predictive Scales (DPS). *Journal of the American Academy of Child & Adolescent Psychiatry, 40*, 443–449.

Marin, G., Sabogal, F., Marin, B., Otero-Sabogal, R., & Perez-Stable, E. J. (1987). Development of a short acculturation scale for Hispanics. *Hispanic Journal of Behavioral Sciences, 9*, 183–205.

Marshal, M. P., & Chassin, L. (2000). Peer influence on adolescent alcohol use: The moderating role of parental support and discipline. *Applied Developmental Science, 4*, 80–88.

Martinez, C. R. (2006). Effects of differential family acculturation on Latino adolescent substance use. *Family Relations, 55*, 306–317.

Martinez, C. R., Jr., & Eddy, J. M. (2005). Effects of culturally adapted parent management training on Latino youth behavioral health outcomes. *Journal of Consulting and Clinical Psychology, 73*, 841–851.

Mummendey, A., Klink, A., & Brown, R. (2001). Nationalism and patriotism: National identification and out-group rejection. *British Journal of Social Psychology, 40*, 159–172.

Newton, N. C., Andrews, G., Teesson, M., & Vogl, L. E. (2009). Delivering prevention for alcohol and cannabis using the Internet: A cluster randomised controlled trial. *Prevention Medicine, 48*, 579–584.

Newton, N. C., Teesson, M., Vogl, L. E., & Andrews, G. (2010). Internet-based prevention for alcohol and cannabis use: Final results of the Climate Schools course. *Addiction, 105*, 749–759.

O'Sullivan, L., Jaramillo, B., Moreau, D., & Meyer-Bahlburg, H. (1999). Mother-daughter communication about sexuality in a clinical sample of Hispanic adolescent girls. *Hispanic Journal of Behavioral Sciences, 21*, 447–469.

Pantin, H., Coatsworth, J. D., Feaster, D. J., Newman, F. L., Briones, E., Prado, G., Schwartz, S. J., & Szapocznik, J. (2003a). Familias Unidas: The efficacy of an intervention to promote parental investment in Hispanic immigrant families. *Prevention Science, 4*, 189–201.

Pantin, H., Schwartz, S. J., Sullivan, S., Coatsworth, J., & Szapocznik, J. (2003b). Preventing substance abuse in Hispanic immigrant adolescents: An ecodevelopmental, parent-centered approach. *Hispanic Journal of Behavioral Sciences, 25*(4), 469–500.

Pantin, H., Schwartz, S. J., Sullivan, S., Prado, G., & Szapocznik, J. (2004). Ecodevelopmental HIV prevention programs for Hispanic adolescents. *The American Journal of Orthopsychiatry, 74*, 545–558.

Pantin, H., Prado, G., Huang, S., Lopez, B., Tapia, M., Sabillon, E., Schwartz, S., & Branchini, J. (2009). A randomized controlled trial of Familias Unidas for Hispanic behavior-problem adolescents. *Psychosomatic Medicine, 71*(9), 987–995.

Parra-Cardon, J., Holtrop, K., Cordova, D., Escobar-Chew, A. R., Horsford, S., Tams, L., et al. (2009). "Queremos aprender": Latino immigrants' call to integrate cultural adaptation with best practice knowledge in a parenting intervention. *Family Process, 48*, 211–231.

Perrino, T., Coatsworth, J. D., Briones, E., Pantin, H., & Szapocznik, J. (2001). Initial engagement in parent-centered preventive interventions: A family systems perspective. *Journal of Primary Prevention, 22*, 21–44.

Pettit, G. S., Bates, J. E., Dodge, K. A., & Meece, D. W. (1999). The impact of after-school peer contact on early adolescent externalizing problems is moderated by parental monitoring, perceived neighborhood safety, and prior adjustment. *Child Development, 70*, 768–778.

Phinney, J. (2003). Ethnic identity and acculturation. In K. Chun, P. Ball, & G. Marin (Eds.), *Acculturation: Advances in theory, measurement, and applied research* (pp. 63–81). Washington, DC: American Psychological Association.

Prado, G., Schwartz, S. J., Pattatucci-Aragon, A., Clatts, M., Pantin, H., Fernandez, M. I., Lopez, B., Briones, E., Amaro, H., & Szapocznik, J. (2006). The prevention of HIV transmission in Hispanic adolescents. *Drug and Alcohol Dependence, 84*, S43–S53.

Prado, G., Pantin, H., Briones, E., Schwartz, S. J., Feaster, D., Huang, S., Sullivan, S., Tapia, M. I., Sabillon, E., Lopez, B., & Szapocznik, J. (2007). A randomized controlled trial of a parent-centered intervention in preventing substance use and HIV risk behaviors in Hispanic adolescents. *Journal of Consulting and Clinical Psychology, 75*, 914–926.

Prado, G., Huang, S., Schwartz, S., Maldonado-Molina, M., Bandiera, F., de la Rosa, M., & Pantin, H. (2009a). What accounts for differences in substance use among U.S. born and immigrant Hispanic adolescents?: Results from a longitudinal prospective cohort study. *Journal of Adolescent Health, 45*, 118–125.

Prado, G., Schwartz, S., Maldonado-Molina, M., Huang, S., Pantin, H., Lopez, B., & Szapocznik, J. (2009b). Ecodevelopmental x intrapersonal risk: Substance use and sexual behavior in Hispanic adolescents. *Health Education & Behavior, 36*, 45–61.

Prado, G., Huang, S., Maldonado-Molina, M., Bandiera, F., Schwartz, S., de la Vega, P., et al. (2010). An empirical test of ecodevelopmental theory in predicting HIV risk behaviors among Hispanic youth. *Health Education & Behavior, 37*, 97–114.

Ramirez, J. R., Crano, W., Quist, R., Burgoon, M., Alvaro, E., & Grandpre, J. (2004). Acculturation, familism, parental monitoring and knowledge as predictors of marijuana and inhalant use in adolescents. *Psychology of Addictive Behaviors, 18*, 3–11.

Redfield, R., Linton, R., & Herskovits, M. J. (1936). Memorandum on the study of acculturation. *American Anthropologist, 38*, 149–152.

Rodgers, K. B. (1999). Parenting processes related to sexual risk taking among adolescent males and females. *Journal of Marriage and the Family, 61*, 99–110.

Rounsaville, B. J., Carroll, K. M., & Onken, L. S. (2001). NIDA's stage model of behavioral therapies research: Getting started and moving on from Stage I. *Clinical Psychology Science and Practice, 8*, 133–142.

Ryder, A., Alden, L., & Paulhus, D. (2000). Is acculturation unidimensional or bidimensional? A head-to-head comparison in the prediction of personality, self-identity, and adjustment. *Journal of Personality and Social Psychology, 79*, 49–65.

Santisteban, D. A., Muir-Malcolm, J. A., Mitrani, V. B., & Szapocznik, J. (2002). Integrating the study of ethnic culture and family psychology intervention science. In H. Liddle, R. Levant, D. A. Santisteban, & J. Bray (Eds.), *Family psychology: Science-based interventions* (pp. 331–352). Washington, DC: American Psychological Association.

Schwartz, S. J., Coatsworth, J. D., Pantin, H., Prado, G., Sharp, E. H., Szapocznik, J., et al. (2006a). The role of ecodevelopmental context and self-concept in depressive and externalizing symptoms in Hispanic adolescents. *International Journal of Behavioral Development, 30*, 359–370.

Schwartz, S. J., Montgomery, M. J., & Briones, E. (2006b). The role of identity in acculturation among immigrant people: Theoretical propositions, empirical questions, and applied recommendations. *Human Development, 49*, 1–30.

Schwartz, S. J., Forthun, L. F., Ravert, R. D., Zamboanga, B. L., Rodriguez, L., Umaña-Taylor, A. J., Filton, B. J., Kim, S. Y., Rodriguez, L., Weisskirch, R. S., Vernon, M., Shneyderman, Y., Williams, M. K., Agocha, V. B., & Hudson, M. (2010). The protective role of identity consolidation against health risk behaviors in college-attending emerging adults. *American Journal of Health Behavior, 34*, 214–224.

Schwinn, T., Schinke, S. P., & Di Nola, J. (2010). Preventing drug abuse among adolescent girls: Outcome data from an Internet-based intervention. *Prevention Science, 11*, 24–32.

Simons, R. L., Lorenz, F. O., Conger, R. D., Wu, C. I., et al. (1992). Support from spouse as mediator and moderator of the disruptive influence of economic strain on parenting. *Child Development, 63*, 1282–1301.

Simons-Morton, B., Crump, A. D., Haynie, D. L., Saylor, K. E., Eitel, P., Yu, K., et al. (1999). Psychosocial, school, and parent factors associated with recent smoking among early-adolescent boys and girls. *Preventive Medicine, 28*, 138–148.

Steinberg, L., & Morris, A. S. (2001). Adolescent development. *Annual Review of Psychology, 52*, 83–110.

Substance Abuse and Mental Health Services Administration. (2005). *Results from the 2004 National Survey on Drug Use and Health: National Findings* (NSDUH series H-28, DHHS publication no. SMA 05–4062). Rockville: Office of Applied Studies.

Szapocznik, J., & Coatsworth, J. D. (1999). An ecodevelopmental framework for organizing the influences on drug abuse: A developmental model of risk and protection. In M. D. Glantz & C. R. Hartel (Eds.), *Drug abuse: Origins & interventions* (pp. 331–366). Washington, DC: American Psychological Association.

Szapocznik, J., & Kurtines, W. M. (1993). Family psychology and cultural diversity: Opportunities for theory, research, and application. *American Psychologist, 48*, 400–407.

Szapocznik, J., & Williams, R. A. (2000). Brief strategic family therapy: Twenty-five years of interplay among theory, research and practice in adolescent behavior problems and drug abuse. *Clinical Child and Family Psychology Review, 3*, 117–134.

Szapocznik, J., Scopetta, M. A., & King, O. E. (1978a). Theory and practice in matching treatment to the special characteristics and problems of Cuban immigrants. *Journal of Community Psychology, 6*, 112–122.

Szapocznik, J., Scopetta, M., Kurtines, W., & Arnalde, M. (1978b). Theory and measurement of acculturation. *Interamerican Journal of Psychology, 12*, 113–130.

Szapocznik, J., Prado, G., Burlew, A. K., Williams, R. A., Santisteban, D. A., et al. (2007). Drug abuse in African American and Hispanic adolescents: Culture, development, and behavior. *Annual Review of Clinical Psychology, 3*, 77–105.

Volkow, N. D. (2011). Physical activity may prevent substance abuse. NIDA Notes. Retrieved from www.drugabuse.com on March 2013.

Volkow, N. D., Tomasi, D., Wang, G. J., Fowler, J. S., Telang, F., Goldstein, R. Z., et al. (2011). Positive emotionality is associated with baseline metabolism in orbitofrontal cortex and in regions of the default network. *Molecular Psychiatry, 16*, 818–825.

Ybarra, M. L., & Bull, S. S. (2007). Current trends in Internet- and cell phone-based HIV prevention and intervention programs. *Current HIV/AIDS Reports, 4*, 201–207.

Yuan, Z., Luo, Z., Volkow, N., Pan, Y., & Congwu, D. (2011). Imaging separation of neuronal from vascular effects of cocaine on rat cortical brain in vivo. *Neuroimage, 54*, 1130–1139.

Chapter 18
Patterns of Risk Behavior Change from Adolescence to Emerging Adulthood: Implications for HIV/STD Racial Disparities

Denise D. Hallfors, Abigail A. Haydon, Carolyn Tucker Halpern, and Bonita J. Iritani

Sexually transmitted infections are a significant health threat to adolescents and young adults. Fifteen to thirty percent of all HIV infections occur in individuals younger than age 25 (Morris et al. 2006). In 2000, youth between the ages of 15–24 accounted for almost half of new cases of sexually transmitted infections (STIs) (Weinstock et al. 2004). Racial and ethnic minorities are disproportionately affected by HIV and other STIs. The Centers for Disease Control and Prevention estimated that African Americans accounted for approximately 45 % of new HIV infections in 2006 and 48 % of persons living with HIV/AIDS in 2007 (CDC 2009). Results from a nationally representative sample of young adults indicate that the prevalences of Chlamydia infection and Trichomoniasis were both approximately six times greater among African American young adults compared with Whites in 2001–2002. Although the overall prevalence of gonorrhea was low compared to other STIs, prevalence was approximately 22 times greater among African American young adults compared with Whites (Miller et al. 2004, 2005).

Some sexual behaviors increase the risk of HIV and STIs, including multiple sexual partners, inconsistent condom use, and early age at first sexual intercourse, although early sexual transition appears to elevate risk only during the adolescent years (Coker et al. 1994; Kaestle et al. 2005; Miller et al. 1999; Upchurch et al. 2004).

D.D. Hallfors (✉)
University of North Carolina at Chapel Hill (UNC-CH), Chapel Hill, NC, USA

PIRE, Chapel Hill, NC, USA
e-mail: hallfors@pire.org

A.A. Haydon
FrameWorks Institute, Washington, DC, USA

C.T. Halpern
University of North Carolina at Chapel Hill (UNC-CH), Chapel Hill, NC, USA

B.J. Iritani
PIRE, Chapel Hill, NC, USA

Y.F. Thomas, L.N. Price (eds.), *Drug Use Trajectories Among Minority Youth*,
DOI 10.1007/978-94-017-7491-8_18

Drug use can also place individuals at risk, either through unsafe patterns of intravenous drug use (CDC 2002) or through effecting changes in sexual behavior such as inconsistent condom use and/or having multiple, potentially high risk, partners (Kotchick et al. 2001). Beyond the clear risk posed by drug–related behaviors such as needle sharing, the implications of substance use for STI risk among adolescents and young adults are not clear. For example, substance use may disinhibit sexual behavior, which could elevate STI risk, but findings about this possible mediating mechanism are ambiguous (Rashad and Kaestner 2004). If substance use plays a central role in STI risk, whether as a distal or proximal factor, we would expect that individuals who exhibit high levels of drug use or especially risky types of use (e.g., sharing needles) would be at elevated risk of infection. However, most work to date has examined single risk behaviors and sociodemographic characteristics such as socioeconomic status, and those elements do not account for the striking racial disparities in HIV and STIs (Ellen et al. 1995, 1998; Santelli et al. 2000; Harawa et al. 2003).

In previous research, we extended the investigation of the contributions of individual risk behaviors to disparities in STIs by examining the covariation of sexual risk taking and substance use. The goal of this work was to understand how ***patterns*** of multiple risk behaviors may contribute to STI risk (Halpern et al. 2004; Hallfors et al. 2007). Examination of patterns of risk behavior is important because if relevant behaviors share causal antecedents, those antecedents could be targeted more effectively in prevention and intervention efforts. Identification of patterns of behavior that map onto infection patterns would also indicate that targeting single risk behaviors is likely to be ineffective. Better documenting patterns of risk-taking related to sexual activity and drug use may be especially important for disparities in STI risk, as there is some evidence that the consistency and magnitude of covariation of these behaviors vary across adolescent sociodemographic groups (Fortenberry 1995). For example, some findings indicate that the association between these behaviors is weaker among African American adolescents than among Whites (Stanton et al. 1993). However, findings have been inconsistent, probably due to reliance on non-representative samples and inadequate controls for confounders. Differences in patterns of risk taking are also evident between male and female adolescents, with females being less likely to engage in multiple risk behaviors (Halpern et al. 2004) but more likely to experience sexually transmitted infections (Halpern et al. 2004; CDC 2008). Thus, it is important to characterize patterns of risk-taking in sexual behavior and drug use using representative samples to determine whether the sociodemographic groups that are most burdened with STIs exhibit distinctive patterns of risk-taking that are linked to infection outcomes over time.

Cluster Analyses and Associations with Self-Reported STIs during Adolescence

To increase our understanding of associations between risk-taking and STIs, we conducted a cluster analysis of 13,998 non-Hispanic African American and White adolescents who participated in Wave I of the National Longitudinal Study of

Adolescent Health (Add Health) (Halpern et al. 2004). Add Health includes a nationally representative sample of adolescents who were in grades 7–12 in the 1994–1995 school year (Harris et al. 2008). At the time of the 2004 analyses, three in-home interviews had been completed, two approximately 1 year apart during high school (1995, 1996) and a third in 2001 when respondents were 18–26 years old.

Cluster Analysis Wave I

Using data from the Wave I in-home interview, we clustered non-Hispanic African American and White adolescents based on multiple behaviors relevant to HIV and other STIs (injection drug use; frequency of binge drinking, use of alcohol, marijuana and other drugs; having vaginal intercourse; condom use at most recent intercourse; using alcohol or other drugs in conjunction with intercourse; having sex for drugs or money; and, for males, having sex with another male in the past 18 months). The clustering approach reflects person-centered analyses of individual behavior patterns. In person-centered analyses, groups of individuals, rather than variables, are sorted into "homogeneous categories with reference to similarities in their profiles based on values for variables relevant to the problem under consideration" (Magnusson and Cairns 1996, page 25). Grouping of individuals is based on the premise that risk behaviors often occur together and interact with each other. Clustering individuals therefore can parsimoniously capture multiple interactions, offering a holistic approach to developmental analysis of populations.

In line with our original project's grant funding from the National Institute on Drug Abuse to examine pathways from adolescence to young adulthood for HIV/AIDS (R01 DA014496; Denise Hallfors, PI), we developed the clusters with risk for HIV and STIs in mind. Using a combination of a priori, theory driven grouping and empirical induction (k-means analysis) defined by modal data patterns, we constructed 16 clusters of adolescents, capturing almost 80 % of the variation in substance use and sexual risk taking patterns in the sample. Each adolescent was assigned to one cluster on the basis of his or her particular pattern of risk behavior; the clusters reflect distinct patterns of activity and gradations of risk behavior. The four a priori clusters (listed in order of construction/priority) were based on: (1) intravenous drug use, (2) males who have sex with males (but have not used intravenously administered drugs), (3) having sex for drugs or money (but no intravenous drug use or, if male, no sex with other males), and (4) abstainers (individuals who reported never having sexual intercourse or drug use). Using an iterative process to produce an optimal number of distinct and homogeneous clusters that captured maximal variation in the behavioral data, we used K-means cluster analysis to inductively derive the remaining 12 clusters. (Scc Halpern et al. 2004 for details about the cluster analysis.)

The Wave I cluster names, their defining behaviors, and the weighted percentages of gender/race groups that fall into each cluster are displayed in Table 18.1 in

Table 18.1 Wave I cluster name, behavioral definition, and weighted percentages of gender/race groups assigned to cluster

		Females		Males	
Cluster name	Behavioral definition	African American (n=2139)	White (n=5079)	African American (n=1891)	White (n=4889)
Substance experimenters (n=3276)	Infrequent or no current use of substances; none have had sex	21.6	24.7	17.2	26.6
Abstainers (n=3125)	None have ever used substances or had sex	24.3	24.6	16.5	22.0
Sex experimenters (n=2398)	All have had sex; median number of partners =1; 60 % used a condom at last sex; infrequent substance use	31.9	10.9	35.4	8.9
Drinkers (n=1020)	All consumed alcohol in past 12 months; 49 % report binge drinking; infrequent/no illicit drug use; none have had sex	3.5	8.6	1.2	8.8
Smokers and sex (n=931)	All smoke cigarettes daily; infrequent use of alcohol/illicit drugs; 62 % have had sex	1.0	10.1	1.4	7.4
Alcohol and sex (n=784)	All drink occasionally; all have had sex; infrequent tobacco/ illicit drug use	8.0	5.7	6.3	4.3
Binge drinkers (n=642)	All binge frequently; infrequent use of other drugs; 45 % have had sex; 60 % binge at least 1 time/week	3.2	3.1	4.1	6.1
Heavy substance use and sex (n=450)	All smoke, drink, and binge drink with moderate frequency; 45 % use marijuana; few use other illicit drugs; 91 % have had sex	0.3	4.2	1.0	4.3
Combination sex and drug use (n=450)	All have had sex; all used alcohol/illicit drug at last sex	2.8	3.5	2.6	3.6
Marijuana users (n=235)	All use marijuana frequently; few have used other illicit drugs; 94 % use alcohol; 79 % smoke cigarettes; 74 % have had sex	0.8	1.4	1.3	2.3

(continued)

Table 18.1 (continued)

		Females		Males	
Cluster name	Behavioral definition	African American (n=2139)	White (n=5079)	African American (n=1891)	White (n=4889)
Multiple partners (n=189)	All report at least 14 sexual partners; 75 % report low or moderate use of substances	0.9	0.7	6.0	0.9
Sex for drugs or money (n=155)	All have had sex for drugs or money; 50 % report low or moderate use of substances; median number of partners=3	0.6	0.6	2.7	1.4
High marijuana use and sex (n=149)	All use marijuana frequently; all have had sex; all used alcohol/other drug at last sex; 82 % have had more than 1 partner	0.4	0.9	2.5	1.2
Marijuana and other drug users (n=64)	95 % report heavy marijuana use; all use other illicit drugs; 68 % have had sex; 28 % used alcohol/other drug at last sex	0.2	0.6	0.1	0.7
Injection-drug users (n=82)	All have injected drugs; 82 % have had sex; median number of partners=4	0.2	0.4	0.4	1.0
Males who have sex with males (n=48)	All are males who have had sex with another male; 78 % have had multiple partners; 40 % used marijuana in past 30 days; 50 % used alcohol at least one time per month; 17 % have had sex for drugs or money	0.0	0.0	1.4	0.5
Total		100.0	100.0	100.0	100.0

Note. Percentages are weighted to yield national probability estimates

order of cluster size. The clusters represent different patterns of sexual activity and drug use, and frequencies of engagement. Descriptive statistics for risk behaviors were examined at each wave for each cluster yielded by the cluster analysis, and modal and other prominent behavioral characteristics were used to name each cluster. In general, larger clusters tend to reflect low to modest levels of apparent risk whereas smaller clusters reflect high risk behavior that relatively few adolescents report. For example, at Wave I, the Sex Experimenters cluster was comprised of adolescents who had all engaged in sexual intercourse (modal number of sexual partners was one) and had infrequent use of substances. Members of the Substance Experimenters cluster reported infrequent or no current substance use and had not had sex. At the other extreme, the 82 injection drug users had all injected drugs and the median number of sex partners was 4. Detailed information about each of the Wave I clusters is shown in Table 18.1.

Although the same cluster structure was evident for African American and White males and females, some behavioral patterns were more or less common within each race/gender group. White males and females tended to show similar patterns of populating the clusters, with membership in Substance Experimenters, Abstainers, Sex Experimenters, Drinkers, Smokers and Sex, and Alcohol and Sex being the most common. Except for Smokers and Sex, these clusters were also the ones African American teens were most likely to populate. However, Whites were more likely than African Americans to be in clusters defined heavily by substance use (such as Drinkers and Smokers and Sex), and African American males and females were more dissimilar from each other than were White males and females. For example, African American females were more likely to be Abstainers and Substance Experimenters than African American males, but White females and males had similar high membership in these two clusters. Differences in the membership of less common but higher risk clusters also clearly demonstrated the different magnitudes, and sometimes different directions, of race differences in cluster membership for males and females. For example, a smaller proportion of African American females, compared with White females, were in the Heavy Substance Use and Sex cluster. In contrast, a higher proportion of African American males, compared with White males, were in the Multiple Partners or Males Who Have Sex with Males cluster.

Figure 18.1 displays the relative risk ratios for cluster membership for the five clusters where there are significant race by gender interactions in membership; ratios are adjusted for age, parental education, and family structure. It may be seen that, except for membership in the Sex Experimenters cluster, African American females are actually less likely than White females to be a member of these risk clusters (compared with being an Abstainer). The picture for males is reversed, with African American males having a much higher relative risk (except for Heavy Substance Use and Sex) of being in the risk cluster (rather than Abstainer cluster) compared to White males. For the remaining 10 clusters, there were no significant race by gender interactions but there were main effects of gender and of race. Males are more likely than females to be members of four (characterized by substance use) of the remaining 10 clusters (Fig. 18.2). African Americans are more likely than Whites to be members of the Sex Experimenters, Alcohol & Sex, and MSM clusters, while Whites are more likely than African Americans to be members of the six remaining clusters which again primarily pertain to substance use (Fig. 18.3).

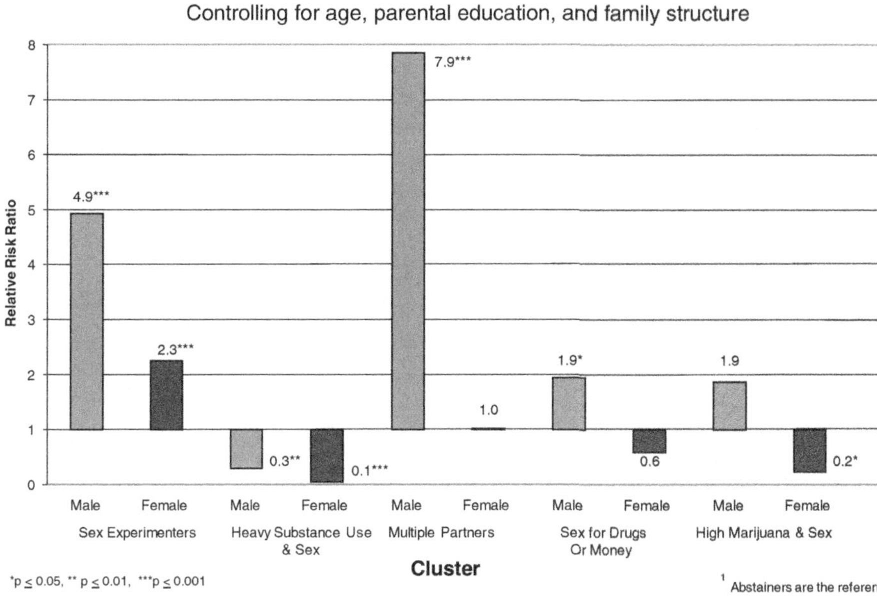

Fig. 18.1 Risk ratios for African Americans relative to Whites, conditional on gender, Wave 1

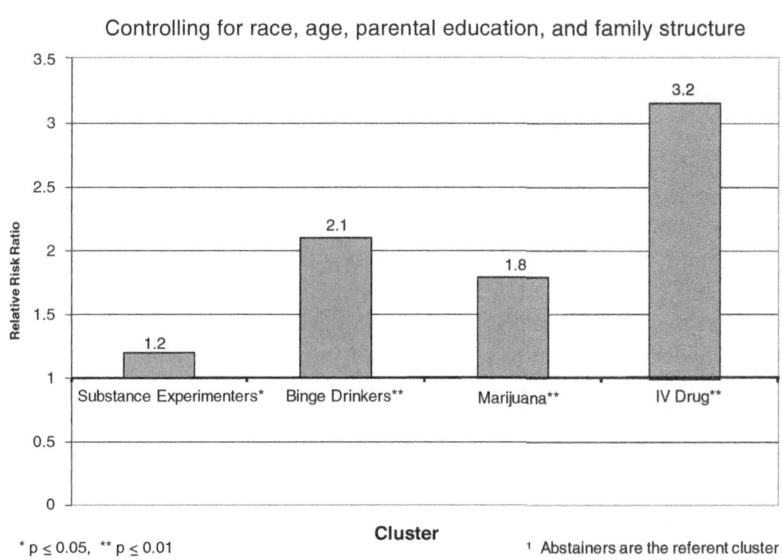

Fig. 18.2 Risk ratios of males relative to females by Wave 1 cluster

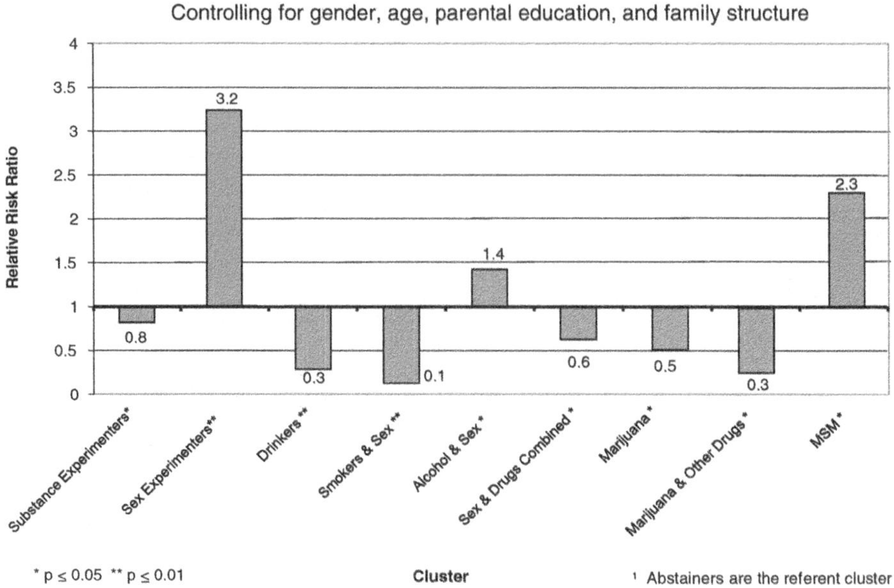

Fig. 18.3 Risk ratios for African Americans relative to Whites by Wave I cluster

Self-Reported STIs & Association with Cluster Membership in Adolescence

We next examined the associations between cluster membership and STIs. At the Wave I adolescent interview respondents were asked whether a doctor or nurse had ever told them that they had chlamydia, syphilis, gonorrhea, HIV or AIDS, genital herpes, genital warts, trichomoniasis or hepatitis B virus. Respondents reporting yes for any of these were coded 1. Six percent of sexually experienced adolescents reported that they had had an STI. African American females were the most likely to report a history of infection (18 %); other prevalence percentages were 6 % of African American males, 6 % of White females, and 2 % of White males.

The clusters that had a high proportion (over 7 %) of STI diagnoses were Multiple partners, Sex for drugs or money, High marijuana use and sex, Injection drug users, and MSM.

However, it was also striking that clusters such as Sex Experimenters, which are characterized by relatively low risk behavior (e.g., median of one partner and very low ATOD), were over-represented among the population reporting STIs. Sex Experimenters constituted 15 % of the sample, but accounted for 29 % of adolescents who reported having had an STI. About a third of African American males and females were members of the Sex Experimenters cluster.

Taken together, these analyses indicated that African American and White teens were differentially likely to display patterns of risk taking related to STIs. African American adolescents were less likely to appear in clusters defined primarily by substance use (Substance Experimenters, Drinkers, Marijuana, and Marijuana & Other Drugs) or by high substance use and sexual risk taking (Heavy Substance Use & Sex, Smokers & Sex, Sex & Drugs Combined). Further, females overall were less likely to appear in clusters defined primarily by substance use (Substance Experimenters, Binge Drinkers, Marijuana, and IV Drug) or high levels of sexual activity. The results also showed that some patterns or clusters (i.e., those defined primarily by sexual risk taking) were more likely to have members who report having had an STI. However, relatively low risk clusters (e.g., Sex Experimenters) accounted for disproportionate numbers of teens with a positive STI history. Further, African American females were heavily represented in low risk clusters, yet had the highest prevalence of a past STI diagnosis according to self-report. One in five sexually experienced African American females reported having received a past STI diagnosis compared with fewer than one in ten White females, who showed similar or even higher risk patterns.

Cluster Analyses and Associations with Tested STIs during Young Adulthood

To better understand the implications of individuals' patterns of risk behavior for vulnerability to STIs, we addressed similar issues in a subsequent study based on Wave III data from Add Health (conducted in 2001–2002 when respondents were 18–26 years old). At Wave III, Add Health conducted biological testing for HIV and three STIs (chlamydia, gonorrhea, and trichomoniasis). We examined the association between patterns of sexual and substance use risk taking, and current, test-identified STIs for 8706 African American and White young adults in the sample (Hallfors et al. 2007).

Cluster Analysis Wave III

Using a cluster construction strategy that paralleled that of our 2004 study, we identified 16 risk patterns that explained 73 % of individual differences in sexual and substance use risk behaviors. The modal cluster was characterized by having had few sexual partners in the past year and low alcohol, tobacco, and other drug use. Table 18.2 presents cluster names, defining behaviors, and gender/race membership for all 16 clusters. As in the previous analysis, within each risk cluster African American and White young adults were essentially matched on their sexual and substance use risk behavior patterns. Figure 18.4 displays the relative risk ratios for

Table 18.2 Wave III cluster name, behavioral definition, and weighted percentages of gender/race groups assigned to cluster

		Females		Males	
Cluster name	Behavioral definition	African American (n=1378)	White (n=3315)	African American (n=1071)	White (n=2942)
Few partners; low ATOD use (n=1673)	Sexually active (almost all had intercourse in the past year) but few partners and low ATOD use; median number of sexual partners=1; 57 % did not use condom at last sex; none regretted AOD use in sexual situation in past year	44.7	15.9	30.2	9.4
Light alcohol consumption and sexual activity (n=1382)	Occasional alcohol consumption but low binge drinking; median number of partners=2; all have had sexual intercourse; very low tobacco and illegal drug use; 59 % did not use condom at last sex	17.6	17.2	11.7	11.6
Low risk behavior (n=1026)	Very low or no current ATOD use; 67 % have not had sex; remainder had not had sex in past year; median number of partners=0	12.4	11.3	11.6	11.4
Tobacco use and sexual activity (n=892)	Daily cigarette smokers; median number of sexual partners=2; infrequent AOD use; 60 % did not use condom at last sex	6.8	15.7	7.2	8.9
Binge drinking (n=689)	Almost all reported binge drinking once a month or more; median number of partners=2; 56 % did not use condom at last sex; 25 % used marijuana	2.3	8.0	5.7	11.7
Substance use and sexual activity (n=583)	Moderate levels of smoking and drinking; one third used marijuana; 89 % had sex in past year; median number of partners=2; 62 % did not use condom at last sex; all smoked in past month; few used other illegal drugs	1.6	7.2	2.2	10.7

(continued)

Table 18.2 (continued)

		Females		Males	
Cluster name	Behavioral definition	African American (n=1378)	White (n=3315)	African American (n=1071)	White (n=2942)
Regret related to AOD with sexual activity; high AOD use (n=521)	All regretted AOD use in sexual situation in past year; almost all engaged in binge drinking at least monthly; median number of partners=4; 59 % did not use condom at last sex; 39 % used marijuana in past month	0.4	5.7	3.6	10.3
Regret related to AOD use with sexual activity; moderate AOD use (n=504)	All regretted AOD use in sexual situation in past year; occasional drinkers with little binge drinking; median number of sexual partners=3; 55 % did not use condom at last sex; 25 % used marijuana in past month	2.7	6.9	5.5	4.8
Marijuana use (n=406)	Frequent marijuana use; 27 % had used other illegal drug in past month; median number of sexual partners=3; 60 % did not use condom at last sex; most smoked cigarettes; almost all drank alcohol; most were occasional drinkers	1.5	3.6	5.2	7.0
Multiple partners (n=287)	All had 8 or more sexual partners in past 6 years; median number of sexual partners=11; 60 % did not use condom at last sex; moderate ATOD use	2.6	3.1	3.4	2.8
High rate of marijuana use and sexual activity (n=212)	All used marijuana frequently and almost all had sex in past year; median number of partners=4; almost all regretted AOD use in sexual situation; 56 % did not use condom at last sex; 43 % reported other illegal drug use	0.2	1.7	3.2	4.1

(continued)

Table 18.2 (continued)

		Females		Males	
Cluster name	Behavioral definition	African American (n=1378)	White (n=3315)	African American (n=1071)	White (n=2942)
Sex for money (n=204)	All reported sex for money in past 5 years; median number of sexual partners=4; 54 % did not use condom at last sex; variations in ATOD use; overall moderate use	6.6	0.8	7.1	1.4
Abstaining (n=106)	Never engaged in any ATOD use and never had sexual intercourse	0.4	1.4	0.7	1.0
Injection drug use (n=83)	All injected drugs in past 6 years; median number of partners=3; 68 % did not use condom at last sex; more than half used marijuana; half used other illegal drug in past month	0.00	0.9	0.8	1.6
Male-male sexual activity (n=82)	All had male-male sexual activity in past 6 years; median number of partners=3; 45 % did not use condom at last sex; 36 % used marijuana; 15 % had used other drug in past month; most were occasional drinkers	N/A	N/A	1.5	2.1
Marijuana and other drug use (n=56)	Most reported heavy marijuana use; all reported other illegal drug use; median number of partners=4; 64 % did not use condom at last sex; one third regretted AOD use in a sexual situation in the past year	0.2	0.6	0.4	1.2
Total		100.0	100.0	100.0	100.0

Note. Percentages are weighted to yield national probability estimates
ATOD Alcohol tobacco and other drug use, *AOD* Alcohol and other drug use

Wave III cluster membership for the six clusters for which there were significant race by gender interactions. As in adolescence, at Wave III White young adults, both male and female, were more likely to be members of clusters defined by substance use (e.g., Binge Drinking) or by a combination of substance use and risky sexual

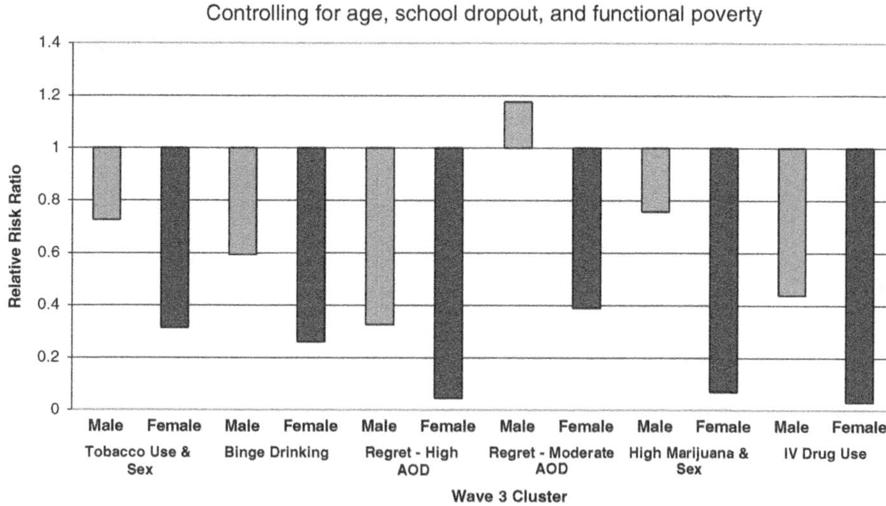

Fig. 18.4 Risk ratios for African Americans relative to Whites, conditional on gender – Wave 3

activity (e.g., Tobacco use and Sexual Activity), although in general these race differences were larger among females. For the remaining clusters, there were main effects of gender and of race. Males were more likely to be in clusters characterized by heavier substance use (see Fig. 18.5). Race differences were more variable than in adolescence (see Fig. 18.6). African Americans were more likely to be in both a low risk cluster (Few Partners, Low ATOD Use, where almost 45 % of African American females and 30 % of African American males appear) and a high risk cluster (Sex for Money). However, African Americans were less likely than Whites to be in three high risk clusters (e.g., Marijuana and Other Drugs) that were primarily characterized by substance use.

Test-Identified STIs & Association with Cluster Membership in Young Adulthood

In young adulthood, the prevalence of current test-identified infection was 6 %. The prevalence of each type of infection was higher among African Americans than Whites, and, as was the case for self-reported STIs during adolescence, the prevalence of test-identified STIs was highest among African American women (almost 23 %), compared with 15 % in African American men, 4 % in White women, and 3 % in White men. Unlike lifetime STI diagnosis reported in adolescence, which tended to be much more common among clearly high risk clusters such as Injection Drug Use and Multiple Partners, associations between cluster membership and current infection tended to be disbursed across clusters in young adulthood for African Americans. Among whites, members of clusters such as Sex for Money and Injection

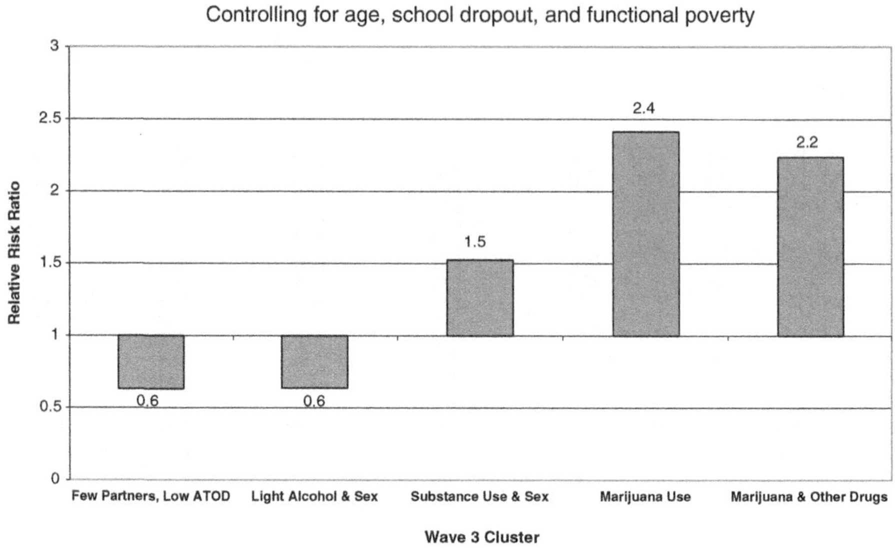

Fig. 18.5 Risk ratios for males relative to females by Wave 3 cluster

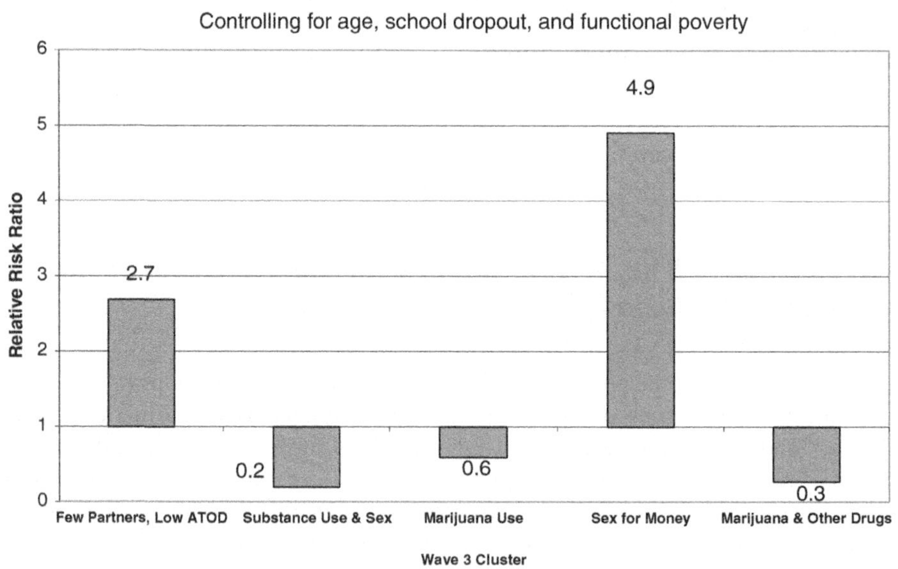

Fig. 18.6 Risk ratios for African Americans relative to Whites by Wave 3 cluster

Drug Use were more likely to have a current infection (9 % and 8 % respectively) compared with lower risk clusters such as Few Partners, Low ATOD Use (3 %). However, among African Americans, levels of infection bore little resemblance to the presumed risk levels of various clusters. For example, 23 % of African Americans in the Sex for Money and Injection Drug Use clusters tested positive for an STI compared with 20 % of African Americans in the Few Partners, Low ATOD Use cluster.

Race differences were even more obvious in examining adjusted odds of testing positive for an STI by cluster membership. African American young adults in the Low Risk Behavior cluster had 25 times the odds of testing positive than Whites in the same cluster. Even in the Injection Drug Use cluster, the odds of African Americans testing positive were almost 19 times greater for African Americans than Whites. These findings, which echoed the patterns seen in adolescence, indicate that African American young adults are at high risk for STIs, as assessed by current test-identified measures, even when their reported behavior is normative and similar to the behavior of White young adults who have a lower STI prevalence.

Trajectories of Risk Behavior & Associations with Test-Identified STIs in Young Adulthood

The two studies described above indicate that cross-sectional analyses of individual behavior do not explain race and gender differences in STI and HIV risk in adolescence or in young adulthood. However, longitudinal trajectories of risk behavior may provide additional explanatory power. According to Life Course Theory, the implications of behavior and life experiences may depend on the developmental stage in which they occur (Elder 1998). Thus the implications of current behavior patterns may vary by more distal patterns of risk taking. Relatively little is known about longitudinal patterns of risk behavior and the ways in which these patterns vary by race and gender. For example, Dariotis et al. used data from the National Survey of Adolescent Males to construct clusters based on sexual risk behaviors, and observed a general pattern of movement from higher-risk clusters in late adolescence (ages 17–22) to lower-risk clusters in young adulthood (ages 21–26) (Dariotis et al. 2008). In one of the few recent analyses of race differences in developmental patterns of sexual risk behavior, Fergus et al. found that although risk behavior among all racial groups accelerated during adolescence, peaked during young adulthood, and then declined, African American youth engaged in more risk behavior during early adolescence but less risk behavior during early adulthood relative to Whites (Fergus et al. 2007). Similarly, using National Longitudinal Survey of Youth data, Murphy et al. (2009) found both African American males and females to be overrepresented in the decreasing HIV risk behavior trajectory group compared to other racial and ethnic groups. Murphy et al. (2009) did not offer an interpretation of this difference, nor did they examine relationships between risk behavior trajectories and STIs.

In contrast to the relative paucity of information on longitudinal patterns of sexual risk behavior, a growing body of literature has examined trajectories of substance use from adolescence to young adulthood. These studies describe several distinct patterns of substance use over time. For example, cigarette, alcohol, and marijuana use is typically characterized by adolescent onset followed by steady increases over time, with peak use occurring during early adulthood (Chen and Kandel 1995). Certain patterns of use, typically identified by early onset and/or persistent high levels of use, are associated with increased risk of poor psychosocial and health outcomes (Windle and Wiesner 2004; Hill et al. 2000). Both overall prevalence and longitudinal patterns of substance use also vary markedly by race and gender, however. In general, African American youth report lower levels of substance use during adolescence compared to White youth (Bachman et al. 1991), and reach peak levels of use at later ages. By young adulthood, race differences in the use and abuse of certain substances are attenuated or even reversed. In a recent longitudinal analysis of alcohol use, for example, African American adolescents increased their level of use at a slower rate compared to White adolescents and reached highest levels of involvement later. Because they desisted from use more gradually, African Americans ultimately reported levels of involvement comparable to those of Whites (Cooper et al. 2008). Prevalence of substance use among African American females is particularly low. Race and gender comparisons of trend data from Monitoring the Future, a nationally representative study of substance use among 8th, 10th, and 12th grade students, indicate that rates of marijuana, alcohol, and cigarette use are consistently lowest among Asian and African American females (Wallace et al. 2003). Taken together, these findings suggest that analyses of the implications of different patterns of substance use and sexual risk behavior over time must consider possible subgroup differences.

We now turn to a longitudinal analysis that integrates information about adolescent and young adult risk taking, and thereby extends our understanding of risk behavior trajectories and their association with HIV/STI outcomes. Because of the patterns seen in our cross-sectional work, we focus in particular on differences by race and gender. The longitudinal analysis, also based on Add Health data, offers several methodological advantages. First, as in our earlier analyses, we capitalize on a nationally representative and sociodemographically diverse sample that provides detailed information on sexual risk behavior and STI-relevant substance use, and includes laboratory tests for STI infection in young adulthood. Second, we continue with a person-centered approach that allows us to examine the interaction and covariation of relevant adolescent and young adult risk behaviors. To date, very few studies have examined trajectories of both sexual risk behavior and substance use, despite evidence that risk behaviors during adolescence covary (Duncan et al. 1999; Graves and Leigh 1995). Third, we use longitudinal data spanning two developmental periods – adolescence and emerging adulthood – thus allowing us to track developmental pathways in STI/HIV risk behavior and their association with STI status in young adulthood.

In this longitudinal analysis we address the following questions:

1. What are the longitudinal trajectories of drug use and sexual behavior from adolescence to young adulthood, and how do they differ by race and gender (biological sex)?
2. What is the prevalence of sexually transmitted infections in young adulthood for each longitudinal trajectory by race and gender?
3. Does the likelihood of sexually transmitted infection among individuals with similar risk behavior trajectories differ by race and gender?

Method

Add Health Data

In Add Health, a sample of 80 high schools and 52 middle schools was selected with unequal probability of selection. Use of systematic sampling methods and implicit stratification in the study design ensured that selected schools were representative of U.S. schools with respect to region of the country, urbanicity, school size, school type and ethnicity. Interviews were conducted using laptop computers; computer-assisted self-interviewing (CASI) technology was used to collect information on sensitive topics such as sexual activity and substance use. Additional information about the Add Health study is available elsewhere (Harris et al. 2008). Present analyses used the sample of non-Hispanic African American and White respondents who participated in Waves I and III, have appropriate sampling weights, and could be assigned to a behavioral risk cluster at Waves I and III (n = 10,737). The response rate for Wave III was 77 % (Harris et al. 2008).

Measures

Sexually Transmitted Infection At Wave III, respondents were asked to provide a specimen of first stream urine for testing. Chlamydia, gonorrhea, and trichomoniasis were assessed by analyzing the specimens for the presence of *Chlamydia trachomatis, Neisseria gonorrhoeae*, and *Trichomonas vaginalis*, respectively. Assays used ligase chain reaction (LCR) to detect the presence of *C. trachomatis*, and *N. gonorrhoeae* DNA. PCR-ELISA tests were used to detect *T. vaginalis* DNA. Of the Wave III participants, 8 % refused to provide a urine specimen, and another 6 % of *N. gonorrhoeae* results were excluded due to manufacturer product recall. Additional reasons for missing test result data were participants' inability to provide a urine sample at the interview (2 %) and shipping and laboratory problems (3 %). Further details about the STI testing conducted in Wave III of Add Health is available elsewhere (Miller et al. 2004, 2005). For the present longitudinal analyses, we used the same dichotomous combined STI measure used for Wave III cross-sectional analysis. It was coded to indicate a positive test on any of the three STIs versus negative results on all three tests. Participants who had missing data on any test were categorized as missing on the composite STI variable.

Cluster Membership and Trajectory Patterns As described earlier, we previously categorized each respondent into one of 16 clusters based on his or her pattern of substance use and sexual risk behavior at Wave I (Halpern et al. 2004). Sixteen clusters were similarly created for Wave III substance use and sexual behavior (Hallfors et al. 2007). For the present longitudinal analysis we examine stability and change in cluster membership from Wave I to Wave III. Given the number of possible transition patterns, examining the movement between 16 clusters in adolescence to 16 clusters in young adulthood required simplification. Thus, for the research presented in this chapter, we collapsed the clusters from 16 to 4 risk group levels. Specifically, we categorized each Wave I and Wave III cluster into one of four risk profile levels ("Very high," "High," "Medium," or "Low") based on extent of sexual risk behavior and associations in the literature with HIV/STD (Table 18.3). We sought to make generally comparable risk assignments across the two waves, but also recognized the need to take into account differences in the risk behaviors of members of Wave I compared to Wave III clusters, as well as differences in the developmental norms of behavior for adolescents versus adults. For example, the Wave I (adolescent) Alcohol and Sex cluster was assigned to the "High" risk category while the Wave III (young adult) Light Alcohol and Sex cluster was assigned to the "Medium" risk category. Wave I Alcohol and Sex members were more likely to binge drink than the Wave III Light Alcohol and Sex cluster. Another difference in these clusters is that the Alcohol and Sex cluster members had a median of two lifetime sexual partners by the time of the Wave I adolescent survey while the Light Alcohol and Sex young adults had a median of two partners in the 6 years prior to the Wave III survey, which is more statistically normative.

A total of 16 combinations of movement are possible from the four Wave I risk levels to the four Wave III risk levels. Patterns with very few members were combined with similar patterns, resulting in 12 patterns of movement from Wave I to Wave III. We then collapsed the 12 patterns of movement into four risk trajectory categories: "Stable Low," "Decreasing," "Increasing," and "Stable High" (see Table 18.4). Again, normativeness of behavior at different ages was taken into account. The "Low to Medium" Wave I to Wave III movement pattern was assigned to the *stable low* risk behavior trajectory group, given that "Medium" risk behavior is normative for young adults. (Ninety percent of young adults had had vaginal intercourse at Wave III [Halpern et al. 2006]. Also, over one-third of whites and almost two-thirds of African Americans were in the "Low" and "Medium" categories at Wave III, marked by Low ATOD.) Likewise, the *decreasing* trajectory group included "Medium to Medium" movement. Our rationale for this grouping is that in adolescence, the Sex Experimenters cluster was a nonnormative group, but sexual activity was normative at Wave III.

Table 18.3 Risk behavior trajectories at Wave I and Wave III

Risk category	Wave I	Wave III
	Drinkers	
Low	Substance experimenters	Low risk behavior
	Abstainers	Abstainers
Medium	Sex experimenters	Few partners, low ATOD[a]
		Light alcohol and sex
High	Marijuana users	Marijuana users
	Heavy substance use and sex	Heavy substance use and sex
	Sex and drugs combined	Regret related to AOD[b], high AOD[b]
	Binge drinkers	Binge drinkers
	Alcohol and sex	Regret related to AOD[b], moderate AOD[b]
	Smokers and sex	Smokers and sex
Very high	MSM[c]	MSM[c]
	IV drug users	IV drug users
	Marijuana and other drugs	Marijuana and other drugs
	Sex for drugs or money	Sex for money
	Multiple partners	Multiple partners
	High Marijuana and sex	High Marijuana and sex

[a]Alcohol, tobacco, and other drugs
[b]Alcohol and other drugs
[c]Men who have sex with men

Table 18.4 Wave I to Wave III movement patterns and risk behavior trajectory categories

Wave I cluster → Wave III cluster (12 Patterns of movement)	Risk behavior trajectory
Low → Low	Stable low
Low → Medium	
Medium → Low	Decreasing
Medium → Medium	
High/Very High → Low	
High/Very High → Medium	
Low → High	Increasing
Low → Very High	
Medium → High/Very High	
High → High	Stable High
High → Very High	
Very High → High/Very High	

Sociodemographics Respondent race (African American or White) was based on self-report. Other variables used in analyses were gender (biological sex), chronological age, marital status (married or not married), functional poverty (whether or not respondents or their households did not pay the full amount of either the rent or mortgage or utility [gas, electricity, or oil] bill because they did not have enough money), and high school dropout status (yes or no), all of which were measured at Wave III.

Statistical Analyses

First, descriptive analyses were conducted assessing the prevalence of each risk behavior trajectory pattern by gender and race of respondent. Race and gender differences and a race by gender interaction in trajectory group membership were tested using multinomial logistic regression controlling for age, marital status, functional poverty, and school dropout. Next, prevalence of STI infection in young adulthood for each risk behavior trajectory was examined by race and gender. Finally, odds of STI infection among African Americans versus Whites were examined in multivariate logistic regressions stratified by both trajectory pattern and gender.

Post-stratification sampling weights were used to yield estimates representative of the national population. Procedures for survey data in Stata (Stata Corp, College Station, TX) were used to account for the complex sampling design. Reported percentages are weighted percentages; sample sizes presented are the unweighted numbers.

Results

Participants

The analytic sample for the present analysis consisted of 8694 participants (White: n=6245; African American: n=2449). Approximately 15 % of participants reported functional poverty (13 % of Whites, 21 % of African Americans), 15 % either dropped out or never attended high school (13 % of Whites, 19 % of African Americans), and 17 % were presently married (18 % of Whites, 11 % of African Americans). Overall, the mean age of participants at Wave III was 21.7 years. Table 18.5 presents descriptive statistics for selected characteristics by race and

Table 18.5 Descriptive statistics for analytic sample, by race and gender

	Females		Males	
	African Americans (n=1381)	Whites (n=3315)	African Americans (n=1068)	Whites (n=2930)
High school dropout	16.0	12.3	22.6	14.5*
Married	12.6	23.2**	9.5	13.4
In functional poverty	24.2	15.1**	16.9	11.5*
First sex at age 14 or younger	23.0	18.9	31.6	14.5**

Note. Percentages are weighted to yield national probability estimates
*p<.01 for the difference between African Americans and Whites within gender
**p<.001 for the difference between African Americans and Whites within gender

gender of participants. Compared to White females, a smaller proportion of African American females were married (13 % vs. 23 %, $F=27.71$, $p<.001$) and a greater proportion lived in functional poverty (24 % vs. 15 %, $F=27.76$, $p<.001$). Compared to White males, a greater proportion of African American males had not completed high school (23 % vs. 15 %, $F=10.38$, $p<.01$), were in functional poverty (17 % vs. 12 %, $F=10.04$, $p<.01$), and had first intercourse at age 14 or younger (32 % vs. 15 %, $F=78.75$, $p<.001$).

Risk Behavior Trajectories from Adolescence to Young Adulthood

Risk behavior increased overall from adolescence to young adulthood (Table 18.6). Thirty-one percent of White females and 40 % of White males moved into higher risk clusters in young adulthood. Thirty-one percent of White females and 25 % of White males were in low risk clusters across the two time points, 14 % of White females and 9 % of White males went from a higher risk cluster to lower risk, and 24 % of White females and 27 % of White males maintained high risk cluster membership. African American males were more evenly distributed across risk behavior trajectories compared to other race/gender groups, with 29 % moving to lower risk clusters, 26 % transitioning to higher risk clusters, 24 % maintaining stable membership in low risk clusters from adolescence to young adulthood, and 20 % continuing membership in high risk clusters. In contrast, the modal behavioral trajectory for African American females was stable membership in low risk clusters in both adolescence and young adulthood (41 %). Only 9 % of African American females maintained membership in clusters characterized by high levels of risk behavior; 16 % reflected movement to higher risk and 34 % moved to lower levels of risk.

Table 18.6 Percentage distribution of participants, by risk behavior pattern, according to race and gender

	Females		Males		
Risk behavior trajectory	African Americans (n=1381)	Whites (n=3315)	African Americans (n=1068)	Whites (n=2930)	Total
Stable low (n=2633)	41.0	31.4	24.4	24.7	29.0
Decreasing (n=1542)	34.1	14.3	29.3	8.7	15.4
Increasing (n=2662)	15.9	30.5	26.0	40.0	32.5
Stable High (n=1857)	9.0	23.8	20.3	26.6	23.1
Total	100	100	100	100	100

Note. Percentages are weighted to yield national probability estimates

Table 18.7 Relative risk ratios showing African American participants' likelihood of belonging to a trajectory group, relative to whites', by Gender

	Females (n=4696)		Males (n=3998)	
Trajectory	RRR	95 % CI	RRR	95 % CI
Decreasing	1.7*	(1.3, 2.2)	3.5*	(2.7, 4.7)
Increasing	0.3*	(0.2, 0.5)	0.6*	(0.4, 0.8)
Stable high	0.2*	(0.1, 0.3)	0.7*	(0.4, 1.0)

Note. Reference category=Stable low. Analyses adjusted for age, marital status, school dropout, and functional poverty
*$p<0.01$

In multinomial logistic regressions predicting trajectory group membership (controlling for age, marital status, functional poverty, and school dropout), there was a significant interaction between race and gender for each trajectory group (Decreasing: RRR=0.5, 95 % CI=0.3, 0.7; Increasing: RRR=0.6, 95 % CI=0.4, 0.8; Stable High: RRR=0.3, 95 % CI=0.2, 0.5). Compared to their White counterparts, African American females were more likely to be in the *decreasing* group, and less likely to be in the *increasing* and *stable high* groups, compared with the *stable low* group (Table 18.7). The same direction of associations between race and likelihood of being in each trajectory pattern was observed among males. However, the positive association between African American race and belonging to the *decreasing* group was larger among males (RRR=3.5) than among females (RRR=1.7), and the negative association between African American race and belonging to the *increasing* and *stable high* trajectory groups was larger among females (RRR=0.3 and 0.2) than among males (RRR=0.6 and 0.7, respectively).

Race and Gender Differences in the Risk of STI Infection, by Risk Behavior Trajectory

As noted in the Wave III cross-sectional analysis, 6 % of respondents tested positive for at least one of the three STIs. White males were least likely to have a positive STI test result (3 %), while African American females were most likely (23 %). Four percent of White females and 14 % of African American males had positive test results.

The STI prevalence within each trajectory pattern, stratified by gender and race, is displayed in Figs. 18.7 (females) and 18.8 (males). Within each trajectory, the percentage of African American respondents with positive STI tests was greater than the percentage of White respondents. This pattern was evident for both African American males and African American females. However, STI prevalence was particularly high among African American females regardless of longitudinal pattern of risk behavior. For example, 33 % of African American females in the *increasing* risk trajectory had a positive STI test result, compared to 4 % of White females in

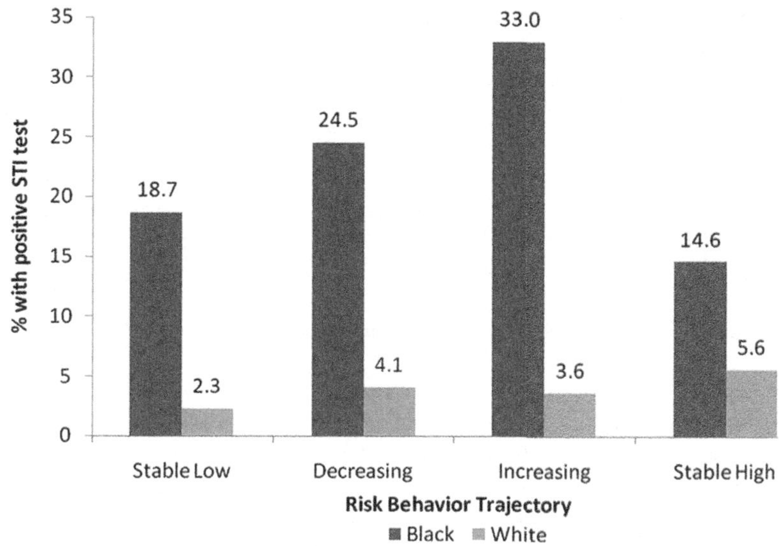

Fig. 18.7 STI prevalence by risk trajectory and race, among females

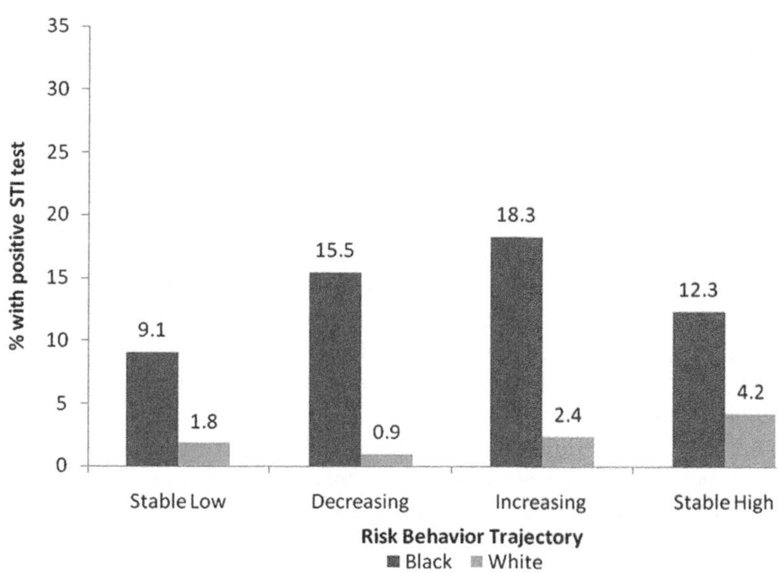

Fig. 18.8 STI prevalence by risk trajectory and race, among males

the same trajectory group (Fig. 18.7). STI risk among African American females extended even to those with low behavioral risk profiles: almost 20 % of African American females in the *stable low* risk trajectory had a positive STI test. In logistic

Table 18.8 Odds of positive STI test among African Americans, compared to Whites, by gender and trajectory pattern

Trajectory	Females (n=4696)		Males (n=3998)	
	OR[a]	95 % CI	OR[a]	95 % CI
Stable low	9.2**	(4.8, 17.8)	5.7**	(2.2, 14.6)
Decreasing	6.9**	(3.6, 13.0)	17.7**	(4.0, 77.8)
Increasing	11.8**	(6.9, 20.1)	8.1**	(4.5, 14.6)
Stable high	2.8*	(1.1, 7.0)	3.0**	(1.5, 6.2)

* p<0.0
** p<0.01
[a]Adjusted for age, marital status, school dropout, and functional poverty

regressions predicting the odds of having a positive STI test for African Americans relative to Whites within each trajectory pattern, controlling for age, marital status, school dropout, and functional poverty, and stratified by gender, both African American females and African American males had significantly greater odds of having a positive STI test in young adulthood compared to their White counterparts within all trajectory groups (Table 18.8).

Discussion

We have summarized our earlier work documenting large racial disparities in both self-reported and test-identified STI/HIV prevalence that could not be explained by contemporaneous risk behavior (Halpern et al. 2004; Hallfors et al. 2007). The present analysis expanded these earlier investigations by applying a longitudinal person-centered analysis of sex and drug use behavior patterns to see whether the implications of behavioral risk taking for STI/HIV racial disparities among young adults vary according to the developmental period in which they begin, and/or according to trajectory over time. We conducted analyses separately by race and gender to identify any differences in longitudinal trajectories that would inform prevention interventions.

Comparing young adult African American men to White men, we found that similar proportions (24–25 %) were in the *stable low* risk category. More African American men, however, were in the *decreasing* risk category (29 % versus 9 %) and more White men were in the *increasing* and *stable high* risk categories (40 % versus 26 %, and 27 % versus 20 % respectively). These prospective self reports of risk-taking suggest that although a larger proportion of African American male youth engage in sexual risk behavior in adolescence, they are more likely to revert to conventional behavior by young adulthood, at which point more White than African American men are engaging in high risk behavior (67 % versus 46 %). In other words, White males tend to start off slower than African American males but are more likely to escalate their risk behavior as they enter adulthood.

Despite these differences in patterns, African American men were much more likely than White men to have a positive STI/HIV test in young adulthood in all trajectories. The difference is especially pronounced in the *decreasing* trajectory; the few White men who demonstrated diminishing risk were notably less likely to test positive for an STI. This suggests that early risk behavior, even with desistance in the transition to adulthood, is associated with a greater likelihood of infection among young adult African American men, but not among their White peers.

On the other hand, young adult African American women were more likely to be consistently low-risk in their behavior compared to White women (41 % versus 31 % in the *stable low* category). Another 34 % of African American women engaged in sexual risk behavior in adolescence but then moved to a lower risk cluster by young adulthood; in comparison, only 14 % of White women were in this *decreasing* category. By young adulthood, only 25 % of African American women were in the *increasing* or *stable high* groups compared to 54 % of white women. Thus, despite a higher prevalence of consistently low risk behavior and desistance from high risk patterns evident during adolescence, the prevalence of STIs among African American women was much higher than for any other gender/race group.

These findings add to mounting evidence that other factors – in addition to individual behavior – contribute to the radically greater likelihood of STIs among African Americans compared to Whites. Many of the risk clusters that we have grouped as High and Very High risk are heavily defined by drug behavior (alcohol, tobacco, and other drugs) as well as greater numbers of sexual partners. We assumed that drug behavior is a potent risk factor for STIs, with a greater likelihood of unprotected sex and sex with partners who are at higher risk for STIs. However, African American males and females, who are more likely to currently have an STI, are less likely to evidence patterns of *stable high* or *increasing* risk behavior as tapped in our measures.

What are we to make of these findings that suggest a disconnect between risk taking and vulnerability to STIs? We can only speculate, but several considerations seem relevant. First, although proper and consistent use of condoms is effective in preventing STIs, it is difficult to measure these qualities of use, making interpretation of available findings unclear. For example, African Americans consistently report higher condom use than Whites; in our previous study we reported that 53 % of African Americans versus 37 % of White young adults who had sex in the past year reported condom use at their most recent sexual intercourse (Hallfors et al. 2007). Further, we found little variation in condom use by cluster membership (authors' calculations). However, use at most recent intercourse does not describe consistent use or proper use in the past 12 months, nor do reports about "typical use" describe use at particular sexual encounters, which may be partly driven by partner characteristics, which in turn cannot be easily systematically assessed. The common necessity of using global self-report condom use measures affords only modest insight into proper and consistent use. Although data are limited, other studies suggest there may be high levels of improper use (Crosby et al. 2005; de Visser and Smith 2000) or errors in reporting (Rose et al. 2009). Thus, better measurement within the context of representative, population-based studies is needed.

A second, related consideration is that personal risk behavior seems to have limited association with STI/HIV status in young adulthood. The majority of Whites, but not African Americans, show high levels of both sex and drug involvement, yet relatively low STI prevalence. Stable high levels of risk behavior are associated with the highest prevalence of STIs for Whites but not for African Americans. These findings suggest that African Americans may recognize their STI/HIV risk stemming from consistently high risk behavior and obtain testing and treatment, while counterpart Whites may not.

Overall, our findings suggest that African American young adults, and especially women, are on average, at very high risk for STIs regardless of current or past behavior when that behavior is considered apart from the context in which it takes place. The prevalent prevention paradigm, to reduce personal risk behavior and increase personal protective behavior (condom use), is not adequate for ameliorating racial STI/HIV disparities. We suggest a more holistic approach to individual risk behavior, simultaneously considering multiple factors that reflect biological, behavioral, and structural/contextual domains.

Sexual partnering patterns are an essential component of the wider context that links individual and environmental characteristics. Propinquity, or proximity with the opportunity to interact, is among the most important factors in personal attraction; other factors include status and similarity (Festinger et al. 1950; Hallinan and Williams 1989). Because racial groups tend to be segregated residentially, and because shared cultural values exist within different groups and differ across groups, individuals are more likely to have relationships with members of their own racial or ethnic groups.

Laumann and Youm (1999) convincingly argued that racial segregation in mating patterns, coupled with a high likelihood for bridging between high and low risk sexual partners among African Americans but not Whites, are major factors in STI/HIV disparities. Data from Add Health Wave III confirm that sexual networks are highly segregated by race for African Americans and Whites. For example, 86 % of non-Hispanic White males reported only White sexual partners and 82 % of White females reported only White sexual partners. Among non-Hispanic African American men, 71 % reported only African American partners; but 90 % of African American women reported only African American partners (authors' calculations). Thus, White males and females are similarly segregated in their sexual partnering, while African American women show the greatest segregation and African American men the least. These data support the racial segregation findings of Laumann and Youm (1999).

Sexual partnering is exceedingly complex. Marriage is related to sexual partnering, but more public and stable, as is the contemporary trend for couples to live together in stable relationships without marriage. In both of these types of more public relationships there is evidence that women choose marriage partners with similar or higher education than themselves; there appears to be less upward economic selection for cohabitating partners (African Americanwell and Lichter 2000).

Moreover, racial segregation appears to be even stronger in marriage and cohabitation relationships (Ford and Norris 1997; Blackwell and Lichter 2000).

Marriage is negatively related to sexual concurrency (i.e., having more than one sexual partner at the same time), a major factor driving STI/HIV racial disparities (Adimora et al. 2006, 2007; Manhart et al. 2002; Rosenberg et al. 1999), and race. Analyses of data from the 1995 and 2002 National Survey of Family Growth (NSFG) indicate that African Americans are much less likely to be married than Whites (25 % versus 54 % for women, and 31 % versus 44 % for men), and unmarried status is associated with partner concurrency (Adimora et al. 2002, 2007). The combination of concurrency, dense sexual networks, and bridging (between high risk and low risk individuals) appears to efficiently transmit STIs and HIV. Forces that inhibit marriage, such as a low ratio of men to women, appear to affect sexual networks and foster widespread concurrent sexual relationships (Guttentag and Secord 1983; Adimora and Schoenbach 2005; Adimora et al. 2006).

U.S. census data indicate that the ratio of men to women is lower among young African American adults compared to Whites (US Census Bureau 2008). This national demographic difference appears to be well known among African Americans. For example, rural African American women reported a huge shortage of marriageable African American men, because most lacked education and a good job, or were incarcerated, deceased, or doing drugs (Adimora et al. 2001). Criminal justice statistics indicate that African American men are six times more likely to be incarcerated than White men in the U.S., with as many as a third of African American men likely to be incarcerated at some point during their lifetime (Bonczar 2003). Illiteracy and high school dropout have consistently been linked to higher rates of crime and incarceration (see e.g., Lochner 2004; Lopoo and Western 2005), and high community incarceration rates, in turn, have been associated with STI prevalence (Thomas and Torrone 2006). In Adimora and colleagues' qualitative work, women thought the shortage of marriageable men led to less committed sexual relationships and concurrency by both men and women (Adimora et al. 2001).

Differences in marriage rates also imply that African American and White youth have different family structures in which to grow. From our Add Health research, we found that White adolescents were twice as likely as African American adolescents to report two biological parents in the home (Cuffee et al. 2007). Two parents in the home and school achievement have been found to be protective factors preventing early sexual debut among African American males (Bakken and Winter 2002). We found that African American males were more than twice as likely (32 % versus 15 % among White males) to report sexual intercourse before age 15. In summary, it appears that there are important associations between family structure, early sexual debut, school achievement, employment, marriage, concurrency, and HIV/STIs. These contextual factors warrant further research to better understand how they relate to each other and to racial STI/HIV disparities.

The above considerations point to the importance of combining investigations of individual behavior with study of contextual factors, both at the level of the couple and the community. In addition to social contextual factors, biological and genetic factors, which may interact with context and experience differentially according to race and gender, should be included in studies of STI/HIV racial disparities. There are a number of factors that could alter the likelihood of infection, given exposure. For example, African American women are more likely than White women to have a condition known as bacterial vaginosis, in which the normal bacterial balance in the vagina is disrupted, and which is associated with greater STI/HIV infection (Cherpes et al. 2008; Peipert et al. 2008). In terms of genetic research, there is some evidence that AIDS restriction genes, most notably the variant CCR5 Δ32, are more likely to be found in Whites, and particularly northern European Whites, compared to African Americans (O'Brien and Nelson 2004; Galvani and Novembre 2005). CCR5 Δ32 effectively blocks HIV-1 infection in homozygous people and slows AIDS progression in heterozygotes, which may account for some of the skewness in geographic HIV/AIDS distribution (Winkler et al. 2004). Despite the concentration of HIV in sub-Saharan Africa, the bulk of human genetic studies have focused on White populations, and therefore very little is known about natural HIV resistance in other at-risk populations living in different social and physical contexts, such as African American Africans (Donfack et al. 2006). Although an adequate review of these lines of research is beyond the scope of the present chapter, we include them to suggest the importance of cross-disciplinary work to bring a multi-faceted approach to efforts to understand and reduce STI/HIV racial disparities.

In summary, longitudinal trajectories of drug use and sexual behavior patterns provide additional evidence that individual behavior alone does not account for racial disparities in STI/HIV. More research is needed to better understand both the biological and social contexts in which such disparities occur, and how those contexts interact to influence the likelihood of infection. The challenge for public health is to explore new paradigms for prevention that will more effectively reduce the significant and disproportionate burden of disease borne by African American men and women.

Authors' Note This chapter is based on research from grant R01 DA014496 (Denise Hallfors, PI) from the National Institute on Drug Abuse. We thank Martha W. Waller for her data analysis work on race of partners that we report in this chapter. This research uses data from Add Health, a program project designed by J. Richard Udry, Peter S. Bearman, and Kathleen Mullan Harris, and funded by a grant P01-HD31921 from the Eunice Kennedy Shriver National Institute of Child Health and Human Development, with cooperative funding from 17 other agencies. Special acknowledgment is due Ronald R. Rindfuss and Barbara Entwisle for assistance in the original design. Persons interested in obtaining Data Files from Add Health should contact Add Health, The University of North Carolina at Chapel Hill, Carolina Population Center, 123 W. Franklin Street, Chapel Hill, NC 27516–2524 (addhealth@unc.edu). No direct support was received from grant P01-HD31921 for this analysis.

References

Adimora, A. A., & Schoenbach, V. J. (2005). Social context, sexual networks, and racial disparities in rates of sexually transmitted infections. *The Journal of Infectious Diseases, 191*, S115–S122.

Adimora, A. A., Schoenbach, V. J., Martinson, F. E., Donaldson, K. H., Fullilove, R. E., & Aral, S. O. (2001). Social context of sexual relationships among rural African Americans. *Sexually Transmitted Diseases, 28*, 69–76.

Adimora, A. A., Shoenbach, V. J., Bonas, D. M., et al. (2002). Concurrent partnerships among women in the US. *Epidemiology, 13*, 320–327.

Adimora, A. A., Schoenbach, V. J., & Doherty, I. A. (2006). HIV and African Americans in the Southern United States: Sexual networks and social context. *Sexually Transmitted Diseases, 33*(7), S39–S45.

Adimora, A. A., Schoenbach, V. J., & Doherty, I. A. (2007). Concurrent sexual partnerships among men in the United States. *American Journal of Public Health, 97*, 2230–2237.

Blackwell, D. L., & Lichter, D. T. (2000). Mate selection among married and cohabiting couples. *Journal of Family Issues, 21*(3), 275–302.

Bachman, J. G., Wallace, J. M., O'Malley, P. M., Johnston, L. D., Kurth, C. L., & Neighbors, H. W. (1991). Racial ethnic differences in smoking, drinking, and illicit drug use among American high school seniors, 1976–89. *American Journal of Public Health, 81*(3), 372–377.

Bakken, R. J., & Winter, M. (2002). Family characteristics and sexual risk behaviors among African American men in the United States. *Perspectives on Sexual and Reproductive Health, 34*, 252–258.

Bonczar, T. P. (2003). *Prevalence of imprisonment in the U.S. population, 1974–2001*. Bureau of Justice Statistics Special Report. Available at http://www.ojp.usdoj.gov/bjs/abstract/piusp01.htm. Accessed 23 Sept 2009.

Centers for Disease Control and Prevention. (2002). *Drug-associated HIV transmission continues in the United States*. Available at http://www.cdc.gov/hiv/resources/factsheets/idu.htm. Accessed 23 Sept 2009.

Centers for Disease Control and Prevention. (2008). *Sexually transmitted disease surveillance, 2007*. Atlanta: U.S. Department of Health and Human Services.

Centers for Disease Control and Prevention. (2009). *HIV/AIDS surveillance report, 2007* (Vol. 19, pp. 1–63). Atlanta: US Department of Health and Human Services. Also available at: http://www.cdc.gov/hiv/topics/surveillance/resources/reports/

Chen, K., & Kandel, D. B. (1995). The natural history of drug use from adolescence to the mid-thirties in a general population sample. *American Journal of Public Health, 85*(1), 41–47.

Cherpes, T. L., Hillier, S. L., Meyn, L. A., Busch, J. L., & Krohn, M. A. (2008). A delicate balance: Risk factors for acquisition of bacterial vaginosis include sexual activity, absence of hydrogen peroxide-producing lactobacilli, African American race, and positive herpes simplex virus type 2 serology. *Sexually Transmitted Diseases, 35*, 78–83.

Coker, A. L., Richter, D. L., Valois, R. F., McKeown, R. E., Garrison, C. Z., & Vincent, M. L. (1994). Correlates and consequences of early initiation of sexual intercourse. *Journal of School Health, 64*(9), 372–377.

Cooper, M. L., Krull, J. L., Agocha, V. B., Flanagan, M. E., Orcutt, H. K., Grabe, S., Dermen, K. H., & Jackson, M. (2008). Motivational pathways to alcohol use and abuse among African American and White adolescents. *Journal of Abnormal Psychology, 117*(3), 485–501.

Crosby, R., Salazar, L. F., DiClemente, R. J., Yarber, W. L., Caliendo, A. M., & Staples-Horne, M. (2005). Accounting for failures may improve precision: Evidence supporting improved validity of self-reported condom use. *Sexually Transmitted Diseases, 32*, 513–515.

Cuffee, J. J., Hallfors, D. D., & Waller, M. W. (2007). Racial and gender differences in adolescent sexual attitudes and longitudinal associations with coital debut. *Journal of Adolescent Health, 41*, 19–26.

Dariotis, J. K., Sonenstein, F. L., Gates, G. J., Capps, R., Astone, N. M., Pleck, J. H., et al. (2008). Changes in sexual risk behavior as young men transition to adulthood. *Perspectives on Sexual and Reproductive Health, 40*(4), 218–225.

de Visser, R. O., & Smith, A. M. A. (2000). When always isn't enough: Implications of the late application of condoms for the validity and reliability of self-reported condom use. *AIDS Care, 12*(2), 221–224.

Donfack, J., Buchinsky, F. J., Post, J. C., & Ehrlick, G. D. (2006). Human susceptibility to viral infection: The search for HIV-protective alleles among Africans by means of genome-wide studies. *AIDS Research and Human Retroviruses, 22*(10), 925–930.

Duncan, S. C., Strycker, L. A., & Duncan, T. E. (1999). Exploring associations in developmental trends of adolescent substance use and risky sexual behavior in a high-risk population. *Journal of Behavioral Medicine, 22*(1), 21–34.

Elder, G. H. (1998). The life course as developmental theory. *Child Development, 69*(1), 1–12.

Ellen, J. M., Kohn, R. P., Bolan, G. A., Shiboski, S., & Krieger, N. (1995). Socioeconomic differences in sexually-transmitted disease rates among African American-and-white adolescents, San-Francisco, 1990 to 1992. *American Journal of Public Health, 85*(11), 1546–1548.

Ellen, J. M., Aral, S. O., & Madger, L. S. (1998). Do differences in sexual behaviors account for the racial/ethnic differences in adolescents' self-reported history of a sexually transmitted disease? *Sexually Transmitted Diseases, 25*(3), 125–129.

Fergus, S., Zimmerman, M. A., & Caldwell, C. H. (2007). Growth trajectories of sexual risk behavior in adolescence and young adulthood. *American Journal of Public Health, 97*(6), 1096–1101.

Festinger, L., Schachter, S., & Back, K. (1950). The spatial ecology of group formation. In L. Festinger, S. Schachter, & K. Back (Eds.), *Social pressures in informal groups* (pp. 33–69). Palo Alto: Stanford University Press.

Ford, K., & Norris, A. (1997). Sexual networks of African American and Hispanic youth. *Sexually Transmitted Diseases, 24*(6), 327–333.

Fortenberry, J. D. (1995). Adolescent substance use and sexually transmitted diseases risk: A review. *Journal of Adolescent Health, 16*(4), 304–308.

Galvani, A. P., & Novembre, J. P. (2005). The evolutionary history of the *CCR5-Δ32* HIV-resistance mutation. *Microbes and Infection, 7*, 301–308.

Graves, K. L., & Leigh, B. C. (1995). The relationship of substance use to sexual activity among young adults in the United States. *Family Planning Perspectives, 27*(1), 18–22, 33.

Guttentag, M., & Secord, P. (1983). *Too many women: The sex ratio question*. Beverly Hills: Sage.

Hallfors, D. D., Iritani, B. J., Miller, W. C., & Bauer, D. J. (2007). Sexual and drug behavior patterns and HIV and STD racial disparities: The need for new directions. *American Journal of Public Health, 97*(1), 125–132.

Hallinan, M., & Williams, R. A. (1989). Interracial friendship choices in secondary school. *American Sociological Review, 54*, 67–78.

Halpern, C. T., Hallfors, D., Bauer, D. J., Iritani, B., Waller, M. W., & Cho, H. (2004). Implications of racial and sex differences in patterns of adolescent risk behavior for HIV and other sexually transmitted diseases. *Perspectives on Sexual and Reproductive Health, 36*(6), 239–247.

Halpern, C. T., Waller, M. W., Spriggs, A., & Hallfors, D. D. (2006). Adolescent predictors of emerging adult sexual patterns. *Journal of Adolescent Health, 39*(6), 926.e1–926.e10.

Harawa, N. T., Greenland, S., Cochran, S. D., Cunningham, W. E., & Visscher, B. (2003). Do differences in relationship and partner attributes explain disparities in sexually transmitted disease among young white and African American women? *Journal of Adolescent Health, 32*(3), 187–191.

Harris, K. M., Halpern, C. T., Entzel, P., Tabor, J., Bearman, P. S., & Udry, J. R. (2008). *The national longitudinal study of adolescent health: Research design* [WWW document]. Retrieved July 27, 2009, from: http://www.cpc.unc.edu/projects/addhealth/design

Hill, K. G., White, H. R., Chung, I. J., Hawkins, J. D., & Catalano, R. F. (2000). Early adult outcomes of adolescent binge drinking: Person- and variable-centered analyses of binge drinking trajectories. *Alcoholism, 24*(6), 892–901.

Kaestle, C. E., Halpern, C. T., Miller, W. C., & Ford, C. A. (2005). Young age at first sexual intercourse and sexually transmitted infections in adolescents and young adults. *American Journal of Epidemiology, 161*(8), 774–780.

Kotchick, B. A., Shaffer, A., Forehand, R., & Miller, K. S. (2001). Adolescent sexual risk behavior: A multi-system perspective. *Clinical Psychology Review, 21*(4), 493–519.

Laumann, E. O., & Youm, Y. (1999). Racial/ethnic group differences in the prevalence of sexually transmitted diseases in the United States: A network explanation. *Sexually Transmitted Diseases, 26*(5), 250–261.

Lochner, L. (2004). Education, work, and crime: A human capital approach. *International Economic Review, 45*(3), 811–843.

Lopoo, L. M., & Western, B. (2005). Incarceration and the formation and stability of marital unions. *Journal of Marriage and the Family, 67*, 721–734.

Magnusson, D., & Cairns, R. B. (1996). Developmental science: Toward a unified framework. In R. B. Cairns, G. H. Elder Jr., & E. J. Costello (Eds.), *Developmental science* (pp. 7–30). Cambridge: Cambridge University Press.

Manhart, L. E., Aral, S., Holmes, K. K., & Foxman, B. (2002). Sex partner concurrency: Measurement, prevalence, and correlates among urban 18-39-year-olds. *Sexually Transmitted Diseases, 29*(3), 133–143.

Miller, H. G., Cain, V. S., Rogers, S. M., Gribble, J. N., & Turner, C. F. (1999). Correlates of sexually transmitted bacterial infections among US women in 1995. *Family Planning Perspectives, 31*(1), 4–9, 23.

Miller, W. C., Ford, C. A., Morris, M., Handcock, M. S., Schmitz, J. L., Hobbs, M. M., Cohen, M. S., Harris, K. M., & Udry, J. R. (2004). Prevalence of chlamydial and gonococcal infections among young adults in the United States. *JAMA, 291*, 2229–2236.

Miller, W. C., Swygard, H., Hobbs, M. M., Ford, C. A., Handcock, M. S., Morris, M., Schmitz, J. L., Cohen, M. S., Harris, K. M., & Udry, J. R. (2005). The prevalence of trichomoniasis in young adults in the United States. *Sexually Transmitted Diseases, 32*, 593–598.

Morris, M., Handcock, M. S., Miller, W. C., Ford, C. A., Schmitz, J. L., Hobbs, M. M., et al. (2006). Prevalence of HIV infection among young adults in the United States: Results from the Add Health study. *American Journal of Public Health, 96*, 1091–1097.

Murphy, D. A., Brecht, M., Herbeck, D. M., & Huang, D. (2009). Trajectories of HIV risk behavior from age 15 to 25 in the national longitudinal survey of youth sample. *Journal of Youth and Adolescence, 38*, 1226–1239.

O'Brien, S., & Nelson, G. (2004). Human genes that limited AIDS. *Nature Genetics, 36*(6), 565–574.

Peipert, J., Lapane, K., Allwsorth, J., Redding, C., Blume, J., & Stein, M. (2008). Bacterial vaginosis, race, and sexually transmitted infections: Does race modify the association? *Sexually Transmitted Diseases, 35*(4), 363–367.

Rashad, I., & Kaestner, R. (2004). Teenage sex, drugs, and alcohol use: Problems identifying the cause of risky behavior. *Journal of Health Economics, 23*, 493–503.

Rose, E., DiClemente, R. J., Wingood, G. M., Sales, J. M., Latham, T. P., Crosby, R. A., Zenilman, J., Melendez, J., & Hardin, J. (2009). The validity of teens' and young adults' self-reported condom use. *Archives of Pediatric and Adolescent Medicine, 163*(1), 61–64.

Rosenberg, M. D., Gurvey, J. E., Adler, N., Dunlop, M. B., & Jonathan, E. (1999). Concurrent sex partners and risk for sexually transmitted diseases among adolescents. *Sexually Transmitted Diseases, 26*(4), 208–212.

Santelli, J. S., Lowry, R., Brener, N. D., & Robin, L. (2000). The association of sexual behaviors with socioeconomic status, family structure, and race/ethnicity among US adolescents. *American Journal of Public Health, 90*(10), 1582–1588.

Stanton, B., Romer, D., Ricardo, I., African American, M., Feigelman, S., & Galbraith, J. (1993). Early initiation of sex and its lack of association with risk behaviors among adolescent African Americans. *Pediatrics, 92*(1), 13–19.

Thomas, J., & Torrone, E. (2006). Incarceration as forced migration: Effects on selected community health outcomes. *American Journal of Public Health, 96*(10), 1762–1765.

Upchurch, D. M., Mason, W. M., Kusunoki, Y., Johnson, M., & Kriechbaum, J. (2004). Social and behavioral determinants of self-reported STD among adolescents. *Perspectives on Sexual and Reproductive Health, 36*(6), 276–287.

U.S. Census Bureau, Population Division. (2008). *U.S. Census Bureau National Population Estimates. Table 4: Annual estimates of the African American or African American alone population by sex and age for the United States: April 1, 2000 to July 1, 2007 and Table 4: Annual estimates of the White alone population by sex and age for the United States: April 1, 2000 to July 1, 2007*. Available at: http://www.census.gov/popest/national/asrh/NC-EST2007-asrh.html. Accessed 2/12/09.

Wallace, J. M., Bachman, J. G., O'Malley, P. M., Schulenberg, J. E., Cooper, S. M., & Johnston, L. D. (2003). Gender and ethnic differences in smoking, drinking and illicit drug use among American 8th, 10th and 12th grade students, 1976–2000. *Addiction, 98*(2), 225–234.

Weinstock, H., Berman, S., & Cates, W., Jr. (2004). Sexually transmitted diseases among American youth: Incidence and prevalence estimates, 2000. *Perspectives on Sexual and Reproductive Health, 36*(1), 6–10.

Windle, M., & Wiesner, M. (2004). Trajectories of marijuana use from adolescence to young adulthood: Predictors and outcomes. *Development and Psychopathology, 16*(4), 1007–1027.

Winkler, C., An, P., & O'Brien, S. (2004). Patterns of ethnic diversity among the genes that influence AIDS. *Human Molecular Genetics, 13*, R9–R19.

Chapter 19
A Contextual-Genetics Approach to Adolescent Drug Use and Sexual Risk Behavior

Gene H. Brody, Steven R.H. Beach, and Robert A. Philibert

In this chapter, we describe a research program that began in 2005 that was designed to refine etiologic models of drug use and sexual risk behavior, as well as the prevention programs the models inform. We were motivated to start this research because inclusive reviews of programs designed to prevent drug use/abuse and sexual risk behavior reveal mixed results. Many prevention programs do not attain their goals, and others are effective for some subgroups but not others (Foxcroft 2006; Foxcroft et al. 2003; Kraemer et al. 2002; White and Pitts 1998), suggesting that the causes of these risk behaviors are not yet well enough understood for prevention efforts to achieve large and reliable effects. This suggested to us a need for new approaches to etiologic models of drug use/abuse and sexual risk behavior, particularly greater articulation of the ways in which interactions among genetic, psychosocial, and developmental processes can inform them. Concurrent advances in both

---

This research was supported by Award Number 3R01AA012768-09S1 from the National Institute on Alcohol Abuse and Alcoholism and by Awards Numbers 5R01DA021736, 5R01DA019230, 3R01DA021736-04S1, 5R01DA010923-04S1, and 1P30DA027827 from the National Institute on Drug Abuse. The content is solely the responsibility of the authors and does not necessarily represent the official views of the National Institute on Alcohol Abuse and Alcoholism, the National Institute on Drug Abuse, or the National Institutes of Health.

G.H. Brody (✉)
Center for Family Research, Institute for Behavioral Research, University of Georgia, Athens, GA 30602, USA

Department of Behavioral Science and Health Education, Rollins School of Public Health, Emory University, Atlanta, GA 30322, USA
e-mail: gbrody@uga.edu

S.R.H. Beach
Center for Family Research, Institute for Behavioral Research, University of Georgia, Athens, GA 30602, USA

R.A. Philibert
Department of Psychiatry, University of Iowa, Iowa City, IA 52242, USA

Y.F. Thomas, L.N. Price (eds.), *Drug Use Trajectories Among Minority Youth*,
DOI 10.1007/978-94-017-7491-8_19

knowledge and technology related to basic genetics and epigenetic processes have created unprecedented opportunities for conceptual integration. We responded to this challenge and this opportunity by initiating a research program that uses findings from gene-environment interplay, developmental, and epigenetic research conducted with rural African Americans and participants in the Iowa Adoption Studies (IAS) to refine etiologic models in ways that increase their predictive utility.

Our goals in this chapter are to chronicle the development of this gene-environment interplay research program to help others who want to establish similar programs, and to suggest future directions that may turn the promise of this approach into practical benefits for the populations we serve. In the following sections, we provide (a) an overview of our research program involving rural African American youths and families, including a description of the study population; (b) the reasons for our examination of gene-environment interplay processes; (c) the development of our transdisciplinary team and our partnerships with rural African American communities; (d) some illustrative findings from gene × environment and epigenetic research; and (e) future research directions.

Research Involving Rural African American Families and Youths

For more than 25 years, we have been systematically investigating rural African American family life. More than 3000 families have taken part in longitudinal, developmental studies and randomized prevention trials. These studies examined (a) normative processes among married-parent and single-parent rural African American families (Brody and Flor 1996, 1997; Brody et al. 1998a, 1999, 2002, 2005a); (b) longitudinal, developmental models of family and contextual processes associated with drug use, behavioral and emotional problems, academic competence, and self-regulation among rural African American children, adolescents, and young adults (Brody et al. 1998b, 2000, 2001, 2003, 2005b, 2006a, 2010a; Brody and Forehand 1993); (c) protective factors that moderate the effects of adversity on child and adolescent development, enabling youths to avoid negative developmental trajectories; and (d) the application of research findings to the development and evaluation of drug use and risk behavior preventive interventions (Brody et al. 2004, 2006b, c, 2010b). Our work in rural communities exemplifies a belief in the importance of creating partnerships with representatives of the population from which participants are to be recruited (Murry and Brody 2004). In 30 rural Georgia counties, we have developed partnerships with rural African American community members who (a) provide ongoing feedback that we apply to the conceptual frameworks on which the projects are based (Brody et al. 2004), (b) evaluate the validity of measures to be used with rural African Americans (Brody and Stoneman 1992), (c) work with scientists on the development of prevention programs, and (d) participate in the Community Liaison system that our group developed for the recruitment and retention of African American families in our projects (Murry and Brody 2004).

These partnerships have been essential in establishing the CFR's credibility with participants, leading to high rates of participant retention and DNA collection rates of more than 90 %.

Demographic Background and Trends Involving Drug Use and Risk Behavior Among Rural African Americans

The African American families on whom our research program focuses live in small towns and communities in which poverty rates are among the highest in the nation and unemployment rates are above the national average (Proctor and Dalaker 2003). Although most of the primary caregivers work an average of 39 h per week, 50 % live below the poverty threshold and another 25 % live within 150 % of it. Their poverty status reflects the dominance of low-wage, resource-intensive industries in those areas (Tickamyer and Duncan 1990). For rural African American families, the challenge of overcoming the environmental obstacles associated with poverty and chronic economic stress is exacerbated by oppressive social structures and racial discrimination (Brody et al. 2006a; Tickamyer and Duncan 1990; Murry et al. 2001). Many rural African American families thus live with severe, chronic economic stress that takes a toll on children, youths, and young adults, increasing their risk for drug use and sexual risk behavior.

Drug Abuse and Sexual Risk Behavior Trends and Consequences Among African Americans

Historically, residing in rural communities has protected African American adolescents and young adults from the drug abuse prevalent in urban areas. Rural African American youths, however, are now using illicit drugs at rates equal to or exceeding those of their inner-city counterparts (Kogan et al. 2006; Levine and Coupey 2003). Although in rural areas African Americans initiate drug use later than do European Americans (Wallace et al. 2002), they experience rapidly escalating use across adolescence that ultimately equals or surpasses European Americans' use (French et al. 2002; Galea and Rudenstine 2005). For example, marijuana use rates at age 15 are lower for African Americans than for other racial/ethnic groups; by age 20, African Americans are more likely than members of other groups to be dependent on marijuana (Reardon and Buka 2002). The consequences of use, including arrests and clinic admissions, also are greater for African American adolescents and adults than for their European American peers (Wallace et al. 2002; Galea and Rudenstine 2005; Herd 1994). Racial/ethnic disparities also exist for access to prevention services (Brody et al. 1997), drug abuse treatment (Schmidt et al. 2006; Kline 1996; Lowman and Le Fauve 2003), and treatment program completion (Jacobson et al. 2007; Stack et al. 2000).

A critical consequence of drug abuse among rural African Americans is its impact on the spread of HIV and other STIs. HIV, once confined mainly to urban centers, has increased drastically among rural African Americans in the Southern United States (Centers for Disease Control and Prevention 1998). As of 2003, the infection rate of 55.6 per 100,000 among African Americans in the Southeast was more than 8 times the rate for European Americans in this region, and the rate among rural African Americans in this area was 3 times that of the United States as a whole (23.2 and 7.3, respectively; [Hall et al. 2005]). Nationally, about 75 % of HIV-positive adolescents 13–19 years of age are African American (Centers for Disease Control and Prevention 2006). Taken together, these data underscore the importance of refining etiologic models of the development and escalation of drug use and sexual risk behavior and the prevention programs they inform, both in general and for rural African Americans in particular.

Why Add Genes and Their Interplay with Contextual Processes to Etiologic Models of Drug Use and Risk Behavior?

Adolescence and, to a lesser extent, young adulthood are often perceived as times of experimentation with drugs and sexual activity. In reality, great variability exists among individuals during these developmental periods. Some youths become entrapped in self-perpetuating patterns of drug use and its consequences, some become transiently involved in such behaviors, and still others pass through these years without so much as sampling a cigarette. Our research program is designed to determine what accounts for these considerable differences in initiation, escalation, and maintenance.

Most research on the etiology of drug use has focused on a range of contextual (e.g., parenting, peers, neighborhood) and intraindividual (e.g., temperament, self-regulation, psychological symptomatology) factors (Pandina and Johnson 1999; Weinberg et al. 1998, other chapters in this volume). Recently, this research has begun to move beyond main effect models to address transactions between individual characteristics and environmental contexts that influence the probability of drug use initiation (Brody et al. 1998b). Until recently, though, genetic attributes that operate in conjunction with contextual factors to forecast drug use across preadolescence, adolescence, and young adulthood have not been considered.

Converging information suggests that genes in concert with contextual processes should play a significant role in determining the etiology of youth drug use/abuse. Exemplars in the behavior genetic literature of highly heritable traits that contextual factors can modify suggest the likely importance of the environment. For example, poverty can reduce IQ (Turkheimer et al. 2003), the heritability of youth smoking can be reduced by family processes (Kendler et al. 2004), inherited tendencies toward depression (Heath et al. 1998) and alcohol consumption (Heath et al. 1989) can be reduced by both marriage and religiosity, and residence in a rural environment can reduce the heritability of drug use (Rose et al. 2001). Behavior genetic findings highlight the contextual nature of heritability; estimates from twin designs are not fixed, but are sensitive to contextual conditions.

Genetic variation also has been shown to influence exposure to life circumstances, such as harsh parenting (Deater-Deckard et al. 1999) or stressful life events (Rutter and Silberg 2002). These genetically influenced effects may arise because individuals' behaviors shape and select their environments and influence others' responses to them; this process is termed *gene-environment correlation* (*r*GE). Gene-environment correlations can masquerade as environmental effects (e.g., the effects of children on their environments rather than effects of rearing experiences on children), suggesting the importance of genetic assessment for appropriate testing of alternative models. Another form of G-E interplay, *gene-environment interaction* (G×E), occurs when genetic variation alters an individual's sensitivity to specific environmental events (e.g., maltreatment in the development of depression); or when environmental events (e.g., participation in a preventive intervention or exposure to life stress) exert differential control over genetic influences (Brody et al. 2009a; Kendler and Eaves 1986). G×E effects imply that genetic variation brings about differences in individuals' resilience or vulnerability to the environmental causes of many disorders, including drug use and sexual risk behavior (Hamer 2002). Prevention scientists have begun to give attention to differences between G×E models that suggest vulnerability and those that suggest differential susceptibility (Belsky and Pluess 2009), a distinction that may be of particular conceptual importance among disadvantaged populations.

From etiologic, prevention science, and public health perspectives, the misspecification of genetically mediated effects and G×E interactions as environmentally mediated effects has the potential to yield weak etiologic models of drug use/abuse and sexual risk behavior that misinform preventive intervention design (Kellam and Van Horn 1997). This research and theory suggest that scientists studying both genetic and contextual effects must recognize that many of these effects are contingent upon several forms of gene-environment co-action.

Gene-Environment Interplay Processes and Rural African Americans

An important limitation of existing genetic research is its almost exclusive focus on populations of northern European descent to the exclusion of other ethnic groups. African Americans are particularly unlikely to participate in genetic studies (Espeland et al. 2006; Green et al. 2006; McQuillan et al. 2003, 2006). This is a problem for prevention and public health scientists because some risk mechanisms (e.g., poverty and racism) disproportionately affect African Americans (Brody et al. 2006a; Gibbons et al. 2004; Simons et al. 2003). Findings from European American samples regarding *r*GE or G×E may not be generalizable to rural African Americans and may exclude relevant environmental risk or protective factors; thus, these findings may have little or no heuristic value for the development of preventive interventions for rural African Americans. In addition, substantial genetic variation exists across ethnic groups at loci known to be important in moderating responses to environmental risk mechanisms. For example, greater diversity is generally found in the African American population in the serotonin transporter gene *SLC6A4* (Disotell 2000). The absence of information regarding this locus in African Americans serves as an additional caveat regarding the inadequacy of generalizing information derived from European American populations to rural African American families. One goal, then, of our research program was to collect data on rural African American youths and their caregivers that included characterizations of contexts (family, peer, and neighborhood), drug use and sexual risk behavior phenotypes, and genetic data across preadolescence, adolescence, and young adulthood. In the next section, we describe the origins of our transdisciplinary collaboration and the work that resulted from the availability of genetically and contextually informed data.

Getting Going: Collaboration, Focus Groups, and Pilot Data Collection

Around 2005, Brody and Beach decided, for reasons described previously, to develop a gene-environment interplay research program. Formation of a transdisciplinary team was necessary to achieve this goal. Traditionally, cooperation between biological and contextual scientists has been encouraged, but few exemplars of successful transdisciplinary endeavors have emerged. Brody and Beach were extremely fortunate to become acquainted with genetics experts Robert Philibert and Alexandre Todorov, forming a team that included a developmental psychologist (Brody), a clinical psychologist (Beach), a psychiatrist and human geneticist (Philibert), and a statistical geneticist (Todorov). To prepare for their collaborative work, the team met regularly over a 2-year period to review the genetic and family process literature, adding collaborators as the effort solidified. These meetings yielded testable hypotheses, workable data collection methods, and strategies for managing genetically and contextually informed data sets.

Two focus groups of rural African Americans were formed, one for parents and one for adolescents. Each group included 10 persons who met for 2 h. Many focus group members wanted a clear explanation of the procedures for obtaining DNA, and they wanted to know how DNA collection would advance knowledge about the development of African American youths. This feedback was incorporated into a brochure, presented in Fig. 19.1, that includes answers to frequently asked questions. The answers address in a straightforward manner the issues the focus groups raised. A copy of the brochure was given to each participating family to provide them with written information that they could consult. Response to the brochure has been very positive, contributing to high rates of participation in the collection of genetic data.

A New Way of Understanding Health and Success in African American Families

For more information call us toll-free at:

1-888-542-3068

and ask to speak to our genetic information specialist.

The Center for Family Research
1095 College Station Road
Athens, Georgia 30602

Center for Family Research
THE UNIVERSITY OF GEORGIA

FREQUENTLY ASKED QUESTIONS:

Why collect information about genes?

Information about how genes and life experiences work together to make us who we are has the potential to change our understanding of health and illness. We are particularly interested in health problems that occur frequently among African Americans, like diabetes, hypertension and heart disease. By including a genetic sample with information about attitudes and beliefs, relationships, home and health, we will be able to answer questions about how our experiences may actually protect us from genetic vulnerability to some illnesses!

What does this have to do with me as an African American?

In order to understand how genes work, we must study people whose ancestors come from all parts of the world. It is important, then, to include people of all races in genetic studies. If African Americans are to benefit from this research, we must learn first-hand how their genes and life experiences work together to influence their health. We do not want to assume that how it works for people of one race is how it works for all races.

What will I be asked to do?

Since genes are in every cell of the body, we can get a sample of your genetic information from your saliva. You will be asked to rinse your mouth with water, and then spit into a small vial called Oragene. When you cap the vial, a seal will be broken and your saliva will mix with the Oragene liquid. The field interviewer will provide instructions and the materials needed. Your participation is completely voluntary.

What will be done with the mouthwash sample?

Your saliva and Oragene sample will be labeled with an identification number rather than your name. It will be stored and analyzed at the University of Iowa, in Iowa City. When our study is completed, the sample will be destroyed.

Will my genetic information be available to anyone?

We go to great lengths to keep all of the information that you share with us confidential. This commitment to your privacy will be upheld with your genetic information as well. It will not be shared with any third parties for any reason, including purposes of employment, insurance, paternity or criminal investigation except with your written permission or as may be required by law.

What are the benefits to me and my family?

The value of participating comes from contributing to a study about African Americans that may reveal information that will help families like yours in the future.

You will not receive any personal genetic information from our researchers because we are not qualified to make medical diagnoses or to provide genetic counseling.

Will I get paid extra for participating in this part of the study?

You will be paid an additional $20 for providing the saliva sample.

What if I have questions?

Please call us toll free (1-888-542-3068) at the Center for Family Research with any questions you may have. Dr. Yvonne Mensa-Wilmot, AIM Project Coordinator, can answer specific questions about the AIM study and our genetic collection procedures.

For many, many years people have wondered what makes us who we are. Is it the way our parents raise us? Is it the experiences that life sends our way? Or is it what we each carry inside the cells of our bodies–the unique genetic blueprint we are born with? Most scientists would agree that it is a mixture of how we are made on the inside (our genes) and what we experience on the outside (our environment) that makes us who we are.

Our Research is Changing

Over the years, the research conducted at the Center for Family Research has looked at the kinds of families, schools and communities that foster health and success in African American youth. Now we are collecting genetic as well as environmental information from the families who participate in our research. By looking at the connections between how we are made (our genes) and what we experience (our environment), we hope to better understand why some people are able to succeed in difficult situations.

Why This Research is Important

Scientists have identified all of the genes in the human body. More than 20,000 of them! But many questions remain about how genes work and why they seem to have different effects on our health and behavior under different circumstances. We need to know more about when and how different versions of our genes are affected by the experiences we have–because these experiences (how much stress we experience, whether or not we take good care of our bodies, what kind of relationships we have with friends and family members) influence our health.

Some health problems like hypertension, diabetes and heart disease occur much too frequently in the African American community. The explanation for this may be partially genetic, but it is also related to stress, diet and health care practices. By studying the fit between genes and environment, we may learn how to improve treatments for these illnesses. In fact, we may discover ways in which families and communities actually protect us from developing health problems!

How It Will Help

This new approach to our research will help us understand how and why certain conditions do or do not develop in African Americans–and what we might do to prevent them. We are looking at genes related to (1) health conditions like hypertension, diabetes, and heart disease, and (2) emotional and behavioral conditions like depression, anxiety, substance abuse, and attention deficit hyperactivity disorder. Our findings will help us develop new ideas about how family life, parenting practices and everyday experiences can encourage health and success in African American children and families.

Fig. 19.1 DNA collection brochure

A pilot study was also conducted to assess the viability of DNA collection from saliva versus whole blood (Philibert et al. 2008a). As predicted, concentrations of DNA were higher in blood than in saliva; the saliva samples, however, contained adequate amounts of DNA to permit genotyping. We concluded that the ease and economy of DNA collection from saliva made it appropriate for the research we planned. We also learned that researchers should prepare and budget for some resampling because some saliva samples may include contaminants such as tobacco and food residues. We are beginning to collect whole blood as well as saliva samples for research involving epigenetics: the influence of contexts on gene expression. Our experience thus far suggests that participants are no more likely to refuse to provide blood samples than saliva samples, as long as phlebotomists are well trained and professional and the overall rationale for the sample collection is clear and well justified.

Gene × Environment Research

In this section, we present some recent findings on G × E interactions that have heuristic value for informing etiologic models of drug use/abuse and sexual risk behavior.

Theoretical Approaches to G × E Interactions

Ecological (Bronfenbrenner and Ceci 1994), systems (Lerner 1991), and resilience approaches to lifespan development share the tenet that biological predispositions transact with contextual processes to create variations in phenotypes over time. These G × E interactions occur when genetic variation alters an organism's sensitivity to specific environmental effects or when environmental features exert differential control over genetic effects (Kendler and Eaves 1986). Although scientists agree that G × E investigations are worthwhile, no one formula captures adequately the ways in which genetic and contextual factors interact during adolescence to produce individual differences in drug use. Three predominant viewpoints shape this research. According to the diathesis-stress model, genetic vulnerabilities are expressed under risk-enhancing conditions (Kahn et al. 2003). From this perspective, the link between selected genes and drug use becomes most evident when youths experience stressors that dysregulate behavior and emotions or when they encounter risk factors that have established links with drug use initiation and escalation. Such risk factors include conflicted family relationships, affiliations with drug-using peers, and disengagement from schooling (Hawkins et al. 1992; Petraitis et al. 1995). According to the resilience model, genetic status is less important when youths' environments include protective processes that impede drug use initiation and escalation. Such processes include parenting characterized by high levels of emotional support, instrumental assistance, and communication; engagement with schooling; and affiliations with prosocial peers (Brody et al. 2006a; Caprara et al. 2000; Wills et al. 2003). According to the differential susceptibility model, individuals vary in their responsiveness to both buffering and risk-enhancing environmental conditions. Preliminary evidence indicates that several genes (5HTT, DRD4, MAOA) may contribute to variations in sensitivity (Belsky et al. 2009). Youths who carry specific copies of alleles in these genes are hypothesized to display a "for better or for worse" pattern that would be expressed, for instance, in a negative association between a candidate gene and drug use in the presence of a protective factor and a positive association between the gene and drug use in the absence of that factor.

Using Longitudinal, Epidemiological Research and Randomized Prevention Programs to Test G × E Interactions

We have used two research designs to test G×E hypotheses: longitudinal, epidemiological research and randomized prevention programs. A primary challenge of this design is the unambiguous identification of a true environmental effect, particularly if exposure occurs over an extended period of time. For many years, we have used longitudinal, epidemiological designs to test buffering and risk-enhancing contextual effects, and we plan to continue using them to understand the ways in which genes and environments interact to create phenotypic differences over time. Nevertheless, we recognize that this design is essentially correlational, opening results to multiple interpretations.

The use of intervention strategies such as randomized prevention trials is one means of determining whether an environmental factor has causal status. Through the implementation of such trials, a causal relationship between an environmental manipulation and the course of a targeted outcome can be identified (Rutter 2005). Randomized prevention designs also rule out an alternative explanation for G×E interactions, namely, *r*GE that occur when genetic influence on participants' likelihood of exposure to environmental factors contaminates environmental measures (Rutter 2007).

Random assignment to an intervention program has the additional advantage of ruling out confounds that, in epidemiological designs, may be taken for environmental effects. These include history (unmeasured events, such as an economic downturn, that co-occur with measured events), maturation (natural change across time, such as the onset of puberty), repeated testing (effects of prior assessments on responses to subsequent assessments as participants become familiar with the instruments), and statistical regression (a subsequent shift toward the population mean following an initial low or high assessment). Finally, the testing of G×E hypotheses using randomized prevention trials enhances statistical power as much as fivefold over epidemiological approaches (McClelland and Judd 1993); consequently, fewer participants are needed to obtain a detectable G×E interaction.

The Candidate Genes

Thus far, our research program has focused primarily on candidate genes, with two genes, 5HTT and DRD4, occupying much of our attention. Although genome-wide association studies (GWAS) are encouraged so that genetic linkages can be identified, they are useful primarily in the identification of genetic main effects. It is much more difficult to use GWAS to examine contextually sensitive, developmental change. By its nature, examination of such processes requires clear theoretical models and well-specified mechanisms of action, rendering such investigations less compatible with the atheoretical methods of the GWAS approach. To the extent that genes and their transactions with contextual processes are relevant to etiology, a focus on particular candidate genes is likely to be necessary. In addition, implications for intervention are most likely to become apparent in the context of gene-environment transactions. As Belsky and Pluess (2009) note, there is also good reason to anticipate that GWAS will not identify some important genetic loci. Specifically, a G×E effect will only manifest as a main effect if it takes the form of a vulnerability, not if it confers greater sensitivity for better or worse. To the extent that many genetic effects of particular developmental importance confer sensitivity rather than vulnerability, it would be inappropriate to require demonstration of a main effect prior to examination of contextually sensitive models.

The serotonin transporter protein 5HTT plays a key role in serotonergic activity by regulating the reuptake of serotonin (5-HT) following synaptic release (Fabre et al. 2000). The most common polymorphism in the 5-HTTLPR region of the serotonin transporter gene SCL6A4 results in two common variants, a long (*l*) and a short (*s*) allele. The *s* variant is associated with lower availability of 5-HTT and reduced efficiency of 5-HT reuptake (Whale et al. 2000); it has been found to be associated with alcohol consumption in college students (Covault et al. 2007; Herman et al. 2003), maltreated youths (Kaufman et al. 2007), and a large representative sample in the United Kingdom (Munafò et al. 2005), as well as with high ethanol tolerance among young adults (Tucker et al. 2003). Other studies involving children and adolescents link low serotonergic function to conduct problems, a correlate and predictor of drug use/abuse. Several of these studies were contemporaneous (Gerra et al. 2005; Haberstick et al. 2006; Halperin et al. 1997; Mitsis et al. 2000; Sakai et al. 2006) and two used prospective designs (Flory et al. 2007; Halperin et al. 2006). The literature is not entirely consistent, however; two contemporaneous analyses (Beitchman et al. 2003; Davidge et al. 2004) and one longitudinal analysis (Sakai et al. 2007) found no association between the *s* allele and conduct problems. Thus, although the literature is somewhat mixed, the preponderance of data suggest that the *s* allele of 5-HTTLPR may be important in understanding the etiology of drug use and conduct problems.

DRD4 is a key regulator of dopamine neurotransmission. The VNTR contains 2–11 repeats; the 4-repeat and 7-repeat alleles are most common. The 7-repeat allele produces less reactive D4 receptors in both *in vitro* and *in vivo* tests of responsiveness, resulting in weaker transmission of intracellular signals for those with the

7-repeat allele than for those with the 4-repeat allele (Levitan et al. 2006). Youths carrying at least one 7-repeat allele have been found to engage in higher lifetime rates of smoking (Laucht et al. 2005, 2008; Shields et al. 1998) and alcohol use (Conner et al. 2010; Laucht et al. 2007; Ray et al. 2008; Skowronek et al. 2006; Vaughn et al. 2009) than similar youths without the allele. Other researchers looked for an association between the 7-repeat allele and novelty seeking, a risk factor for substance use (Cloninger et al. 1993). Some found an association among the DRD4 VNTR, novelty seeking, and alcohol use (Laucht et al. 2005, 2007; Ray et al. 2008); in one of these studies, novelty seeking mediated the association between DRD4 and alcohol use (Ray et al. 2008). The literature is not entirely consistent, though; some studies found no associations between DRD4 variation and indicators of youth substance use (Hopfer et al. 2005; Luciano et al. 2004; Rodríguez et al. 2006) or novelty seeking (Luciano et al. 2004). The preponderance of studies, however, suggest that the 7-repeat allele variant of DRD4 may forecast increases in youths' substance use. In our research, therefore, we contrasted drug use among youths with at least one 7-repeat allele with use among those with two 4-repeat alleles (Ding et al. 2002).

Findings from the Longitudinal, Epidemiological Research

In this section, we present results from G×E studies that demonstrate resilience, diathesis-stress, and differential susceptibility effects. We begin with a study that examined parenting as a moderator of a genetic vulnerability factor.

Resilience Effects of Involved-Supportive Parenting

Using a prospective design, we investigated a moderation effect in the association between the genetic vulnerability factor in 5-HTTLPR and increases in drug use among African American youths age 11–14 years (Brody et al. 2009b). We predicted that involved-supportive parenting would attenuate the link between one or two copies of the *s* allele in 5-HTTLPR and longitudinal increases in drug use. Among African American parents, involved-supportive parenting is characterized by high levels of emotional support, instrumental assistance, and communication (Brody et al. 2006a). Research has consistently indicated that these parenting practices have protective moderation effects for African American youths and reduce the impact of risk factors on youth drug use (Brody et al. 2004; Wills et al. 2003; DiClemente et al. 2001).

The results indicated that 5-HTTLPR status was linked positively with the development of drug use, but no association emerged between the presence of the *s* allele and the development of drug use when youths received high levels of involved-supportive parenting. The risk conferred by 5-HTTLPR status on youths whose

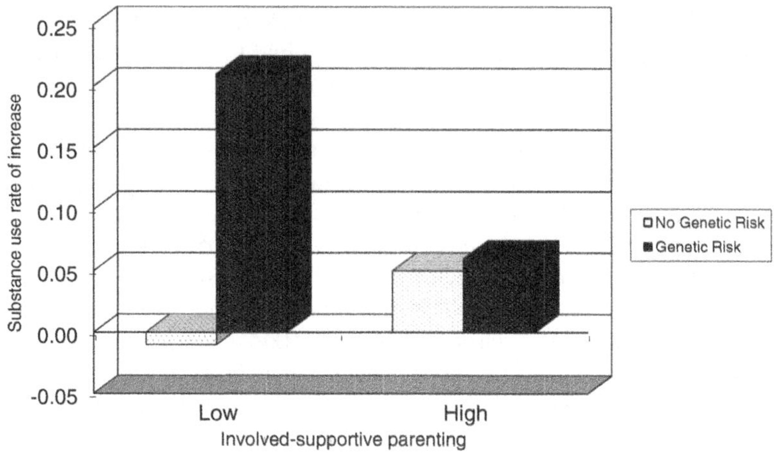

Fig. 19.2 Mean rate of increase on the substance use index for Genetic Risk × Parenting combinations. Groups classified as low on involved–supportive parenting include families at or below the 30th percentile for the study sample; groups classified as high include those at or above the 70th percentile

parents engaged in low levels of involved-supportive parenting was three times as large as the coefficient among youths whose caregivers provided high levels of such parenting. These findings are depicted in Fig. 19.2.

Diathesis-Stress Effects Involving 5-HTTLPR

One of the most frequently cited risk factors for HIV-related sexual risk behavior during middle and late adolescence is drug use (Leigh and Stahl 1993). Drug users in early adolescence have been found to be more likely than abstainers to report unprotected sexual intercourse, intercourse with multiple partners, intoxication during intercourse, and pregnancy (Stueve and O'Donnell 2005). Others researchers have found prospective links between drug use in early adolescence and inconsistent condom use, intercourse with multiple partners, and intoxication during intercourse (Guo et al. 2002; Ellickson et al. 2001). Tubman et al. (1996), however, reported that more than half of youths who used substances in early adolescence did not engage in sexual risk behavior 2 years later. To account for the considerable variation in these findings, we conjectured that 5-HTTLPR status would moderate the link between drug use and subsequent sexual risk behavior during early adolescence. Consistent with our hypotheses, drug use among rural African Americans at age 14 was related strongly to sexual risk behavior 2 years later if they carried a copy of the *s* allele at 5HTT, but not otherwise. In the absence of this genetic vulnerability, no significant prospective association emerged between drug use and sexual risk behavior. Figure 19.3 graphically presents the G × E interaction.

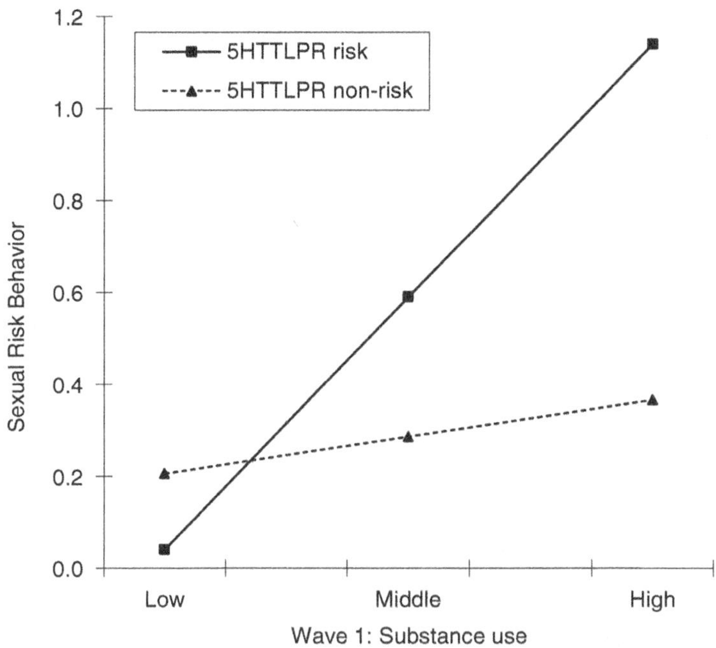

Fig. 19.3 5-HTTLPR status moderates the association between substance use and sexual risk behavior

Resilience, Diathesis-Stress, and Differential Susceptibility Effects Involving DRD4

Using a prospective research design, we investigated moderation effects in the association between a genetic vulnerability factor, a variable nucleotide repeat polymorphism (VNTR) in the coding region of DRD4, and increases in drug use among African American youths from 13 to 16 years of age. The primary purpose of this study, however, was to determine whether experiences in important areas of the adolescents' lives—the quality of the parenting they receive, their connection and commitment to schooling, and the type of peers with whom they affiliate—condition the hypothesized link between DRD4 status and the development of drug use. We were particularly interested in whether the moderation effects conformed to a resilience, diathesis-stress, or differential susceptibility model.

We found that DRD4 status was linked positively with the development of drug use across adolescence, and this association was conditioned by high levels of involved-supportive parenting, high levels of deviant peer affiliations, and both high and low levels of school engagement. Figures 19.4a, 19.4b, and 19.4c present each of the G×E interactions. To increase the precision with which we could detect the form of the G×E interactions, we used the Johnson-Neyman (J-N) technique

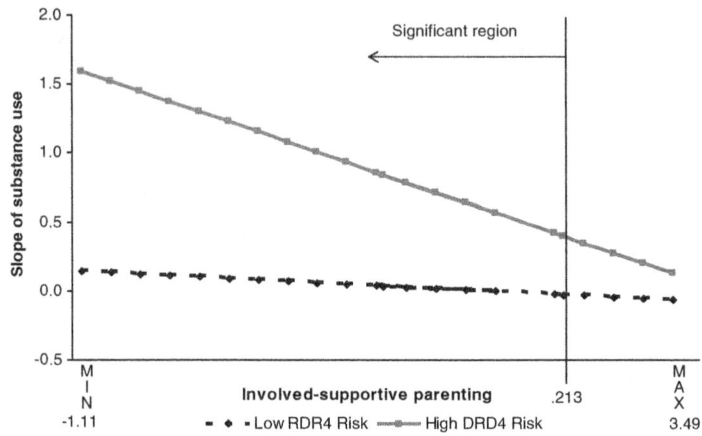

Fig. 19.4a DRD4 moderates the relationship between involved-supportive parenting and slope of substance use

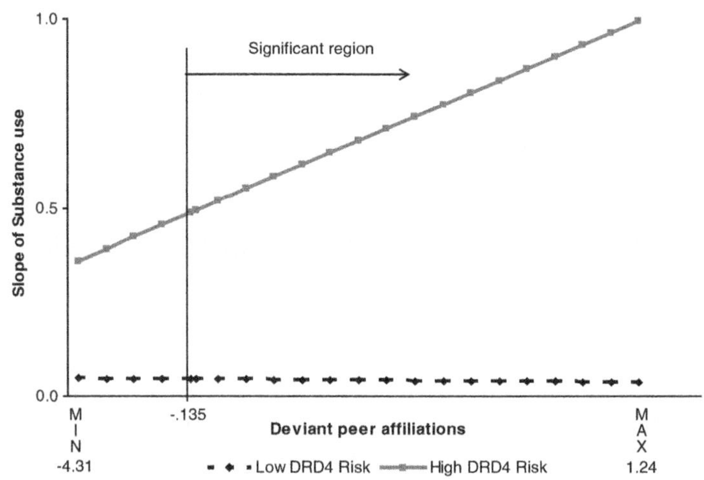

Fig. 19.4b DRD4 moderates the relationship between deviant peer affiliation and slope of substance use

(Johnson and Fay 1950; Johnson and Neyman 1936). The J-N technique allows researchers to determine exactly where on the continuous distribution of the moderator variable that moderation is detected. The results of this analysis revealed a resilience pattern for involved-supportive parenting, a diathesis-stress pattern for deviant peer affiliations, and a differential susceptibility pattern for school engagement. For school engagement, high levels ameliorated the link between DRD4 and drug use (resilience effects) and low levels were associated with a more robust association between DRD4 and drug use. This differential susceptibility pattern demonstrates the powerful effects that schooling has on individual differences in

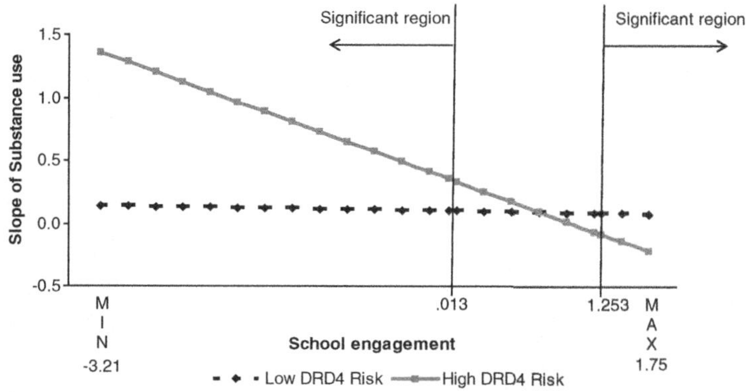

Fig. 19.4c DRD4 moderates the relationship between school engagement and slope of substance use

adolescents' drug use. The moderators' various effects on youth drug use suggest the following proposition. Protective environments that include involved-supportive parenting and high levels of school engagement facilitate the development of self-regulation; the development of self-regulation constrains proclivities that DRD4 status may foster. Conversely, deviant peer affiliations and low levels of school engagement may be symptoms of poor self-regulation, which promotes the expression of DRD4-related behaviors by encouraging exposure to circumstances that provide both opportunities and reinforcement for drug use.

Overview of G × E Findings Using Randomized Prevention Trial Designs

Prevention and 5-HTTLPR G × E Effects

The Strong African American Families (SAAF) program was designed to prevent the initiation of a cluster of risk behaviors that included alcohol use, marijuana use, and sexual activity. Evaluations of SAAF have confirmed its efficacy in preventing the initiation of risk behaviors (Brody et al. 2004, 2006b, c). The primary purpose of this study was to test the G × E hypothesis that random assignment to the SAAF prevention group versus a control group would interact with genetic risk to predict youths' risk behavior initiation over a period of 2 years and 5 months. Risk behavior was measured using an index that indicated whether a youth had ever used alcohol, smoked marijuana, or had sexual intercourse. Specifically, we predicted that (a) youths at genetic risk from the presence of the *s* allele at 5-HTTLPR who were assigned randomly to the control condition would initiate more risk behaviors compared with youths at genetic risk assigned randomly to the SAAF prevention condition; (b) youths at genetic risk assigned to the control condition would initiate more

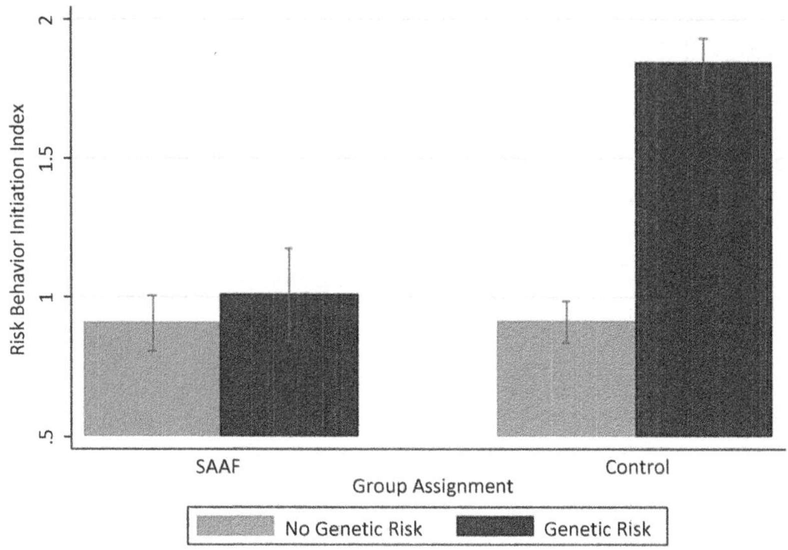

Fig. 19.5 Mean risk behavior initiation for each Prevention Group × Genetic Risk condition at long-term follow-up, adjusted for pretest values

risk behaviors than would youths without genetic risk assigned randomly to either the prevention or control condition; and (c) youths at genetic risk assigned to the prevention condition would not initiate more risk behaviors than would youths without genetic risk assigned to either condition.

These hypotheses were based on Rutter's (1985) thesis that protective processes have their greatest effects on youths at highest risk; presumably, protective processes augment at-risk youths' inhibitory controls and foster competencies that occasion positive developmental outcomes. We conjectured that enhancing these protective caregiving practices via SAAF would be particularly important for those youths carrying the *s* allele of 5-HTTLPR by decreasing the likelihood that youths would encounter opportunities to initiate and engage in risk behaviors. The results supported the hypotheses, indicating that youths with the *s* allele who were assigned to the control condition initiated risk behavior at higher rates than did youths with the *s* allele in the SAAF condition and youths with two copies of the *l* allele in either the SAAF or control condition. These results are presented in Fig. 19.5.

Prevention × DRD4 Effects

After determining that participation in a preventive intervention could condition the risk that 5-HTTLPR confers for risk behavior onset, we extended this line of research by examining the possibility that SAAF participation could ameliorate the

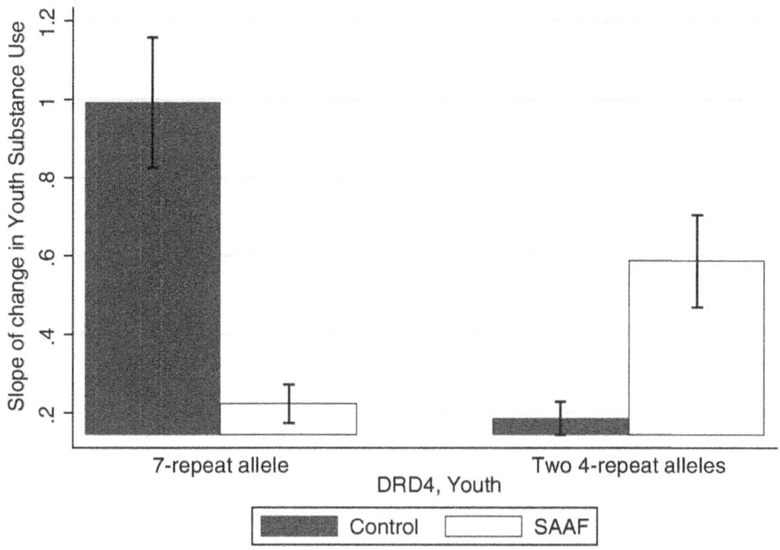

Fig. 19.6 The slope of change in youth past-month substance use as a function of SAAF and target genotype

risk that variation in DRD4 is hypothesized to confer on drug use (Beach et al. 2010a). We focused on DRD4 for reasons explained previously that support its contribution to individual variation in drug use. In this analysis, we contrasted substance use among youths with at least one 7-repeat allele versus those with two 4-repeat alleles (Ding et al. 2002).

The results supported the hypothesis that youths who took part in SAAF who carried the DRD4 7-repeat allele would be more responsive to the intervention than would youths with two copies of the 4-repeat allele. Among youths with the 7-repeat allele, those in the control group increased their past-month substance use substantially more than did those who took part in SAAF. These results are presented in Fig. 19.6. These results, like those presented previously for 5-HTTLPR, support the differential susceptibility hypothesis (Belsky and Pluess 2009), which suggests that some genes, such as DRD4 and 5-HTT, increase sensitivity to contextual influences whether those influences are protective or risk promoting. Applied to our results, youths with the 7-repeat DRD4 allele or the *s* allele of 5-HTTLPR were more sensitive to the enhanced family environments that prevention programming provided as well as to opportunities for risk behavior that youth in the control condition, experiencing life as usual, encountered. These results support the idea that, for genetic reasons, some youths may be particularly responsive to contexts that do or do not promote resilience.

Epigenetic Regulation

In addition to the "hard-wired" variation in gene activity associated with particular genetic polymorphisms, we are also beginning to explore biological mechanisms by which G×E interactions confer long-term risks for disorder. In mammals, epigenetic effects occur via three major mechanisms: DNA methylation, histone modification, and non-coding RNA interference. Acting separately or together, these three mechanisms can alter mammalian tissue structure or behavior. The mechanism most readily available for investigation is methylation of promoter regions. Indeed, some environments appear to change gene functioning by contributing to the methylation of the gene's promoter region, typically resulting in reduced transcription at that locus and perhaps setting the stage for vulnerability to drug use/abuse and related disorders. Accordingly, methylation of promoter regions of various genes is a potentially important mechanism of "epigenesis," defined as the production of stable changes in DNA expression that do not result from a change in DNA sequence.

Increases in the amount of methylation in promoter-associated CpG islands usually, but not always, results in the down-regulation of RNA transcription from downstream genes. Because CpG methylation can also recruit factors associated with chromatin remodeling, the methylation of these promoter CpG residues also can act synergistically with other regulatory mechanisms, such as histone modification, to encourage further transcriptional down-regulation. In turn, these changes in regulation of gene transcription further up-regulate or down-regulate the effects of genetic polymorphisms.

If CpG methylation can serve as a mechanism for the cellular fine-tuning of gene expression in response to extracellular events, measuring the extent of DNA methylation, may make it possible to define more precisely how environmental risk mechanisms contribute to drug use/abuse. A potential impediment to research linking serious childhood stressors, such as sexual abuse, to CpG methylation is that no technologies exist to assess the function of these regulatory elements in the intact human CNS. As a result, we have used lymphoblast cell lines to explore these elements, because they are more readily obtained and are increasingly accepted as a valuable tool for deciphering the effects of methylation on gene expression (Bradley et al. 2005) and other outcomes (Feinberg 2008). In addition, lymphoblast cell lines derived from human subjects appear to retain transcriptional signatures that reflect the clinical status of their donors (Philibert et al. 2008b), and low-pass lymphoblast lines of the sort used in the current study retain the methylation signature of lymphocytes (Grafodatskaya et al. 2010). Hence, by using *in vitro* cell lines from well-characterized subjects as surrogates for *in vivo* CNS cells, it may be possible to delineate the identity of the regulatory elements and determine whether epigenetic effects on these elements affect their function in neuropsychiatric illness.

Can Environmental Events Produce Changes in Methylation at Key Loci?

Supporting the potential impact of childhood experiences in humans on epigenetic change, methylation differences have been demonstrated in the postmortem hippocampus obtained from suicide victims with a history of childhood abuse relative to those from either suicide victims with no childhood abuse or non-suicide controls (McGowan et al. 2008, 2009). Because it seems likely that there is a role for both genetic and environmental factors in models of drug use with a probable role for family environment during development, it is important to examine possible biological processes that may mediate these long-term impacts on adult psychopathology. For behavioral scientists, changes in methylation produced by clear environmental stressors are of particular interest because alterations in methylation, in the form of changes in CpG motifs, may remain stable over a relatively long time in humans (Eckhardt et al. 2006), making individual differences in methylation a potential biological marker of environmental contributions to the phenotypic divergence of those with similar genetic endowments (Fraga et al. 2005). In addition, because CpG motifs are potentially modifiable by environmental factors, they provide a plausible physical substrate by which environmental events may have lasting effects on behavior.

Currently, the best evidence that epigenetic effects may result from experiences in rearing environments and then subsequently affect important aspects of behavior is found in studies using non-human models. For example, poor maternal care by rat dams of their pups within the first 10 days of life has been shown to influence gene expression by decreasing RNA expression in the hippocampus, resulting in increased sensitivity to stress that lasts over the entire lifetime of the maltreated pups (Liu et al. 1997).

What Has Our Initial Work on Epigenetics Found?

In an initial test of the possibility that child abuse might influence the methylation profile at 5HTT, we found that physical abuse, harsh parenting, and sexual abuse were associated with overall hypermethylation of the *5HTT* promoter region (Beach et al. 2010b). A significant association also emerged in that sample between sexual abuse alone and overall methylation of the CpG island at *5HTT* among female participants, $r(82)=360$, $p<001$. To replicate the effect of sexual abuse on methylation in women, we subsequently examined a larger sample only of women. Extending the prior report, we also examined symptomatic outcomes for adult ASPD and the potential role of methylation in modifying the impact of genetic risk alleles. We simultaneously controlled for the potential impact of parental diagnostic status on both symptoms and level of methylation to rule out a range of potential third-variable explanations for observed associations. As such, this investigation provides a model for epigenetic effects on behavioral outcomes with clear relevance for adult risk behavior. We found a significant effect of childhood sexual abuse on methylation of the *5HTT* promoter region among women. In addition, a significant effect of methylation at *5HTT* on symptoms of ASPD among women emerged. Significant associations remained with the significant effect of biological parent psychopathology controlled. Accordingly, we concluded that child sexual abuse may create long-lasting changes in methylation of the promoter region of *5HTT* in women, that hypermethylation may be one mechanism linking childhood sexual abuse to changes in risk for adult antisocial behavior in women, and that better understanding of the methylome may prove critical in understanding the role of childhood environments on long-term psychiatric sequelae.

Future Directions

Recent advances in human genetic and epigenetic processes offer unprecedented opportunities to generate new knowledge that has the potential to enhance preventive interventions. By examining the role of genetic and epigenetic change, we may better understand the specific environmental factors that contribute to long-term change as well as identifying the most cost-effective interventions that could interrupt environmental pathogenesis. Prevention scientists need to identify protective contextual mechanisms in youths' lives that override or regulate genetic risks and sensitivities to ensure that these mechanisms are targeted in prevention programs.

Recently we have described a "microtrials" approach to improve utilization of G×E research as we tailor existing prevention programs and develop new ones (Howe et al. 2010). We believe that existing data have great heuristic value, but must be translated in several ways before they have direct implications for preventive intervention. For example, the data imply that an interaction between the social environment and particular genetic diatheses may be critical for the correct specification of the etiology of some important outcomes, such as the prevention of SUDs. In many cases, however, understanding how to specify these environments for humans and the extent to which the relevant processes are malleable for humans requires additional direct examination. We anticipate that randomized microtrials will build on genetic moderation observed in animal models or hypothesized on the basis of basic research and establish similar effects in human populations. Through the implementation of microtrials, a cause and effect relationship is easily established and rival hypotheses can be eliminated, thereby guiding theoretical development in prevention science.

On the basis of our existing and future work, we hope to develop a genetically informed research base that increases the accuracy and predictive utility of causative models of drug use among rural African Americans. Predictive models may also generalize to other groups, and both replications and failures to replicate with other groups have the potential to be theoretically informative. We plan to test hypotheses that address the ways in which genes, *r*GE, and G×E at various developmental stages inform the selection of intervention targets in drug use prevention programs.

We also plan to use data from prevention and longitudinal, epidemiological studies to inform basic genetic science regarding epigenetic processes and gene expression and to address questions about the ways in which environments affect responses to prevention programs. In particular, by measuring epigenetic processes (e.g., DNA methylation) as well as sequence variation at a given locus, it may be possible to define more precisely the set of biological influences on gene expression, vulnerability, and resilience, explicating some of the physical bases of G×E and *r*GE effects on drug use/abuse as well as refining predictions of differential responses to preventive interventions.

References

Beach, S. R. H., Brody, G. H., Gunter, T. D., Packer, H., Wernett, P., & Philibert, R. A. (2010a). Child maltreatment moderates the association of MAOA with symptoms of depression and antisocial personality disorder. *Journal of Family Psychology, 24*(1), 12–20.

Beach, S. R. H., Brody, G. H., Todorov, A., Gunter, T. D., & Philibert, R. A. (2010b). Methylation at SLC6A4 is linked to family history of child abuse: An examination of the Iowa Adoptee sample. *American Journal of Medical Genetics Part B: Neuropsychiatric Genetics, 153B*(2), 710–713.

Beitchman, J. H., Davidge, K. M., Kennedy, J. L., Atkinson, L., Lee, V., Shapiro, S., et al. (2003). The serotonin transporter gene in aggressive children with and without ADHD and nonaggressive matched controls. *Annals of the New York Academy of Sciences, 1008*, 248–251.

Belsky, J., & Pluess, M. (2009). Beyond diathesis stress: Differential susceptibility to environmental influences. *Psychological Bulletin, 135*(6), 885–908.

Belsky, J., Jonassaint, C. R., Pluess, M., Stanton, M., Brummett, B., & Williams, R. (2009). Vulnerability genes or plasticity genes? *Molecular Psychiatry, 14*(8), 746–754.

Bradley, S. P., Rastellini, C., da Costa, M. A., Kowalik, T. F., Bloomenthal, A. B., Brown, M., et al. (2005). Gene silencing in the endocrine pancreas mediated by short-interfering RNA. *Pancreas, 31*(4), 373–379.

Brody, G. H., & Flor, D. L. (1996). Coparenting, family interactions, and competence among African American youths. In J. P. McHale & P. A. Cowan (Eds.), *Understanding how family-level dynamics affect children's development: Studies of two-parent families* (pp. 77–91). San Francisco: Jossey-Bass.

Brody, G. H., & Flor, D. L. (1997). Maternal psychological functioning, family processes, and child adjustment in rural, single-parent African American families. *Developmental Psychology, 33*(6), 1000–1011.

Brody, G. H., & Forehand, R. (1993). Prospective associations among family form, family processes, and adolescents' alcohol and drug use. *Behaviour Research and Therapy, 31*(6), 587–593.

Brody, G. H., & Stoneman, Z. (1992). Child competence and developmental goals among rural Black families: Investigating the links. In I. E. Sigel, A. V. McGillicuddy-DeLisi, & J. J.

Goodnow (Eds.), *Parental belief systems: The psychological consequences for children* (2nd ed., pp. 415–431). Hillsdale: Erlbaum.
Brody, G. H., Neubaum, E., Boyd, G. M., & Dufour, M. (1997). Health consequences of alcohol use in rural America. In E. B. Robertson, Z. Sloboda, G. M. Boyd, L. Beatty, & N. J. Kozel (Eds.), *Rural substance abuse: State of knowledge and issues* (pp. 137–174). Rockville: U.S. Department of Health and Human Services.
Brody, G. H., Flor, D. L., & Neubaum, E. (1998a). Coparenting processes and child competence among rural African-American families. In M. Lewis & C. Feiring (Eds.), *Families, risk, and competence* (pp. 227–243). Mahwah: Erlbaum.
Brody, G. H., Flor, D. L., Hollett-Wright, N., & McCoy, J. K. (1998b). Children's development of alcohol use norms: Contributions of parent and sibling norms, children's temperaments, and parent–child discussions. *Journal of Family Psychology, 12*(2), 209–219.
Brody, G. H., Flor, D. L., & Gibson, N. M. (1999). Linking maternal efficacy beliefs, developmental goals, parenting practices, and child competence in rural single-parent African American families. *Child Development, 70*(5), 1197–1208.
Brody, G. H., Ge, X., Katz, J., & Arias, I. (2000). A longitudinal analysis of internalization of parental alcohol-use norms and adolescent alcohol use. *Applied Developmental Science, 4*(2), 71–79.
Brody, G. H., Ge, X., Conger, R. D., Gibbons, F. X., Murry, V. M., Gerrard, M., et al. (2001). The influence of neighborhood disadvantage, collective socialization, and parenting on African American children's affiliation with deviant peers. *Child Development, 72*(4), 1231–1246.
Brody, G. H., Murry, V. M., Kim, S., & Brown, A. C. (2002). Longitudinal pathways to competence and psychological adjustment among African American children living in rural single-parent households. *Child Development, 73*(5), 1505–1516.
Brody, G. H., Ge, X., Kim, S. Y., Murry, V. M., Simons, R. L., Gibbons, F. X., et al. (2003). Neighborhood disadvantage moderates associations of parenting and older sibling problem attitudes and behavior with conduct disorders in African American children. *Journal of Consulting and Clinical Psychology, 71*(2), 211–222.
Brody, G. H., Murry, V. M., Gerrard, M., Gibbons, F. X., Molgaard, V., McNair, L. D., et al. (2004). The Strong African American Families program: Translating research into prevention programming. *Child Development, 75*(3), 900–917.
Brody, G. H., Murry, V. M., McNair, L. D., Chen, Y.-f., Gibbons, F. X., Gerrard, M., et al. (2005a). Linking changes in parenting to parent–child relationship quality and youth self-control: The Strong African American Families program. *Journal of Research on Adolescence, 15*(1), 47–69.
Brody, G. H., Kim, S., Murry, V. M., & Brown, A. C. (2005b). Longitudinal links among parenting, self-presentations to peers, and the development of externalizing and internalizing symptoms in African American siblings. *Development and Psychopathology, 17*(1), 185–205.
Brody, G. H., Chen, Y.-f., Murry, V. M., Ge, X., Simons, R. L., Gibbons, F. X., et al. (2006a). Perceived discrimination and the adjustment of African American youths: A five-year longitudinal analysis with contextual moderation effects. *Child Development, 77*(5), 1170–1189.
Brody, G. H., Murry, V. M., Gerrard, M., Gibbons, F. X., McNair, L. D., Brown, A. C., et al. (2006b). The Strong African American Families program: Prevention of youths' high-risk behavior and a test of a model of change. *Journal of Family Psychology, 20*(1), 1–11.
Brody, G. H., Murry, V. M., Kogan, S. M., Gerrard, M., Gibbons, F. X., Molgaard, V., et al. (2006c). The Strong African American Families program: A cluster-randomized prevention trial of long-term effects and a mediational model. *Journal of Consulting and Clinical Psychology, 74*(2), 356–366.
Brody, G. H., Beach, S. R. H., Philibert, R. A., Chen, Y.-f., & Murry, V. M. (2009a). Prevention effects moderate the association of 5-HTTLPR and youth risk behavior initiation: Gene × environment hypotheses tested via a randomized prevention design. *Child Development, 80*, 645–661.

Brody, G. H., Beach, S. R. H., Philibert, R. A., Chen, Y.-f., Lei, M.-K., Murry, V. M., et al. (2009b). Parenting moderates a genetic vulnerability factor in longitudinal increases in youths' substance use. *Journal of Consulting and Clinical Psychology, 77*(1), 1–11.

Brody, G. H., Chen, Y.-f., & Kogan, S. M. (2010a). A cascade model connecting life stress to risk behavior among rural African American emerging adults. *Development and Psychopathology, 22*(3), 667–678.

Brody, G. H., Chen, Y.-f., Kogan, S. M., Murry, V. M., & Brown, A. C. (2010b). Long-term effects of the Strong African American Families program on youths' alcohol use. *Journal of Consulting and Clinical Psychology, 78*(2), 281–285.

Bronfenbrenner, U., & Ceci, S. J. (1994). Nature-nurture reconceptualized in developmental perspective: A bioecological model. *Psychological Review, 101*(4), 568–586.

Caprara, G. V., Barbaranelli, C., Pastorelli, C., Bandura, A., & Zimbardo, P. G. (2000). Prosocial foundations of children's academic achievement. *Psychological Science, 11*(4), 302–306.

Centers for Disease Control and Prevention. (1998). Risks for HIV infection among persons residing in rural areas and small cities: Selected sites, southern United States, 1995–1996. *Morbidity and Mortality Weekly Report, 47*, 974–978.

Centers for Disease Control and Prevention. (2006). *HIV/AIDS surveillance report* (p. 2008). Washington, DC: U.S. Department of Health and Human Services.

Cloninger, C. R., Svrakic, D. M., & Przybeck, T. R. (1993). A psychobiological model of temperament and character. *Archives of General Psychiatry, 50*(12), 975–990.

Conner, B. T., Hellemann, G. S., Ritchie, T. L., & Noble, E. P. (2010). Genetic, personality, and environmental predictors of drug use in adolescents. *Journal of Substance Abuse Treatment, 38*(2), 178–190.

Covault, J., Tennen, H., Armeli, S., Conner, T. S., Herman, A. I., Cillessen, A. H. N., et al. (2007). Interactive effects of the serotonin transporter 5-HTTLPR polymorphism and stressful life events on college student drinking and drug use. *Biological Psychiatry, 61*(5), 609–616.

Davidge, K. M., Atkinson, L., Douglas, L., Lee, V., Shapiro, S., Kennedy, J. L., et al. (2004). Association of the serotonin transporter and 5HT1Dbeta receptor genes with extreme, persistent and pervasive aggressive behaviour in children. *Psychiatric Genetics, 14*(3), 143–146. URL.

Deater-Deckard, K., Fulker, D. W., & Plomin, R. (1999). A genetic study of the family environment in the transition to early adolescence. *Journal of Child Psychology and Psychiatry, 40*(5), 769–775.

DiClemente, R. J., Wingood, G. M., Crosby, R., Sionean, C., Cobb, B. K., Harrington, K., et al. (2001). Parental monitoring: Association with adolescents' risk behaviors. *Pediatrics, 107*(6), 1363–1368.

Ding, Y.-C., Chi, H.-C., Grady, D. L., Morishima, A., Kidd, J. R., Kidd, K. K., et al. (2002). Evidence of positive selection acting at the human dopamine receptor D4 gene locus. *Proceedings of the National Academy of Sciences of the United States of America, 99*(1), 309–314.

Disotell, T. R. (2000). Human genomic variation. *Genome Biology, 1*(5), comment2004.1–3.

Eckhardt, F., Lewin, J., Cortese, R., Rakyan, V. K., Attwood, J., Burger, M., et al. (2006). DNA methylation profiling of human chromosomes 6, 20 and 22. *Nature Genetics, 38*(12), 1378–1385.

Ellickson, P. L., Tucker, J. S., & Klein, D. J. (2001). High-risk behaviors associated with early smoking: Results from a 5-year follow-up. *Journal of Adolescent Health, 28*(6), 465–473.

Espeland, M. A., Dotson, K., Jaramillo, S. A., Kahn, S. E., Harrison, B., Montez, M., et al. (2006). Consent for genetics studies among clinical trial participants: Findings from Action for Health in Diabetes (Look AHEAD). *Clinical Trials, 3*(5), 443–456.

Fabre, V., Boutrel, B., Hanoun, N., Lanfumey, L., Fattaccini, C. M., Demeneix, B., et al. (2000). Homeostatic regulation of serotonergic function by the serotonin transporter as revealed by nonviral gene transfer. *Journal of Neuroscience, 20*(13), 5065–5075.

Feinberg, A. P. (2008). Epigenetics at the epicenter of modern medicine. *Journal of the American Medical Association, 299*(11), 1345–1350.

Flory, J. D., Newcorn, J. H., Miller, C., Harty, S., & Halperin, J. M. (2007). Serotonergic function in children with attention-deficit hyperactivity disorder: Relationship to later antisocial personality disorder. *British Journal of Psychiatry, 190*, 410–414.

Foxcroft, D. R. (2006). *Alcohol misuse prevention for young people: A rapid review of recent evidence*. Oxford: Oxford Brookes University.

Foxcroft, D. R., Ireland, D., Lister-Sharp, D. J., Lowe, G., & Breen, R. (2003). Longer-term primary prevention for alcohol misuse in young people: A systematic review. *Addiction, 98*(4), 397–411.

Fraga, M. F., Ballestar, E., Paz, M. F., Ropero, S., Setien, F., Ballestar, M. L., et al. (2005). Epigenetic differences arise during the lifetime of monozygotic twins. *Proceedings of the National Academy of Sciences of the United States of America, 102*(30), 10604–10609.

French, K., Finkbiner, R., & Duhamel, L. (2002). *Patterns of substance use among minority youth and adults in the United States: An overview and synthesis of national survey findings*. Fairfax: Department of Health and Human Services. URL.

Galea, S., & Rudenstine, S. (2005). Challenges in understanding disparities in drug use and its consequences. *Journal of Urban Health, 82*(2), iii5–iii12. Suppl. 3.

Gerra, G., Garofano, L., Castaldini, L., Rovetto, F., Zaimovic, A., Moi, G., et al. (2005). Serotonin transporter promoter polymorphism genotype is associated with temperament, personality traits and illegal drugs use among adolescents. *Journal of Neural Transmission, 112*(10), 1397–1410.

Gibbons, F. X., Gerrard, M., Cleveland, M. J., Wills, T. A., & Brody, G. H. (2004). Perceived discrimination and substance use in African American parents and their children: A panel study. *Journal of Personality and Social Psychology, 86*(4), 517–529.

Grafodatskaya, D., Choufani, S., Ferreira, J. C., Butcher, D. T., Lou, Y., Zhao, C., et al. (2010). EBV transformation and cell culturing destabilizes DNA methylation in human lymphoblastoid cell lines. *Genomics, 95*, 73–83.

Green, D., Cushman, M., Dermond, N., Johnson, E. A., Castro, C., Arnett, D., et al. (2006). Obtaining informed consent for genetic studies: The multiethnic study of atherosclerosis. *American Journal of Epidemiology, 164*(9), 845–851.

Guo, J., Chung, I.-J., Hill, K. G., Hawkins, J. D., Catalano, R. F., & Abbott, R. D. (2002). Developmental relationships between adolescent substance use and risky sexual behavior in young adulthood. *Journal of Adolescent Health, 31*(4), 354–362.

Haberstick, B. C., Smolen, A., & Hewitt, J. K. (2006). Family-based association test of the 5HTTLPR and aggressive behavior in a general population sample of children. *Biological Psychiatry, 59*(9), 836–843.

Hall, H. I., Li, J., & McKenna, M. T. (2005). HIV in predominantly rural areas of the United States. *The Journal of Rural Health, 21*(3), 245–254.

Halperin, J. M., Newcorn, J. H., Schwartz, S. T., Sharma, V., Siever, L. J., Koda, V. H., et al. (1997). Age-related changes in the association between serotonergic function and aggression in boys with ADHD. *Biological Psychiatry, 41*(6), 682–689.

Halperin, J. M., Kalmar, J. H., Schulz, K. P., Marks, D. J., Sharma, V., & Newcorn, J. H. (2006). Elevated childhood serotonergic function protects against adolescent aggression in disruptive boys. *Journal of the American Academy of Child and Adolescent Psychiatry, 45*, 833–840.

Hamer, D. (2002). Rethinking behavior genetics. *Science, 298*, 71–72.

Hawkins, J. D., Catalano, R. F., & Miller, J. Y. (1992). Risk and protective factors for alcohol and other drug problems in adolescence and early adulthood: Implications for substance abuse prevention. *Psychological Bulletin, 112*(1), 64–105.

Heath, A. C., Jardine, R., & Martin, N. G. (1989). Interactive effects of genotype and social environment on alcohol consumption in female twins. *Journal of Studies on Alcohol, 50*(1), 38–48. URL.

Heath, A. C., Eaves, L. J., & Martin, N. G. (1998). Interaction of marital status and genetic risk for symptoms of depression. *Twin Research, 1*(3), 119–122.

Herd, D. (1994). Predicting drinking problems among Black and White men: Results from a national survey. *Journal of Studies on Alcohol, 55*, 61–71.

Herman, A. I., Philbeck, J. W., Vassilopoulos, N. L., & Depetrillo, P. B. (2003). Serotonin transporter promoter polymorphism and differences in alcohol consumption behavior in a college student population. *Alcohol and Alcoholism, 38*(5), 446–449.

Hopfer, C. J., Timberlake, D., Haberstick, B. C., Lessem, J. M., Ehringer, M. A., Smolen, A., et al. (2005). Genetic influences on quantity of alcohol consumed by adolescents and young adults. *Drug and Alcohol Dependence, 78*, 187–193.

Howe, G. W., Beach, S. R. H., & Brody, G. H. (2010). Microtrial methods for translating gene-environment dynamics into preventive interventions. *Prevention Science, 11*(4), 343–354.

Jacobson, J. O., Robinson, P. L., & Bluthenthal, R. N. (2007). Racial disparities in completion rates from publicly funded alcohol treatment: Economic resources explain more than demographics and addiction severity. *Health Services Research, 42*(2), 773–794.

Johnson, P. O., & Fay, L. C. (1950). The Johnson-Neyman technique: Its theory and application. *Psychometrika, 15*, 349–367.

Johnson, P. O., & Neyman, J. (1936). Tests of certain linear hypotheses and their applications to some educational problems. *Statistical Research Memoirs, 1*, 57–93.

Kahn, R. S., Khoury, J., Nichols, W. C., & Lanphear, B. P. (2003). Role of dopamine transporter genotype and maternal prenatal smoking in childhood hyperactive-impulsive, inattentive, and oppositional behaviors. *Journal of Pediatrics, 143*(1), 104–110.

Kaufman, J., Yang, B.-Z., Douglas-Palumberi, H., Crouse-Artus, M., Lipschitz, D., Krystal, J. H., et al. (2007). Genetic and environmental predictors of early alcohol use. *Biological Psychiatry, 61*, 1228–1234.

Kellam, S. G., & Van Horn, Y. V. (1997). Life course development, community epidemiology, and preventive trials: A scientific structure for prevention research. *American Journal of Community Psychology, 25*(2), 177–188.

Kendler, K. S., & Eaves, L. (1986). Models for the joint effect of genotype and environment on liability to psychiatric illness. *American Journal of Psychiatry, 143*(3), 279–289.

Kendler, K. S., Aggen, S. H., Prescott, C. A., Jacobson, K. C., & Neale, M. C. (2004). Level of family dysfunction and genetic influences on smoking in women. *Psychological Medicine, 34*(7), 1263–1269.

Kline, A. (1996). Pathways into drug user treatment: The influence of gender and racial/ethnic identity. *Substance Use and Misuse, 31*(3), 323–342.

Kogan, S. M., Berkel, C., Chen, Y.-F., Brody, G. H., & Murry, V. M. (2006). Metro status and African-American adolescents' risk for substance use. *Journal of Adolescent Health, 38*(4), 454–457.

Kraemer, H. C., Wilson, G. T., Fairburn, C. G., & Agras, W. S. (2002). Mediators and moderators of treatment effects in randomized clinical trials. *Archives of General Psychiatry, 59*(10), 877–883.

Laucht, M., Becker, K., El-Faddagh, M., Hohm, E., & Schmidt, M. H. (2005). Association of the DRD4 exon III polymorphism with smoking in fifteen-year-olds: A mediating role for novelty seeking? *Journal of the American Academy of Child and Adolescent Psychiatry, 44*(5), 477–484.

Laucht, M., Becker, K., Blomeyer, D., & Schmidt, M. H. (2007). Novelty seeking involved in mediating the association between the dopamine D4 receptor gene exon III polymorphism and heavy drinking in male adolescents: Results from a high-risk community sample. *Biological Psychiatry, 61*(1), 87–92.

Laucht, M., Becker, K., Frank, J., Schmidt, M. H., Esser, G., Treutlein, J., et al. (2008). Genetic variation in dopamine pathways differentially associated with smoking progression in adolescence. *Journal of the American Academy of Child and Adolescent Psychiatry, 47*(6), 673–681.

Leigh, B. C., & Stahl, R. (1993). Substance use and risky sexual behavior for exposure to HIV: Issues in methodology, interpretation, and prevention. *American Psychologist, 48*(10), 1035–1045.

Lerner, R. M. (1991). Changing organism-context relations as the basic process of development: A developmental context and perspective. *Developmental Psychology, 27*(1), 27–32.

Levine, S. B., & Coupey, S. M. (2003). Adolescent substance use, sexual behavior, and metropolitan status: Is "urban" a risk factor? *Journal of Adolescent Health, 32*, 350–355.

Levitan, R. D., Masellis, M., Lam, R. W., Kaplan, A. S., Davis, C., Tharmalingam, S., et al. (2006). A birth-season/DRD4 gene interaction predicts weight gain and obesity in women with seasonal affective disorder: A seasonal thrifty phenotype hypothesis. *Neuropsychopharmacology, 31*(11), 2498–2503.

Liu, D., Diorio, J., Tannenbaum, B., Caldji, C., Francis, D., Freedman, A., et al. (1997). Maternal care, hippocampal glucocorticoid receptors, and hypothalamic-pituitary-adrenal responses to stress. *Science, 277*(5332), 1659–1662.

Lowman, C., & Le Fauve, C. E. (2003). Health disparities and the relationship between race, ethnicity, and substance abuse treatment outcomes. *Alcoholism, Clinical and Experimental Research, 27*(8), 1324–1326.

Luciano, M., Zhu, G., Kirk, K. M., Whitfield, J. B., Butler, R., Heath, A. C., et al. (2004). Effects of dopamine receptor D4 variation on alcohol and tobacco use and on novelty seeking: Multivariate linkage and association analysis. *American Journal of Medical Genetics Part B: Neuropsychiatric Genetics, 124B*(1), 113–123.

McClelland, G. H., & Judd, C. M. (1993). Statistical difficulties of detecting interactions and moderator effects. *Psychological Bulletin, 114*(2), 376–390.

McGowan, P. O., Sasaki, A., Huang, T. C. T., Unterberger, A., Suderman, M., Ernst, C., et al. (2008). Promoter-wide hypermethylation of ribosomal RNA gene promoter in the suicide brain [electronic article]. *PLoS ONE, 3*(5). e2085.

McGowan, P. O., Sasaki, A., D'Alessio, A. C., Dymov, S., Labonté, B., Szyf, M., et al. (2009). Epigenetic regulation of the glucocorticoid receptor in human brain associates with childhood abuse. *Nature Neuroscience, 12*(3), 342–348.

McQuillan, G. M., Porter, K. S., Agelli, M., & Kington, R. (2003). Consent for genetic research in a general population: The NHANES experience. *Genetics in Medicine, 5*(1), 35–42.

McQuillan, G. M., Pan, Q., & Porter, K. S. (2006). Consent for genetic research in a general population: An update on the National Health and Nutrition Examination Survey experience. *Genetics in Medicine, 8*(6), 354–360.

Mitsis, E. M., Halperin, J. M., & Newcorn, J. H. (2000). Serotonin and aggression in children. *Current Psychiatry Reports, 2*(2), 95–101.

Munafò, M. R., Lingford-Hughes, A. R., Johnstone, E. C., & Walton, R. T. (2005). Association between the serotonin transporter gene and alcohol consumption in social drinkers. *American Journal of Medical Genetics. Part B, Neuropsychiatric Genetics, 135B*(1), 10–14.

Murry, V. M., & Brody, G. H. (2004). Partnering with community stakeholders: Engaging rural African American families in basic research and the Strong African American Families preventive intervention program. *Journal of Marital and Family Therapy, 30*(3), 271–283.

Murry, V. M., Brown, P. A., Brody, G. H., Cutrona, C. E., & Simons, R. L. (2001). Racial discrimination as a moderator of the links among stress, maternal psychological functioning, and family relationships. *Journal of Marriage and Family, 63*(4), 915–926.

Pandina, R. J., & Johnson, V. L. (1999). Why people use, abuse, and become dependent on drugs: Toward a heuristic model. In M. D. Glantz & C. R. Hartel (Eds.), *Drug abuse: Origins and interventions* (pp. 119–147). Washington, DC: American Psychological Association.

Petraitis, J., Flay, B. R., & Miller, T. Q. (1995). Reviewing theories of adolescent substance use: Organizing pieces in the puzzle. *Psychological Bulletin, 117*(1), 67–86.

Philibert, R. A., Zadorozhnyaya, O., Beach, S. R. H., & Brody, G. H. (2008a). Comparison of the genotyping results using DNA obtained from blood and saliva. *Psychiatric Genetics, 18*(6), 275–281.

Philibert, R. A., Sandhu, H., Hollenbeck, N., Gunter, T., Adams, W., & Madan, A. (2008b). The relationship of 5HTT (SLC6A4) methylation and genotype on mRNA expression and liability to major depression and alcohol dependence in subjects from the Iowa Adoption Studies. *American Journal of Medical Genetics Part B: Neuropsychiatric Genetics, 147B*(5), 543–549.

Proctor, B. D., & Dalaker, J. (2003)*Poverty in the United States: 2002* (Current population reports, P60-222). Washington, DC: U.S. Bureau of the Census.

Ray, L. A., Bryan, A., MacKillop, J., McGeary, J., Hesterberg, K., & Hutchison, K. E. (2008). The dopamine D4 Receptor (DRD4) gene exon III polymorphism, problematic alcohol use and novelty seeking: Direct and mediated genetic effects. *Addiction Biology, 14*(2), 238–244.

Reardon, S. F., & Buka, S. L. (2002). Differences in onset and persistence of substance abuse and dependence among Whites, Blacks, and Hispanics. *Public Health Reports, 117*(Suppl 1), S51–S59.

Rodríguez, S., Huang, S., Chen, X.-h., Gaunt, T. R., Syddall, H. E., Gilg, J. A., et al. (2006). A study of TH01 and IGF2-INS-TH haplotypes in relation to smoking initiation in three independent surveys. *Pharmacogenetics and Genomics, 16*(1), 15–23.

Rose, R. J., Dick, D. M., Viken, R. J., & Kaprio, J. (2001). Gene–environment interaction in patterns of adolescent drinking: Regional residency moderates longitudinal influences on alcohol use. *Alcoholism, Clinical and Experimental Research, 25*(5), 637–643.

Rutter, M. L. (1985). Resilience in the face of adversity: Protective factors and resistance to psychiatric disturbance. *British Journal of Psychiatry, 147*, 598–611.

Rutter, M. L. (2005). Environmentally mediated risks for psychopathology: Research strategies and findings. *Journal of the American Academy of Child and Adolescent Psychiatry, 44*(1), 3–18.

Rutter, M. L. (2007). Proceeding form observed correlation to causal inference: The use of natural experiments. *Perspectives on Psychological Science, 2*(4), 377–395.

Rutter, M. L., & Silberg, J. (2002). Gene-environment interplay in relation to emotional and behavioral disturbance. *Annual Review of Psychology, 53*(1), 463–490.

Sakai, J. T., Young, S. E., Stallings, M. C., Timberlake, D., Smolen, A., Stetler, G. L., et al. (2006). Case–control and within-family tests for an association between conduct disorder and 5HTTLPR. *American Journal of Medical Genetics. Part B, Neuropsychiatric Genetics, 141B*(8), 825–832.

Sakai, J. T., Lessem, J. M., Haberstick, B. C., Hopfer, C. J., Smolen, A., Ehringer, M. A., et al. (2007). Case–control and within-family tests for association between 5HTTLPR and conduct problems in a longitudinal adolescent sample. *Psychiatric Genetics, 17*(4), 207–214.

Schmidt, L., Greenfield, T., & Mulia, N. (2006). Unequal treatment: Racial and ethnic disparities in alcoholism treatment services. *Alcohol Research & Health, 29*(1), 49–54.

Shields, P. G., Lerman, C., Audrain, J., Bowman, E. D., Main, D., Boyd, N. R., et al. (1998). Dopamine D4 receptors and the risk of cigarette smoking in African-Americans and Caucasians. *Cancer Epidemiology, Biomarkers and Prevention, 7*(6), 453–458.

Simons, R. L., Chen, Y.-f., Stewart, E. A., & Brody, G. H. (2003). Incidents of discrimination and risk for delinquency: A longitudinal test of strain theory with an African American sample. *Justice Quarterly, 20*(4), 827–854.

Skowronek, M. H., Laucht, M., Hohm, E., Becker, K., & Schmidt, M. H. (2006). Interaction between the dopamine D4 receptor and the serotonin transporter promoter polymorphisms in alcohol and tobacco use among 15-year-olds. *Neurogenetics, 7*(4), 239–246.

Stack, K., Cortina, J., Samples, C., Zapata, M., & Arcand, L. F. (2000). Race, age, and back pain as factors in completion of residential substance abuse treatment by veterans. *Psychiatric Services, 51*(9), 1157–1161.

Stueve, A., & O'Donnell, L. N. (2005). Early alcohol initiation and subsequent sexual and alcohol risk behaviors among urban youths. *American Journal of Public Health, 95*(5), 887–893.

Tickamyer, A. R., & Duncan, C. M. (1990). Poverty and opportunity structure in rural America. *Annual Review of Sociology, 16*, 67–86.

Tubman, J. G., Windle, M., & Windle, R. C. (1996). Cumulative sexual intercourse patterns among middle adolescents: Problem behavior precursors and concurrent health risk behaviors. *Journal of Adolescent Health, 18*(3), 182–191.

Tucker, J. S., Orlando, M., & Ellickson, P. L. (2003). Patterns and correlates of binge drinking trajectories from early adolescence to young adulthood. *Health Psychology, 22*(1), 79–87.

Turkheimer, E., Haley, A., Waldron, M., D'Onofrio, B., & Gottesman, I. I. (2003). Socioeconomic status modifies heritability of IQ in young children. *Psychological Science, 14*(6), 623–628.

Vaughn, M. G., Beaver, K. M., DeLisi, M., Howard, M. O., & Perron, B. E. (2009). Dopamine D4 receptor gene exon III polymorphism associated with binge drinking attitudinal phenotype. *Alcohol, 43*(3), 179–184.

Wallace, J. M., Jr., Bachman, J. G., O'Malley, P. M., Johnston, L. D., Schulenberg, J. E., & Cooper, S. M. (2002). Tobacco, alcohol, and illicit drug use: Racial and ethnic differences among U.S. high school seniors, 1976–2000. *Public Health Reports, 117*(Suppl 1), S67–S75.

Weinberg, N. Z., Rahdert, E., Colliver, J. D., & Glantz, M. D. (1998). Adolescent substance abuse: A review of the past 10 years. *Journal of the American Academy of Child and Adolescent Psychiatry, 37*, 252–261.

Whale, R., Quested, D. J., Laver, D., Harrison, P. J., & Cowen, P. J. (2000). Serotonin transporter (5-HTT) promoter genotype may influence the prolactin response to clomipramine. *Psychopharmacology, 150*(1), 120–122.

White, D., & Pitts, M. (1998). Educating young people about drugs: A systematic review. *Addiction, 93*(10), 1475–1487.

Wills, T. A., Gibbons, F. X., Gerrard, M., Murry, V. M., & Brody, G. H. (2003). Family communication and religiosity related to substance use and sexual behavior in early adolescence: A test for pathways through self-control and prototype perceptions. *Psychology of Addictive Behaviors, 17*(4), 312–323.

The manufacturer's authorised representative in the EU is Springer Nature Customer Service Centre GmbH, Europaplatz 3, 69115 Heidelberg, Germany. If you have any concerns regarding our products, please contact ProductSafety@springernature.com

Printed and bound by CPI Group (UK) Ltd, Croydon, CR0 4YY
28/09/2026
02224061-0005